Principles of Food Science

Principles of Food Science

Edited by Lisa Jordan

SYRAWOOD
PUBLISHING HOUSE

New York

Published by Syrawood Publishing House,
750 Third Avenue, 9th Floor,
New York, NY 10017, USA
www.syrawoodpublishinghouse.com

Principles of Food Science
Edited by Lisa Jordan

International Standard Book Number: 978-1-68286-549-1 (Hardback)

Cataloging-in-Publication Data

Principles of food science / edited by Lisa Jordan.
    p. cm.
Includes bibliographical references and index.
ISBN 978-1-68286-549-1
1. Food. 2. Nutrition. 3. Food industry and trade.
I. Jordan, Lisa.
TX353 .P75 2018
641.3--dc23

TABLE OF CONTENTS

PREFACE

Food science studies the nature, processing, and composition of food. The process of food processing is related to principles of food chemistry, food microbiology, food preservation, molecular gastronomy, etc. Food science helps in the effectual utilization of food resources and correct preservation methods. This book attempts to understand the multiple branches that fall under the discipline of food science and how such concepts have practical applications. It aims to serve as a resource guide for experts as well as students associated with this field.

This book is a comprehensive compilation of works of different researchers from varied parts of the world. It includes valuable experiences of the researchers with the sole objective of providing the readers (learners) with a proper knowledge of the concerned field. This book will be beneficial in evoking inspiration and enhancing the knowledge of the interested readers.

In the end, I would like to extend my heartiest thanks to the authors who worked with great determination on their chapters. I also appreciate the publisher's support in the course of the book. I would also like to deeply acknowledge my family who stood by me as a source of inspiration during the project.

Editor

Maternal Food Provisioning in a Substrate-Brooding African Cichlid

Kazutaka Ota*, Masanori Kohda

Department of Biology and Geosciences, Osaka City University, Sumiyoshi, Osaka, Japan

Abstract

Fish demonstrate the greatest variety of parental care strategies within the animal kingdom. Fish parents seldom provision food for offspring, with some exceptions predominantly found in substrate-brooding Central American cichlids and mouth-brooding African cichlids. Here, we provide the first evidence of food provisioning in a substrate-brooding African cichlid *Neolamprologus mondabu*. This fish is a maternal substrate-brooding cichlid endemic to Lake Tanganyika, and feeds on benthic animals using unique techniques–individuals typically feed on the surface of sandy substrates, but also expose prey by digging up substrates with vigorous wriggling of their body and fins. Young also feed on benthos on the substrate surface, but only using the first technique. We observed that feeding induced by digging accounted for 30% of total feeding bouts in adult females, demonstrating that digging is an important foraging tactic. However, parental females fed less frequently after digging than non-parental females, although both females stayed in pits created by digging for approximately 30 s. Instead, young gathered in the pit and fed intensively, suggesting that parental females provision food for young by means of digging. We tested this hypothesis by comparing the feeding frequency of young before and after digging that was simulated by hand, and observed that young doubled their feeding frequency after the simulated digging. This suggests that parental females engage in digging to uncover food items that are otherwise unavailable to young, and provision food for them at the expense of their own foraging. This behavior was similar to what has been observed in Central American cichlids.

Editor: Walter Salzburger, University of Basel, Switzerland

Funding: This work was supported by Grants-in-Aid for Young Scientists (B) (to KO) and Grants-in-Aid for Overseas Scientific Research (to MK) from the Ministry of Education, Culture, Sports, Science and Technology, Japan. The funders had no role in study design, data collection and analysis, decision to publish, or preparation of the manuscript.

Competing Interests: The authors have declared that no competing interests exist.

* E-mail: otkztk@gmail.com

Introduction

Organisms experience heightened vulnerability to threats such as disease and predation, early in their life history [1,2]. Many species have evolved parental care strategies to increase offspring survival during this period [3–5]. In general, parental care includes preparation of a nest, egg care (e.g. incubation and fanning), brood guarding and nourishment. Parental nourishment largely influences the early growth and development of offspring, and thus their survivorship and future reproductive performance [6–9]. Lecithotrophy (yolk-only provisioning) is the most prevalent form of parental nourishment among oviparous species [10]. In addition, parents may provide nutrition to their offspring even after hatching, as offspring have not developed traits useful to search, detect, and handle food; this is especially prevalent in mammals, many altricial birds and some insects [3].

Teleost fish demonstrate the greatest variety of parental care among the animal kingdom, although parental care is typically uncommon [11–13]. Fish parents generally do not provision nutrition to offspring, other than yolk, but a handful of exceptions mainly exist in cichlids that show an unrivalled diverse array of parental care strategies [14–16]. For example, in the mouth-brooding African cichlids *Tropheus duboisi*, *T. moorii* [17] and *Cyphotilapia frontosa* [18], part or all of the food taken by female parents are ingested by their young held in their buccal cavity. In the substrate-brooding Central American cichlids *Symphysodon*

discus [19] and *Amphilophus citrinellus* (formerly *Cichlasoma citrinellum*) [20], young feed on epidermal mucus on parents' bodies. Furthermore, in other substrate-brooding Central American cichlids, *Amatitlania siquia* (formerly *Archocentrus nigrofasciatum*) [21] and *Rocio octofasciata* (formerly *Archocentrus octofasciatum*) [22], parents increase food availability for their offspring by means of fin digging, where they stir up loose substrate with a short bout of vigorous, rapid beating of their pectoral fins.

The substrate-brooding African cichlid *Neolamprologus mondabu* is known to engage in fin-digging for its own feeding [23]. This cichlid is endemic to Lake Tanganyika in East Africa, and preferentially inhabits sandy substrates in rocky areas, where it feeds on benthic animals [24]. Its prey occur within the substrates as well as on the substrate surface, and *N. mondabu* feed using the following four methods [23]: (1) picking–moving slowly or darting a short distance and picking up prey; (2) thrusting–sucking prey while thrusting the mouth into loose substrate with a short dash, then ejecting inedible matter through the gills and mouth; (3) flipping–picking up prey exposed by lifting and flipping over pebbles or small vacant shells; and (4) digging–intensively feeding on prey exposed in pits that are dug in loose substrate using vigorous wriggling of the body, pectoral fin and caudal fin. Fry spread horizontally around nest burrows and peck at substrate to feed on the similar benthic animals as their parents, although they also feed on plankton in the water column in earlier postlarval stages [25,26]. Fry do not use digging and therefore have no access to benthic animals that remain in the substrate and under pebbles. We witnessed several events during which parental females engaged in digging, but appeared to feed less frequently in the fin-dug pits while their young gathered and foraged there. We propose that parental females are provisioning food for their offspring through digging. If this is the case, we predicted that parental females would demonstrate a reduced feeding frequency, and that young would increase their feeding frequency following parental digging. We tested these predictions in the field through observations and experimental manipulations.

Materials and Methods

Study Species

Both sexes of adult *N. mondabu* (>60 mm standard length [SL]) defend territories against same-sex rivals and food competitors [27]. Males are polygynous and their territories encompass 1–6 female territories [27,28]. Spawning takes place in a burrow that the female digs under a rock within her territory. Females care for eggs and embryos in the nest burrows and subsequently guard free-swimming young until independence at *ca.* 10 weeks of age, during which time the young grow rapidly [29]. Females spawn 140 eggs on average, but the survival rate is considerably low among Lake Tanganyika cichlids [29]. Males pay no regard to their offspring while periodically visiting female territories [27,28].

Field Study

To test our hypotheses, field surveys were conducted in September 2013 at depths of 4–10 m at Nkumbula Island (8°75′ S, 31°09′ E), near Mpulungu, Zambia. Although *N. mondabu* is common in the littoral zone, brood-guarding females are present at a low density (0.002–0.004/m² [Ota unpublished data]). To quantify female feeding behaviors, we looked for parental females in a large area (*ca.* 5000 m²) and non-parental females within a subset (*ca.* 900 m²); the observation area consisted of rocks and sandy substrate that this fish prefers. We found a total of 17 females ($n_{non-parental} = 8$, $n_{parental} = 9$) and recorded the behaviors of each fish for 30 min (two 15-min periods in a row) using a digital camera. Pecking at the substrate or rock surface was classified as a feeding bout, but we could not monitor all feeding events within the 30 min as individuals occasionally moved under or behind rocks (range: 0–384 s, $n = 17$). Counts of feeding events were therefore limited to unobstructed 15-min periods. However, visual obstruction did not affect our count of digging events because individuals have insufficient space to dig when under rocks. Therefore, we counted the number of digging events within 30 min and recorded the amount of time that females spent in the dug pit after each digging event. We observed 65 digging events from 17 females during the observations. To examine the effect of brood care on feeding, we also counted the number of attacks against approaching fish for unobstructed 15 min.

To examine the effect of digging on feeding by young, we compared their feeding frequency between before- and after-digging periods using a repeated-measures design. Since digging is infrequent and unpredictable, spontaneous digging could not be used to collect sufficient samples. Instead, we simulated digging by quivering a hand for approximately 2 s, and compared the feeding frequency of young before and after digging. We performed this experiment with 11 clutches; three young from each clutch were randomly selected. To avoid cofounding observations of individual offspring feeding before and after simulated digging, we performed manipulations separately for each young; we counted the feeding bouts of an individual young over five min, then simulated digging

near it and counted feeding events within the next five min. The latter observations began after 30 s of digging. The simulated digging successfully induced feeding in the dug pits, although young usually fled from the hand briefly before returning. To avoid multiple observations of the same individuals, the first young was captured immediately following post-digging observations and observations of the third individual began immediately following the second individual's experiment. Captured young were held in a plastic bag and released at the end of the experiment.

Our study complied with the current laws of Zambia and Japan, and was approved by the Zambian Ministry of Agriculture, Food and Fisheries for fish research in Lake Tanganyika.

Statistical Analysis

The frequencies of attacks, feeding and digging were compared between parental and non-parental females using generalized linear models (GLMs). Because digging was measured repeatedly in some females, time spent in dug pits, the number of feeding events in pits and frequency of feeding were compared using generalized linear mixed models (GLMMs); a female identifier included as a random factor. The feeding frequency of young before and after simulated digging was compared using a GLMM with two random factors (young and clutch identifiers) included. Difference in stages of young was not considered because of inadequate sample size. We fitted a Gaussian distribution to the models when data were normally distributed, while other count and frequency data were fitted to a Poisson or negative binomial (if overdispersion was observed) distribution. All analyses were performed using R version 2.15.2.

Results

Non-parental females were observed feeding much more frequently than parental females for 15 min (Table 1). In non-parental females, feeding induced by digging accounted for 29.5% (mean; SD = 19.5, range: 3.8–63.9%, $n = 8$) of total feeding bouts; although the success of every feeding bout could not be confirmed, this indicates that digging was a significant behavior for obtaining nutrition. Parental and non-parental females spent similar amounts of time following digging in the fin-dug pit, but parental females fed less frequently during the period than non-parental females (Table 1, see Movies S1 and S2). Parental females performed digging marginally less frequently than non-parental females, but practiced aggressive attacks more frequently than non-parental females (Table 1). There was a tendency of negative correlation between these behaviors across females (negative binomial GLM, $\chi^2 = 3.01$, $df = 1$, $p = 0.08$).

Free-swimming young foraged at a mean frequency of 26.8 times per 5 min (SD = 14.9, $n = 33$ young). They gathered in the hand-dug pits within 38 s following simulated digging, on average (SD = 45.7, $n = 33$), and doubled their frequency of feeding during the 5 min following digging compared to their feeding frequency during the 5 min before digging (GLMM, $F = 39.07$, $df = 122.8$, $p < 0.001$; Fig. 1).

Discussion

Non-parental females relied on digging for 30% of total feeding bouts. Conversely, parental females marginally decreased digging frequency and significantly decreased the frequency of feeding following digging, compared to non-parental females. We found a negative correlation between the numbers of digging behaviors and aggressive attacks, suggesting that the decrease in digging frequency is attributable to an increase in vigilance for brood

Table 1. Differences in feeding activities between parental and non-parental *N. Mondabu* females.

Variables	parental females	non-parental females	statistics		
			F/χ^2	df	p
Number of attacks (/15min)[†]	14.7±9.5 (9)	7.4±4.5 (8)	3.92	1,15	0.066
Number of feedings (/15min)[†]	76.4±31.2 (9)	134.4±40.5 (8)	11.01	1,15	0.005
Time spent in dug pit after diggings (sec)[‡]	28.2±21.2 (9)	32.0±14.4 (8)	1.18	1	0.278
Number of feeding in dug pit after digging[§]	4.7±2.4 (9)	12.7±7.4 (8)	4.70	1	0.030
Frequency of feeding in dug pit after digging (/sec)[§]	0.20±0.13 (9)	0.36±0.07 (8)	9.04	1	0.002
Number of digging (/12min)[‖]	2.7±2.6 (9)	5.1±4.7 (8)	2.50	1	0.11

[†]LM; [‡]GLMM; [§]Poisson GLMM; [‖]negative binomial GLM.
Values are means ± SD. Sample sizes are in parentheses.

defense. However, brood defense is an unlikely cause for the decreased feeding observed after digging because the time spent in the dug pit did not differ between parental and non-parental females. Our experiment showed that young increased their feeding frequency after hand-simulating digging. These results suggest that parental females sacrifice their own foraging opportunity to provision food for their young by digging to uncover food items that would otherwise be unavailable to them. Young foraged twice as much after parental digging relative to normal foraging, suggesting that food provisioning contributed considerably to their nourishment. The contribution from parental digging would be actually much greater than the current estimate from our simulated-digging experiment, given that young immediately enter pits dug by parental females or sometimes gather around females prior to digging but did not necessarily rush into pits immediately following simulated digging. The improved accessibility to food would enhance the growth rate of young, and thus contribute to increased female fitness. For example, an increase in growth rate would increase size at independence, and thus survival of young due to increase in ability to escape predation and future reproductive performance [6–9,22,30–32]. Enhanced growth rate would also decrease the time required to reach independence, which may enable females to shorten the interval to the next mating [32–34]. In a mouth-brooding Lake Tanganyika cichlid *Tropheus moorii*, young that are provisioned by parents grow larger, and thus have superior competitive and predator-avoidance

abilities, compared to unprovisioned young [35]. Increased growth through improved accessibility to food by parents would be beneficial for *N. mondabu*, especially considering young of this fish experience high mortality rates compared to other Lake Tanganyika cichlids [29].

To our knowledge, this is the first study to show parental food provisioning in substrate-brooding African cichlids. Why does only *N. mondabu* provision food for its offspring among them? The reason is probably related to limited access to food of their young. In fish, young are generally capable of feeding after yolk sac absorption (i.e., precocial). For example, young of many substrate-brooding Lake Tanganyika cichlids begin feeding on plankton immediately after yolk sac absorption under parental care, and access to food is not restricted by their own abilities. However, young *N. mondabu* feed on benthos, as adult *N. Mondabu* do, but only by means of picking because of insignificant power to dig and expose prey; they only supplement their diet with plankton in earlier postlarval stages. This suggests that food accessibility in this fish is limited by individuals' abilities, and thus their growth may be highly restricted without parental provisioning. Likewise, parents in some mouth-brooding African cichlids provision food for their young that are mainly held in females' buccal cavities, and thus have limited access to food [17,18,35]. In addition, female territoriality is likely also partly responsible for the limited access to food by young. The prey of benthivorous *N. mondabu* are distributed discretely on substrate, and females are highly aggressive against both conspecific and heterospecific food competitors to defend feeding territories [27]. The energetic demand of these defensive behaviors would require increased foraging within territories, likely at the expense of constant care and attention to their young. Feeding on plankton in predator-vulnerable water columns should be risky for young under loose parental care, and as such, feeding in the water column is limited to the earlier postlarval stages. Food provisioning may be a compromise between parental females' own feeding and parental care.

Food provisioning is expected to impose costs on parental female *N. mondabu* because they partly overlap food items with their own young in their territories [29], and they sacrifice their own feeding for provisioning offspring. Therefore, parents and young are expected to be in conflict [32]; we observed that young typically rush in whenever parental females approach nest burrows, but the females do not always respond (Ota pers. obs.). We do not know whether or how young actively solicit food provisioning from their parent or what motivates parents to provision food. In Central American cichlids, parents provide food

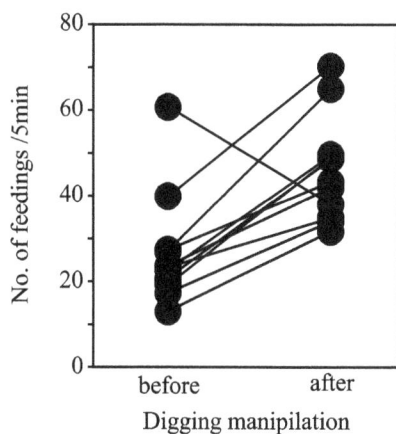

Figure 1. Differences in feeding frequency by young between periods before and after hand-simulated digging. Each plot is a mean value in each clutch.

to their offspring according to the parents' saturation status [21,22]. Parents may also be motivated to provision in response to behavioral cues from young, such as rushing by young (e.g., begging in bird species [36]) or chemical cues, such as odors concentrated by aggregating young (e.g., chemical solicitation in insect species [37]). Further studies will be conducted to examine the optimization of provisioning and solicitation strategies.

This study offers one of the few scarce examples in parental care in fishes where parents seldom nourish their offspring after hatching. Notably and surprisingly, the form of food provisioning in *N. mondabu* is similar to what is observed in substrate-breeding Central American cichlids, whereby both sexes (but more frequently females) engage in digging and provisioning [15,16]. Given the phylogenetic and biogeographic distances between African and Central American cichlids [16,38] and the lack of food provisioning in most other cichlids, these similar forms of food provisioning should evolved independently.

Supporting Information

Movie S1 Non-parental female digging. Females are observed intensively feeding on the freshly dug pit.

Movie S2 Parental female digging. Young are observed rushing toward their mother as she digs, and intensively feeding in the freshly dug pit.

Acknowledgments

We thank the staff of the Lake Tanganyika Research Unit at the Fisheries Research Institute of Zambia for their assistance during fieldwork. We are grateful to two anonymous referees for their useful comments and suggestions on an earlier version of the manuscript.

Author Contributions

Conceived and designed the experiments: KO. Performed the experiments: KO. Analyzed the data: KO. Wrote the paper: KO MK.

References

1. Sogard SM (1997) Size-selective mortality in the juvenile stage of teleost fishes: a review. Bull Mar Sci 60: 1129–1157.
2. Taborsky B, Dieckmann U, Heino M (2003) Unexpected discontinuities in life-history evolution under size-dependent mortality. Proc R Soc Lond B 270: 713–721.
3. Clutton-Brock TH (1991) The evolution of parental care. Princeton: Princeton University Press. 352 p.
4. Royle NJ, Smiseth PT, Kölliker M (2012) The evolution of parental care. Oxford: Oxford University Press. 356 p.
5. Klug H, Bonsall MB (2010) Life history and the evolution of parental care. Evolution 64: 823–835.
6. Einum S, Fleming IA (1999) Maternal effects of egg size in brown trout (*Salmo trutta*): norms of reaction to environmental quality. Proc R Soc Lond B 266: 2095–2100.
7. Donelson JM, Munday PL, McCormick MI (2009) Parental effects on offspring life histories: when are they important? Biol Lett 5: 262–265.
8. Segers FH, Taborsky B (2011) Egg size and food abundance interactively affect juvenile growth and behaviour. Funct Ecol 25: 166–176.
9. Monteith K, Andrews C, Smiseth P (2012) Post-hatching parental care masks the effects of egg size on offspring fitness: a removal experiment on burying beetles. J Evol Biol 25: 1815–1822.
10. Blackburn DG (1992) Convergent evolution of viviparity, matrotrophy, and specializations for fetal nutrition in reptiles and other vertebrates. Am Zool 32: 313–321.
11. Gross MR, Sargent RC (1985) The evolution of male and female parental care in fishes. Am Zool 25: 807–822.
12. Reynolds JD, Goodwin NB, Freckleton RP (2002) Evolutionary transitions in parental care and live bearing in vertebrates. Philos Trans R Soc Lond B 357: 269–281.
13. Mank JE, Promislow DE, Avise JC (2005) Phylogenetic perspectives in the evolution of parental care in ray-finned fishes. Evolution 59: 1570–1578.
14. Keenleyside MH (1991) Cichlid fishes: Behaviour, ecology and evolution. London: Springer. 378 p.
15. Barlow GW (2002) The cichlid fishes: nature's grand experiment in evolution. New York: Basic Books. 335 p.
16. Goodwin NB, Balshine-Earn S, Reynolds JD (1998) Evolutionary transitions in parental care in cichlid fish. Proc R Soc Lond B 265: 2265–2272.
17. Yanagisawa Y, Sato T (1990) Active browsing by mouthbrooding females of *Tropheus duboisi* and *Tropheus moorii* (Cichlidae) to feed the young and/or themselves. Env Biol Fish 27: 43–50.
18. Yanagisawa Y, Ochi H (1991) Food intake by mouthbrooding females of *Cyphotilapia frontosa* (Cichlidae) to feed both themselves and their young. Env Biol Fish 30: 353–358.
19. Hildemann W (1959) A cichlid fish, *Symphysodon discus*, with unique nurture habits. Am Nat: 27–34.

20. Noakes DL, Barlow GW (1973) Ontogeny of parent-contacting in young *Cichlasoma citrinellum* (Pisces, Cichlidae). Behaviour: 221–255.
21. Wisenden BD, Lanfranconi-Izawa TL, Keenleyside MH (1995) Fin digging and leaf lifting by the convict cichlid, *Cichlasoma nigrofasciatum*: examples of parental food provisioning. Anim Behav 49: 623–631.
22. Zworykin DD (1998) Parental fin digging by *Cichlasoma octofasciatum* (Teleostei: Cichlidae) and the effect of parents' satiation state on brood provisioning. Ethology 104: 771–779.
23. Hori M (1983) Feeding ecology of thirteen species of *Lamprologus* (Teleostei; Cichlidae) coexisting at a rocky shore of Lake Tanganyika. Physiol Ecol Jpn 20: 129–149.
24. Takemon Y, Nakanishi K (1998) Reproductive success in female *Neolamprologus mondabu* (Cichlidae): influence of substrate types. Env Biol Fish 52: 261–269.
25. Gashagaza MM, Makoto N (1986) Comparative study on the food habits of six species of *Lamprologus* (Osteichthyes: Cichlide). Afr Stud Monogr 6: 37–44.
26. Nagoshi M (1983) Distribution, abundance and parental care of the genus *Lamprologus* (Cichlidae) in Lake Tanganyika. Afr Stud Monogr 3: 39–47.
27. Ota K, Awata S, Morita M, Kohda M (2014) Sneaker males are not necessarily similar to females in colour in a sexually monochromatic cichlid. J Zool 1: 63–70.
28. Gashagaza M (1991) Diversity of breeding habits in lamprologine cichlids in Lake Tanganyika. Physiol Ecol Jpn 28: 29–65.
29. Nagoshi M (1985) Growth and survival in larval stage of the genus *Lamprologus* (Cichlidae) in Lake Tanganyika. Verh Internat Verein Limnol 22: 2663–2670.
30. Reznick DN (1990) Plasticity in age and size at maturity in male guppies (*Poecilia reticulata*): An experimental evaluation of alternative models of development. J Evol Biol 3: 185–203.
31. Lindström J (1999) Early development and fitness in birds and mammals. Trends Ecol Evol 14: 343–348.
32. Trivers RL (1974) Parent-offspring conflict. Am Zool 14: 249–264.
33. Taborsky B, Foerster K (2004) Female mouthbrooders adjust incubation duration to perceived risk of predation. Anim Behav 68: 1275–1281.
34. Smith C, Wootton RJ (1994) The cost of parental care in *Haplochromis 'argens'* (Cichlidae). Env Biol Fish 40: 99–104.
35. Schürch R, Taborsky B (2005) The functional significance of buccal feeding in the mouthbrooding cichlid *Tropheus moorii*. Behaviour 142: 265–281.
36. Kilner R, Johnstone RA (1997) Begging the question: are offspring solicitation behaviours signals of need? Trends Ecol Evol 12: 11–15.
37. Mas F, Kölliker M (2008) Maternal care and offspring begging in social insects: chemical signalling, hormonal regulation and evolution. Anim Behav 76: 1121–1131.
38. Friedman M, Keck BP, Dornburg A, Eytan RI, Martin CH, et al. (2013) Molecular and fossil evidence place the origin of cichlid fishes long after Gondwanan rifting. Proc R Soc Lond B 280: 20131733.

Bridging Developmental Boundaries: Lifelong Dietary Patterns Modulate Life Histories in a Parthenogenetic Insect

Alison M. Roark[1]*, Karen A. Bjorndal[2]

1 Department of Biology, Furman University, Greenville, South Carolina, United States of America, 2 Department of Biology, University of Florida, Gainesville, Florida, United States of America

Abstract

Determining the effects of lifelong intake patterns on performance is challenging for many species, primarily because of methodological constraints. Here, we used a parthenogenetic insect (*Carausius morosus*) to determine the effects of limited and unlimited food availability across multiple life-history stages. Using a parthenogen allowed us to quantify intake by juvenile and adult females and to evaluate the morphological, physiological, and life-history responses to intake, all without the confounding influences of pair-housing, mating, and male behavior. In our study, growth rate prior to reproductive maturity was positively correlated with both adult and reproductive lifespans but negatively correlated with total lifespan. Food limitation had opposing effects on lifespan depending on when it was imposed, as it protracted development in juveniles but hastened death in adults. Food limitation also constrained reproduction regardless of when food was limited, although decreased fecundity was especially pronounced in individuals that were food-limited as late juveniles and adults. Additional carry-over effects of juvenile food limitation included smaller adult size and decreased body condition at the adult molt, but these effects were largely mitigated in insects that were switched to ad libitum feeding as late juveniles. Our data provide little support for the existence of a trade-off between longevity and fecundity, perhaps because these functions were fueled by different nutrient pools. However, insects that experienced a switch to the limited diet at reproductive maturity seem to have fueled egg production by drawing down body stores, thus providing some evidence for a life-history trade-off. Our results provide important insights into the effects of food limitation and indicate that performance is modulated by intake both within and across life-history stages.

Editor: Joshua B. Benoit, University of Cincinnati, United States of America

Funding: This project was funded by a National Science Foundation Doctoral Dissertation Improvement Grant (project number 0508592). The funders had no role in study design, data collection and analysis, decision to publish, or preparation of the manuscript.

Competing Interests: The authors have declared that no competing interests exist.

* Email: alison.roark@furman.edu

Introduction

Many morphological, physiological, and life-history traits are shaped by when and to what extent resources are acquired, assimilated, and allocated to various functions. Even in the absence of genetic variation, the expression of these traits can differ among individuals of a species [1]. Such phenotypic plasticity is particularly apparent when resource availability varies spatially or temporally.

Resource scarcity restricts the capacity for growth, maintenance, and reproduction. Even when resources are plentiful, upper limits on the rates of nutrient intake and uptake constrain performance [2]. According to the principle of allocation, these extrinsic and intrinsic constraints should prevent animals from simultaneously maximizing the allocation of nutrients to all traits that influence fitness, such that increased use of resources for one function necessarily decreases allocation to a different function [3,4,5]. Resources should therefore be allocated to functions including growth, development, reproduction, and survival according to priority rules [6,7,8].

The principle of allocation has generated a number of testable predictions about how animals optimize their use of limited resources according to these priority rules. These predictions almost always include the existence of negative correlations (or trade-offs) among traits. For example, it has been suggested that forgoing reproduction during periods of food scarcity allows animals to divert available resources into maintenance and storage, thereby increasing starvation resistance and the probability of survival until conditions are more conducive to reproduction [9,10]. As predicted, some food-limited adults do suppress reproduction and survive longer [11,12].

Trade-offs are also apparent when food limitation occurs early in life. For example, food-limited juveniles often delay developmental transitions to extend the time available for growth [13]. Individuals thus mature at older ages, potentially shortening the reproductive lifespan and lengthening generation time. Alternatively, size thresholds for life-history transitions may shift downward in food-limited juveniles to reduce the demographic costs of extended development (e.g., predation or starvation of juveniles) [14], thus yielding smaller, potentially less fecund, adults

[15]. On the other hand, when resources are plentiful, growth and development are accelerated such that both age at maturity and generation time are minimized while adult size, fecundity, and relative fitness are maximized [16,17].

On the surface, this evidence supports the notions that a) resources allocated to growth or maintenance necessarily diminish the resources available for reproduction, and b) reproduction is inherently costly with respect to survival [18,19]. Presumably, then, food limitation extends lifespan by inhibiting reproduction. However, preventing oogenesis or vitellogenesis does not necessarily improve lifespan [20,21,22], and lifespan can be extended by food restriction even in post-reproductive adults [23]. Furthermore, several studies have reported either no correlation [24] or a positive correlation between longevity and fecundity [25,26,27].

Whether traits are negatively or positively correlated may also be determined by the lag between acquisition and allocation. When these processes are asynchronous, the magnitude and direction of trait correlations may differ from when these processes occur concomitantly. For example, adult food limitation may have very different effects on the relationship between fecundity and longevity in "income" breeders that provision offspring with nutrients acquired concurrently than in "capital" breeders that rely on stored nutrients for egg production [6,28,29,30]. Additionally, nutritional stress experienced early in life can constrain subsequent reproductive potential, even if conditions improve for adults [31,32]. Previously food-limited animals may also be more vulnerable to starvation, illness, or predation because of their small size and limited energy reserves [33,34,35,36]. In these cases, the carry-over effects of previous food limitation can lead to positive correlations between body size, fecundity, and survival.

Clearly, resource acquisition and allocation play a significant role in dictating life histories. Because acquisition and allocation patterns change with age [37,38,39], and because life-history traits may be expressed only once (e.g., age at maturity), multiple times (e.g., body size at each molt), or continuously throughout life (e.g., metabolic rate) [40], determining the full range of responses to differences in food availability requires manipulation of food intake throughout the entire lifespan. This approach is particularly challenging for long-lived animals with complex life histories. Even in tractable animal models, studies in which food availability is manipulated are complicated by a lack of consistency in the protocols used for feeding and housing of animals.

In feeding trials, diet is manipulated by altering either the quantity or the quality of food offered. In the former case, which is more common in vertebrate feeding trials, intake can be restricted by pair-feeding, intermittent feeding, or reducing the mass of food offered [41,42]. In insect feeding trials, these approaches are typically used only for adults [43,44,45]. As a result, the effects of quantitative, juvenile food limitation in insects are largely unknown. In the latter case, animals are offered ad libitum quantities of chemically defined diets that vary in nutrient density and/or protein:carbohydrate ratio [24,46,47,48]. In many such studies, particularly in insects, individual intake is not quantified [49] (although see [48]), despite the need for such information when assessing trade-offs [7]. Studies in which individual intake has been quantified have provided exciting information about how insects regulate intake to achieve specific "nutrient targets" that maximize survival or fecundity [49,50], but such regulation may not be possible on more natural diets. Additionally, evaluating the effects of food limitation on fitness requires that females of sexual species be allowed to mate. However, co-housing individuals complicates the quantification of individual intake and can influence longevity and fecundity due to the effects of crowding

[51,52]. To avoid such problems, the production of eggs by virgin females is often used as a measure of fitness despite the fact that mating enhances egg production [53,54,55,56]. For these reasons, life-history responses to lifelong patterns of intake of natural diets are largely unknown, particularly in invertebrates.

To overcome these obstacles, we used a novel approach by evaluating the effects of quantitative food limitation during different life-history stages in a parthenogenetic animal. The Indian stick insect, *Carausius morosus* (Br.) (Phasmatodea, Lonchodinae) reproduces via apomictic parthenogenesis [57], thus permitting us to manipulate the intake of individually housed females while still allowing them to reproduce. Indian stick insects are hemimetabolous and, unlike many insects, do not undergo an ontogenetic diet shift. For this reason, we were able to use the same food source in either limited (L) or unlimited (U) quantities throughout the entire lifespan and thus to test the effects of constant low and high food availabilities across developmental boundaries. We also switched some insects from L to U or from U to L during the juvenile stage or at reproductive maturity to evaluate the effects of a changing environment. We were then able to determine whether the direction and magnitude of the responses to intake changed with age and ontogeny. The specific questions we addressed were:

1. How does the lifelong pattern of intake affect growth, development, and age and size at life-history transitions? Do these effects differ depending on when food is limited?

2. What morphological and physiological parameters are correlated with longevity and fecundity in reproductively active females with different intake histories? Are these measures of performance negatively or positively correlated with each other?

3. Does juvenile intake have carry-over effects in the adult stage, or is adult performance independent of previous intake history?

Throughout the study, we determined how patterns of both juvenile and adult intake affect physiological and life-history parameters including rates of growth and development, age and size at critical life-history transitions, fecundity, and various measures of lifespan. Our expectations were that a) ad libitum-fed insects should grow and develop faster into larger adults that reproduce more and die sooner than food-limited insects, regardless of when food is limited; b) the influence of adult intake on the expression of adult traits will be constrained by carry-over effects of juvenile intake on age, size, and body condition at the adult molt; and c) we should observe life-history trade-offs between pairs of traits including longevity and fecundity, egg size and number, and current and future reproduction.

Material and Methods

Animal care

Insects were housed in a quarantine facility in the Department of Biology, University of Florida. Lights were on a 12 h:12 h light:dark cycle. Room temperature averaged 22.5–24.5°C (Fig. S1), and relative humidity averaged 45–55% (Fig. S2) throughout the trial.

Twenty adult female Indian stick insects (*Carausius morosus*) were obtained from the Exploratorium in San Francisco, California. Eggs oviposited by these females were individually incubated in plastic well plates until hatching. The resulting offspring ($n = 86$) were systematically assigned to treatment groups such that the offspring produced by each adult female were distributed among groups.

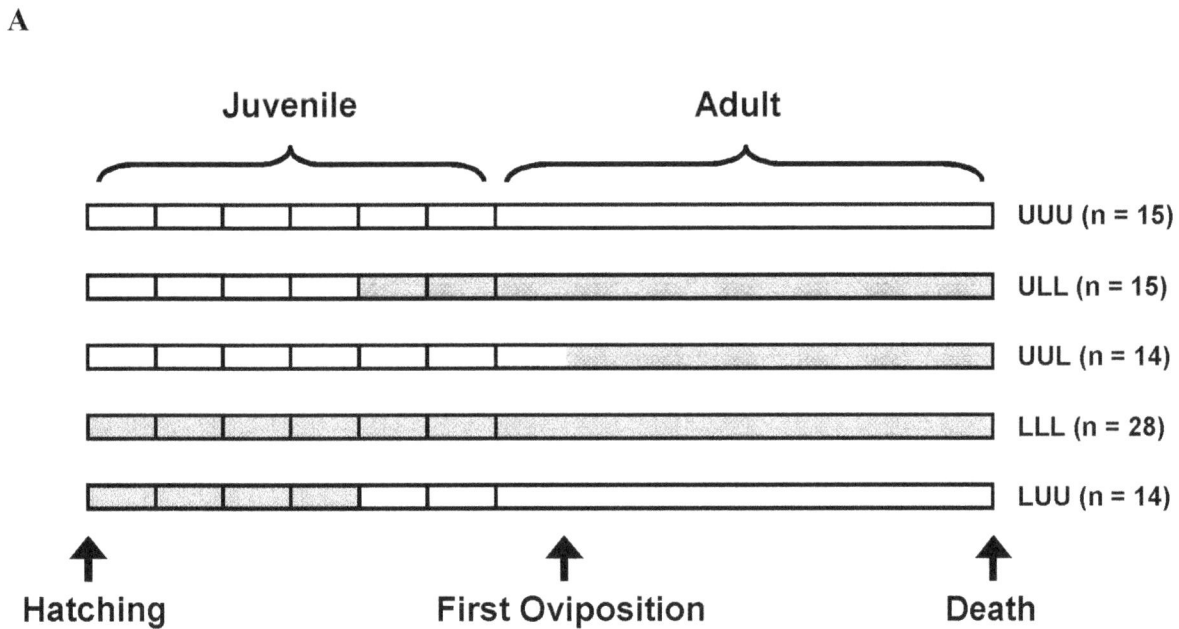

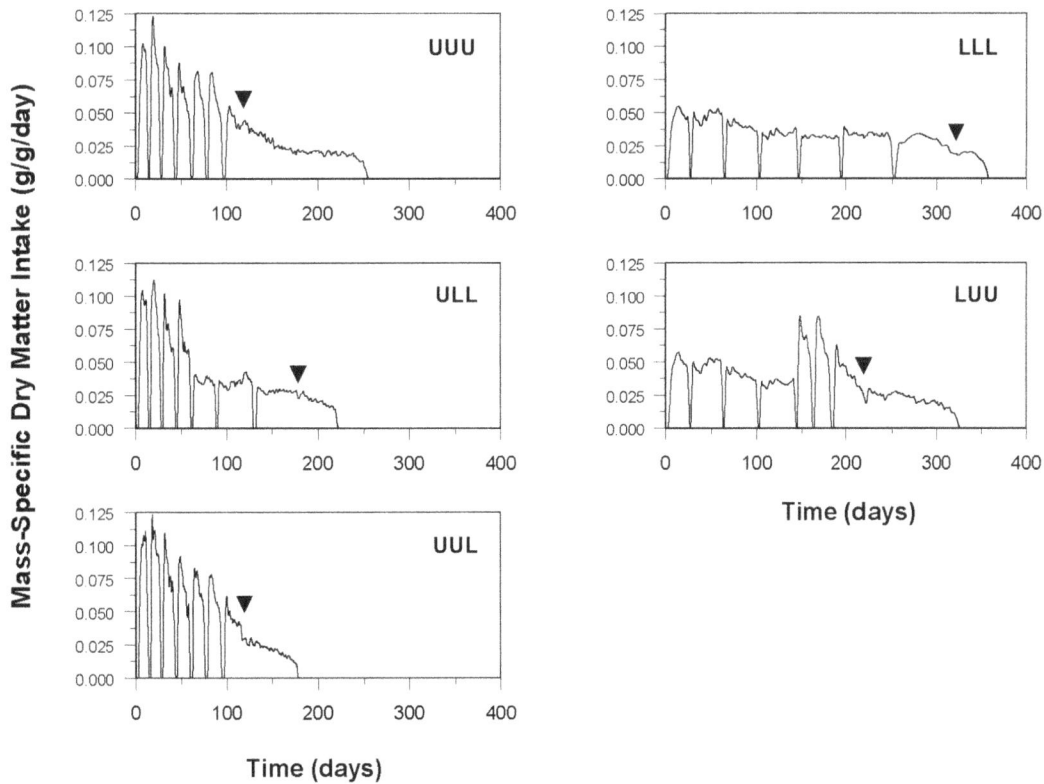

Figure 1. Treatment groups and mass-specific intake. (A) Experimental design. Lifespans are represented by horizontal bars divided into six instars and an adult stage. Time is not to scale, and differences in timing of life-history transitions between groups are not graphically presented. Vertical lines in juvenile stages denote ecdyses. White bars represent life stages when food was offered ad libitum (U, unlimited access to food). Shaded bars represent life stages when food was limited (L) to 60% of the amount of food consumed by insects in group UUU on a percent body

mass basis. Because survival to first oviposition was low for insects that were food-limited for the duration of juvenile development, we were unable to test the effects of a diet switch from L to U at first oviposition (LLU). Sample sizes reflect the number of individuals present in each treatment group at the beginning of the trial. (B) Daily mass-specific dry matter intake (g/g/day). Curves were constructed by scaling the duration of each stage for each insect to the average duration of that stage for each treatment group and fitting a loess smoothing function to these data. Points where mass-specific intake declined to zero correspond to ecdyses. The first six time intervals represent juvenile stages; the final time interval represents the adult stage. Arrowheads denote the average age at first oviposition. Mass-specific intake for UUU insects declined after first oviposition. The amount of food offered to food-limited adults after first oviposition was decreased proportionally to match this decline. Sample sizes: UUU $n = 13$, ULL $n = 13$, UUL $n = 13$, LLL $n = 19$ juveniles and 7 adults, LUU $n = 12$.

Insects were maintained individually in plastic cages (29.5 cm × 19 cm × 19 cm) that were misted daily with deionized water to provide drinking water. Insects were fed discs cut from leaves of English ivy (*Hedera helix*) daily. Biopsy punches (Miltex Instrument Co., Inc.) were used to create discs of different diameters: 2 mm for first instar insects, 3 mm for second instars, 4 mm for third instars, 5 mm for fourth instars, 6 mm for fifth instars, and 8 mm for sixth instars and adults. Samples of leaf discs of each size were dried daily to constant mass at 60°C and weighed, and the dry mass per disc was calculated (Fig. S3).

Initially, leaf discs were all punched from leaves of ivy grown in culture. English ivy was first obtained from a commercial supplier (Benchmark Foliage, Inc., Plymouth, FL). These plants were maintained under a metal halide grow lamp (Sunmaster Cool Deluxe) on a 12 h:12 h light:dark cycle at approximately 21°C. Ivy was watered weekly with deionized water and fertilized monthly using Peter's Professional all-purpose plant food (20% total N, 20% available phosphate, 20% soluble potash). Ivy was cultured by taking cuttings twice per month. Cuttings were allowed to root in deionized water for two weeks before being planted in Bayer Advanced Garden multi-purpose potting mix. Beginning in week 17 (prior to the first oviposition of all insects) and continuing until the end of the study, leaves used to cut 8-mm discs were obtained from a private wooded lot near the University of Florida. Runners of *H. helix* were collected weekly from this lot and maintained in Bayer Advanced Garden multi-purpose potting mix under the same conditions used for cultured ivy. Leaf discs smaller than 8 mm in diameter were always cut from cultured ivy leaves.

Insects were offered either more leaf discs than they could consume within 24 hours (ad libitum or unlimited food, U) or a limited number of discs (L) equal to 60% of the average daily mass-specific intake by continuously ad libitum-fed insects in the same life-history stage. Stages included each of six juvenile instars (although one insect progressed through a supernumerary instar), adult prior to first oviposition (pre-ov adult), and adult after first oviposition (post-ov adult). Correcting intake for body size on a percent body mass basis is appropriate because metabolic rate scales proportionally with body mass in *C. morosus* [58]. Because mass-specific intake by continuously ad libitum-fed insects declined after first oviposition, the amount of food offered to food-limited adults after first oviposition was decreased proportionally to match this decline. Food-limited insects in all life-history stages almost always consumed all of the discs they were offered each day, except on days immediately preceding a molt.

Leaf discs were offered according to five treatment schedules (Fig. 1A). Acronyms for group names refer to the feeding treatment in three distinct life-history periods: hatch to the end of the fourth instar, the beginning of the fifth instar to first oviposition, and first oviposition to death. For example, the ULL group was fed ad libitum until the end of the fourth instar and was then switched to the limited diet for the remainder of its lifespan. Although the feeding trial initially included a LLU group, we could not test this treatment because survival to reproductive maturity was low (25%) for continuously food-limited juveniles. To

ensure a sufficient sample size in the LLL group, all insects ($n = 7$) that were continuously food-limited as juveniles (including those initially in the LLU group) and successfully oviposited as adults were maintained on the limited diet throughout adulthood in group LLL. Unless otherwise noted, the sample sizes indicated in each table and figure include only those individuals that survived through the end of the sixth instar. Given the small number of insects in the LLL group that successfully oviposited, data presented for adult endpoints measured on insects in this group are presented and evaluated with caution.

Physiological and life-history data

Daily intake by each insect was calculated by determining the number of discs consumed. Partially eaten discs were pressed between microscope slides and scanned, and the surface area of each fragment as a proportion of an uneaten disc was determined using ImageJ (1.37 v). Daily dry matter intake was calculated as the product of discs consumed and dry mass per disc. Daily mass-specific dry matter intake was calculated using estimates of daily body mass computed from periodic body mass measurements, as described below. Frass was collected at the end of each life-history stage.

Samples of each size of leaf disc offered each week were ground with dry ice in a mill (C.W. Brabender Instruments, Inc., South Hackensack, NJ) and dried to constant mass at 60°C. Frass samples were dried to constant mass at 60°C and ground using a mortar and pestle. Nitrogen content of leaf (Fig. S4) and frass samples was determined using a Carlo Erba NA 1500 CNS Elemental Analyzer. Assimilated nitrogen in each life-history stage was calculated as consumed N – frass N, and apparent nitrogen assimilation efficiency (NAE) was then calculated as assimilated N*100/consumed N during each life-history stage.

Insects were weighed weekly and at the end of each life-history stage. The end of an instar was defined as the day on which no leaf discs were consumed prior to a molt. Insects were also photographed at the end of each life-history stage, and body lengths were determined using ImageJ. Measurements of body size at the end of each instar for juvenile UUU insects were then fitted to the allometric equation $\ln(y) = \ln(a) + b\ln(x)$, where $y = $ body mass and $x = $ body length. Relative mass (as an index of body condition) of insects in all treatment groups at the adult molt was calculated as the ratio between actual body mass and body mass predicted by the allometric equation [59]. Specific growth rate (SGR) in each life-history stage was calculated as $SGR = 100*(\ln BM_f - \ln BM_i)/t$, where BM_f is body mass at the end of a stage, BM_i is body mass at the beginning of a stage, and t is the time in that stage.

All eggs oviposited by each insect were weighed. Fecundity was measured as the number of eggs oviposited, and reproductive investment was calculated as the sum wet mass of all eggs oviposited. Reproductive lifespan was calculated as the time between the oviposition of first and last eggs. After death, the number of ovarioles in each insect was determined by dissection. Unfulfilled reproductive potential was measured as the number of eggs in the ovaries after death, and potential fecundity was calculated as unfulfilled reproductive potential + fecundity.

Statistical analyses

Analysis of variance (ANOVA) was used to test for differences among treatment groups. Data were first tested for normality (Shapiro-Wilk's test) and homogeneity of variances (Levene's test) and transformed, if necessary. Pairwise comparisons were evaluated using Tukey's Honestly Significant Difference post hoc test (if variances were homogeneous) or Tamhane's T2 post hoc test (if variances were not homogeneous). Data that could not be normalized were analyzed using a Kruskal-Wallis test, and pairwise comparisons were evaluated using Mann-Whitney U tests with α set at 0.005 to account for multiple comparisons. Fecundity was analyzed using ANOVA and also using analysis of covariance with body mass at first oviposition as the covariate. Spearman's rank correlation test was used to evaluate the strength of the relationship between relative mass at the adult molt and adult lifespan.

Stepwise linear regression was used to determine the factors that best explained variance in fecundity, early egg output (the number of eggs oviposited during the first six days of the reproductive lifespan), and longevity. For these analyses, the dependent variables were fecundity and potential fecundity, number of eggs oviposited in the first six days of the reproductive lifespan, adult lifespan, reproductive lifespan, and total lifespan. Data for seven potential independent variables (body mass, age, stage duration, specific growth rate, mass-specific and total intake, and assimilated nitrogen) in each of the six instars and the pre-oviposition adult stage were tested, along with mass-specific metabolic rates in each life-history stage beginning in the fourth instar (Table S1). Because some insects require a threshold level of food consumption or body stores to initiate reproductive processes [60,61], the ratio of actual to predicted body mass at the adult molt (as an index of body condition) and both total intake and assimilated nitrogen during a number of multi-stage intervals were also tested as potential independent variables. Variables had to meet a 0.05 significance level to enter a model, and variables with a variance inflation factor (VIF) greater than 10 were excluded from analysis [62]. Normality of the standardized residuals for the most significant of each set of models was confirmed using Shapiro-Wilk's test.

Least squares linear regression was used to examine the relationships between fecundity and both total and adult lifespan and also to evaluate the strength of the relationship between fecundity and total intake during the reproductive lifespan. Kaplan-Meier survivorship curves were constructed for the entire lifespan ($n = 86$, including all insects used in the study) and for adult lifespan ($n = 70$, including only those individuals that survived through the adult molt). Pairwise comparisons among groups were evaluated using log-rank tests with α set at 0.005 to account for multiple comparisons.

Data were analyzed using SPSS for Windows (Release 11.0.0). S-Plus (Version 7.0) was used for graphing smoothing functions and Kaplan-Meier curves.

Results

Intake and nitrogen assimilation efficiency differed among treatments

Diet treatments yielded different mass-specific intake trajectories among groups. Food-limited insects almost always consumed all of the discs offered each day, which amounted to approximately 60% of the mass-specific intake of ad libitum-fed insects in group UUU in the same life-history stage (Fig. 1B). Total dry matter consumed during each life-history stage except the first instar differed among groups (Table S2). In the second instar, food-limited insects consumed more total dry matter despite being smaller than insects feeding ad libitum because of the longer duration of the instar (Fig. 2 and Table S3). In the third and fourth instars, food-limited insects consumed less total dry matter than insects feeding ad libitum because, although the instar duration was longer, the difference in body mass between food-limited and ad libitum-fed insects was much greater and the limited diet was offered on a mass-specific basis. In subsequent life-history stages, total dry matter consumption was dependent on intake history.

Intake history also affected nitrogen assimilation efficiency (NAE) in all life-history stages. Food-limited insects demonstrated lower NAE than ad libitum-fed insects (Fig. S5), either because of lower nitrogen digestibility or higher nitrogen excretion relative to nitrogen intake than in ad libitum-fed insects.

Development and growth rates affected age and size at each life-history transition

The duration of each life-history stage differed among groups (Fig. S6). Food-limited insects generally progressed more slowly through each stage than ad libitum-fed insects. Previous intake history affected the duration of the fifth and sixth instars and the pre-oviposition adult stage for ULL and LUU insects, as individuals in these groups progressed through these stages more rapidly than continuously food-limited individuals but more slowly than continuously ad libitum-fed individuals. Insects experiencing food limitation during adulthood prior to first oviposition laid their first eggs later in the adult stage than insects feeding ad libitum during this time. However, duration of adulthood after first oviposition was significantly shorter for insects that experienced food limitation than for insects that were feeding ad libitum during this time, regardless of when the food limitation was initiated. The senescent lifespan, or the duration of the post-reproductive adult stage, did not differ significantly among groups ($F_{4,53} = 0.179$, $p = 0.948$). Differences in stage duration are reflected in the percentage of the total lifespan spent as juveniles and adults (Fig. S7).

Food-limited insects also grew more slowly than insects feeding ad libitum (Fig. S8). After a switch from limited to ad libitum feeding at the beginning of the fifth instar, specific growth rates of LUU insects through the final two instars were comparable to those of continuously ad libitum-fed insects. However, during adulthood prior to first oviposition, growth of LUU insects was slower than that of UUU and UUL insects but faster than that of LLL and ULL insects. Insects that experienced a switch from ad libitum to limited feeding grew faster in both the fifth and sixth instars than insects that were continuously food-limited. All insects lost body mass between first oviposition and death. UUL insects lost proportionally more body mass than UUU insects, but all other pairwise comparisons of adult growth rates after first oviposition were not significant.

Body mass did not differ among groups at hatching ($F_{4,65} = 1.00$, $p = 0.414$). Mean body mass and mean age at the end of each life-history stage differed among groups (Fig. 2 and Table S3). Body mass was greater and molting occurred at younger ages in ad libitum-fed insects than in food-limited insects in each of the first four instars. At the end of subsequent instars and at first oviposition, body mass and age differed for all groups except UUU and UUL. At death, body mass of UUL insects was not different from body mass of UUU and LUU insects, but all other pairwise comparisons of size were significant. Mean age at death differed among all groups except LLL and LUU.

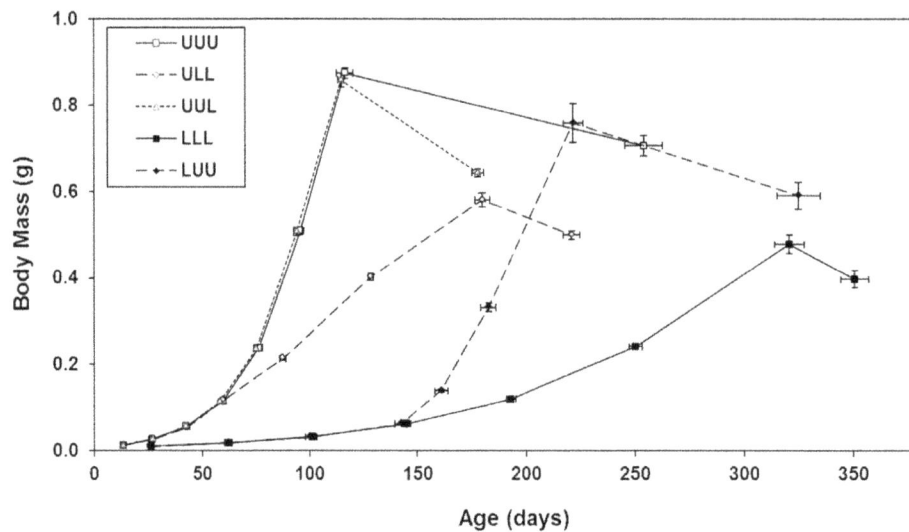

Figure 2. Age and size at each life-history transition. Points represent means ($\pm$ standard errors) at the end of each instar, at first oviposition, and at death. U = unlimited access to food, L = limited access to food. Sample sizes: UUU $n = 13$, ULL $n = 13$, UUL $n = 13$, LLL $n = 19$ juveniles and 7 adults, LUU $n = 12$. See Table S3 for statistical results for size and age at each point.

Body condition at the adult molt was correlated with adult lifespan

Least squares regression of body mass (y) and length (x) for UUU insects at the end of each instar yielded the equation $\ln(y) = 2.7112*\ln(x)-12.018$ ($F_{1,76} = 14077.66$, $p<0.0001$, $R^2 = 0.995$). This equation was used to calculate predicted body masses for insects in all treatment groups using actual body lengths at the adult molt. The ratio of actual to predicted body mass (as an index of body condition) at the adult molt differed among groups (Table S4). Insects that were food limited in the final two instars (groups LLL and ULL) had significantly smaller relative body masses at the adult molt than those that were ad libitum-fed in the final two instars.

Relative but not absolute body mass at the adult molt was significantly and positively correlated with adult lifespan ($\rho = 0.269$, $p = 0.041$ for relative body mass; $\rho = -0.036$, $p = 0.790$ for absolute body mass). Because the diet was switched at first oviposition for UUL insects, these individuals may have experienced a mismatch among body condition at the adult molt, early egg output, and intake during reproductive activity. To determine whether this mismatch confounded the influence of body condition on adult lifespan, we excluded these insects and re-analyzed the correlations between adult lifespan and both relative and absolute body mass at the adult molt. In these cases, the p-values decreased and both relationships were significant ($\rho = 0.523$, $p<0.001$ for the correlation between adult lifespan and relative body mass; $\rho = 0.323$, $p = 0.030$ for the correlation between adult lifespan and absolute body mass).

Juvenile and adult diets influenced fecundity and longevity

Fecundity differed among groups ($F_{4,53} = 50.31$, $p<0.0001$, Fig. 3A). These differences appeared to result both from differences in reproductive lifespan ($F_{4,53} = 41.70$, $p<0.0001$, Fig. 3A) and from differences in early egg output (Fig. 3B). The low fecundity of LLL insects was compounded by low survival to first oviposition. Differences in fecundity did not simply result from differences in body size, as analysis of covariance revealed differences in adjusted mean fecundity when body mass at first

oviposition was used as a covariate (Table S4). Insects with different diet histories also laid eggs that differed in size ($F_{4,53} = 8.195$, $p<0.0001$, Fig. 3B). Differences in fecundity and egg size did not result from differences in number of ovarioles (Table S4).

Intake history affected the number of eggs remaining in the ovaries at death ($F_{4,53} = 5.286$, $p = 0.001$), with LLL insects having a higher unfulfilled reproductive potential than ULL insects (Table S4). All other pairwise comparisons of unfulfilled reproductive potential were not significant. Groups also differed in potential fecundity ($F_{4,53} = 71.62$, $p<0.0001$) and total reproductive investment ($F_{4,53} = 49.57$, $p<0.0001$). The patterns of potential fecundity and total reproductive investment were comparable to that of fecundity.

An event history diagram (sensu [63]) demonstrates the variation in life histories induced by diet treatments (Fig. 4). Pairwise log-rank tests of survival indicated that all groups except LLL and LUU differed in total lifespan (Fig. 5A). This result parallels the ANOVA results for age at death (Table S3). Pairwise log-rank tests of adult lifespan (Fig. 5B) indicated that adult longevity was greater for UUU and LUU insects than for all insects feeding at a limited rate during adulthood.

Intake and growth determined reproductive performance and longevity

We used stepwise multiple linear regression to identify the most significant predictors of fecundity, early egg output, adult lifespan, reproductive lifespan, and total lifespan (Table 1). Dry matter consumed during the reproductive lifespan was the primary predictor of reproductive output selected by the model, explaining 82.8% of the variance in fecundity (Table 1, model A1, Fig. S9) and 84.7% of the variance in potential fecundity. In addition, growth rate during adulthood prior to first oviposition, average mass-specific intake during the reproductive lifespan, nitrogen assimilated between the beginning of the first instar and first oviposition, and SGR during the fifth instar were also selected as variables in a model that explained a total of 95.3% of the variance in fecundity (Table 1, model A5, $F_{5,52} = 229.84$, $p<0.0001$).

A

B

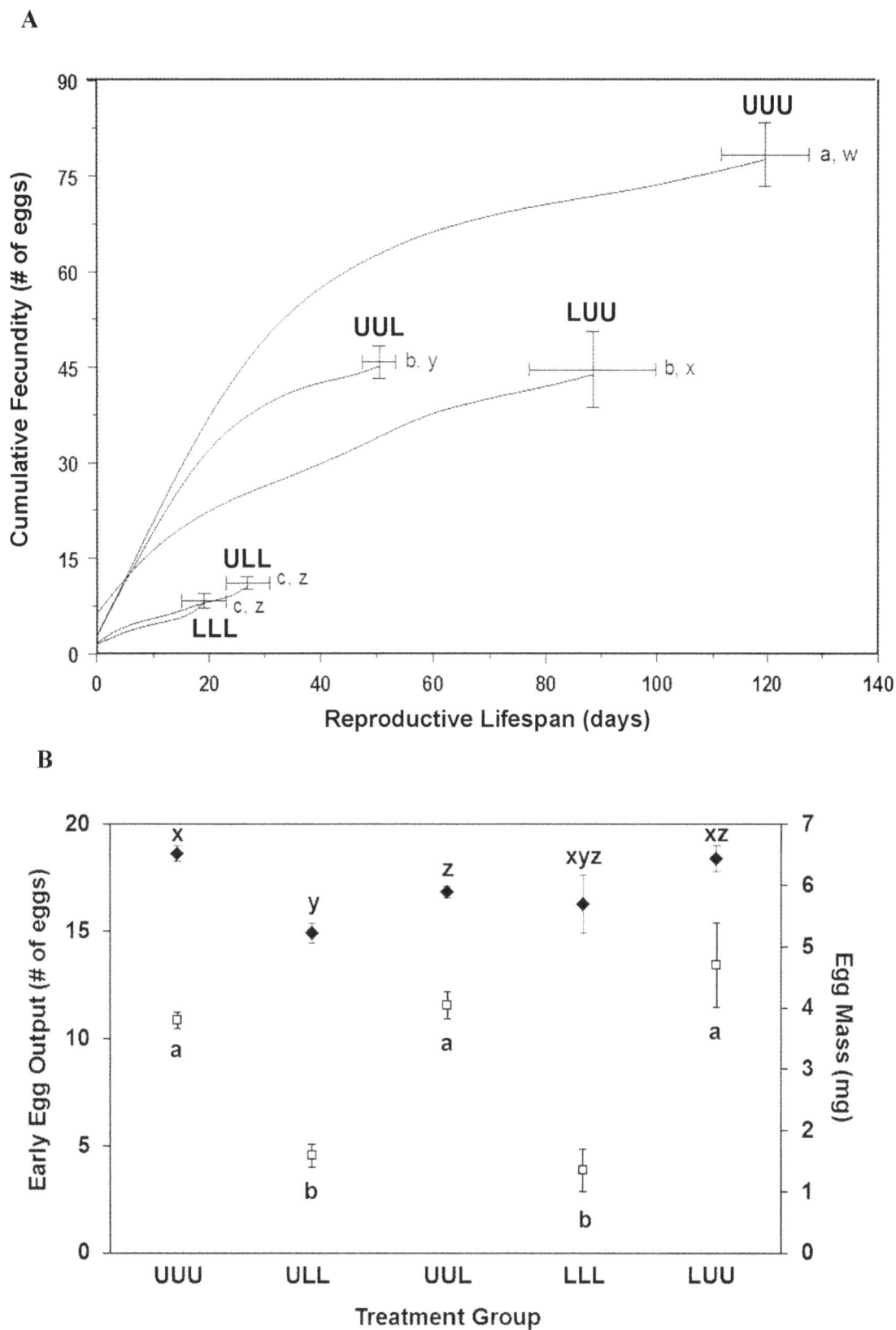

Figure 3. Reproductive performance. (A) Cumulative fecundity of insects in each of five treatment groups. The x-axis represents days of the reproductive lifespan. Each curve terminates at a point corresponding to the mean duration (± standard error) of reproductive activity and the mean fecundity (± standard error) for each group. Curves were constructed by scaling the reproductive lifespan of each insect to the mean reproductive lifespan for that group, determining the mean cumulative fecundity of all insects in that group on each day of the scaled reproductive lifespan, and fitting a smooth spline (df = 7) to the resulting means. U = unlimited access to food, L = limited access to food. Sample sizes: UUU n = 13, ULL n = 13,

UUL $n = 13$, LLL $n = 7$, LUU $n = 12$. Different letters to the right of each point indicate significantly different means for fecundity (a, b, and c) and reproductive lifespan (w, x, y, and z) among treatment groups. (B) Metrics of egg production. Early egg output (white squares) was measured as the number of eggs oviposited during the first six days of the reproductive lifespan, and mean egg mass (black diamonds) was measured in mg (means $\pm$ standard errors). U = unlimited access to food, L = limited access to food. Sample sizes: UUU $n = 13$, ULL $n = 13$, UUL $n = 13$, LLL $n = 7$, LUU $n = 12$. Different letters indicate significantly different means for early egg output (a, b, and c) and mean egg mass (x, y, and z) among treatment groups.

Early egg output (the number of eggs oviposited during the first six days of the reproductive lifespan) was selected *a posteriori* as a dependent variable because of its apparent dependence on diet and its relationship with both survival and fecundity. Stepwise multiple linear regression identified nitrogen assimilated between the beginning of the sixth instar and first oviposition, body length at first oviposition, average mass-specific intake during adulthood prior to first oviposition, nitrogen assimilated during the fifth instar, and metabolic rate in the fourth instar as significant independent variables in a model that explained a total of 80.9% of the variance in early egg output (Table 1, model B7, $F_{5,52} = 49.37$, $p < 0.0001$).

Adult lifespan was best predicted by a model (Table 1, model C2, $F_{2,55} = 8.956$, $p < 0.001$, $R^2 = 0.218$) that included SGR in the fifth instar and mass-specific intake in the third instar as significant independent variables. Reproductive lifespan was best predicted by a model (Table 1, model D2, $F_{2,55} = 27.63$, $p < 0.0001$, $R^2 = 0.483$) that included SGR during adulthood prior to first oviposition and SGR in the fifth instar as significant independent variables. Total lifespan was best predicted by a model (Table 1, model E1, $F_{1,56} = 139.03$, $p < 0.0001$, $R^2 = 0.708$) that included a single, negatively correlated variable (SGR in the third instar). The SGRs and mass-specific intakes we tested as potential variables for

these models were calculated within but not across life-history stages (Table S1), such that multi-stage patterns of growth and mass-specific intake were not represented as potential independent variables. For this reason, the influence of growth and intake during specific instars on subsequent lifespan should not be overstated. Rather, performance during the third and fifth instars likely represents overall patterns of growth and/or intake prior to or after the juvenile diet switch, respectively.

Life histories provide little evidence of trade-offs

We found little evidence for a trade-off between reproduction and survival. Total lifespan was not significantly associated with fecundity or potential fecundity (Fig. S10) when data for all groups were combined ($p > 0.5$) or when each group was analyzed individually ($p > 0.3$ in all cases). On the other hand, adult lifespan was significantly and positively associated with fecundity when data for all treatments were combined (fecundity: $F_{1,56} = 25.67$, $p < 0.0001$, $R^2 = 0.302$, Fig. S11; potential fecundity: $R^2 = 0.343$). However, individual regressions of fecundity versus adult lifespan for each group were all insignificant ($p > 0.2$ in all cases), indicating that no relationship existed between these variables within individual groups.

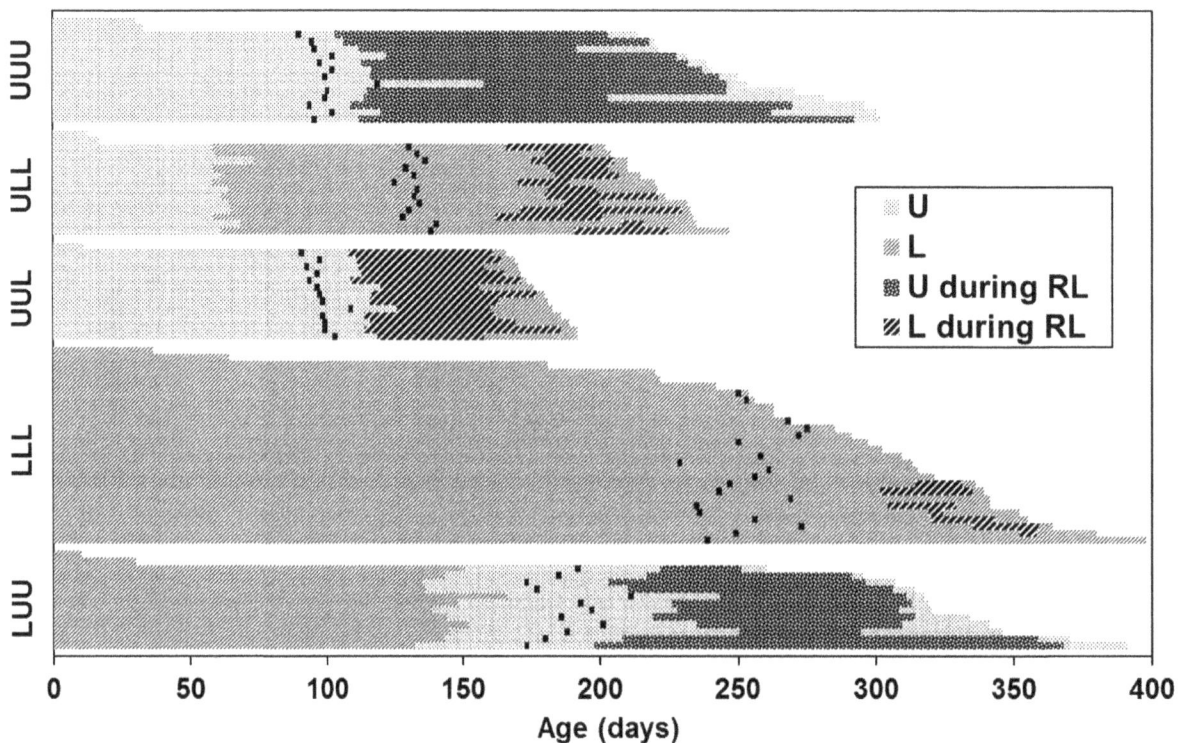

Figure 4. Event history diagram for individual insects maintained on five diet treatments. Each horizontal line represents the lifespan of one individual, with insects in each group arranged in order (top to bottom within a treatment group) from shortest to longest lifespan. U = unlimited access to food, L = limited access to food, RL = reproductive lifespan. The adult molt is indicated by a vertical black line. Data for insects that died during the juvenile stages are included in this diagram but were not included in any analyses except for survivorship curves (Fig. 5). Sample sizes: UUU $n = 15$, ULL $n = 15$, UUL $n = 14$, LLL $n = 28$, LUU $n = 14$.

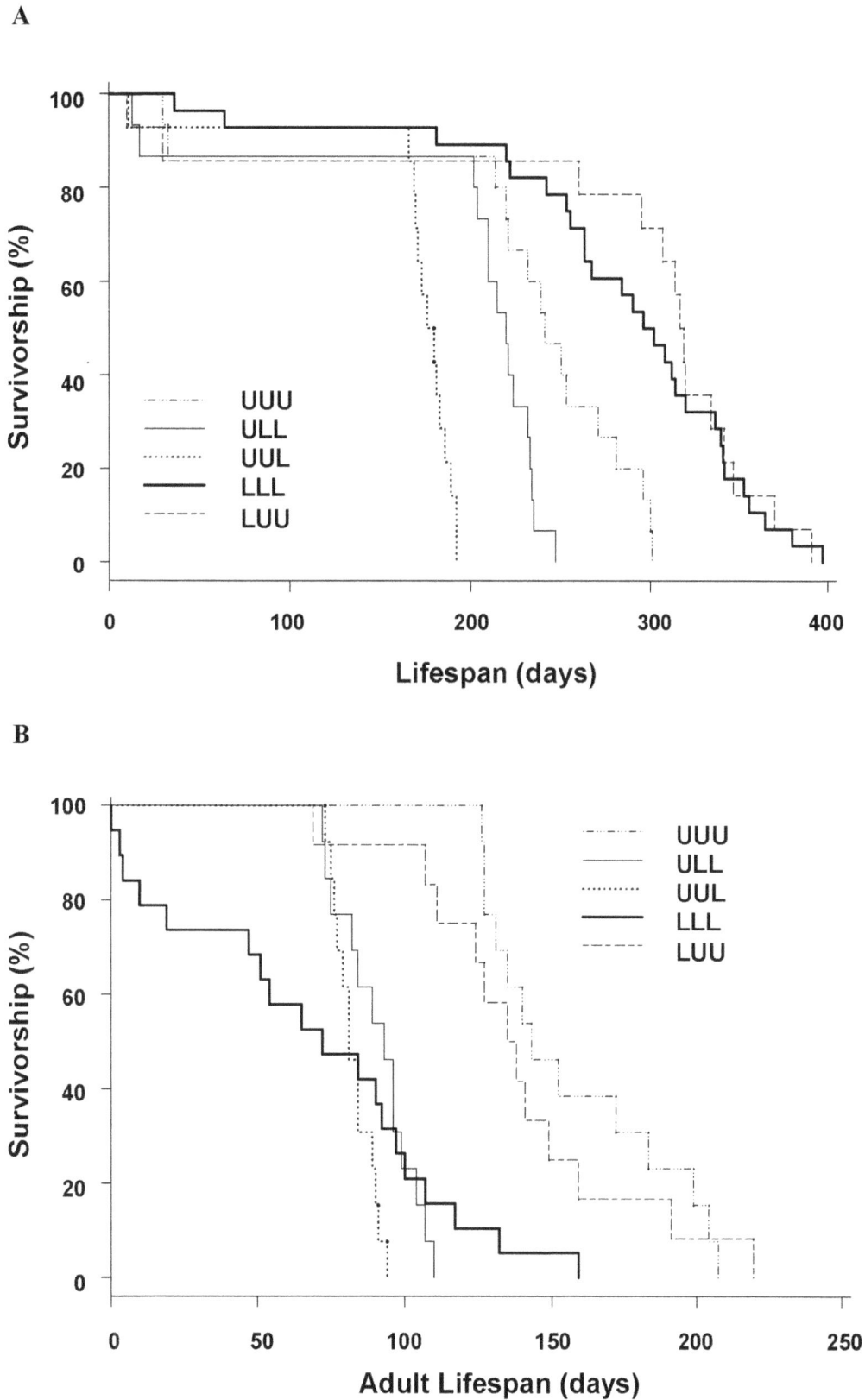

Figure 5. Kaplan-Meier survivorship curves for (A) the entire lifespan and (B) the adult lifespan. Abbreviations: U = unlimited access to food, L = limited access to food. For A, sample sizes are the same as in Figure 4, including insects that died prior to the adult molt (UUU $n = 15$, ULL $n = 15$, UUL $n = 14$, LLL $n = 28$, LUU $n = 14$). For B, only insects that survived to adulthood are included, and sample sizes are UUU $n = 13$, ULL $n = 13$, UUL $n = 13$, LLL $n = 19$ (7 of which oviposited), LUU $n = 12$. For graph A, pairwise log-rank tests indicated that all groups except LLL and LUU differed significantly in longevity. For graph B, UUU and LUU insects had significantly enhanced adult longevity compared to ULL, UUL, and LLL insects.

Table 1. Stepwise multiple linear regression models predicting fecundity (models A1 to A5), initial egg output (models B1 to B4), adult lifespan (models C1 and C2), reproductive lifespan (models D1 and D2), and total lifespan (model E1).

	y	x_1	x_2	x_3	x_4	x_5		intercept	B_1	B_2	B_3	B_4	B_5	B_6	R^2
A1	Fecund	Int RL						9.038	28.05						0.828
A2	Fecund	Int RL	Ad SGR					−4.363	19.09	11.28					0.904
A3	Fecund	Int RL	Ad SGR	MS Int RL				−44.30	17.23	9.583	2051.1				0.929
A4	Fecund	Int RL	Ad SGR	MS Int RL	Nit 1-Ov			−61.89	16.49	6.883	1823.9	1366.5			0.945
A5	Fecund	Int RL	Ad SGR	MS Int RL	Nit 1-Ov	SGR 5		−63.69	17.22	9.875	1651.9	2030.4	−3.837		0.953
B1	6d Eggs	Nit 6-Ov						−9.951	1222.4						0.677
B2	6d Eggs	Nit 6-Ov	SGR 1					−7.575	1264.6	−0.413					0.721
B3	6d Eggs	Nit 6-Ov	SGR 1	Len Ov				−22.44	978.7	−0.926	0.316				0.754
B4	6d Eggs	Nit 6-Ov	SGR 1	Len Ov	MS Int Ad			−21.11	865.6	−0.860	0.258	114.5			0.770
B5	6d Eggs	Nit 6-Ov	SGR 1	Len Ov	MS Int Ad	Nit 5		−24.18	847.1	−0.546	0.371	166.8	−3240.2		0.787
B6	6d Eggs	Nit 6-Ov	SGR 1	Len Ov	MS Int Ad	Nit 5	MR 4	−20.84	777.8	−0.302	0.423	189.0	−3887.9	−0.019	0.811
B7	6d Eggs	Nit 6-Ov	Len Ov	MS Int Ad	Nit 5	MR 4		−17.89	816.8	0.383	212.7	−4840.4	−0.022		0.809
C1	Ad Life	SGR 5						78.00	9.37						0.127
C2	Ad Life	SGR 5	MS Int 3					118.37	11.17	−761.8					0.218
D1	RL	Ad SGR						3.333	29.51						0.455
D2	RL	Ad SGR	SGR 5					−10.93	16.15	10.37					0.483
E1	Tot Life	SGR 3						379.9	−29.89						0.708

Notes: Abbreviations: Fecund = fecundity (number of eggs), 6d Eggs = early egg production (number of eggs oviposited during the first six days of the reproductive lifespan), Ad Life = adult lifespan (days), RL = reproductive lifespan (days), Tot Life = total lifespan (days), Int RL = total dry matter intake (g) during the reproductive lifespan, Nit 6-Ov = nitrogen (g) assimilated between the beginning of the sixth instar and first oviposition, SGR 5 = specific growth rate for body mass (per day) during the fifth instar, Ad SGR = specific growth rate for body mass (per day) during the adult stage prior to first oviposition, SGR 3 = specific growth rate for body mass (per day) during the third instar, SGR 1 = specific growth rate for body mass (per day) during the first instar, Len Ov = body length (mm) at first oviposition, MS Int 3 = average mass-specific intake (g/g) during the third instar, MS Int RL = average mass-specific intake (g/g) during the reproductive lifespan, MS Int Ad = average mass-specific intake (g/g) during adult stage prior to first oviposition, Nit 1-Ov = nitrogen (g) assimilated between the beginning of the first instar and first oviposition, Nit 5 = nitrogen (g) assimilated during the fifth instar, MR 4 = mass-specific metabolic rate (µL/g/hr) in the fourth instar. See Table S1 for the independent variables tested with each dependent variable. Significant independent variables are listed in the order in which they were selected by the models. All models are significant at $p < 0.0001$ with $n = 58$ insects.

We also found no evidence for a trade-off between fecundity and egg size. There was neither a significant interaction between average egg mass and fecundity ($p>0.2$) nor an effect of fecundity on average egg mass ($p>0.3$).

Discussion

Despite the importance of nutrition in determining life histories, few studies have controlled food availability across developmental boundaries to evaluate the effects of lifelong intake patterns. Here, we used a hemimetabolous parthenogen to determine the life-history responses to limited and unlimited availability of a natural diet. Using a hemimetabolous insect that does not undergo an ontogenetic diet shift permitted the use of a homogeneous diet throughout the study, and using a parthenogen allowed us to house our insects individually while monitoring reproductive output.

Our results indicate that life histories are plastic in response to both juvenile and adult food limitation, although the magnitude and direction of the responses differed ontogenetically. Food limitation during developmental and reproductive life-history stages had opposite effects on the rate of progression through each stage, with stage duration increasing in response to food limitation during juvenile and pre-oviposition adult stages and stage duration decreasing in response to food limitation after reproductive maturity. Although developmental delays provided additional time for growth, they were not sufficient to allow full compensation of body size by food-limited juveniles. As a result, insects that experienced juvenile food limitation were both smaller and older at each molt and during adulthood than continuously well fed insects. Adjusting size thresholds downward and age thresholds upward represents a compromise between the need to maximize body size (because of its potential effects on fitness) and the need to minimize the demographic costs of extended development [14].

Although growth slowed in response to food limitation, juvenile insects that experienced a switch from ad libitum to limited feeding grew faster in both the fifth and sixth instars than insects that were continuously food-limited. These insects (in group ULL) also had metabolic rates that were marginally lower than those of LLL insects in the fifth instar [58]. Thus they may have been slightly more efficient in converting their limited incoming nutrients to growth, allowing them to compensate partially for the effects of food limitation.

Food limitation also had drastic negative effects on reproduction, particularly when food was limited during the late juvenile and early adult stages. Plasticity of size thresholds alone does not explain these results, as they persisted even when fecundity was corrected for body mass. Furthermore, differences in fecundity were substantial even among insects of similar size that ate at different rates as adults. Ovary morphology also does not explain these results, as groups did not differ in ovariole number. Rather, reproductive output was likely modulated by a combination of factors including both cumulative and mass-specific adult intake, growth rates during late juvenile and early adult stages (which indirectly influence adult intake through effects on adult body size), and nitrogen assimilated prior to reproductive activity. Of these factors, the total dry mass of food consumed during the reproductive lifespan most significantly influenced reproductive output, with 83% of the variation in fecundity explained by this factor alone.

Intake predicts fecundity; growth predicts survival

This strong, positive correlation between fecundity and total dry matter intake during the reproductive lifespan indicates an "income" breeding strategy [6,28,29,30], in which the resources allocated to reproduction are acquired primarily during the reproductive period. An income breeding strategy is appropriate for a species like *C. morosus*, in which oogenesis and vitellogenesis are non-cyclic and continuous and in which the ovaries contain primarily immature oocytes immediately after the adult molt [64,65]. Furthermore, income breeders commonly demonstrate both a positive correlation between longevity and fecundity [66,67] and a pattern of decreased consumption and production with age [68], both of which were seen in this study. Although income breeders could conceivably mitigate the effects of food limitation on reproductive rate by extending the duration of reproductive activity, food-limited adults in our study had shortened, rather than protracted, reproductive lifespans. Thus, the severely diminished fecundities of LLL and ULL insects likely resulted from both limited access to food and hastened senescence, the combination of which constrained both the resources and time available to generate eggs.

While fecundity was positively correlated with intake during the reproductive lifespan, adult survival was positively correlated with growth rates during late juvenile and immature adult stages and with body composition at the adult molt. Thus, insects that ate and grew the fastest before reproducing were larger, had proportionally greater body stores as adults, and also survived longer as adults than food-limited insects. We therefore demonstrated that adult food limitation had a direct and negative impact on fecundity, whereas juvenile food limitation had a direct and negative impact on adult survival. These data, like those of [43] and [69], suggest that adult-derived "income" fuels egg production while "capital" acquired and stored prior to the onset of reproductive activity fuels subsequent survival, at least in some phytophagous insects.

Effects of intake on early egg output

Body stores present in early adulthood may also function as an index of food availability that entrains the rate and pattern of egg production at the onset of reproductive activity [60,70,71,72]. Modulating reproductive output in response to internal state is a particularly appropriate strategy for a phytophagous insect whose performance is likely nitrogen-limited [73]. In our study, early egg output was predicted primarily by the quantity of nitrogen assimilated during the late juvenile and early adult stages, although factors such as body size at the onset of reproduction, mass-specific intake during the pre-reproductive adult stage, and juvenile nitrogen intake and metabolic rate were also influential. Given their greater consumption rates, insects feeding ad libitum as late juveniles and pre-reproductive adults had higher body condition scores and likely accumulated proportionally more reserves (including nitrogen) prior to first oviposition, thus permitting a higher early egg output than those feeding at a limited rate.

Prospectively matching reproductive activity to the environment is a logical strategy for an income breeder, but only if the assumption that past and future conditions are correlated holds true. If, however, environmental conditions differ for juveniles and adults, then intake and reproductive rates may be mismatched. In our study, for example, the incongruity between high early egg output and low food availability for UUL insects may have forced these individuals to supplement incoming resources with body stores that could otherwise have been allocated to survival. The exhaustion of these stores would explain the shortened adult lifespan of UUL insects relative to continuously well fed individuals. Two specific results lend support to this suggestion:

all UUL individuals were more fecund than expected given their total intake during the reproductive lifespan, and the relationship between relative body mass at the adult molt and adult lifespan for all insects was stronger when UUL insects (the only insects to experience a diet switch as adults) were excluded from the correlation analysis. Thus, we suggest that a mismatch between juvenile and adult intake led to a partial decoupling of the relationship between relative body mass and longevity for insects that experienced a diet switch as adults. In this specific case, a tradeoff between longevity and fecundity may have existed, as resources were preferentially allocated away from survival and toward reproductive output when resource availability changed suddenly at reproductive maturity.

Food limitation and life histories

Food availability clearly modulates life histories through its effects on growth, development, storage, and reproduction. Whether these effects carry over from one life-history stage to another determines how well individuals compensate for a changing environment. In our study, insects that experienced a switch from low to high food availability during development compensated somewhat for their poor start in life by growing faster, molting at larger sizes, accumulating more body stores, and producing more eggs than insects that were continuously food-limited. However, compensation was incomplete and was not accompanied by catch-up growth or an extended reproductive lifespan relative to continuously ad libitum-fed insects. Thus, life histories were moderately flexible in response to environmental variation, allowing insects to mitigate (albeit only partially) a poor early start once conditions improved. On the other hand, insects that experienced a switch from high to low food availability during development were less able to mitigate the effects of poor conditions. Their slow growth, short reproductive lifespans, limited body stores, and diminished reproductive output are evidence that poor conditions experienced late in development compromise performance more profoundly than similar conditions experienced earlier in life.

The fact that food limitation impaired both reproduction and adult survival indicates that insects feeding ad libitum as adults did not incur mortality costs simply because they reproduced more than food-limited insects. This result contradicts the assumption that food limitation elicits a trade-off between reproduction and survival [18,19]. Furthermore, we detected no evidence of a trade-off between current and future reproduction or between early fecundity and adult lifespan [74,75]. Insects feeding ad libitum as pre-reproductive adults exhibited both higher early egg outputs and higher fecundities than those insects that were food-limited during this period.

Reproductive costs (such as increased mortality risk) may differ among species depending on the relative timing of resource acquisition and allocation to reproduction [6]. The existence of a longevity-fecundity trade-off requires that the resources allocated to reproduction and maintenance are derived from a common resource pool and that the utilization of resources from this pool for egg production necessarily decreases the availability of resources for survival [7,76]. Because insects in our study seem to have allocated body stores primarily to maintenance and diverted incoming resources to egg production, these processes likely did not compete for resources and therefore were not negatively correlated [30].

Our study demonstrates that the life-history responses to intake depend to a large extent on the timing of nutritional stress. Food limitation experienced at any point during life led to decreased fecundity, such that reproductive output was maximized when

intake throughout life was also maximized. Total lifespan was maximized when intake and correspondingly growth were limited early in life, but only because food limitation extended development. On the surface, these results seem to provide support for a longevity-fecundity trade-off. However, although insects that were continuously food-restricted experienced both increased longevity and decreased reproductive output relative to continuously ad libitum-fed insects, those individuals that were switched to the limited diet either as juveniles or at reproductive maturity demonstrated deficits in both longevity and fecundity. We also found that intake and growth during juvenile and immature adult stages were positively associated with adult and reproductive lifespans but negatively associated with total lifespan. These data support the contention that a "grow fast and die young" strategy [77] can help to maximize reproductive output when food is readily available. When food is limited, however, ontogeny-dependent costs influence fitness-related traits such as body size, reproductive lifespan, and fecundity. Performance is therefore modulated by nutritional conditions experienced both within and across life-history stages.

Supporting Information

Figure S1 Air temperature in the quarantine facility.

Figure S2 Relative humidity in the quarantine facility.

Figure S3 Leaf disc dry mass.

Figure S4 Leaf disc nitrogen content.

Figure S5 Apparent nitrogen assimilation efficiency.

Figure S6 Duration of each life-history stage.

Figure S7 Percent of the total lifespan comprised of juvenile and adult stages.

Figure S8 Specific growth rates.

Figure S9 Regression of fecundity and intake during the reproductive lifespan.

Figure S10 Relationship between fecundity and total lifespan.

Figure S11 Relationship between fecundity and adult lifespan.

Table S1 The five dependent and 91 independent variables tested in stepwise linear regression analyses.

Table S2 Total leaf dry mass consumed during each life-history stage.

Table S3 Statistical results for body mass and age at each life-history transition.

Table S4 Body condition at the adult molt, mass-corrected fecundity, number of ovarioles, and unfulfilled reproductive potential.

Acknowledgments

Conduct of this study complied with the United States Department of Agriculture (USDA, PPQ-526 permit #69292). We thank L. Nong, S. Porter, and M. Thomas of the USDA for assistance with the construction of an insect quarantine facility; C. Carlson and A. Armendariz of the Exploratorium (San Francisco, CA) for donating adult insects; and Benchmark Foliage, T. Edwards, and B. Moore for donating English ivy.

A. Bolten, L. Guillette, D. Hahn, D. Julian, and J. Sivinski assisted with project development, and D. Hahn also provided comments on the manuscript. J. Curtis, K. Johnson, A. Mazor, S. Strul, and D. Yasova contributed to various aspects of the study. B. Bolker, A. Bolten, and J. Gillooly provided valuable assistance with statistical analyses.

Author Contributions

Conceived and designed the experiments: AMR KAB. Performed the experiments: AMR. Analyzed the data: AMR. Contributed reagents/materials/analysis tools: AMR KAB. Contributed to the writing of the manuscript: AMR KAB.

References

1. Olijnyk AM, Nelson WA (2013) Positive phenotypic correlations among life-history traits remain in the absence of differential resource ingestion. Funct Ecol 27: 165–172.
2. Speakman JR, Król E (2005) Limits to sustained energy intake IX: A review of hypotheses. J Comp Physiol, B 175: 375–394.
3. Cody ML (1966) A general theory of clutch size. Evolution 20: 174–184.
4. Levins R (1968) Evolution in changing environments. New Jersey: Princeton University Press.
5. Gadgil M, Bossert WH (1970) Life historical consequences of natural selection. Am Nat 104: 1–24.
6. Boggs CL (1992) Resource allocation: Exploring connections between foraging and life history. Funct Ecol 6: 508–518.
7. Zera AJ, Harshman LG (2001) The physiology of life history trade-offs in animals. Annu Rev Ecol Syst 32: 95–126.
8. Glazier DS (2002) Resource-allocation rules and the heritability of traits. Evolution 56: 1696–1700.
9. Holliday R (1989) Food, reproduction and longevity: Is the extended lifespan of calorie-restricted animals an evolutionary adaptation? Bioessays 10: 125–127.
10. Masoro EJ, Austad SN (1996) The evolution of the antiaging action of dietary restriction: A hypothesis. J Gerontol 51A: B387–B391.
11. Austad SN (1989) Life extension by dietary restriction in the bowl and doily spider, Frontinella pyramitela. Exp Gerontol 24: 83–92.
12. Chippindale AK, Leroi AM, Kim SB, Rose MR (1993) Phenotypic plasticity and selection in Drosophila life-history evolution. I. Nutrition and the cost of reproduction. J Evol Biol 6: 171–193.
13. Blanckenhorn WU (2006) Divergent juvenile growth and development mediated by food limitation and foraging in the water strider Aquarius remigis (Heteroptera: Gerridae). J Zool 268: 17–23.
14. Rowe L, Ludwig D (1991) Size and timing of metamorphosis in complex life cycles: time constraints and variation. Ecology 72: 413–427.
15. Honěk A (1993) Intraspecific variation in body size and fecundity in insects: A general relationship. Oikos 66: 483–492.
16. Nylin S, Gotthard K (1998) Plasticity in life-history traits. Annu Rev Entomol 43: 63–83.
17. Day T, Rowe L (2002) Developmental thresholds and the evolution of reaction norms for age and size at life-history transitions. Am Nat 159: 338–350.
18. Roff DA (1992) The evolution of life histories: Theory and analysis. New York: Chapman and Hall.
19. Stearns SC (1992) The evolution of life histories. New York: Oxford University Press.
20. Mair W, Sgrò CM, Johnson AP, Chapman T, Partridge L (2004) Lifespan extension by dietary restriction in female Drosophila melanogaster is not caused by a reduction in vitellogenesis or ovarian activity. Exp Gerontol 39: 1011–1019.
21. Barnes AI, Boone JM, Jacobson J, Partridge L, Chapman T (2006) No extension of lifespan by ablation of germ line in Drosophila. Proc R Soc London, Ser B 273: 939–947.
22. Medeiros MN, Ramos IB, Oliveira DMP, da Silva RCB, Gomes FM, et al. (2011) Microscopic and molecular characterization of ovarian follicle atresia in Rhodnius prolixus Stahl under immune challenge. J Insect Physiol 57: 945–953.
23. Kaeberlein TL, Smith ED, Tsuchiya M, Welton KL, Thomas JH, et al. (2006) Lifespan extension in Caenorhabditis elegans by complete removal of food. Aging Cell 5: 487–494.
24. Gribble KE, Welch DBM (2013) Life-span extension by caloric restriction is determined by type and level of food reduction and by reproductive mode in Brachionus manjavacas (Rotifera). J Gerontol 68: 349–358.
25. Smith JNM (1981) Does high fecundity reduce survival in song sparrows? Evolution 35: 1142–1148.
26. Messina FJ, Fry JD (2003) Environment-dependent reversal of a life history trade-off in the seed beetle Callosobruchus maculatus. J Evol Biol 16: 501–509.
27. Agarwala BK, Yasuda H, Sato S (2008) Life history response of a predatory ladybird, Harmonia axyridis (Pallas) (Coleoptera: Coccinellidae), to food stress. Appl Entomol Zool 43: 183–189.
28. Sibly RM, Calow P (1984) Direct and absorption costing in the evolution of life cycles. J Theor Biol 111: 463–473.
29. Sibly RM, Calow P (1986) Physiological ecology of animals: An evolutionary approach. Oxford: Blackwell Scientific Publications.
30. Jönsson KI (1997) Capital and income breeding as alternative tactics of resource use in reproduction. Oikos 78: 57–66.
31. Barrett ELB, Hunt J, Moore AJ, Moore PJ (2009) Separate and combined effects of nutrition during juvenile and sexual development on female life-history trajectories: the thrifty phenotype in a cockroach. Proc R Soc London, Ser B 276: 3257–3264.
32. Dmitriew C, Rowe L (2011) The effects of larval nutrition on reproductive performance in a food-limited adult environment. PLoS One 6: e17399.
33. Lindström J (1999) Early development and fitness in birds and mammals. Trends Ecol Evol 14: 343–348.
34. Morgan IJ, Metcalfe NB (2001) Deferred costs of compensatory growth after autumnal food shortage in juvenile salmon. Proc R Soc London, Ser B 268: 295–301.
35. Lummaa V, Clutton-Brock T (2002) Early development, survival and reproduction in humans. Trends Ecol Evol 17: 141–147.
36. McMillen IC, Robinson JS (2005) Developmental origins of the metabolic syndrome: prediction, plasticity, and programming. Physiol Rev 85: 571–633.
37. Perrin N (1992) Optimal resource allocation and the marginal value of organs. Am Nat 139: 1344–1369.
38. Perrin N, Sibly RM (1993) Dynamic models of energy allocation and investment. Annu Rev Ecol Syst 24: 379–410.
39. Dudycha JL, Lynch M (2005) Conserved ontogeny and allometric scaling of resource acquisition and allocation in the Daphniidae. Evolution 59: 565–576.
40. Nussey DH, Wilson AJ, Brommer JE (2007) The evolutionary ecology of individual phenotypic plasticity in wild populations. J Evol Biol 20: 831–844.
41. Weindruch R, Walford RL (1988) The retardation of aging and disease by dietary restriction. Springfield, IL: Charles C. Thomas.
42. Anson RM, Guo Z, de Cabo R, Iyun T, Rios M, et al. (2003) Intermittent fasting dissociates beneficial effects of dietary restriction on glucose metabolism and neuronal resistance to injury from calorie intake. Proc Natl Acad Sci USA 100: 6216–6220.
43. Boggs CL, Ross CL (1993) The effect of adult food limitation on life history traits in Speyeria mormonia (Lepidoptera: Nymphalidae). Ecology 74: 433–441.
44. Carey JR, Liedo P, Harshman L, Zhang Y, Müller H-G, et al. (2002) Life history response of Mediterranean fruit flies to dietary restriction. Aging Cell 1: 140–148.
45. Cooper TM, Mockett RJ, Sohal BH, Sohal RS, Orr WC (2004) Effect of caloric restriction on life span of the housefly, Musca domestica. FASEB J 18: 1591–1593.
46. Partridge L, Piper MDW, Mair W (2005) Dietary restriction in Drosophila. Mech Ageing Dev 126: 938–950.
47. Bass TM, Grandison RC, Wong R, Martinez P, Partridge L, et al. (2007) Optimization of dietary restriction protocols in Drosophila. J Gerontol 62A: 1071–1081.
48. Fanson BG, Taylor PW (2012) Protein:carbohydrate ratios explain life span patterns found in Queensland fruit fly on diets varying in yeast:sugar ratios. Age 34: 1361–1368.
49. Lee KP, Simpson SJ, Clissold FJ, Brooks R, Ballard JWO, et al. (2008) Lifespan and reproduction in Drosophila: New insights from nutritional geometry. Proc Natl Acad Sci USA 105: 2498–2503.
50. Behmer ST (2009) Insect herbivore nutrient regulation. Annu Rev Entomol 54: 165–187.
51. Joshi A, Wu W-P, Mueller LD (1998) Density-dependent natural selection in Drosophila: adaptation to adult crowding. Evol Ecol 12: 363–376.
52. Ban S, Tenma H, Mori T, Nishimura K (2009) Effects of physical interference on life history shifts in Daphnia pulex. J Exp Biol 212: 3174–3183.
53. Wheeler D (1996) The role of nourishment in oogenesis. Annu Rev Entomol 41: 407–431.
54. De Clercq P, Degheele D (1997) Effects of mating status on body weight, oviposition, egg load, and predation in the predatory stinkbug Podisus maculiventris (Heteroptera: Pentatomidae). Ann Entomol Soc Am 90: 121–127.

55. Foster SP, Howard AJ (1999) The effects of mating, age at mating, and plant stimuli, on the lifetime fecundity and fertility of the generalist herbivore *Epiphyas postvittana*. Entomol Exp Appl 91: 287–295.

56. Meats A, Leighton SM (2004) Protein consumption by mated, unmated, sterile and fertile adults of the Queensland fruit fly, *Bactrocera tryoni* and its relation to egg production. Physiol Entomol 29: 176–182.

57. Pijnacker LP (1966) The maturation divisions of the parthenogenetic stick insect *Carausius morosus* Br. (Orthoptera, Phasmidae). Chromosoma 19: 99–112.

58. Roark AM, Bjorndal KA (2009) Metabolic rate depression is induced by caloric restriction and correlates with rate of development and lifespan in a parthenogenetic insect. Exp Gerontol 44: 413–419.

59. Perrin N, Bradley MC, Calow P (1990) Plasticity of storage allocation in *Daphnia magna*. Oikos 59: 70–74.

60. Juliano SA, Olson JR, Murrell EG, Hatle JD (2004) Plasticity and canalization of insect reproduction: testing alternative models of life history transitions. Ecology 85: 2986–2996.

61. Hatle JD, Waskey T Jr, Juliano SA (2006) Plasticity of grasshopper vitellogenin production in response to diet is primarily a result of changes in fat body mass. J Comp Physiol, B 176: 27–34.

62. Belsley DA, Kuh E, Welsch RE (1980) Regression diagnostics: Identifying influential data and sources of collinearity. New York: John Wiley.

63. Carey JR, Liedo P, Müller H-G, Wang J-L, Vaupel JW (1998) A simple graphical technique for displaying individual fertility data and cohort survival: case study of 1000 Mediterranean fruit fly females. Funct Ecol 12: 359–363.

64. Bradley JT, Masetti M, Cecchettini A, Giorgi F (1995) Vitellogenesis in the allatectomized stick insect *Carausius morosus* (Br.) (Phasmatodea: Lonchodinae). Comp Biochem Physiol, Part B: Biochem Mol Biol 110B: 255–266.

65. Jervis MA, Boggs CL, Ferns PN (2005) Egg maturation strategy and its associated trade-offs: a synthesis focusing on Lepidoptera. Ecol Entomol 30: 359–375.

66. Bauerfeind SS, Fischer K (2008) Maternal body size as a morphological constraint on egg size and fecundity in butterflies. Basic Appl Ecol 9: 443–451.

67. Walker PW, Allen GR (2010) Mating frequency and reproductive success in an income breeding moth, *Mnesampela privata*. Entomol Exp Appl 136: 290–300.

68. Dixon AFG, Agarwala BK (2002) Triangular fecundity function and ageing in ladybird beetles. Ecol Entomol 27: 433–440.

69. Boggs CL, Freeman KD (2005) Larval food limitation in butterflies: effects on adult resource allocation and fitness. Oecologia 144: 353–361.

70. McNamara JM, Houston AI (1996) State-dependent life histories. Nature 380: 215–221.

71. Moehrlin GS, Juliano SA (1998) Plasticity of insect reproduction: testing models of flexible and fixed development in response to different growth rates. Oecologia 115: 492–500.

72. Hatle JD, Borst DW, Juliano SA (2003) Plasticity and canalization in the control of reproduction in the lubber grasshopper. Integr Comp Biol 43: 635–645.

73. Price PW, Denno RF, Eubanks MD, Finke DL, Kaplan I (2011) Insect ecology: Behavior, populations and communities. New York: Cambridge University Press.

74. Reznick D (1985) Costs of reproduction: an evaluation of the empirical evidence. Oikos 44: 257–267.

75. Reznick D (1992) Measuring the costs of reproduction. Trends Ecol Evol 7: 42–45.

76. van Noordwijk AJ, de Jong G (1986) Acquisition and allocation of resources: their influence on variation in life history tactics. Am Nat 128: 137–142.

77. Metcalfe NB, Monaghan P (2003) Growth versus lifespan: perspectives from evolutionary ecology. Exp Gerontol 38: 935–940.

3

The Association between Selenium and Other Micronutrients and Thyroid Cancer Incidence in the NIH-AARP Diet and Health Study

Thomas J. O'Grady[1]*, **Cari M. Kitahara**[2], **A. Gregory DiRienzo**[1], **Margaret A. Gates**[1]

1 University at Albany, School of Public Health, Rensselaer, New York, United States of America, 2 Division of Cancer Epidemiology and Genetics, National Cancer Institute, National Institutes of Health, Rockville, Maryland, United States of America

Abstract

Background: Selenium is an essential trace element that is important for thyroid hormone metabolism and has antioxidant properties which protect the thyroid gland from oxidative stress. The association of selenium, as well as intake of other micronutrients, with thyroid cancer is unclear.

Methods: We evaluated associations of dietary selenium, beta-carotene, calcium, vitamin D, vitamin C, vitamin E, folate, magnesium, and zinc intake with thyroid cancer risk in the National Institutes of Health – American Association of Retired Persons Diet and Health Study, a large prospective cohort of 566,398 men and women aged 50–71 years in 1995–1996. Multivariable-adjusted Cox proportional hazards regression was used to examine associations between dietary intake of micronutrients, assessed using a food frequency questionnaire, and thyroid cancer cases, ascertained by linkage to state cancer registries and the National Death Index.

Results: With the exception of vitamin C, which was associated with an increased risk of thyroid cancer ($HR_{Q5\ vs\ Q1}$, 1.34; 95% CI, 1.02–1.76; P_{trend}, <0.01), we observed no evidence of an association between quintile of selenium ($HR_{Q5\ vs\ Q1}$, 1.23; 95% CI, 0.92–1.65; P_{trend}, 0.26) or other micronutrient intake and thyroid cancer.

Conclusion: Our study does not suggest strong evidence for an association between dietary intake of selenium or other micronutrients and thyroid cancer risk. More studies are needed to clarify the role of selenium and other micronutrients in thyroid carcinogenesis.

Editor: Javier S. Castresana, University of Navarra, Spain

Funding: This work was supported in part by the Intramural Research Program of the National Cancer Institute, National Institutes of Health. The funders had no role in study design, data collection and analysis, decision to publish, or preparation of the manuscript. The authors received no other funding of any type for the preparation of this manuscript.

Competing Interests: The authors have declared that no competing interests exist.

* Email: togrady@albany.edu

Introduction

The past three decades have seen a rapid increase in thyroid cancer incidence in the United States (U.S.) and other countries [1–5]. Increased medical surveillance and diagnostic scrutiny are likely responsible for some, but not all, of this trend [6]. In addition to established risk factors such as ionizing radiation exposure and benign thyroid nodules [7], recent studies have focused on modifiable etiologic factors such as diet and obesity. Findings from these studies indicate that obesity and excessive weight gain during adulthood [8–10] and dietary nitrate and nitrite intake [11,12] are associated with increased thyroid cancer risk while eating various fruits and vegetables [13–17], having adequate iodine intake and consuming fish [18], and a Polynesian-style diet [19] may be protective. Still, there is no consensus as to what dietary factors contribute to or inhibit thyroid carcinogenesis, as others have reported that a traditional Western diet [19,20] and fruit and vegetable consumption [21] are unassociated with thyroid cancer risk.

While there has been research into dietary patterns as a whole, less work has been done to assess specific dietary constituents such as micronutrients and their impact on thyroid cancer risk. Additional research into the association between micronutrients and thyroid cancer may help further the understanding of the biological mechanisms involved in thyroid carcinogenesis. Selenium is an essential trace element that is found at higher concentrations in the thyroid gland than in other organs [22]. Selenium is important in the metabolism of thyroid hormones triiodothyronine (T3) and thyroxine (T4). Specifically, type I 5'-deiodinase is a selenium-containing protein which assists in activating naturally occurring T4 (which has little biological activity) into biologically active T3. Additionally, selenium has antioxidant properties which may help protect the thyroid gland from H_2O_2 and reactive oxygen species [23–26]. While selenium has been shown to have an inverse association with other cancers [27] this relationship has not been investigated for thyroid cancer. A recent meta-analysis of the association between supplemental micronutrients and thyroid cancer showed both that there is limited research on the topic of micronutrients and thyroid cancer and that the results of these prior studies are largely inconclusive [28].

We therefore examined the association between dietary intake of micronutrients and thyroid cancer risk in the U.S. National Institutes of Health-American Association of Retired Persons (NIH-AARP) Diet and Health Study, a large prospective cohort of 566,398 men and women ages 50–71 years at baseline. The primary micronutrient of interest was selenium because of its established importance for proper function of the thyroid gland, its potential antioxidant properties, and the possible inverse association between selenium and other cancers [23,24,27–29]. To our knowledge this is the first prospective evaluation of dietary selenium intake in relation to thyroid cancer risk. Additionally, we evaluated the associations between dietary intake of beta-carotene, calcium, folate, magnesium, vitamin C, vitamin D, vitamin E, and zinc with thyroid cancer risk to expand upon recent studies investigating supplemental intake of these micronutrients and thyroid cancer [28].

Methods

Study Population

The NIH-AARP Diet and Health study began in 1995–1996 with the mailing of an extensive baseline questionnaire to 3.5 million AARP members aged 50–71 years old and residing in six U.S. states (California, Florida, Louisiana, North Carolina, New Jersey, and Pennsylvania) or two U.S. metropolitan areas (Atlanta, Georgia and Detroit, Michigan). Information ascertained by the questionnaire included usual dietary intake over the past twelve months, use of individual and multivitamin supplements, alcohol intake, smoking history, height and weight at baseline, and other factors. Of the 566,398 individuals who were deemed to have satisfactorily completed the baseline questionnaire we excluded, in the following order, participants with proxy respondents (n = 15,760), those who reported poor health or end stage renal disease (n = 9,134), participants with a previous diagnosis of cancer other than non-melanoma skin cancer (n = 53,195), those with extreme or missing values for total energy intake (n = 4,279) and individuals with extreme values for daily selenium intake (n = 1,223). The remaining analytic cohort included 482,807 participants (287,944 men and 194,863 women). The NIH-AARP Diet and Health Study was approved by the Special Studies Institutional Review Board of the U.S. National Cancer Institute and the Institutional Review Boards from all participating institutions approved the use of these data.

Cancer Ascertainment

Participants accrued time in the study from the date of baseline questionnaire completion to the date of any cancer other than non-melanoma skin cancer, the date when a person moved out of the registry ascertainment area, death, or December 31, 2006, whichever occurred first. Incident thyroid cancers (International Classification of Disease for Oncology, Third Edition (ICD-O-3), topography code C73) [30] were identified through December 31, 2006, via linkage of the NIH-AARP cohort membership to state cancer registries and the National Death Index [31]. The state cancer registries are certified by the North American Association of Central Cancer Registries as being at least 90% complete within two years of the close of the diagnosis year [32]. A validation study comparing linkage to state cancer registries with self-report and subsequent medical record conformation of incident cancers estimated that 90% of all cancer cases identified in the NIH-AARP Diet and Health Study were valid [31]. Subtypes of thyroid cancer were defined by ICD-O-3 morphology codes (papillary, 8050, 8052, 8130, 8260, 8340–8344) and (follicular, 8290, 8330–8332, 8335).

Dietary Intake

The baseline questionnaire had a dietary component, which included questions about the frequency of dietary consumption during the past twelve months of 124 food items and the corresponding portion sizes of 100 of those food items. Intake frequency was recorded as one of ten categories ranging from "never" to "2+ times per day" for foods and "never" to "6+ times per day" for beverages. Additionally, each item included three possible portion size responses. The methods of Subar et al [33] along with national dietary data from the U.S. Department of Agriculture's 1994–1996 Continuing Survey of Food Intake by Individuals (CSFII) [34] were used to construct the food items, portion sizes, nutrient database, and pyramid food servings database. A recipe file was used by the Pyramid Servings Database to disaggregate food mixtures into component ingredients and assign the components to food groups. The FFQs used by the NIH-AARP Diet and Health Study have been validated using two 24-hour recalls in a subset of the cohort [35]. The daily micronutrient consumption of an individual was determined by multiplying the frequency of consumption of each line item by its micronutrient content (determined from CSFII) and summing over all line items.

Statistical Analysis

Cox proportional hazards regression [36] with person-years as the underlying time metric was used to estimate the cause-specific hazard ratios (HR) and corresponding 95% confidence intervals (CI) for thyroid cancer by quintiles of selenium and other micronutrient intake for thyroid cancer overall and the papillary and follicular subtypes. Quintiles of micronutrient intake were assessed first using the same cut points for men and women, resulting in an equal number of participants in each quintile when summed across men and women but not within sex. We then ran the analysis using separate, sex-specific, cut points for men and women. We assessed and verified, using cumulative sums of martingale residuals [37], that there was no violation of the proportional hazards assumption. We tested for linear trend by including the median value of each micronutrient category as a continuous variable in the model and assessing the significance of the Wald Chi-square p-value.

Micronutrient intake was adjusted for total energy intake using the nutrient residual method [38], which computes nutrient intake by first removing both calorie and micronutrient outliers, separately by sex, and then taking residuals from the regression model with total caloric intake as the independent variable and absolute nutrient intake as the dependent variable and adding a fixed constant (mean caloric intake by sex) of the study population.

Our minimal model was adjusted for age (continuous) and sex in the overall analysis and age in the sex specific analysis. Potential confounding variables were identified based on a review of the literature and previous studies on thyroid cancer using the NIH-AARP Diet and Health Study. Potential confounders were assessed using a backward elimination method in which we

Table 1. Baseline characteristics of the NIH-AARP Diet and Health Study cohort by quintiles of residual adjusted selenium intake.

	Male N = 287,944					Female N = 194,683				
	Q1	Q2	Q3	Q4	Q5	Q1	Q2	Q3	Q4	Q5
N* =	22,396	39,795	60,176	77,287	88,290	74,165	56,767	36,385	19,275	8,271
Number of thyroid cancer cases	19	25	49	68	96	119	106	66	29	15
Number of papillary thyroid cancer cases	12	14	33	49	56	83	77	49	21	12
Number of follicular thyroid cancer cases	4	8	8	14	23	21	14	14	5	2
Energy adjusted Selenium quintile median (mcg/day Residual Method)	7.1	7.6	8	8.4	8.9	7.1	7.6	8	8.4	8.9
Unadjusted Selenium quintile median (mcg/day)	47	68.4	86.6	108.6	150.1	47	68.4	86.6	108.6	150.1

Parameter	Characteristics									
	Q1	Q2	Q3	Q4	Q5	Q1	Q2	Q3	Q4	Q5
Age[+]	62.4 (5.4)	62.5 (5.3)	62.4 (5.3)	62.1 (5.3)	61.7 (5.4)	62.1 (5.4)	61.9 (5.4)	61.7 (5.4)	61.5 (5.4)	61.5 (5.4)
Race White (%)	88.2	91.4	92.7	93.4	93.5	86.9	90.8	91.3	91.1	91.6
Black (%)	5.6	3.4	2.5	2.2	2.1	7.8	4.4	4	3.9	3.8
Other (%)	4.5	3.9	3.7	3.4	3.4	3.5	3.5	3.4	3.6	3.3
Currently Married (%)	79.6	83.9	85.6	86.4	85.7	41.7	45.9	46.9	46.4	43.9
BMI (kg/m^2)	26.8 (4.2)	26.8 (4.1)	27.0 (4.1)	27.3 (4.2)	27.7 (4.5)	26.3 (5.7)	26.8 (5.9)	27.2 (6.1)	27.5 (6.3)	27.8 (6.4)
Smoking History Never Smoker (%)	26.7	28.9	29.6	30	29.6	44.9	45	44	41.8	39.4
Former Smoker (%)	53.6	55	56.2	56.6	57.8	35.1	38.5	40.1	42.7	45.1
Current Smoker (%)	15.3	12	10.3	9.7	8.8	16.3	13.3	12.5	11.9	11.1
Education College Grad/PostGrad (%)	37.6	40.7	43.8	45.9	48.4	26.5	30.9	32.7	34.1	34.4
20 minutes Physical activity 5 or more times per week (%)	21.1	21.3	20.9	21.0	22.5	16.2	15.9	16.1	16.8	19.3
Post- menopause (% Yes)	n/a	n/a	n/a	n/a	n/a	94.5	93.9	93.8	93.5	93.2
Hormone Use (% Yes Ever)	n/a	n/a	n/a	n/a	n/a	51.3	55.2	55.2	54.5	51.5
Currently taking hormones (% yes)	n/a	n/a	n/a	n/a	n/a	42.2	46.1	46.2	45.5	43.2
Dietary Intakes (Vitamins Residually Adjusted) Calories	2,226.0 (1,018.8)	1,995.4 (875.0)	1,935.1 (800.6)	1,944.1 (781.5)	2,064.9 (812.5)	1,568.6 (677.7)	1,522.9 (616.0)	1,561.0 (623.4)	1,623.0 (647.0)	1,702.1 (648.5)
Vitamin C (mg/day)	9.7 (2.6)	9.6 (2.1)	9.5 (1.9)	9.4 (1.7)	9.2 (1.7)	9.5 (2.0)	9.2 (1.6)	9.1 (1.6)	9.0 (1.6)	8.9 (1.7)
Vitamin D (mcg/day)	1.2 (0.8)	1.5 (0.8)	1.6 (0.7)	1.7 (0.7)	1.8 (0.6)	1.2 (0.8)	1.4 (0.7)	1.4 (0.6)	1.5 (0.6)	1.7 (0.7)
Vitamin E (mg/day)	2.2 (0.5)	2.3 (0.4)	2.4 (0.4)	2.4 (0.4)	2.4 (0.3)	2.1 (0.4)	2.2 (0.3)	2.2 (0.3)	2.2 (0.3)	2.2 (0.3)
Beta-carotene (mcg/day)	9.5 (1.2)	9.7 (1.1)	9.8 (1.0)	9.9 (1.0)	10.0 (1.0)	9.9 (1.1)	10.1 (1.0)	10.1 (1.0)	10.2 (1.0)	10.2 (1.1)
Calcium (mg/day)	7.3 (0.6)	7.5 (0.5)	7.5 (0.5)	7.6 (0.5)	7.5 (0.4)	7.3 (0.5)	7.4 (0.5)	7.4 (0.5)	7.3 (0.4)	7.3 (0.4)
Folate (mg/day)	12.8 (1.7)	13.3 (1.2)	13.4 (1.1)	13.6 (1.1)	13.8 (1.0)	12.5 (1.3)	12.8 (1.0)	12.9 (1.0)	13.0 (1.0)	13.1 (1.0)

Table 1. Cont.

	Male N = 287,944					Female N = 194,683				
	Q1	Q2	Q3	Q4	Q5	Q1	Q2	Q3	Q4	Q5
Magnesium (mg/day)	10.8 (1.0)	11.2 (0.7)	11.3 (0.7)	11.4 (0.6)	11.5 (0.6)	10.6 (0.8)	10.8 (0.6)	10.8 (0.6)	10.9 (0.6)	11.0 (0.6)
Zinc (mg/day)	2.4 (0.4)	2.7 (0.4)	2.9 (0.3)	3.0 (0.3)	3.1 (0.3)	2.4 (0.4)	2.6 (0.3)	2.6 (0.3)	2.7 (0.3)	2.7 (0.3)

*The same quintile cut points were used for men and women. In an analysis using sex-specific cut points, the results presented in the manuscript were unchanged.
†Mean and (standard deviation).

removed the least significant covariate in the model and assessed whether this changed the main exposure HR by more than 10%. We assessed for confounding by age (continuous), sex, body mass index (BMI; <18.5, 18.5–24.99, 25–29.99, >30), total calories (continuous), education (high school or less, some college, college or post graduate, unknown), physical activity (<1–2 times per week, 3–4 times per week, 5+ times per week, unknown), race (White, Black, other, unknown), smoking status (never, current, former, unknown), marital status (yes, no, unknown), alcohol intake (≤1, 2, 3, 4+ drinks/day), and micronutrient intake (continuous intake of vitamin C, vitamin E, beta-carotene, and folate). Effect modification was assessed using the likelihood ratio test comparing a model with the cross-product terms to one without.

Finally, we tested the assumption that the risk of thyroid cancer was log-linearly associated with selenium and other micronutrient intake by comparing the linear model with a non-parametric regression curve obtained with restricted cubic splines [39]. We used a stepwise selection process to identify the number and location of knots for each micronutrient analyzed. The likelihood ratio test was used to fit the restricted cubic splines. SAS software version 9.3 (SAS Institute, Cary, NC) was used for all statistical analyses. All reported p-values are based on two-sided tests and an alpha level of 0.05.

Results

Over a total of 4,406,634 person-years of follow-up we identified 592 incident thyroid cancer cases (257 in men and 335 in women) of which 406 were of the papillary histologic subtype (164 in men and 242 in women) and 113 (57 in men and 56 in women) were of the follicular histologic subtype. Select characteristics of the study population are presented by quintile of dietary selenium intake (Table 1). Participants in the highest quintile of selenium intake were more likely to be married, never or former smokers, and college educated when compared to the lowest quintile of selenium intake. The mean and standard deviation for selenium intake was 94.0 ± 42.9 mcg/day (other micronutrient mean and standard deviation intakes presented in Table 2). The five largest contributors to selenium intake were breads & rolls (13.9%), pasta (6.3%), tuna (4.8%), fish – not fried (4.1%), and eggs (4.0%). Contributors to vitamin C intake were orange & grapefruit juices (29.1%), oranges & tangelos (7.9%), broccoli (7.3%), other juices (5.1%), and grapefruit (4.5%). The dietary contributors of each micronutrient studied were similar for men and women. Table 2 provides a detailed summary of the major dietary contributors of each micronutrient in our study.

Table 3 presents the association between dietary micronutrient intake and thyroid cancer risk. After controlling for potential confounders there were no statistically significant associations between increasing quintile of selenium intake and risk of thyroid cancer ($HR_{Q5 \text{ vs } Q1}$, 1.23; 95% CI, 0.92–1.65; P_{trend}, 0.26). There was a statistically significant increase in risk of thyroid cancer for the highest versus lowest quintile of vitamin C intake in our multivariable model prior to adjusting for antioxidant intake ($HR_{Q5 \text{ vs } Q1}$, 1.34; 95% CI, 1.02–1.76; P_{trend}, <0.01), but not after ($HR_{Q5 \text{ vs } Q1}$, 1.35; 95% CI, 0.99–1.84; P_{trend}, 0.09). Although we observed evidence of a statistically significant positive linear trend for vitamin C, the HRs were highest for the fourth versus first quintile and were slightly attenuated for the highest quintile. For the remaining micronutrients in our study there was no clear evidence of a positive or negative association, or of a linear trend, between intake of any micronutrient and thyroid cancer risk.

Table 2. Top Five Primary Dietary Sources for Micronutrients in NIH-AARP Diet and Health Study for Men and Women Combined.

Micronutrient[+] Mean (SD) Intake	Primary Source (%)	Secondary Source (%)	Tertiary Source (%)	Fourth Source (%)	Fifth Source (%)
Selenium: 94±42.9	Bread/Rolls (13.9)	Pasta (6.7)	Tuna (4.8)	Fish - Not Fried (4.1)	Eggs (4.0)
Vitamin C: 157.0±103.7	Orange/Grapefruit Juice (29.1)	Orange/Tangelos (7.9)	Broccoli (7.3)	Other Juice (5.1)	Grapefruit (4.5)
Beta-carotene: 4,234.3±3,767.6	Carrots (36.8)	Sweet Potatoes (12.4)	Spinach/Greens (9.8)	Vegetable Medley (5.3)	Broccoli (2.9)
Calcium: 766.5±429.1	Milk –1 & 2% (11.2)	Milk - Skim (10.8)	Milk - In Cereal (9.0)	Bread/Rolls (5.4)	Cereal (3.7)
Folate: 411.4±182.2	Cereal (14.5)	Orange/Grapefruit Juice (10.8)	Lettuce (4.9)	Bread/Rolls (4.7)	Spinach/Greens (4.1)
Magnesium: 326.2±129.7	Coffee (11.3)	Bread/Rolls (5.6)	Cereal (5.4)	Orange/Grapefruit Juice (4.2)	Bananas (3.5)
Vitamin E: 8.8±4.6	Cereal (10.2)	Salad Dressing (7.5)	Oils (3.8)	Nuts/Seed -Whole (3.7)	Tomato Sauces w/Meat (3.6)
Zinc: 10.4±4.8	Cereal (11.0)	Beef - Steak (4.8)	Bread/Rolls (4.4)	Beef - Burger (4.2)	Beef - Meatball (3.3)

[+]Micronutrients measured as followed: Selenium (mcg/day), Betacarotene (mcg/day), Calcium (mg/day), Folate (mcg/day), Magnesium (mg/day), Vitamin C (mg/day), Vitamin D (mcg/day), Vitamin E (mg/day), Zinc (mg/day). Vitamin D food sources not available.

We also evaluated the associations between micronutrient intake and thyroid cancer separately by sex because of the higher incidence of thyroid cancer in women and suspected differences in the etiology of thyroid cancer by sex [40] (Tables S1 and S2). Although for some micronutrients, such as vitamin C, the strength and direction of the association appeared to differ by sex there were no statistically significant interactions by sex. Results restricted to papillary or follicular thyroid cancer, the two most common subtypes of thyroid cancer, were largely similar to the overall analysis (Tables S3–S8).

When we restricted the analysis to only participants with complete covariate information the results did not vary substantially from the presented data with an indicator variable for unknown values. There was also no indication that additionally adjusting for intake of the antioxidants vitamin C, vitamin E, beta-carotene, and folate in our multivariable model had an impact on the association with selenium or any other micronutrient (multivariable model 2 vs. 3). Finally there was no evidence that using sex-specific quintiles of micronutrient intake in men and women had a substantial impact on the results presented.

Discussion

Using a large prospective cohort study, we observed no evidence of an association between quintile of selenium intake and incidence of total thyroid cancer or the papillary and follicular subtypes. This was the first prospective study on this topic. We observed a suggestion of a positive linear relationship between increasing quintile of vitamin C and the risk of total thyroid cancer and the papillary and follicular subtypes, but no evidence of an association between thyroid cancer risk and calcium, folate, vitamin E, vitamin D, magnesium, or zinc intakes.

Selenium is a biologically important micronutrient shown to have a preventive effect for cancers other than thyroid [27]. Selenium assists in the production and proper function of thyroid hormones and has antioxidant properties [23–26]. Several limitations of our study may have contributed to the null results observed for selenium. Using an FFQ to determine dietary selenium intake may have resulted in exposure misclassification because the selenium content of soil varies largely by geographic region [41,42], resulting in a difference in the accumulation of

selenium in animals and plant products and in the selenium content of foods. An additional limitation of our study is that measurement error in dietary micronutrient intake, which was likely non-differential, may have attenuated the associations of interest. Use of bio-specimens in future studies would allow for more accurate measurement of an individual's selenium intake. Measuring plasma selenium is a good indicator of recent intake [43,44] although plasma selenium concentrations are not useful for determining long term intake and recent infections can influence plasma selenium levels [45]. Toenail clippings have been used to indicate long term selenium intake and are useful in investigations between selenium and chronic diseases [45,46].

Another limiting factor of using an FFQ in our study population was that the participants were not asked about every possible contributor to dietary selenium. For example, dietary consumption of Brazilian nuts, the food with the highest selenium content [47], was not assessed. Intake of fresh halibut and sardines, two other food sources with high selenium content were also not assessed. A study of the major food sources of antioxidants in U.S. adults [48] showed that Brazilian nuts, halibut, and sardines are not major dietary contributors of selenium in the U.S. diet and that the dietary contributors of selenium included in our study are representative of the major sources of selenium in the U.S. diet. Therefore, the impact of not having information on Brazilian nut consumption in our study was likely minimal. Further, the bio-availability of selenium from different food sources varies, and wheat, which was the highest contributor of dietary selenium in our study, has a high bio-availability [49].

In our analysis of other micronutrients and their association with thyroid cancer we saw evidence for an association with intake of vitamin C only. Higher intake of vitamin C appeared to be positively associated with thyroid cancer risk. Biologically, vitamin C has been shown to improve and mediate abnormalities seen in the levels of thyroid hormones and thyroid stimulating hormone in the serum of humans [50] and rats [51]. Previous studies have indicated both a positive and negative association between increased vitamin C intake and thyroid cancer [28]. A recent case-control study by Jung et al. indicated that vitamin C intake and citrus consumption in controls was higher than in individuals with either benign or malignant thyroid cancer, although the difference was not statistically significant [13]. A positive

Table 3. Hazard Ratios (HRs) and corresponding 95% confidence intervals (CIs) for total thyroid cancer by quintile of micronutrient intake in men and women combined in The NIH-AARP Diet and Health Study.

Selenium	Q1	Q2	Q3	Q4	Q5	P trend
Median Intake	7.05	7.64	8.03	8.41	8.93	
Number of Cases	138	131	115	97	111	
Age-adjusted HR[1] (95% CI)	1.00 (ref)	0.96 (0.75, 1.22)	0.85 (0.66, 1.09)	0.72 (0.56, 0.94)	0.83 (0.65, 1.07)	0.03
Multivariable HR[2] (95% CI)	1.00 (ref)	1.00 (0.79, 1.28)	0.99 (0.76, 1.29)	1.00 (0.75, 1.33)	1.23 (0.92, 1.65)	0.26
Multivariable HR[3] (95% CI)	1.00 (ref)	1.07 (0.83, 1.36)	1.07 (0.81, 1.40)	1.10 (0.82, 1.48)	1.35 (0.99, 1.84)	0.09
Vitamin C	Q1	Q2	Q3	Q4	Q5	P trend
Median Intake	7	8.41	9.36	10.28	11.67	
Number of Cases	36	43	46	67	63	
Age-adjusted HR[1] (95% CI)	1.00 (ref)	1.19 (0.77, 1.86)	1.27 (0.82, 1.96)	1.80 (1.20, 2.70)	1.57 (1.04, 2.36)	0.01
Multivariable HR[2] (95% CI)	1.00 (ref)	1.03 (0.78, 1.36)	1.07 (0.81, 1.42)	1.43 (1.09, 1.86)	1.34 (1.02, 1.76)	<0.01
Multivariable HR[3] (95% CI)	1.00 (ref)	1.03 (0.77, 1.37)	1.11 (0.82, 1.49)	1.47 (1.09, 1.98)	1.46 (1.05, 2.04)	<0.01
Betacarotene	Q1	Q2	Q3	Q4	Q5	P trend
Median Intake	8.67	9.38	9.89	10.43	11.3	
Number of Cases	111	116	107	123	132	
Age-adjusted HR[1] (95% CI)	1.00 (ref)	1.03 (0.80, 1.34)	0.95 (0.73, 1.24)	1.09 (0.84, 1.40)	1.16 (0.90, 1.50)	0.2
Multivariable HR[2] (95% CI)	1.00 (ref)	1.00 (0.77, 1.30)	0.87 (0.67, 1.15)	1.01 (0.78, 1.32)	1.07 (0.83, 1.39)	0.6
Multivariable HR[3] (95% CI)	1.00 (ref)	0.93 (0.71, 1.22)	0.78 (0.59, 1.04)	0.87 (0.66, 1.16)	0.89 (0.65, 1.20)	0.38
Calcium	Q1	Q2	Q3	Q4	Q5	P trend
Median Intake	8.67	9.38	9.89	10.43	11.3	
Number of Cases	111	116	107	123	132	
Age-adjusted HR[1] (95% CI)	1.00 (ref)	1.03 (0.80, 1.34)	0.95(0.73, 1.24)	1.09 (0.84, 1.40)	1.16(0.90, 1.50)	0.2
Multivariable HR[2] (95% CI)	1.00 (ref)	0.94 (0.71, 1.23)	0.92 (0.69, 1.23)	1.15 (0.85, 1.56)	1.08 (0.74, 1.58)	0.63
Multivariable HR[3] (95% CI)	1.00 (ref)	0.92 (0.69, 1.21)	0.90 (0.66, 1.21)	1.11 (0.81, 1.53)	1.00 (0.67, 1.48)	0.94
Folate	Q1	Q2	Q3	Q4	Q5	P trend
Median Intake	11.7	12.58	13.17	13.78	14.72	
Number of Cases	121	115	129	110	117	
Age-adjusted HR[1] (95% CI)	1.00 (ref)	0.94 (0.73, 1.22)	1.06 (0.83, 1.36)	0.90 (0.70, 1.17)	0.96 (0.75, 1.24)	0.7
Multivariable HR[2] (95% CI)	1.00 (ref)	0.94 (0.73, 1.22)	1.17 (0.91, 1.51)	1.04 (0.79, 1.36)	1.27 (0.97, 1.67)	0.06
Multivariable HR[3] (95% CI)	1.00 (ref)	0.94 (0.73, 1.22)	1.17 (0.91, 1.51)	1.04 (0.79, 1.36)	1.27 (0.97, 1.67)	0.06
Vitamin E	Q1	Q2	Q3	Q4	Q5	P trend
Median Intake	1.85	2.09	2.26	2.43	2.71	
Number of Cases	125	133	126	108	97	
Age-adjusted HR[1] (95% CI)	1.00 (ref)	1.06 (0.83, 1.35)	1.01 (0.79, 1.29)	0.87 (0.67, 1.12)	0.78 (0.60, 1.02)	0.03
Multivariable HR[2] (95% CI)	1.00 (ref)	1.09 (0.84, 1.39)	1.12 (0.87, 1.45)	1.05 (0.80, 1.37)	0.99 (0.75, 1.31)	0.92
Multivariable HR[3] (95% CI)	1.00 (ref)	1.07 (0.83, 1.38)	1.11 (0.85, 1.44)	1.02 (0.77, 1.35)	0.93 (0.69, 1.25)	0.58
Vitamin D	Q1	Q2	Q3	Q4	Q5	P trend
Median Intake	0.58	1.14	1.51	1.89	2.46	
Number of Cases	114	116	135	117	109	
Age-adjusted HR[1] (95% CI)	1.00 (ref)	1.04 (0.80, 1.35)	1.19 (0.91, 1.57)	0.99 (0.73, 1.34)	0.90 (0.62, 1.31)	0.77
Multivariable HR[2] (95% CI)	1.00 (ref)	1.10 (0.84, 1.44)	1.24 (0.94, 1.65)	1.09 (0.79, 1.49)	1.03 (0.70, 1.50)	0.72
Multivariable HR[3] (95% CI)	1.00 (ref)	1.08 (0.82, 1.43)	1.25 (0.93, 1.66)	1.08 (0.78, 1.49)	1.03 (0.70, 1.53)	0.71
Magnesium	Q1	Q2	Q3	Q4	Q5	P trend
Median Intake	10.1	10.72	11.11	11.49	12.03	
Number of Cases	145	115	116	113	100	
Age-adjusted HR[1] (95% CI)	1.00 (ref)	0.79 (0.62, 1.01)	0.80 (0.63, 1.02)	0.79 (0.61, 1.00)	0.70 (0.54, 0.90)	<0.01

Table 3. Cont.

Magnesium	Q1	Q2	Q3	Q4	Q5	P_{trend}
Multivariable HR[2] (95% CI)	1.00 (ref)	0.85 (0.66, 1.09)	0.96 (0.74, 1.23)	1.03 (0.79, 1.34)	1.00 (0.76, 1.33)	0.66
Multivariable HR[3] (95% CI)	1.00 (ref)	0.79 (0.61, 1.03)	0.86 (0.66, 1.14)	0.88 (0.66, 1.19)	0.80 (0.57, 1.13)	0.33
Zinc	**Q1**	**Q2**	**Q3**	**Q4**	**Q5**	P_{trend}
Median Intake	2.24	2.54	2.75	2.95	3.24	
Number of Cases	146	138	106	104	98	
Age-adjusted HR[1] (95% CI)	1.00 (ref)	0.95 (0.75, 1.20)	0.74 (0.57, 0.94)	0.73 (0.57, 0.94)	0.69 (0.54, 0.90)	<0.01
Multivariable HR[2] (95% CI)	1.00 (ref)	0.96 (0.76, 1.22)	0.85 (0.66, 1.11)	0.95 (0.72, 1.25)	0.98 (0.73, 1.32)	0.96
Multivariable HR[3] (95% CI)	1.00 (ref)	0.94 (0.74, 1.21)	0.82 (0.62, 1.08)	0.87 (0.64, 1.18)	0.89 (0.63, 1.25)	0.36

[1]Adjusted for entry age.
[2]Adjusted for entry age, sex (overall), calories, smoking status, race, education, BMI, and physical activity.
[3]Additionally adjusted for vitamin C, vitamin E, beta-carotene, and folate.

association between citrus consumption and thyroid cancer was also seen by Xiao et al. and appeared to be driven by the intake of orange and grapefruit juice [52]. Orange and grapefruit juices were the primary contributors of vitamin C intake in our study. Xiao et al. suggest some individuals may have included artificially flavored drinks in their report of orange and grapefruit juice consumption, leading to misclassification and potentially a false positive association.

It is also possible that residual confounding may have contributed to our observed association between increasing quintile of vitamin C intake and thyroid cancer risk. Participants in the highest quintile of vitamin C intake had higher education, which has been associated with healthcare utilization and increased rates of thyroid cancer diagnosis [53,54]. Additionally, individuals in the highest quintile of vitamin C intake had greater physical activity and lower caloric intake, which are characteristics associated with a health conscious lifestyle. A healthier lifestyle, again, may correspond to greater healthcare utilization and increased opportunity for thyroid cancer detection. Although we did control for these factors in our analysis it is possible that they were measured imperfectly or that we were unable to fully capture aspects of a healthy lifestyle or health consciousness that could influence detection. It seems unlikely, however, that increased detection would occur only in men and impact only the results for vitamin C.

Our study has several strengths. While previous studies have utilized case-control designs this study is the first to examine the association between selenium, and other micronutrient, intake and thyroid cancer using a large prospective cohort design, greatly reducing the possibility of recall and selection biases [55]. Furthermore, there was generally complete follow-up of study participants to ascertain outcomes and a relatively large number of thyroid cancer cases, which allowed us to analyze thyroid cancer by histologic subtype. Additionally we had covariate information that allowed us to adjust for potential confounders in our multivariable analysis of this study population. Using quintiles of selenium intake allowed for a natural comparison group, as the lowest unadjusted quintile of intake for selenium (less than 47 mcg/day) was the only group which fell below the recommended daily intake for men and women of 55 mcg/day [56].

It is possible that exposure to essential micronutrients earlier in life, or over the course of an individual's lifetime, may be more important in determining thyroid cancer risk. Although the NIH-AARP study utilized follow-up questionnaires that asked about

diet at different points in the participant's life, the follow-up questions were not designed to assess selenium and therefore were not used in this analysis. Future studies with information on dietary intake over the span of a person's life could add valuable information to dietary studies on thyroid cancer risk. Finally, because the FFQ in this study was not designed to measure iodine consumption, we did not have information on daily iodine intake, which is important for thyroid function and may be a potential important confounding variable. However, iodine deficiency and the subsequent impact of not controlling for iodine intake on our study results is likely minimal, as iodine fortification of salt and other foods has occurred in the United States since the 1920's [57].

In conclusion we observed no evidence of an association between thyroid cancer and quintile of selenium intake. Given the unexpected positive association observed for vitamin C, further evaluation of the association between vitamin C and thyroid cancer in other prospective cohorts is warranted. Valuable follow-up studies of selenium and thyroid cancer risk could evaluate selenium intake and intake of other micronutrients more objectively, such as by use of biomarkers.

Supporting Information

Table S1 Hazard Ratios (HRs) and corresponding 95% confidence intervals (CIs) for total thyroid cancer by quintile of micronutrient intake among men in The NIH-AARP Diet and Health Study.

Table S2 Hazard Ratios (HRs) and corresponding 95% confidence intervals (CIs) for total thyroid cancer by quintile of micronutrient intake among women in The NIH-AARP Diet and Health Study.

Table S3 Hazard Ratios (HRs) and corresponding 95% confidence intervals (CIs) for papillary thyroid cancer by quintile of micronutrient intake among men and women combined in The NIH-AARP Diet and Health Study.

Table S4 Hazard Ratios (HRs) and corresponding 95% confidence intervals (CIs) for papillary thyroid cancer

by quintile of micronutrient intake among men in The NIH-AARP Diet and Health Study.

Table S5 Hazard Ratios (HRs) and corresponding 95% confidence intervals (CIs) for papillary thyroid cancer by quintile of micronutrient intake among women in The NIH-AARP Diet and Health Study.

Table S6 Hazard Ratios (HRs) and corresponding 95% confidence intervals (CIs) for total thyroid cancer by quintile of micronutrient intake among men in The NIH-AARP Diet and Health Study.

Table S7 Hazard Ratios (HRs) and corresponding 95% confidence intervals (CIs) for follicular thyroid cancer

by quintile of micronutrient intake among men in The NIH-AARP Diet and Health Study.

Table S8 Hazard Ratios (HRs) and corresponding 95% confidence intervals (CIs) for follicular thyroid cancer by quintile of micronutrient intake among women in The NIH-AARP Diet and Health Study.

Author Contributions

Conceived and designed the experiments: TJO CMK MAG. Performed the experiments: TJO. Analyzed the data: TJO AGD. Contributed to the writing of the manuscript: TJO. Revised the manuscript critically for important intellectual content: CMK AGD MAG.

References

1. Kilfoy BA, Zheng T, Holford TR, Han X, Ward MH, et al. (2009) International patterns and trends in thyroid cancer incidence, 1973–2002. Cancer Causes Control 20: 525–531.
2. Colonna M, Grosclaude P, Remontet L, Schvartz C, Mace-Lesech J, et al. (2002) Incidence of thyroid cancer in adults recorded by French cancer registries (1978–1997). Eur J Cancer 38: 1762–1768.
3. dos Santos Silva I, Swerdlow AJ (1993) Thyroid cancer epidemiology in England and Wales: time trends and geographical distribution. Br J Cancer 67: 330–340.
4. Enewold L, Zhu K, Ron E, Marrogi AJ, Stojadinovic A, et al. (2009) Rising thyroid cancer incidence in the United States by demographic and tumor characteristics, 1980–2005. Cancer Epidemiol Biomarkers Prev 18: 784–791.
5. Montanaro F, Pury P, Bordoni A, Lutz JM, Swiss Cancer Registries N (2006) Unexpected additional increase in the incidence of thyroid cancer among a recent birth cohort in Switzerland. Eur J Cancer Prev 15: 178–186.
6. Cramer JD, Fu P, Harth KC, Margevicius S, Wilhelm SM (2010) Analysis of the rising incidence of thyroid cancer using the Surveillance, Epidemiology and End Results national cancer data registry. Surgery 148: 1147–1152; discussion 1152–1143.
7. Ron E, Schneider A (2006) Thyroid Cancer. In Shottenfeld, D.; Fraumeni, JF., Jr, editors. Cancer Epidemiology and Prevention. New York: Oxford University Press.
8. Leitzmann MF, Brenner A, Moore SC, Koebnick C, Park Y, et al. (2010) Prospective study of body mass index, physical activity and thyroid cancer. Int J Cancer 126: 2947–2956.
9. Kitahara CM, Platz EA, Freeman LE, Hsing AW, Linet MS, et al. (2011) Obesity and thyroid cancer risk among U.S. men and women: a pooled analysis of five prospective studies. Cancer Epidemiol Biomarkers Prev 20: 464–472.
10. Kitahara CM, Platz EA, Park Y, Hollenbeck AR, Schatzkin A, et al. (2012) Body fat distribution, weight change during adulthood, and thyroid cancer risk in the NIH-AARP Diet and Health Study. Int J Cancer 130: 1411–1419.
11. Ward MH, Kilfoy BA, Weyer PJ, Anderson KE, Folsom AR, et al. (2010) Nitrate intake and the risk of thyroid cancer and thyroid disease. Epidemiology 21: 389–395.
12. Kilfoy BA, Zhang Y, Park Y, Holford TR, Schatzkin A, et al. (2011) Dietary nitrate and nitrite and the risk of thyroid cancer in the NIH-AARP Diet and Health Study. Int J Cancer 129: 160–172.
13. Jung SK, Kim K, Tae K, Kong G, Kim MK (2013) The effect of raw vegetable and fruit intake on thyroid cancer risk among women: a case-control study in South Korea. Br J Nutr 109: 118–128.
14. Dal Maso L, Bosetti C, La Vecchia C, Franceschi S (2009) Risk factors for thyroid cancer: an epidemiological review focused on nutritional factors. Cancer Causes Control 20: 75–86.
15. Bosetti C, Negri E, Kolonel L, Ron E, Franceschi S, et al. (2002) A pooled analysis of case-control studies of thyroid cancer. VII. Cruciferous and other vegetables (International). Cancer Causes Control 13: 765–775.
16. Markaki I, Linos D, Linos A (2003) The influence of dietary patterns on the development of thyroid cancer. Eur J Cancer 39: 1912–1919.
17. Galanti MR, Hansson L, Bergstrom R, Wolk A, Hjartaker A, et al. (1997) Diet and the risk of papillary and follicular thyroid carcinoma: a population-based case-control study in Sweden and Norway. Cancer Causes Control 8: 205–214.
18. Clero E, Doyon F, Chungue V, Rachedi F, Boissin JL, et al. (2012) Dietary iodine and thyroid cancer risk in French Polynesia: a case-control study. Thyroid 22: 422–429.
19. Clero E, Doyon F, Chungue V, Rachedi F, Boissin JL, et al. (2012) Dietary patterns, goitrogenic food, and thyroid cancer: a case-control study in French Polynesia. Nutr Cancer 64: 929–936.
20. Mack WJ, Preston-Martin S, Bernstein L, Qian D (2002) Lifestyle and other risk factors for thyroid cancer in Los Angeles County females. Ann Epidemiol 12: 395–401.
21. George SM, Park Y, Leitzmann MF, Freedman ND, Dowling EC, et al. (2009) Fruit and vegetable intake and risk of cancer: a prospective cohort study. Am J Clin Nutr 89: 347–353.
22. Kohrle J (1999) The trace element selenium and the thyroid gland. Biochimie 81: 527–533.
23. Hard GC (1998) Recent developments in the investigation of thyroid regulation and thyroid carcinogenesis. Environ Health Perspect 106: 427–436.
24. Dora JM, Machado WE, Rheinheimer J, Crispim D, Maia AL (2010) Association of the type 2 deiodinase Thr92Ala polymorphism with type 2 diabetes: case-control study and meta-analysis. Eur J Endocrinol 163: 427–434.
25. Kohrle J (2013) Pathophysiological relevance of selenium. J Endocrinol Invest 36: 1–7.
26. Beckett GJ, Arthur JR (2005) Selenium and endocrine systems. J Endocrinol 184: 455–465.
27. Vinceti M, Dennert G, Crespi CM, Zwahlen M, Brinkman M, et al. (2014) Selenium for preventing cancer. Cochrane Database Syst Rev 3: CD005195.
28. Zhang LR, Sawka AM, Adams L, Hatfield N, Hung RJ (2013) Vitamin and mineral supplements and thyroid cancer: a systematic review. Eur J Cancer Prev 22: 158–168.
29. Drutel A, Archambeaud F, Caron P (2013) Selenium and the thyroid gland: more good news for clinicians. Clin Endocrinol (Oxf) 78: 155–164.
30. Fritz A, Percy C, Jack A, Shanmugaratnam K, Parkin DM, et al. (2000) International Classification of Diseases for Oncology (ICD-O), 3rd Edition. WHO, Geneva.
31. Michaud DS, Midthune D, Hermansen S, Leitzmann M, Harlan L, et al. (2005) Comparison of Cancer Registry Case Ascertainment with SEER Estimates and Self-reporting in a Subset of NIH-AARP Diet and Health Study. Journal of Registry Management 32: 70–75.
32. North American Association of Central Cancer Registries (NAACCR). North American Association of Central Disease Registries. Standards for Cancer Registries Vol. 3. Standards for completeness, quality, analysis, and management of data. 2008.
33. Subar AF, Midthune D, Kulldorff M, Brown CC, Thompson FE, et al. (2000) Evaluation of alternative approaches to assign nutrient values to food groups in food frequency questionnaires. Am J Epidemiol 152: 279–286.
34. Friday JB (2006) MyPyramid Equivalents Database for USDA Survey Food Codes, 1994–2002 USDA, ARS, Community Nutrition Research Group.
35. Thompson FE, Kipnis V, Midthune D, Freedman LS, Carroll RJ, et al. (2008) Performance of a food-frequency questionnaire in the US NIH-AARP (National Institutes of Health-American Association of Retired Persons) Diet and Health Study. Public Health Nutr 11: 183–195.
36. Cox DR (1972) Regression Models and Life-Tables. Journal of the Royal Statistical Society 34: 187–220.
37. Lin DY, Wei LJ, Ying Z (1993) Checking the Cox model with cumulative sums of martingale-based residuals. Biometrika 80: 557–572.
38. Willett W, Stampfer MJ (1986) Total energy intake: implications for epidemiologic analyses. Am J Epidemiol 124: 17–27.
39. Durrleman S, Simon R (1989) Flexible regression models with cubic splines. Stat Med 8: 551–561.
40. Rahbari R, Zhang L, Kebebew E (2010) Thyroid cancer gender disparity. Future Oncol 6: 1771–1779.
41. Sunde RA. Selenium. In: Bowman B RR, eds. Present Knowledge in Nutrition. 9th ed. Washington, DC: International Life Sciences Institute; 2006: 480–97.
42. Sunde RA. Selenium. In: Ross AC CB, Cousins RJ, Tucker KL, Ziegler TR, eds. Modern Nutrition in Health and Disease. 11th ed. Philadelphia, PA: Lippincott Williams & Wilkins; 2012: 225–37.
43. Duffield AJ, Thomson CD, Hill KE, Williams S (1999) An estimation of selenium requirements for New Zealanders. Am J Clin Nutr 70: 896–903.

44. Persson-Moschos M, Alfthan G, Akesson B (1998) Plasma selenoprotein P levels of healthy males in different selenium status after oral supplementation with different forms of selenium. Eur J Clin Nutr 52: 363–367.

45. Longnecker MP, Stampfer MJ, Morris JS, Spate V, Baskett C, et al. (1993) A 1-y trial of the effect of high-selenium bread on selenium concentrations in blood and toenails. Am J Clin Nutr 57: 408–413.

46. van den Brandt PA, Goldbohm RA, van 't Veer P, Bode P, Dorant E, et al. (1993) A prospective cohort study on toenail selenium levels and risk of gastrointestinal cancer. J Natl Cancer Inst 85: 224–229.

47. U.S. Department of Agriculture, Agricultural Research Service. USDA National Nutrient Database for Standard Reference, Release 25. Nutrient Data Laboratory Home Page, 2012.

48. Chun OK, Floegel A, Chung SJ, Chung CE, Song WO, et al. (2010) Estimation of antioxidant intakes from diet and supplements in U.S. adults. J Nutr 140: 317–324.

49. Mutanen M (1986) Bioavailability of selenium. Ann Clin Res 18: 48–54.

50. Jubiz W, Ramirez M (2014) Effect of vitamin C on the absorption of levothyroxine in patients with hypothyroidism and gastritis. J Clin Endocrinol Metab: jc20134360.

51. Ambali SF, Orieji C, Abubakar WO, Shittu M, Kawu MU (2011) Ameliorative effect of vitamin C on alterations in thyroid hormones concentrations induced by subchronic coadministration of chlorpyrifos and lead in wistar rats. J Thyroid Res 2011: 214924.

52. Xiao Q, Park Y, Hollenbeck AR, Kitahara CM (2014) Dietary Flavonoid Intake and Thyroid Cancer Risk in the NIH-AARP Diet and Health Study. Cancer Epidemiol Biomarkers Prev.

53. Sprague BL, Warren Andersen S, Trentham-Dietz A (2008) Thyroid cancer incidence and socioeconomic indicators of health care access. Cancer Causes Control 19: 585–593.

54. Roche LM, Niu X, Pawlish KS, Henry KA (2011) Thyroid cancer incidence in New Jersey: time trend, birth cohort and socioeconomic status analysis (1979–2006). J Environ Public Health 2011: 850105.

55. Key TJ, Schatzkin A, Willett WC, Allen NE, Spencer EA, et al. (2004) Diet, nutrition and the prevention of cancer. Public Health Nutr 7: 187–200.

56. (2000) Institute of Medicine, Food and Nutrition Board. Dietary Reference Intakes: Vitamin C, Vitamin E, Selenium, and Carotenoids. National Academy Press, Washington, DC.

57. Leung AM, Braverman LE, Pearce EN (2012) History of U.S. iodine fortification and supplementation. Nutrients 4: 1740–1746.

4

Effects of Changes in Food Supply at the Time of Sex Differentiation on the Gonadal Transcriptome of Juvenile Fish. Implications for Natural and Farmed Populations

Noelia Díaz[¤], Laia Ribas, Francesc Piferrer*

Institut de Ciències del Mar, Consejo Superior de Investigaciones Científicas (CSIC), Barcelona, Spain

Abstract

Background: Food supply is a major factor influencing growth rates in animals. This has important implications for both natural and farmed fish populations, since food restriction may difficult reproduction. However, a study on the effects of food supply on the development of juvenile gonads has never been transcriptionally described in fish.

Methods and Findings: This study investigated the consequences of growth on gonadal transcriptome of European sea bass in: 1) 4-month-old sexually undifferentiated fish, comparing the gonads of fish with the highest vs. the lowest growth, to explore a possible link between transcriptome and future sex, and 2) testis from 11-month-old juveniles where growth had been manipulated through changes in food supply. The four groups used were: i) sustained fast growth, ii) sustained slow growth, iii) accelerated growth, iv) decelerated growth. The transcriptome of undifferentiated gonads was not drastically affected by initial natural differences in growth. Further, changes in the expression of genes associated with protein turnover were seen, favoring catabolism in slow-growing fish and anabolism in fast-growing fish. Moreover, while fast-growing fish took energy from glucose, as deduced from the pathways affected and the analysis of protein-protein interactions examined, in slow-growing fish lipid metabolism and gluconeogenesis was favored. Interestingly, the highest transcriptomic differences were found when forcing initially fast-growing fish to decelerate their growth, while accelerating growth of initially slow-growing fish resulted in full transcriptomic convergence with sustained fast-growing fish.

Conclusions: Food availability during sex differentiation shapes the juvenile testis transcriptome, as evidenced by adaptations to different energy balances. Remarkably, this occurs in absence of major histological changes in the testis. Thus, fish are able to recover transcriptionally their testes if they are provided with enough food supply during sex differentiation; however, an initial fast growth does not represent any advantage in terms of transcriptional fitness if later food becomes scarce.

Editor: Balasubramanian Senthilkumaran, University of Hyderabad, India

Funding: Research was supported by the Spanish Ministry of Science projects "Epigen-Aqua" (AGL2010-15939) and "Aquagenomics" (CDS2007-0002) to FP. ND was supported by a scholarship from the Government of Spain (BES-2007-14273) and then by a contract by the Epigen-Aqua project. LR was supported by an Aquagenomics postdoctoral contract and Epigen-Aqua project. The funders had no role in study design, data collection and analysis, decision to publish, or preparation of the manuscript.

Competing Interests: The authors have declared that no competing interests exist.

* Email: piferrer@icm.csic.es

¤ Current address: Max Planck Institute for Molecular Biomedicine, Regulatory Genomics Lab, Münster, Germany

Introduction

Food availability and energetic demands fluctuate in most habitats. Animals are capable of sensing their inner energy levels and the external energy availability and thus act accordingly by long-term investments in processes like growth, immune functions or reproduction when food availability is not a problem, or by ensuring survival when food is scarce [1]. In fish, there is a tight relationship between food availability and reproduction [1,2] since it can alter the timing and duration of spawning, fecundity and egg size [3,4], or the timing of the reproductive cycles [5]. Favorable feeding conditions produce early maturation of individuals [6] while a decrease in food availability causes a decrease in energy transfer to the gonads [7], but this relationship may present important differences between species since fish constitute a vast phylogenetic group with different behaviors and reproduction types [8].

The European sea bass (*Dicentrarchus labrax*) is a gonochoristic species with a polygenic sex determining system [9] presenting a long sexually undifferentiated process with sexual dimorphism at the time of sex differentiation onset (~150 days post hatch, dph, for females and ~180 dph for males) [10–13]. However, this dimorphism is more related to the attained length than to age [14]. The relationship between growth and sex differentiation has been previously studied in sea bass [9,12–19]. There is a relationship between body weight and sex since not only sea bass females are larger than males, but also both males and females in batches with higher percent females were bigger than males and females of batches with a lower female percent [9]. Further, early size-gradings of the population (at 66 and 123–143 dph, [13]; at 70 dph, [15]; or at 82 dph, [17]) selecting for the largest fish resulted in ~90% of females, but the opposite, i.e., selecting for the smallest fish, produced only ~65% males at one year of age, meaning that while the largest fish are essentially all females, among the smallest fish there are both males and females [13].

Recently, two experiments on growth rate alteration by manipulating food supply during the sex differentiation period were conducted with the European sea bass in our laboratory [19]. The first experiment showed that transiently but severely reducing food supply starting towards the end, middle or even at the beginning of the sex differentiation period – and thus negatively affecting growth – did not affect resulting sex ratios, indicating that gender was already fixed before the above mentioned period started [19]. The second experiment involved four groups of fish, which, through controlling food supply, were made to experience different growth rates during the sex differentiation period. Two groups, one fast-growing and the other slow-growing, originated from the fast-growing fish at 127 dph. The other two groups, also one fast-growing and the other slow-growing, originated from the slow-growing fish at 127 dph. In this case, there were differences in the final sex ratio of the population as fast-growing-derived groups presented more number of females (~40%) than the slow-growing-derived groups (~10%). Thus, the differences in the final sex ratio were not related to the growth rate during the sex differentiation but to whether fish derived from the fast- or slow-growing fish at 127 dph. These results confirmed the results of the first experiment and indicated that before the first signs of sex differentiation appear the relationship between growth and sex is already established, confirming that in the European sea bass larger sizes are associated with female development [13,19].

Partition of consumed energy into growth, energy storage and gonads according to temporal food availability, metabolic demands and reproductive needs have been studied since a long time ago [20]. Recently, with the advent of new technologies, the underlying mechanisms including associated changes in global gene expression can be investigated. However, transcriptomic analyses in fish have traditionally addressed nutrition and reproduction topics separately. Hence, while growth studies have put efforts towards the effects of diet substitutions [21–24], stocking density and food ration [25–27] and comparing domesticated vs. transgenic fish [28], on the other hand, reproduction-related transcriptome analyses have focused on describing changes associated with gonad maturation and differences between sexes [29–32], environmental effects [33] or hormonal treatment effects [34,35].

However, a study directly analyzing the effects of food supply on reproduction and, particularly, on the development of juvenile gonads has never been described in fish. In mitten crab (*Eriocheir sinensis*) during early development, when crabs store significant amounts of energy in the hepatopancreas, Jiang and collaborators [36] found four genes in the hepatopancreas and 13 genes in testis related to nutritional control, and three genes in the hepatopancreas and eight in the testes related to regulation of reproduction. Among the former, arginine kinase, zinc-finger proteins or leptin were upregulated in the hepatopancreas transcriptome as a sign of energy storage for further energy-demand of the reproductive

processes. Genes involved in the regulation of reproduction, such as cyclins, kinetochore spindle formation or the heat shock protein 70, were upregulated in testis and promoted an increase in cell division during spermatogenesis. In rats, dietary energy intake changes (restrictions and excesses) but also food availability had profound effects in gonads from both sexes at different levels (biochemical, endocrine, behavioral and genetic) [37]. Moreover, a transcriptomic analysis of the gonads of these rats facing diet restriction or excess showed how females were more affected by ration changes than males. Males were also better adapted to an intermittent fasting by increasing the probability of an eventual fertilization, while females were able to sense the food restriction and behaved as sub-fertile females [38].

The present study is based on samples collected in experiment 2 of Díaz et al. [19] described above and had two objectives: 1) to analyze the transcriptional differences in sexually differentiating European sea bass gonads from the naturally fastest growing vs. the slowest growing fish at 127 dph, i.e., before the first histological signs of sex differentiation at 150 dph [13], but after the first signs of molecular sex differentiation at 120 dph [39], to explore a possible link of transcriptomic signatures with future sex; and 2) the consequences of food availability between 133–337 dph (juvenile growth) on the subsequent testes transcriptome by analyzing the effects of growth acceleration and deceleration.

Materials and Methods

Animals and rearing conditions

As stated above, the fish that were transcriptomically analyzed in this study are the same fish described in Experiment 2 of Díaz et al. [19]. Briefly, European sea bass eggs obtained from a commercial hatchery were collected at one day post fertilization (dpf) on March 2009, transported to our experimental aquarium facilities and hatched following established procedures for this species [40] with minor changes, as previously described [18,19]. Fish were reared under natural conditions of photoperiod, pH (~7.9), salinity (~37.8 ppt), oxygen saturation (85–100%) and with a water renewal rate of 30% vol·h^{-1}. In order to avoid temperature influences on the sex ratio, the thermal regime used and previously described [19] included egg spawning at 13–14°C and larval rearing at 16±1°C until 60 dpf. Then, temperature was increased to 21°C at a rate of 0.5–1°C·day^{-1} and maintained until the first fall, when it was let to follow the natural temperature. The rearing density was kept low to avoid any possible distorting effect on sex ratios (details in [19]). Fish were manually fed three times a day with artemia AF, then artemia EG enriched with amino acid (INVE Aquaculture, Belgium) and dry food (ProAqua, S.A., Spain) of the appropriate pellet size as fish grew. Unless otherwise stated, juveniles and adults were fed *ad libitum* two times a day.

Fish were treated in agreement with the European Convention for the Protection of Animals used for Experimental and Scientific Purposes (ETS Nu 123, 01/01/91). Our facilities are approved for animal experimentation by the Ministry of Agriculture and Fisheries (certificate number 08039-46-A) in accordance of Spanish law (Real Decreto 223 of March 1988) and the experimental protocol was approved by the Spanish National Research Council (CSIC) Ethics Committee within project AGL2010-15939. Animals were sacrificed by an overdose of 2-phenoxyethanol (2PE) followed by severing of the spinal cord.

Experimental design

Details of the experimental design can also be found in Díaz et al. [19]. Briefly, fish were individually size-graded at 127 dph, i.e., at ~4 cm standard length (SL), before the histological process of sex differentiation started (∼8 cm SL), and separated into two groups according to the SL they had attained and comprising the two extremes of the normal curve distribution: a fast growth group (group F), with a mean size of 5.0 cm SL (range 4.2–6.4 cm), and a slow growth group (group S), with a mean size of 3.5 cm SL (range 2.6–3.7 cm). After checking that fish of each group was of the desired size, then at 133 dph (time 1, T1, i.e., when fish were 4 months old) each group was further subdivided into two tanks (n = 79 fish per tank). On one hand, the F group was subdivided into two groups with initial similar mean SL and BW: the fast-fast group (FF), in which growth rates from that moment onwards were as before, and the fast-slow group (FS), in which the growth rate was reduced to match what had been the growth rate of group S until then. On the other hand, the S group was also subdivided into two groups with initial similar mean SL and BW: the slow-fast group (SF), in which the growth rate was increased to match what had been the growth rate of group F until then, and the slow-slow group (SS), in which the growth rates from that moment onwards were as before (see Fig. 1A for a diagram of the experimental design). Food supply changed as follows: prior to T1, all fish were fed *ad libitum* three times a day. Then, groups FF and SF (accelerated growth) were fed *ad libitum* four times a day, with an amount of food per day equivalent of 3–6% of their mean BW. On the other hand, groups FS and SS (decelerated growth) were feed only two times a day with an amount of food per day equivalent of 1.5–3% of their mean BW. The growth rate of all groups was carefully monitored by periodic samplings and the amount of food adjusted if necessary. Animals were sacrificed when they were 337 dph juveniles (T2, i.e., when fish were 11 months old) (range of fish per tank at that moment: 52–70). There was no mortality associated to treatments.

Samplings

Details on the follow-up of growth, including sexual growth dimorphism, gonadosomatic, hepatosomatic and carcass indices, as well as sex ratio and the degree of gonadal development of these fish have been previously described [19]. Fish used for objective 1 were sexually undifferentiated. For objective 2, when possible, sex was visually determined and confirmed histologically if necessary [19]. Only males were considered for objective 2 since the goal was to study the relationship between growth during sex differentiation on the subsequent juvenile testis transcriptome. Histological results indicated that testis contained no spermatozoa. We did not do a similar study on females because some of the resulting populations were highly male-biased (∼90% males) and thus we did not have enough individual females in all groups. The number of fish used for each group and the biometry is shown in Table S1. Here, we focus only on the RNA extraction for transcriptomic analysis of gene expression and for microarray validation by qRT-PCR.

RNA extraction and cDNA synthesis

Total RNA was extracted from sexually undifferentiated gonads (mean SL ∼4 cm) at 133 dph (T1) and sexually differentiated juvenile testis at 337 dph (T2).

A classical chloroform-isopropanol-ethanol RNA extraction protocol after a Trizol (Live Technologies, Scotland, UK) homogenization was used. RNA quality and concentration were measured by a ND-1000 spectrophotometer (NanoDrop Technologies) and checked on a 1% agarose/formaldehyde gel. RNA integrity was measured by a Bioanalyzer 2100 (RNA 6000 Nano LabChip kit Agilent, Spain). Samples with a 100–200 ng/μl RNA concentration and RIN>7 were used for microarray hybridizations.

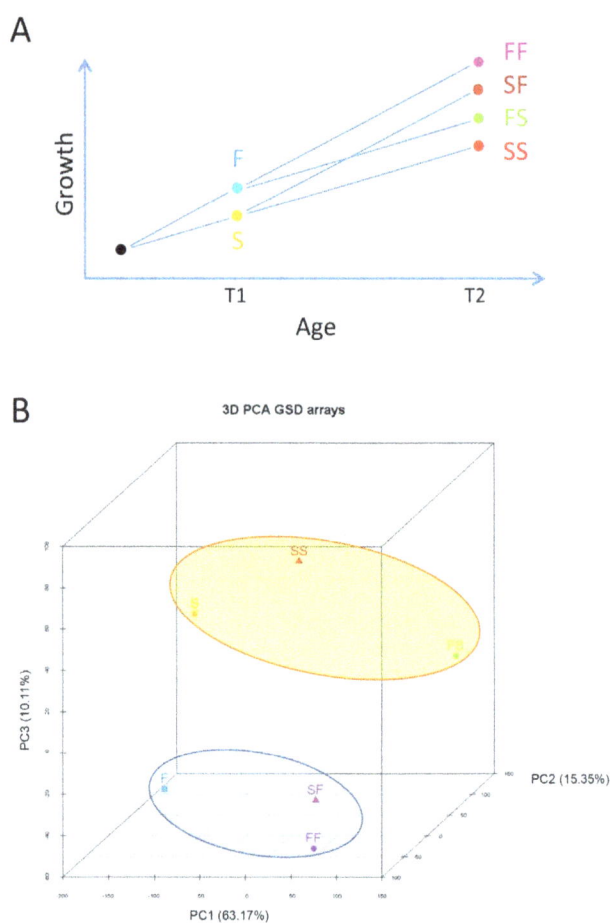

Figure 1. Experimental design and overall transcriptomic results. A) Experimental design involving European sea bass subjected to food restriction at different times during the sex differentiation period. B) Principal component analysis representation of transcriptomic results of the six groups. Time 1 fast- (F) and slow- (S) growing groups, and time 2 F- (FF and FS) and S- (SF and SS) derived groups.

In parallel, 200 ng of total RNA were treated by *E. coli* DNAse H and retrotranscribed (100 ng) to cDNA using SuperScript III RNase Transcriptase (Invitrogen, Spain) and Random hexamer (Invitrogen, Spain) following manufacturer's instructions.

Microarray. Experimental design

Hybridizations were performed at the Universitat Autònoma de Barcelona (UAB, Spain). The experiment included, on one hand, the comparison of 15 undifferentiated gonads from two groups (seven individual gonads from the F group against eight individual gonads from the S group) at 133 dph (T1), to explore transcriptomic differences between two groups with different growth rates since they were selected from the opposite extremes of the normal distribution curve. On the other hand, individual testes from each one of the four groups (groups FF, SF, FS and SS) differentially fed and sampled at 337 dph (T2), were analyzed to investigate the growth acceleration or growth deceleration effects on their transcriptome. Thus, 35 microarrays (one per fish) were used to analyze the gonadal transcriptome of the six groups considered in this study. To avoid batch effects samples were evenly distributed among the slides.

Microarray. RNA sample preparation and array hybridization

RNA labelling, hybridizations, and scanning were performed according to the manufacturer's instructions. Briefly, total RNA (100 ng) was amplified and Cy3-labeled with Agilent's One-Color Microarray-Based Gene Expression Analysis (Low Input Quick Amp Labelling kit) along with Agilent's One-Color RNA SpikeIn Kit. Then cRNA was purified with RNeasy mini spin columns (Qiagen), quantified with the Nanodrop ND–1000 and verified using the Bioanalyzer 2100. Each sample (1.65 µg) was hybridized to the *Dicentrarchus labrax* array (Agilent ID 023790) at 65°C for 17 h using Agilent's GE Hybridization Kit. Washes were conducted as recommended by the manufacturer using Agilent's Gene Expression Wash Pack with stabilization and drying solution. Arrays were scanned with Agilent Technologies Scanner, model G2505B. Spot intensities and other quality control features were extracted with Agilent's Feature Extraction software version 10.4.0.0. The experiment has been submitted to Gene Expression Omnibus (GEO)-NCBI database (GSE54362) and the platform that validates the microarray has the accession number (GPL13443).

Quantitative real time PCR (qRT-PCR)

Microarray validations were carried out by qRT-PCR analysis. Two genes from each one of the six possible microarray comparisons (see Table S2) were used for qRT-PCR validation, including one up- and one downregulated gene. Primers were designed using Primer 3 Plus (http://www.bioinformatics.nl/cgi-bin/primer3plus/primer3plus.cgi) against the annotated gene sequences directly from the European sea bass genome (Tine et al., 2014, unpublished), always trying to design the primers between exons to avoid DNA contamination (Table S2). Primer amplification efficiencies were tested by linear regression analysis from a cDNA dilution series and by running a melting curve (95°C for 15 s, 60°C for 15 s and 95°C for 15 s). Efficiency ($E = 10^{(-1/slope)}$, with values between 1.80 and 2.20, standard curves ranging from −2.9 to −3.9 and linear correlations (R^2) higher than 0.92 were recorded (Table S2). cDNA was diluted 1:10 for all the target genes and 1:500 for the endogenous control (the housekeeping gene *r18S*).

qRT-PCR was analyzed by an ABI 7900HT (Applied Biosystems) under standard cycling conditions. Briefly, an initial UDG decontamination cycle (50°C for 2 min), an activation step (95°C for 10 min) was followed by 40 cycles of denaturation (95°C for 15 s) and one annealing/extension step (60°C for 1 min). A final dissociation step was also added (95°C for 15 s and 60°C for another 15 s). Each sample was run in triplicate in 384-well plates in a final 10 µl volume (2 µl of 5x PyroTaq EvaGreen qPCR Mix Plus (ROX) from Cultek Molecular Bioline, 4 µl distilled water, 2 µl primer mix at a 10 µM concentration and 2 µl of cDNA). Negative controls were run per duplicate and *r18S* was used to calculate intra- and inter-assay variations. SDS 2.3 software (Applied Biosystems) was used to collect raw data and RQ Manager 1.2 (Applied Biosystems) was used to calculate gene expression. qRT-PCR data was analyzed by adjusting for E and normalizing to the *r18S* reference gene [41].

Statistical analysis of data. Microarray raw data normalization

Feature Extraction output data was corrected for background using normexp method [42] and was quantile normalized [43]. Reliable probes showed raw foreground intensity at least two times higher than the respective background intensity and were not saturated nor flagged by the Feature Extraction software. Our sea bass custom-made microarray presents most of the probes (64.7%) in duplicate but also with more than three identical probes for some genes. Median intensities per gene were calculated and a probe was considered reliable when at least half of its replicates were reliable as defined above. An empirical Bayes approach on linear models (Limma) [44] was used to perform a differential expression analysis. A False Discovery Rate (FDR) method was used to correct for multiple testing. Differentially expressed (DE) genes were filtered by fixing an absolute fold change (FC) of 1.5 and an adjusted *P*-value <0.01. MA data analyses were performed with the Bioconductor project (http://www.bioconductor.org/) in the R statistical environment (http://cran.rproject.org/) [45].

Statistical analysis of data. qRT-PCR statistics

Quantitative qRT-PCR statistical analysis was performed using 2DCt from the processed data [41]. 2DCt results were then checked for normality, homoscedasticity of variance and a one-way ANOVA test was used to assess differences between treatments using SPSS v.19 software.

Gene annotation and enrichment analysis

Gene data (names, abbreviations, synonyms and functions) were determined using Genecards (http://www.genecards.org/) and Uniprot (http://www.uniprot.org/). The web based tool AMIGO (http://amigo.geneontology.org/cgi-bin/amigo/go.cgi) [46] was used to look for the sequences of the DE genes found at the MA. After obtaining these sequences, Blast2GO software (http://www.blast2go.com) [47] was used to enrich GO term annotation and to analyze the subsequent altered KEGG pathways (http://www.genome.jp/kegg/), which were also further explored by DAVID (http://david.abcc.ncifcrf.gov/; [48,49]). Completing the analysis, Blast2GO with Fisher's Exact Test with Multiple Testing Correction of the False Discovery Rate [50] was used to analyze our DE genes using the custom-made microarray as background.

Protein names from the DE genes were then uploaded to the web-based tool STRING v9.1 (http://string-db.org/) [51] to analyze physical and functional protein interactions. Furthermore, an FDR test was applied to determine if the protein list was enriched (higher values meaning higher significances). A Mean Linkage Clustering (MLC or UPGMA), a simple agglomerative hierarchical clustering included in STRING was performed to group proteins. This method clusters proteins based on pairwise similarities in relevant descriptor variables.

Results

Overall assessment of transcriptomic results

Visualization of the spatial distribution of the microarray data of the six studied groups along the three major axis of the PCA is shown in Figure 1B. Component 1 contributed to 63.17% of the variation while the first three components together explained 88.63% of the variation. Two clusters could be observed, one containing group F and the F-derived groups with an accelerated growth (groups FF and SF), and the other formed by group S and the S-derived groups with growth deceleration (groups SS and FS).

The number of DE genes found in the only possible comparison at T1 as well as in the six possible comparisons between the four groups at T2 is shown in Table 1. The comparison with larger number of genes was FS vs. SS, while the FF vs. SF comparison gave no DE genes. From each one of the comparisons with DE genes, the most upregulated and the most downregulated genes (a total of twelve) were selected for a qRT-PCR validation (see details and quality control data of the designed primers in Table S2). All

Table 1. Differentially expressed genes in the different comparisons.

Group comparisons	Total # of genes	# Upregulated genes			# Downregulated genes		
		Total	Real	NA	Total	Real	NA
F vs. S	76	41	20	11	35	20	6
FF vs. SS	155	71	43	9	114	70	30
FS vs. FF	1092	662	316	47	431	153	111
SF vs. SS	94	42	26	1	53	37	9
SF vs. FS	938	507	184	162	604	303	40
FF vs. SF	0	0	0	0	0	0	0
FS vs. SS	2014	1452	717	108	562	261	202

NA, non annotated genes.

genes tested showed the same fold change tendency, thus validating the microarray results (Table 2). Among the tested genes four of them (*cct6a*, *rps15*, *fabp3* and *rpl9*) showed statistical differences ($P<0.05$) when analyzed by qRT-PCR. In the comparisons containing DE genes, analysis of the associated GO terms related to biological processes (BP), molecular function (MF) and cell component (CC) provided further information on the molecular signatures of each treatment (Table S3). Seven selected BP subcategories based on prior knowledge that they take place in the gonads are shown in Figure 2. Metabolic process, response to stimulus and signaling were, in that order, the most represented subcategories. Regarding the MF and CC subcategories, no clear differences were seen among the different comparisons. The most represented MF subcategories among the comparisons were binding and catalytic activities.

Transcriptome of sexually undifferentiated gonads of initial fast-growing vs. initial slow-growing fish (group F vs. group S comparison)

All fish from the F group clustered together and all but one fish from the S group did the same as shown in the heatmap (Figure 3A). Of the total 40 DE genes, among the 20 upregulated there were genes related to transcription, immune response or cytoskeleton structure, whereas among the 20 downregulated ones

there were genes mainly related to mitochondrial functions (Table S4).

Further analyzing the BP subcategories for the up- and downregulated GO terms (Figure 2A and 2B, respectively) showed how the number of GO terms for all the subcategories was always low when compared with T2 group comparisons. Tyrosine-protein kinase gene, a gene involved in male gonad development, was upregulated in the F group. In general, genes related to the response to stimulus and metabolic processes were downregulated. This applied also to genes related to growth such as growth hormone (*gh*) and adrenomedullin, which is related to male gonad development and response to stimulus.

DAVID analysis of DE genes showed two gene clusters within the upregulated genes related to cytoskeleton organization and lumen (enrichment scores 2.6 and 1.14, respectively), and five clusters within the downregulated genes mainly related to mitochondrion, binding, membrane structure and ion binding (enrichment scores 3.36, 1.56, 1.47, 0.31 and 0.22, respectively). KEGG pathway analysis of DE genes showed nine affected pathways: three were upregulated and included T-cell receptor signaling and linoleic acid metabolism, and six were downregulated and mainly related to the metabolism of xenobiotics and amino acid degradation (Table S5).

Table 2. Microarray validation by qRT-PCR.

Comparison	Gene	Microarray FC	Microarray adj. P-value	qRT-PCR FC	qRT-PCR SEM	qRT-PCR Student t-test
F vs. S	*flna*	2.85	0.004	1.801	0.509	0.332
	tspan1	−5.90	0.007	−2.47	0.185	0.356
FF vs. SS	*cct6a*	2.33	0.001	2.98	0.551	0.009
	rps15	−13.28	0.000	−1.46	0.407	0.001
FS vs. FF	*ggps1*	11.58	0.006	22.63	16.673	0.284
	fabp3	−15.34	0.007	−7.51	0.125	0.001
SF vs. SS	*rpl9*	2.61	0.001	18.79	7.023	0.047
	pcca	−14.03	0.000	−42.13	0.009	0.970
SF vs. FS	*lpl*	13.93	0.006	2.23	0.440	0.364
	tspan13	−10.27	0.006	−240.52	0.002	0.204
FS vs. SS	*ca1*	36.70	0.004	2.49	0.591	0.631
	agpat5	−13.38	0.000	−2.38	0.292	0.849

A

B

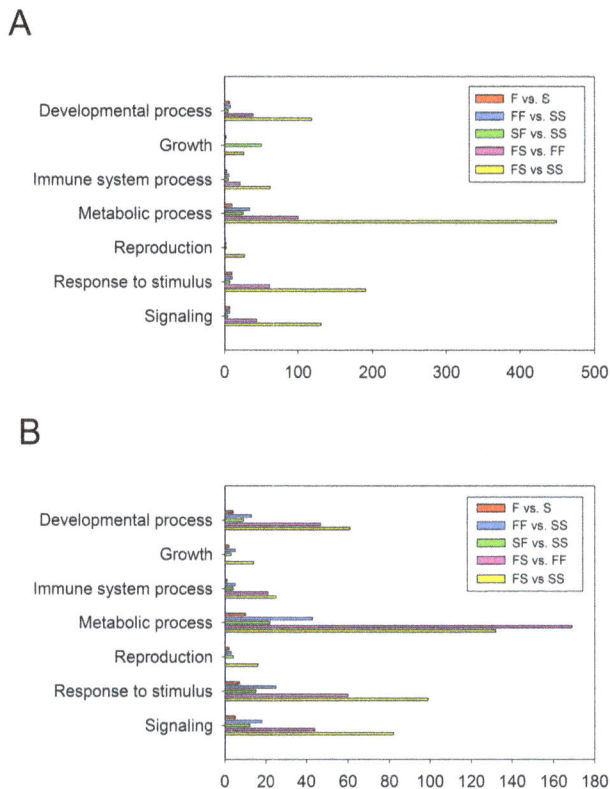

Figure 2. The seven selected GO subcategories for the Biological process (BP) category. A) Upregulated GO-terms and B) downregulated GO terms in the different studied comparisons.

Only seven interactions were found among the proteins coded by these DE genes; however, they were enriched in interactions (P<0.001). Proteins from the upregulated DE genes were required for the 60S ribosomal subunit biogenesis and mRNA stability and included proteins such as Ilf3, Nop56, Nop58 and Noc2l, with combined scores of protein-protein interactions ranging from 0.573 to 0.924 (a value of 1 represents the highest possible relationship). Nevertheless, when analyzing the proteins from the downregulated DE genes, three clusters of relationships were observed and related to: 1) respiratory electron transport (Uqcrc2 and Etfa; combined score: 0.969), 2) amino acid degradation (Bckdha and Ivd; combined score: 0.915), and 3) glutathione-mediated detoxification pathway (Gstk1 and Ggh; combined score: 0.899).

Transcriptome of juvenile testes of sustained fast-growing vs. sustained slow-growing fish (group FF vs. group SS comparison)

There were 113 DE genes when comparing the testis of FF and SS groups (43 up- and 70 downregulated genes; see Table S6 for a detailed list). A heatmap visualization of the data (Figure 3B) clearly separated individuals according to their group of origin.

The three most regulated GO terms in the BP category were related to metabolic processes and response to stimulus, followed by developmental process in the upregulated GO terms, and related to signaling for the downregulated GO terms (Figure 2A and 2B, respectively).

DAVID analysis of the DE genes yielded seven up- and 20 downregulated gene clusters mainly related to the Rps and Rpl

ribosomal protein families. KEGG analysis showed twelve altered pathways: three that were upregulated in the FF fish and showed an opposite behavior to what had been observed for the F vs. S comparison (for example, the drug metabolism and the xenobiotics and glutathione metabolisms). There were also nine downregulated pathways related to accelerated growth and metabolism (see Table S7 for a detailed list of the pathways). Although low representation of sequences was found for each pathway, after a FDR test, ribosome was the most enriched pathway among both the up- and downregulated pathways, while proteasome was also highly enriched among the downregulated DE genes.

A Fisher's Exact Test with Multiple Testing Correction of FDR analysis of the most specific terms showed that eight biological processes, three molecular functions and three cell components GO terms were over-represented when using the whole microarray as a background (see Table S8 for further details). Most of the GO terms were related to the ribosome structure and the translation process.

Protein-protein interaction analysis showed that upregulated proteins clustered in three different groups, where the largest one was related to the Rps (six different Rps) and Rpl (seven different Rpl) ribosomal protein families. These groups of proteins are found at the small and large ribosomal subunits, respectively (combined scores ranging from 0.401 to 0.999; Figure 4A). The other two clusters were conformed by the Iars2 and Cct6a proteins, which are related to translation and folding, as well as the 60S ribosomal subunit biogenesis-related proteins (Ube2a, Nop58, Sf3b1 and Cpsf1). On the other hand, downregulated protein interactions clustered in four groups, being the largest formed by the Rpl protein family (nine different Rpl proteins), but also forming part of the small ribosomal subunit and of the proteasome accessory complex (Figure 4B). The other three clusters were conformed by: 1) Agpat2 and Agpat5, which are involved in phospholipid metabolism; 2) Psmd13, Psmd8 and Psmc1, which are involved in ubiquitinated protein degradation; and 3) Prl and Ren, which are mainly involved in growth regulation and apoptosis.

The effects of accelerating growth: Transcriptome of juvenile testes of growth-accelerated fish vs. sustained slow-growing fish (group SF vs. group SS comparison)

Despite significant differences (P<0.01) in SL and BW in favor of fish from group SF when compared to fish of the SS group (Table S1), the two groups had a similar sex ratio with a clear male bias (90.6 and 92.2% males, respectively; reported in [19]). However, the transcriptional comparison of the SF vs. the SS group had a low or moderate number of DE genes in the testis transcriptome. A heatmap analysis (Figure 5A) visually representing the 63 DE genes, 26 up- and 37 downregulated genes (see Table S9 for further details), showed that these two groups clustered separately.

BP subcategories were analyzed for the up- and downregulated GO terms (Figure 2A and 2B, respectively). Metabolic process GO terms were the most upregulated and contained five genes that were mainly related to amino acid metabolism (ren, psme1, psmc1, trdmt1 and agpat5). However, renin and prolactin (prl), genes involved in positive regulation of growth, male gonad development, response to hormone stimulus and signaling (hormone-mediated or through G-protein coupled receptors) were downregulated.

DAVID analysis of the data with the highest stringency showed seven clusters for the upregulated genes (enrichment score of 23.68 to 0.42), being protein biosynthesis and translational elongation the most enriched ones. Among the downregulated genes, four

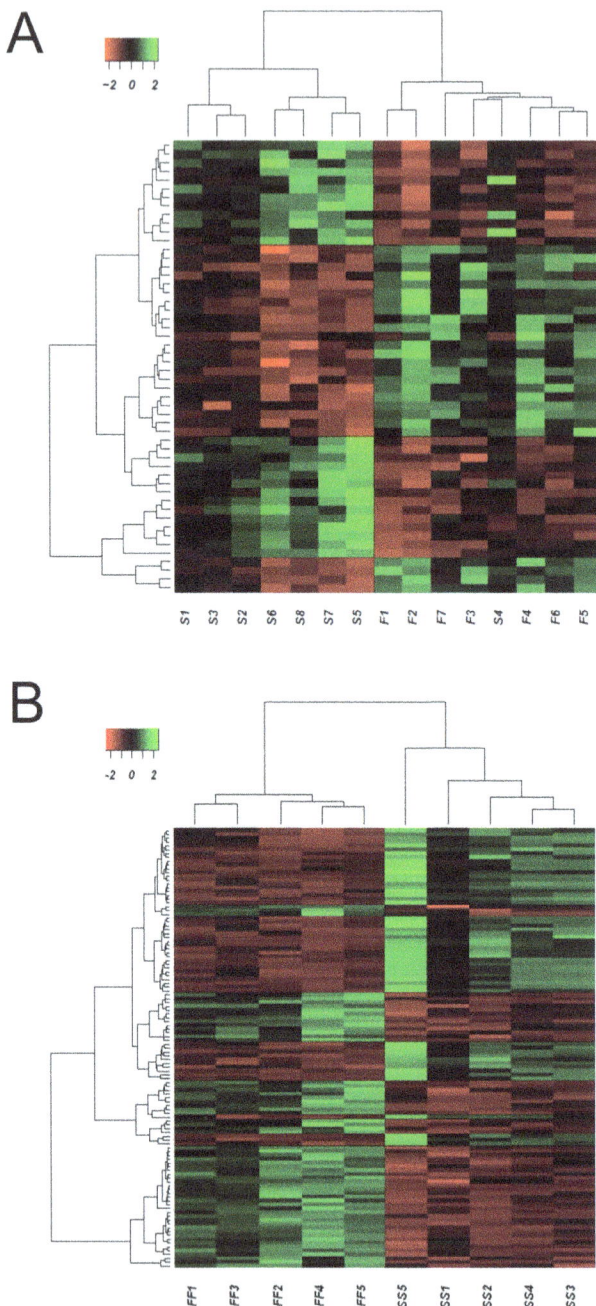

Figure 3. Individual heatmap representation of the transcriptome analysis. A) undifferentiated gonads of F versus S growing fish. B) differentiated testis of sustained fast (FF) or slow (SS) growing fish. Only DE genes are represented in the figure. High to low expression is shown by a degradation color from green to red, respectively. The scale bar shows Z-score values for the heatmap.

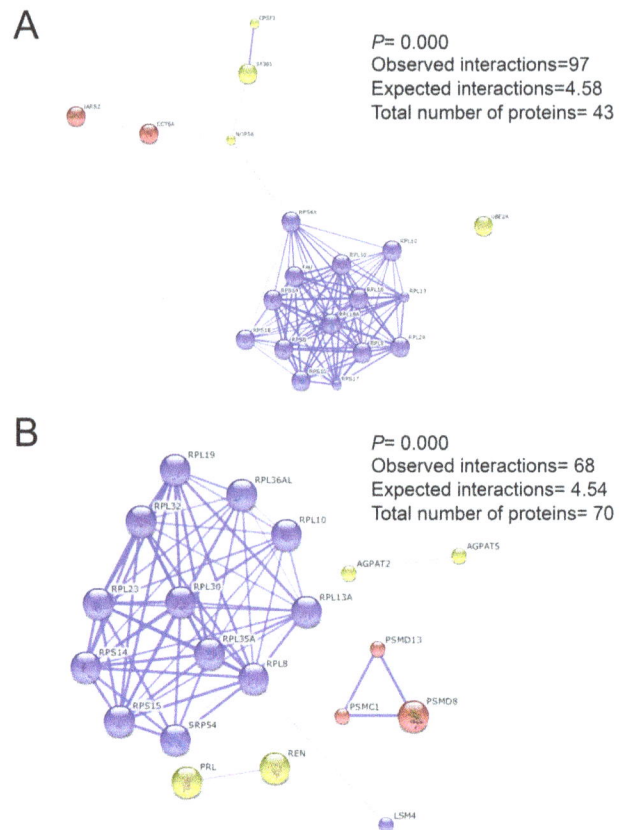

Figure 4. Protein-protein predicted confidence interactions for the FF vs. SS group comparison. A) The 43 proteins from the upregulated DE genes. B) The 70 proteins from the downregulated DE genes.

clusters (enrichment scores from 1.48 to 0.06) were present and mostly related to peptidase activity. KEGG analysis showed 16 pathways altered due to the growth acceleration (three up- and twelve downregulated KEGG pathways; Table S10) that were mostly related to amino acids, glutathione and lipid metabolism. Fisher's Exact Test with Multiple Testing Correction of FDR of the most specific terms showed eight biological processes, three molecular functions and two cellular components that were over-represented when compared against our microarray background and were mainly related to ribosome structure and translational elongation and termination (Table S11).

Protein-protein interaction analysis with an MLC clustering method showed proteins from the upregulated DE genes clustering together and enriched in interactions ($P < 0.001$). These proteins were either ribosomal (Rpl and Rps ribosomal protein families) or ribosome associated proteins (e.g., Ef1a1, Ubc or Fau). On the contrary, proteins from the downregulated DE genes were not enriched in interactions ($P = 0.067$) since just one interaction was present between Prl and Cort proteins (data not shown), which are known to be involved in growth control and signaling.

The effects of accelerating growth: Transcriptome of juvenile testes of growth-accelerated fish vs. sustained fast-growing fish (group SF vs. group FF comparison)

Transcriptional analysis of the SF vs. FF group returned zero DE (Table 1) even when we looked for genes with a lower P-value (0.05) and lower FC (1.2). The two groups had different sex ratios (90.6% and 67.6% males, respectively) due to their initially different growth rates before size-grading, and FF fish were bigger in BW but not in SL at the time of sampling. However, from a transcriptional point of view, they had no differences, indicating a full recovery from the early naturally slow growth rates.

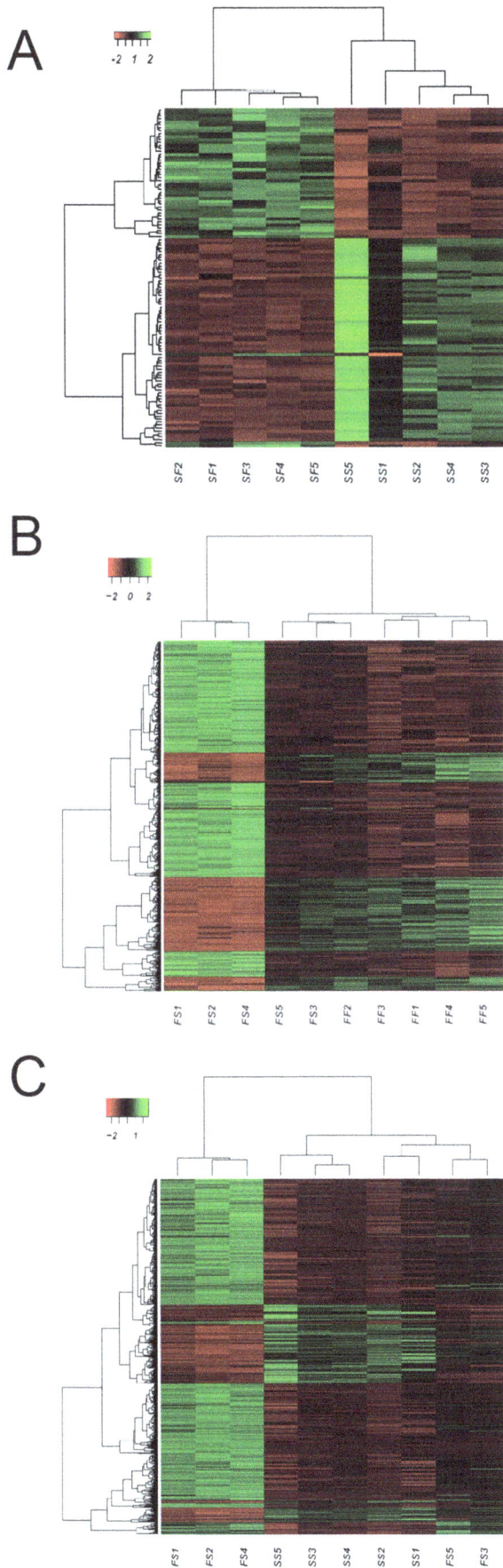

Figure 5. Individual heatmap representation of the transcriptome analysis of European sea bass one-year old testis. A) SF vs. SS comparison. B) FS vs. FF comparison. C) FS vs. SS comparison. Only DE genes are represented in the figure. High to low expression is shown by a degradation color from green to red, respectively. The scale bar shows Z-score values for the heatmap.

The effects of decelerating growth: Transcriptome of juvenile testes of growth-decelerated fish vs. sustained fast-growing fish (group FS vs. group FF comparison)

Fish that experienced the same initial fast-growing rate also had a similar sex ratio (67.6% and 61.4% males, respectively) when compared to the S-derived groups ($P<0.001$), which were highly male-biased (>90%). However, when comparing growth between decelerated fish (FS) vs. sustained fast-growing fish (FF) there were differences in the final growth due to the different feeding regimes (FF>FS in SL and BW) during the sex differentiation period.

Differences at the transcriptomic level were found (469 DE genes: 316 up- and 153 downregulated genes; Table S12). A heatmap visualization of the data (Figure 5B), showed that two FS individuals (FS3 and FS5) shared a transcriptomic pattern with those of the FF group.

The three most regulated GO terms in the BP category were related to metabolic processes, response to stimulus and developmental process in the upregulated GO terms while signaling was for the downregulated subcategory (Figure 2A and 2B, respectively).

DAVID analysis showed 37 clusters from the upregulated genes (enrichment scores from 3.66 to 0.07) and functions were mainly related to proteolysis, regulation of ubiquitin, proteasome and protein modifications processes. On the contrary, downregulated genes (37 clusters; enrichment score from 1.82 to 0.0) had functions mostly related to biosynthesis of phospho- and glycerolipids, anabolic processes and RNA processing and splicing. These DE genes were part of 56 affected pathways (41 upregulated and 15 downregulated; Table S13). Upregulated pathways were the most altered ones after filtering for high stringency and were related to pyrimidine metabolism ($P<0.001$), RNA polymerase ($P<0.05$), oxidative phosphorylation ($P<0.05$), terpenoid backbone biosynthesis ($P<0.05$), epithelial cell signaling ($P<0.05$), purine metabolism ($P<0.05$), glutathione metabolism ($P<0.05$), glycosylphophatidylinositol (GPI)-anchor biosynthesis ($P<0.05$). With this high stringency filtering criteria, only proteasome ($P<0.001$) and ubiquitin mediated proteolysis ($P<0.05$) appeared as being affected among the downregulated pathways.

The Fisher's Exact Test with Multiple Testing Correction of FDR of the most specific terms showed twelve biological processes, eight molecular functions and three cellular components that were over-represented when compared against our microarray as a background and were related to mitochondria and transport activity, while receptor activity was found under-represented (Table S14).

The protein-protein interaction analysis showed that proteins corresponding to both DE up- (four different clusters; Figure S1) and downregulated (ten different clusters; Figure S2) genes were enriched in interactions ($P<0.001$). Upregulated protein clusters were conformed by: 1) proteasome-related proteins (e.g., Psma, Cct6a, Skp1 or Ube2v2), 2) signaling and cholesterol storage-proteins (e.g., Dmd, Mtor or Lpl), 3) transcription regulator proteins (e.g., Max, Pdcd10 or Itgb4), and 4) mitochondrial membrane respiratory chain (e.g., Mt-co1, Mt-nd1 or Mt-nd4). Downregulated proteins clustered in ten different groups but containing a few proteins, with the most enriched ones playing a role in: 1) signaling and protein degradation (e.g., Mapk14, Htra2),

2) translation initiation and protein folding (e.g., Eif4g1, Pdfn1) or 3) transcription (e.g., Polr2h, Gtf3a).

The effects of decelerating growth: Transcriptome of juvenile testes of growth-decelerated fish vs. sustained slow-growing fish (group FS vs. group SS comparison)

The analysis of testes from fish that suffered from a growth deceleration during the sex differentiating period (FS) compared to fish with a sustained slow growth (SS group) showed differences ($P<0.001$) in the final sex ratio (61.4% and 92.2%, respectively) and in the final growth (FS>SS for both SL and BW). These results were further corroborated by the large transcriptomic differences found (978 DE genes: 717 up and 261 downregulated genes; Table S15). Heatmap visualization of data (Figure 5C) showed that two fish from the FS group (FS3 and FS5) clustered with the SS fish.

Analysis of the BP subcategories from the up- and downregulated GO terms (Figure 2A and 2B, respectively) showed how decelerating growth caused the highest changes in all subcategories. None of the genes from these GO terms of the FS vs. SS comparison coincided with those of the other decelerating comparison (FS vs. FF).

Further analysis of the data revealed that these 978 DE genes classified in 71 altered KEGG pathways (62 up- and nine downregulated), and were mostly related to RNA translation and elongation (Table S16). Moreover, clustering analysis of these genes with the highest stringency yielded 82 clusters for the up- and 48 clusters for the downregulated genes. The most enriched upregulated gene clusters were mainly related to the peroxisome, RNA splicing or nucleotide biosynthetic process, while the most enriched downregulated clusters were mostly related to protein catabolism processes, DNA modifications such as methylation, and response to nutrients.

A Fisher's Exact Test with multiple corrections for FDR of the most specific terms gave two BP, one MF and three CC GO terms that were over-represented when comparing to our custom microarray. These GO terms were mainly related to the ribosome structure and translation (Table S17). On the other hand, there was one GO term under-represented and related to the regulation of the immune system.

The highest representation of protein-protein interactions for this comparison (FS vs. SS) showed after a MLC clustering an enrichment in interactions ($P<0.001$), and presented six clusters for the upregulated proteins (Figure S3) related to: 1) ribosomal protein families (Rpl and Rps), 2) post-replication repair of damaged DNA and proteasome (e.g., Rad18 and Psm6, respectively), 3) response to stress (e.g., Tp53, Apex1, Ing1), 4) 60S ribosomal biogenesis and mRNA synthesis (e.g., Nop58, Nop16, Polr2f) and 5) respiratory chain (e.g., Ndufa1, Nduf53, Atp5g1). The twelve clusters for the downregulated proteins (Figure S4) were mainly related to: 1) regulation of metabolic pathways and chromosome stability (e.g., Csnk2b, Mapre1, Tubgcp2), 2) translation initiation (e.g., Eif4a2, Eif4e, Eif3d) and 3) regulation of cell responses (e.g., Prl, Irf1, Wipi1).

The comparison between fast- and slow-growing groups vs. the FS group, a group that shows high transcriptomic activity but still some transcriptomic similarities with both groups (FF and SS), showed 253 shared DE genes (Figure 6). These were mainly grouped by five main functions: positive regulation of ubiquitination, RNA splicing and mRNA processing, regulation of apoptosis, glycerolipid and phospholipid metabolic process, and regulation of phosphorylation. Among these common 253 DE genes, there were just five genes that showed an opposite pattern of expression: *atf4*, *prelid1*, *rps17*, *psma6* and *dmd*, which were more expressed in the

F-derived groups (FF>FS>SS) and mainly related to the proteasome complex and ribosomal structure. Renin (*ren*) and prolactin (*prl*), two genes involved in the positive regulation of growth, growth hormone and G-protein coupled receptor signaling pathways as well as in male gonad development were downregulated in both comparisons (FS vs. SS and FF; Figure 6) with a low expression of these genes in the initial fast-growing groups. This inhibition was also observed in fish with forced accelerated growth when compared to the slow-growing fish (SF vs. SS). These results indicated that these pathways are inhibited when the food availability is altered. Regarding the signaling function, apart from *ren* and *prl*, there were four coincident genes with the same pattern of expression in both comparisons (FS vs. SS and FF). Two of them, *atf4* and *ppkp2*, were upregulated in the FS group and involved in unfolded protein response and cell-cell signaling, while the other two, *errb* and *fkbp14*, were downregulated in the FS group and involved in steroid hormone- mediated signaling pathway and also in unfolded protein response.

Discussion

The relationship between growth and sex has long been known for the European sea bass at the time of gonadal differentiation, where the largest fish are essentially all females whereas both sexes are found among the smallest fish, although males predominate. Early size-grading experiments (between 66–143 dph) have confirmed this [13,15,17] by obtaining ∼90% females among the largest selected fish. Moreover, in a previous study we found that altering growth rates during the sex differentiation period in both size-graded and non-size-graded populations did not alter the sex ratios [19]. Here, it is presented for the first time a microarray analysis of undifferentiated gonads from 4-month-old sea bass with opposite growing rates just after size-grading (T1), and on differentiated testis (11-month-old juveniles, T2). To the best of our knowledge, the question of whether naturally occurring differences in somatic growth are somehow translated in observed transcriptomic differences in the gonads during sex differentiation has never been explored in fish. Nevertheless, what could be called a related type of work was performed in mitten crabs [36], where it was separately analyzed the relationship between nutrition and reproduction, by examining the hepatopancreas and testes transcriptomes, respectively. Interestingly, and regardless the differences among experimental designs and the model organisms used, some traits found in the study with crabs, as the differential expression of some heat shock proteins, cell death suppressors, RNA-dependent DNA polymerases or controllers of splicing, were also found in our study. Similarly to what has been previously reported in fish liver and muscle transcriptomes [22,24,26,52], it is then clear that the juvenile testis is also affected by changes in food supply. Interestingly, there are then common transcriptomic responses with the above mentioned tissues, but not with the brain transcriptomic responses [52].

Fast growing vs. slow growing fish before the onset of SD

Differences between naturally fast- vs. slow-growing (F vs. S) European sea bass of the same family early in development (T1) were not reflected in major transcriptomic changes in their undifferentiated gonads since only 40 DE genes were found. Translation was an active process in F fish gonads since genes coding for ribosomal structure, protein translation and folding were highly expressed. Immune response (positive regulation of apoptosis) and reproduction-related pathways were upregulated, although the only gene directly related to reproduction, the

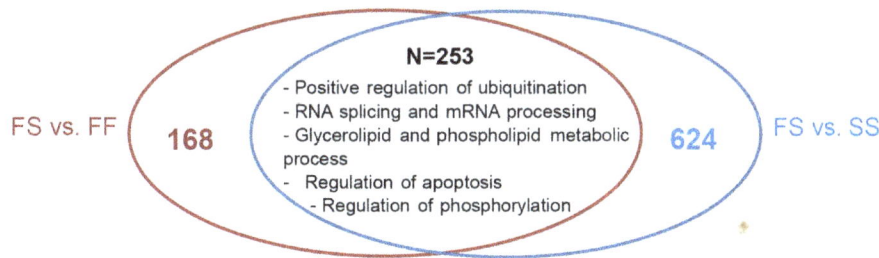

Figure 6. Venn diagram analysis of the DE genes by comparing (FS vs. FF) vs. (FS vs. SS). N represents the total number of common genes and the main categories in which genes are clustered.

tyrosine-protein kinase-like (*ptk*), which is associated with male gonad development function, is also present in other biological functions like cytoskeleton reorganization or cell proliferation. In contrast, gonads from fish that showed the slowest growth before size-grading (group S) were undergoing catabolism processes and protein recycling, since pathways related to negative regulation of growth, protein and amino acid catabolism, protein modification and fatty acid biosynthesis were highly expressed. The lack of abundance in reproduction-related genes among the DE genes present in the F vs. S comparison may be because the custom-made microarray used in this study was more enriched for immunology and growth-related terms rather than for reproduction related terms as it was based on the availability of the public sequences at the time. However, as can be seen in the results of this study, essentially most of the important genes related to sex differentiation found in this and other piscine species are present in this array, which ensures that it fulfills the requirements for such a type of study.

These results indicate that large intrafamily differences in somatic growth within a group of 4-month-old European sea bass due to natural variation are not necessarily translated in a large number of DE genes in their sexually undifferentiated gonads. This is relevant because occurs despite the fact that the groups made selecting the largest fish contain more future females than the groups made selecting the smallest fish, as shown before [13,15,17], and as evidenced by actual subsequent differences in the sex ratios of these groups, since the number of females in the F-derived groups was ~40% while in the S-derived groups it was only ~10% [19]. Thus, there is indeed a clear relationship between growth at the beginning of sex differentiation at 4 months and future sex ratio. Since prior to 4 months growth was not manipulated, growth and future sex can be related, but not food supply and future sex at 4 months. What is also not possible is to relate a given transcriptomic profile at the time of sex differentiation with future sex on an individual fish basis; that is impossible because in order to analyze the transcriptome the fish must be sacrificed. Changes at 4 months probably precede subsequent changes that would account for the differences in sex ratio between the F- and S-derived groups observed when juveniles.

Effects of unrestricted growth on the juvenile testis transcriptome

Food intake is one of the main factors influencing growth rates in aquatic production [25] and food restriction is directly associated with reduced growth rates in fish including the European sea bass [19]. The lack of transcriptomic differences between the SF and FF groups highlights the balance between protein synthesis and degradation, i.e., protein turnover, as one of the most important active processes in the gonads. Protein turnover relies on proteins mainly obtained from the diet, since high protein contributions from diet or low protein turnover (catabolism) are translated into higher growth rates [53]. In fact, we found lower expression of genes related to catabolism in the accelerated growth group (SF) when compared to the group with sustained growth (FF). The genes related to protein turnover, together with genes involved in the immune system, were also downregulated in the juvenile testis, as it had been found before in the liver of Atlantic salmon (*Salmon salar*) fed with a supplemental diet [26], and it was suggested that this is so because they are involved in the regulation of the decrease in whole body metabolic demands, resulting in less energy wastage and an enhancement in growth performance [24].

Groups with unrestricted growth during the sex differentiation period (FF and SF) showed in common an increased expression of genes related to high protein translation and folding, mainly of proteins related to ribosome structure. This, together with the lack of histological differences between groups, shows how gonads from slow-growing fish can still recover after a period of slow growth if food supply is not a limiting factor thereafter. This is remarkable since it shows the plasticity of the gonads during the sex differentiation period to environmental effects, since fish present a capacity of exploiting a situation of food abundance and recover from initial slow growth.

Effects of restricted food supply on the juvenile testis transcriptome

As also found in the Atlantic salmon liver [26] and white muscle transcriptome [24,54] after food deprivation, protein synthesis and degradation decreased in European sea bass juvenile testes, since both processes are very demanding in terms of energy requirements [53]. It is known that defective or damaged proteins (proteolysis process) are constantly degraded by the proteasome following two main pathways: lysosome or ubiquitin-proteasome pathways [55,56]. Our results show that in European sea bass gonads proteolysis was mainly achieved by the ubiquitin-proteasome pathway rather than the lysosome pathway, as described before in rainbow trout and gilthead sea bream, *Sparus aurata* [22,54,57], since we observed a larger representation of genes involved in the ubiquitin-proteasome pathway. The ubiquitin-proteasome pathway, mainly responsible for protein degradation, was downregulated in the group FS when compared to the group FF and contrasting with the SF and SS groups. Also, food supply restriction was accompanied by a downregulation of genes related to protein synthesis and degradation, and with the immune system, in agreement with previous observations made in the Atlantic salmon liver [26] and in white muscle [24] transcriptomes after food deprivation, although some genes of the complement system [58] were still upregulated. Moreover, a decrease in the lipoprotein levels, in transcription, in tRNA synthesis, and in protein synthesis and elongation in group FS, as a consequence of food deprivation, was coincident with previous results in fasted cod (*Gadus morhua*) as an energy-conserving mechanism [59], and is a common and strong downregulated response of the energy-generating processes in the adipose tissue [60]. This may be because during food deprivation hormonal signals such as growth hormone or insulin levels are translating food restriction signals into a protein turnover change by decreasing the protein synthesis and increasing catabolism [61] to save energy [1,62].

Several studies in fish have analyzed the responses to starvation by measuring catch-up growth [63] or analyzing the effects on the muscle and liver transcriptomes [22,64] but, to our knowledge, this is the first time that a similar study is performed in fish gonads. Two individuals from group FS (FS3 and FS5) clustered with individuals from the FF and SS groups, showing transcriptome similarities. This suggest that these FS fish still conserved some traits of the pre-T1 period but also traits related to the T1–T2 period when food supply was reduced, reflecting the existence of an adaptation to sudden feeding changes [1]. Most of these changes were related to protein turnover. This may be due to the adaptive capacity of fish to sense environmental fluctuations that in turn drive changes in their metabolism, since protein turnover is a highly energy-demanding process [24,53].

The comparison of groups FS vs. FF or SS evidenced some common features of expression where processes such as apoptosis, ubiquitin catabolism, peroxisome, kinase activity or regulation of cellular growth were increased. On the other hand, processes such as proteolysis, regulation of protein modifications, RNA processing, regulation of transcription factors, chromatin assembly, response to nutrients or gamete generation and reproduction were decreased, opposite to what has been found for sea bream heart transcriptome, where transcription was enhanced and transcription inhibitors downregulated [22]. These observations indicate that FS fish had to cope with the dramatic reduction of food intake by saving energy at different levels [1,19,24,26]. This is also supported by the fact that pathways related to metabolism such as lipid mobilization, or purine and pyrimidine metabolism were upregulated, as well as the pathways related to stress response such as the mTOR signaling pathway, which is involved in DNA damage and nutrient deprivation. In contrast, amino acid metabolism, xenobiotic removal or glucose metabolism were downregulated when comparing the FS to the FF and SS groups, showing how the naturally fast-growing fish fed with a non-restrictive diet and later subjected to food restriction did not obtain enough dietary energy and therefore had to start mobilizing lipids and activating gluconeogenesis. Moreover, in agreement with what has been found in both white and red muscle of gilthead sea bream under food restriction [22], mitochondria and ATP transport GO terms were enriched in juvenile testes, a fact that has been proposed as a link between food restriction and stress response mediated by cortisol [22].

No matter which one of the extreme groups the FS group was compared to, processes related to translation and protein regulation such as unfolded protein response, proteasome and postregulation of damaged DNA were highly active. Moreover, processes such as translation initiation, protein folding, transcription and cell-cell signaling were also taking place. Together, these results indicated that although food was scarce and growth was decelerated the transcriptional and the translational machineries of the testis were still active in the FS group. Furthermore, the steroid biosynthesis pathway was downregulated in the FS group when compared to the SF and SS groups, suggesting that the adaptation to the growth decrease could be affecting the energy

dedicated to future gonad maturation, although without apparent major consequences, since there were no histological differences when fish were sampled. Reproduction-related processes such as steroid biosynthesis, steroid hormone-mediated signaling and cholesterol storage were affected by growth deceleration, since they were downregulated in the FS group when compared to FF group. However, the FS group showed an increase in GO terms related to spermatogenesis/male gamete generation. This suggests that the FS fish, although being the most different group from a transcriptomic point of view due to food restriction during the sex differentiation period, were still allocating some of the energy in preparation for gonad maturation, which in farmed European sea bass takes place during the second year of life.

Conclusions

To the best of our knowledge, this is the first study evaluating the effects of different growth rates on gonadal development in fish with a transcriptomic approach.

The transcriptome of sexually-undifferentiated gonads was not drastically affected by initial natural differences in growth rates (fish from the opposite sides of the normal distribution curve). In addition, regardless the maturation status of the gonad (T1 and T2), as it has also been shown previously for liver and muscle, the slow-growing fish transcriptome showed an altered protein turnover with a higher catabolism, represented by a reduction in transcription and translation, a decreased immunological response, and a metabolism based on lipids and gluconeogenesis. On the other hand, the transcriptome of fast-growing fish reflects an enhancement of anabolic processes such as transcription, translation, protein synthesis and elongation and a metabolism based on glucose.

In differentiated juvenile gonads, the highest effects on the testis transcriptome were observed when forcing a naturally fast-growing fish to decelerate its growth through food restriction, since those fish showed high transcriptomic differences when compared to the sustained fast-growing fish and even more differences when compared to the fish with sustained slow-grow. These results suggest that food availability during the sex differentiation period is indeed able to modulate the testis transcriptome.

Interestingly, individuals with an initial slow grow but later with an accelerated growth due to increased food supply during sex differentiation showed a recovered transcriptome. These results suggest that fish are able to recover transcriptionally their testes if they are provided with enough food supply during the sex differentiation period. Nevertheless, the opposite is not true, since a natural initial fast growth does not ensure any advantage in terms of transcriptional fitness if later food becomes scarce. These results have implications for natural fish populations subjected to fluctuating food supply in a scenario of global change, as well as for populations or a part thereof of farmed fish under suboptimal feeding regimes, since they provide information on the possible consequences that these situations may have for the reproductive physiology of fish.

Supporting Information

Figure S1   Protein-protein predicted confidence interactions for the FS vs. FS group comparison. The interactions of 266 proteins from the upregulated DE genes are shown. The expected and observed interactions are shown with the significance level.

Figure S2   Protein-protein predicted confidence interactions for the FS vs. FF group comparison. The interactions of 129 proteins from the downregulated DE genes are shown. The expected and observed interactions are shown with the significance level.

Figure S3   Protein-protein predicted confidence interactions for the FS vs. SS group comparison. The interactions of 602 proteins from the upregulated DE genes are shown. The expected and observed interactions are shown with the significance level.

Figure S4   Protein-protein predicted confidence interactions for the FS vs. SS group comparison. The interactions of 206 proteins from the downregulated DE genes are shown. The expected and observed interactions are shown with the significance level.

Table S1   Biometric data of the individuals used for the transcriptomic analysis.

Table S2   Quantitative RT-PCR primer characteristics.

Table S3   List of the number of GO terms found for each category for all the comparisons studied.

Table S4   DE gene list for the F vs. S group comparison.

Table S5   Affected KEGG pathways in the F vs. S group comparison.

Table S6   DE gene list for the FF vs. SS group comparison.

Table S7   Affected KEGG pathways in the FF vs. SS group comparison.

Table S8   Two-tails Fisher's exact test with Multiple Testing Corrections of FDR results for the FF vs. SS group comparison.

Table S9   DE gene list for the SF vs. SS group comparison.

Table S10   Affected KEGG pathways in the SF vs. SS group comparison.

Table S11   Two-tails Fisher's exact test with Multiple Testing Corrections of FDR results for the SF vs. SS group comparison.

Table S12   DE gene list for the FS vs FF group comparison.

Table S13   Affected KEGG pathways in the FS vs. FF group comparison.

Table S14   Two-tails Fisher's exact test with Multiple Testing Corrections for FDR results for the FS vs. FF group comparison.

Table S15   DE gene list for the FS vs. SS group comparison.

Table S16   Affected KEGG pathways in the FS vs. SS group comparison.

Table S17   Two-tails Fisher's exact test with Multiple Testing Correction for FDR results for the FS vs. SS group comparison.

Acknowledgments

Thanks are due to S. Joly for technical assistance and to the staff of our experimental aquarium facilities (ZAE) for assistance with fish rearing.

Author Contributions

Conceived and designed the experiments: FP. Performed the experiments: ND LR. Analyzed the data: ND LR FP. Contributed reagents/materials/analysis tools: FP. Wrote the paper: ND LR FP.

References

1. Schneider JE (2004) Energy balance and reproduction. Physiology and Behavior 81: 289–317.
2. Castellano JM, Roa J, Luque RM, Diéguez C, Aguilar E, et al. (2009) KiSS-1/ kisspeptins and the metabolic control of reproduction: physiologic roles and putative physiopathological implications. Peptides 30: 139–145.
3. Volkoff H, Xu M, MacDonald E, Hoskins L (2009) Aspects of the hormonal regulation of appetite in fish with emphasis on goldfish, Atlantic cod and winter flounder: Notes on actions and responses to nutritional, environmental and reproductive changes. Comparative Biochemistry and Physiology, Part: A 153: 8–12.
4. Morgan MJ, Wright PJ, Rideout MN (2013) Effect of age and temperature of two gadoid species. Fisheries Research 138: 42–51.
5. Yoneda M, Wright PJ (2005) Effects to varying temperature and food availability on growth and reproduction in first-time spawning female Atlantic cod. Journal of Fish Biology 67: 1225–1241.
6. Kjesbu OS (1994) Time of start of spawning in Atlantic cod (*Gadus morhua*) females in relation to vitellogenic oocyte diameter, temperature, fish length and condition. Journal of Fish Biology 45: 719–735.
7. Marshall CT, Yaragina NA, Lambert Y, Kjesbu OS (1999) Total lipid energy as a proxy for total egg production by fish stocks. Nature 402: 288–290.
8. Devlin RH, Nagahama Y (2002) Sex determination and sex differentiation in fish: an overview of genetic, physiological, and environmental influences. Aquaculture 208: 191–364.
9. Vandeputte M, Dupont-Nivet M, Chavanne H, Chatain B (2007) A polygenic hypothesis for sex determination in the European sea bass – *Dicentrarchus labrax*. Genetics 176: 1049–1057.
10. Roblin C, Bruslé J (1983) Gonadal ontogenesis and sex differentiation in the sea bass, *Dicentrarchus labrax*, under fish-farming conditions. Reproduction, Nutrition and Development 23: 115–127.
11. Blázquez M, Piferrer F, Zanuy S, Carrillo M, Donaldson EM (1995) Development of sex control techniques for European sea bass (*Dicentrarchus labrax* L.) aquaculture: Effects of dietary 17 alpha-methyltestosterone prior to sex differentiation. Aquaculture 135: 329–342.
12. Mylonas CC, Anezaki L, Divanach P, Zanuy S, Piferrer F, et al. (2005) Influence of rearing temperature during the larval and nursery periods on growth and sex differentiation in two Mediterranean strains of *Dicentrarchus labrax*. Journal of Fish Biology 67: 652–668.
13. Papadaki M, Piferrer F, Zanuy S, Maingot E, Divanach P, et al. (2005) Growth, sex differentiation and gonad and plasma levels of sex steroids in male- and female-dominant populations of *Dicentrarchus labrax* obtained through repeated size grading. Journal of Fish Biology 66: 938–956.
14. Blazquez M, Carrillo M, Zanuy S, Piferrer F (1999) Sex ratios in offspring of sex-reversed sea bass and the relationship between growth and phenotypic sex differentiation. Journal of Fish Biology 55: 916–930.
15. Koumoundouros G, Pavlidis M, Anezaki L, Kokkari C, Sterioti K, et al. (2002) Temperature sex determination in the European sea bass, *Dicentrarchus labrax* (L., 1758) (Teleostei, Perciformes, Moronidae): Critical sensitive ontogenetic phase. Journal of Experimental Zoology 292: 573–579.
16. Saillant E, Chatain B, Menu B, Fauvel C, Vidal MO, et al. (2003) Sexual differentiation and juvenile intersexuality in the European sea bass (*Dicentrarchus labrax*). Journal of Zoology 260: 53–63.

17. Saillant E, Fostier A, Haffray P, Menu B, Laureau S, et al. (2003) Effects of rearing density, size grading and parental factors on sex ratios of the sea bass (*Dicentrarchus labrax* L.) in intensive aquaculture. Aquaculture 221: 183–206.

18. Navarro-Martín L, Blázquez M, Viñas J, Joly S, Piferrer F (2009) Balancing the effects of rearing at low temperature during early development on sex ratios, growth and maturation in the European sea bass (*Dicentrarchus labrax*). Limitations and opportunities for the production of highly female-biased stocks. Aquaculture 296: 347–358.

19. Díaz N, Ribas L, Piferrer F (2013) The relationship between growth and sex differentiation in the European sea bass (Dicentrarchus labrax). Aquaculture 408–409: 191–202.

20. Adams S, McLean R, Parrotta J (1982) Energy partiotining in largemouth bass under conditions of seasonally fluctuating prey availability transactions. American Fisheries Society 111: 549–558.

21. Geay F, Ferraresso S, Zambonino-Infante JL, Bargelloni L, Quentel C, et al. (2011) Effects of the total replacement of fish-based diet with plant-based diet on the hepatic transcriptome of two European sea bass (*Dicentrarchus labrax*) half-sib families showing different growth rates with the plant-based diet. BMC Genomics 12: 522.

22. Calduch-Giner JA, Sitja-Bobadilla A, Davey GC, Cairns MT, Kaushik S, et al. (2012) Dietary vegetable oils do not alter the intestine transcriptome of gilthead sea bream (*Sparus aurata*), but modulate the transcriptomic response to infection with *Enteromyxum leei*. BMC Genomics 13: 470.

23. Campos C, Valente LMP, Borges P, Bizuayehu T, Fernandes JMO (2010) Dietary lipid levels have a remarkable impact on the expression of growth-related genes in Senegalese sole (*Solea senegalensis* Kaup). The Journal of Experimental Biology 213: 200–209.

24. Tacchi L, Bickerdike R, Douglas A, Secombes CJ, Martin SAM (2011) Transcriptomic responses to functional feeds in Atlantic salmon (*Salmo salar*). Fish and Shellfish Immunology 31: 704–715.

25. Salas-Leiton E, Anguis V, Martín-Antonio B, Crespo D, Planas JV, et al. (2010) Effects of stocking density and feed ration on growth and gene expression in the Senegalese sole (*Solea senegalensis*): Potential effects on the immune response. Fish and Shellfish Immunology 28: 296–302.

26. Martin SAM, Douglas A, Houlihan DF, Secombes CJ (2010) Starvation alters the liver transcriptome of the innate immune response in Atlantic salmon (*Salmo salar*). BMC Genomics 11: 418.

27. Yi SK, Gao ZX, Zhao HH, Zeng C, Luo W, et al. (2013) Identification and characterization of microRNAs involved in growth of blunt snout bream (*Megalobrama amblycephala*) by Solexa sequencing. BMC Genomics 14: 754.

28. Overtuf K, Skhrani D, Devlin RH (2010) Expression profile for metabolic and growth-related genes in domesticated and transgenic coho salmon (*Oncorhynchus kisutch*) modified for increased growth hormone production. Aquaculture 307: 111–122.

29. Sun F, Liu S, Gao X, Jiang Y, Perera D, et al. (2013) Male-biased genes in catfish as revealed by RNA-seq analysis of the testis transcriptome. PLoS ONE 8: e68452.

30. Ravi P, Jiang J, Liew WC, Orban L (2014) Small-scale transcriptomics reveals differences among gonadal stages in Asian seabass (*Lates calcarifer*). Reproductive Biology and Endocrinology 12: 5.

31. Tao W, Yuan J, Zhou L, Sun L, Sun Y, et al. (2013) Characterization of gonadal transcriptomes from Nile Tilapia (*Oreochromis niloticus*) reveals differentially expressed genes. Plos ONE 8: e63604.

32. Rolland AD, Lareyre JJ, Goupil AS, Montfort J, Ricordel MJ, et al. (2009) Expression profiling of rainbow trout testis development identifies evolutionary conserved genes involved in spermatogenesis. BMC Genomics 10: 546.

33. Bozinovic G, Oleksiak MF (2011) Omics and environmental science genomic approaches with natural fish populations from polluted environments. Environmental Toxicology and Chemistry 30: 283–289.

34. Schiller V, Wichmann A, Kriehuber R, Muth-Kohne E, Giesy JP, et al. (2013) Studying the effects of genistein on gene expression of fish embryos as an alternative testing approach for endocrine disruption. Comparative Biochemistry and Physiology C-Toxicology and Pharmacology 157: 41–53.

35. Schiller V, Wichmann A, Kriehuber R, Schafers C, Fischer R, et al. (2013) Transcriptome alterations in zebrafish embryos after exposure to environmental estrogens and anti-androgens can reveal endocrine disruption. Reproductive Toxicology 42: 210–223.

36. Jiang H, Yin Y, Zhang X, Hu S, Wang Q (2009) Chasing relationships between nutrition and reproduction: A comparative transcriptome analysis of hepato-pancreas and testis from Eriocheir sinensis. Comparative Biochemistry and Physiology, Part D: Genomics and Proteomics 4: 227–234.

37. Martin P, Kohlmann K, Scholtz G (2007) The parthenogenetic Marmorkrebs (*Marbled crayfish*) produces genetically uniform offspring. Naturwissenschaften 94: 843–846.

38. Martin B, Pearson M, Brenneman R, Golden E, Wood W III, et al. (2009) Gonadal Transcriptome alterations in response to dietary energy intake: sensing the reproductive environment. PloS One 4: e4146.

39. Blazquez M, Navarro-Martin L, Piferrer F (2009) Expression profiles of sex differentiation-related genes during ontogenesis in the European sea bass

acclimated to two different temperatures. Journal of Experimental Zoology, Part B: Molecular and Developmental Evolution 312B: 686–700.

40. Moretti A, Pedini M, Cittolin G, Guidastri R (1999) Manual on hatchery production of seabass and gilthead seabream. FAO, Roma.

41. Schmittgen TD, Livak KJ (2008) Analyzing real-time PCR data by the comparative CT method. Nature Protocols 3: 1101–1108.

42. Ritchie ME, Silver J, Oshlack A, Holmes M, Diyagama D, et al. (2007) A comparison of background correction methods for two-colour microarrays. Bioinformatics 23: 2700–2707.

43. Bolstad B (2001) Probe level quantile normalization of high density oligonucleotide array data. Unpublished manuscript from the Division of Biostatistics, University of California, Berkely.

44. Smyth G (2005) Limma: linear models for microarray data. In: Gentleman R, Carey VJ, Huber W, Irizarry RA, Dudoit S, editors. Bioinformatics and computational biology solutions using R and Bioconductor. Springer New York. pp. 397–420.

45. Gentleman R, Carey V, Bates D, Bolstad B, Dettling M, et al. (2004) Bioconductor: open software development for computational biology and bioinformatics. Genome Biology 5: R80.

46. Carbon S, Ireland A, Mungall CJ, Shu S, Marshall B, et al. (2009) AmiGO: online access to ontology and annotation data. Bioinformatics 25: 288–289.

47. Conesa A, Gotz S, Garcia-Gomez JM, Terol J, Talon M, et al. (2005) Blast2GO: a universal tool for annotation, visualization and analysis in functional genomics research. Bioinformatics 21: 3674–3676.

48. Huang DW, Sherman BT, Lempicki RA (2009) Systematic and integrative analysis of large gene lists using DAVID Bioinformatics Resources. Nature Protocols 4: 44–57.

49. Huang DW, Sherman BT, Lempicki RA (2009) Bioinformatics enrichment tools: paths toward the comprehensive functional analysis of large gene lists. Nucleic Acids Research 37: 1–13.

50. Benjamini Y, Hochberg Y (1995) Controlling the false discovery rate: a practical and powerful approach to multiple testing. Journal of the Royal Society Series B 57: 289–300.

51. Franceschini A, Szklarczyk D, Frankild S, Kuhn M, Simonovic M, et al. (2013) STRING v9.1: protein-protein interaction networks, with increased coverage and integration. Nucleic Acids Research 41: D808–815.

52. Drew RE, Rodnick KJ, Settles ML, Wacyk J, Churchill EJ, et al. (2008) Effect of starvation on the transcriptomes of the brain and liver in adult female zebrafish (*Danio rerio*). Physiological Genomics 35: 283–295.

53. Houlihan DF, Carter CG, McCarthy ID (1995) Protein synthesis in fish; Hochachka M, editor. Amsterdam: Elsevier.

54. Martin SAM, Blaney S, Bowman AS, Houlihan DF (2002) Ubiquitin-proteasome-dependent proteolysis in rainbow trout (*Oncorhynchus mykiss*), effect of food deprivation. European Journal of Applied Physiology 445: 257–266.

55. Tanaka K, Chiba T (1998) The proteasome. a protein/destroying machine. Genes to Cells 3: 499–510.

56. Craiu A, Akopian T, Goldberg A, Rock KL (1997) Two distinct proteolytic processes in the generation of a major histocompatibility complex class I-presented peptide. Proceedings of the National Academy of Sciences 94: 10850–10855.

57. Palstra A, Beltran S, Burgerhout E, Brittijn S, Magnoni L, et al. (2013) Deep RNA sequencing of the skeletal muscle transcriptome in swimming fish. PLoS ONE 8: e53171.

58. Boshra H, GElman AE, Sunyer JO (2004) Structural and functional characterization of complement C4 and C1s/like molecules in teleost fish. Insights into the evolution of classical and alternative pathways. Journal of Immunology 171: 349–359.

59. Kjaer MA, Vegusdal A, Berge GM, Galloway TF, Hillestad M, et al. (2009) Characterisation of lipid transport in Atlantic cod (*Gadus morhua*) when fasted and fed high or low fat diets. Aquaculture 288: 325–336.

60. Higami Y, Pugh T, Page G, Allison D, Prolla T, et al. (2004) Adipose tissue energy metabolism: altered gene expression profile of mice subjected to long-term caloric restriction. FASEB Journal 18: 415–417.

61. Gabillard JC, Kamangar BB, Monstserrat N (2006) Coordinated regulation of the GH/IGF system genes during refeeding in rainbow trout (*Oncorhynchus mykiss*). Journal of Endocrinology 191: 15–24.

62. Dobly A, Martin SAM, Blaney SC, Houlihan DF (2004) Protein growth rate in rainbow trout (*Oncorhynchus mykiss*) is negatively correlated to liver 20S proteasome activity. Comparative Biochemistry and Physiology, Part A: Molecular and Integrative Physiology 137: 75–85.

63. Rescan PY, Montfort J, Ralliere C, Le Cam A, Esquerre D, et al. (2007) Dynamic gene expression in fish muscle during recovery growth induced by a fasting-refeeding schedule. BMC Genomics 8: 438.

64. Salem M, Kenney PB, Rexroad CE, Yao JB (2006) Molecular characterization of muscle atrophy and proteolysis associated with spawning in rainbow trout. Comparative Biochemistry and Physiology, Part D: Genomics and Proteomics 1: 227–237.

5

Dietary Patterns of Korean Adults and the Prevalence of Metabolic Syndrome

Hae Dong Woo[1], Aesun Shin[1,2], Jeongseon Kim[1]*

1 Molecular Epidemiology Branch, National Cancer Center, Goyang-si, Korea, **2** Department of Preventive Medicine, Seoul National University College of Medicine, Seoul, Republic of Korea

Abstract

The prevalence of metabolic syndrome has been increasing in Korea and has been associated with dietary habits. The aim of our study was to identify the relationship between dietary patterns and the prevalence of metabolic syndrome. Using a validated food frequency questionnaire, we employed a cross-sectional design to assess the dietary intake of 1257 Korean adults aged 31 to 70 years. To determine the participants' dietary patterns, we considered 37 predefined food groups in principal components analysis. Metabolic syndrome was defined according to the National Cholesterol Education Program Adult Treatment Panel III. The abdominal obesity criterion was modified using Asian guidelines. Prevalence ratios and 95% confidence intervals for the metabolic syndrome were calculated across the quartiles of dietary pattern scores using log binomial regression models. The covariates used in the model were age, sex, total energy intake, tobacco intake, alcohol consumption, and physical activity. The prevalence of metabolic syndrome was 19.8% in men and 14.1% in women. The PCA identified three distinct dietary patterns: the 'traditional' pattern, the 'meat' pattern, and the 'snack' pattern. There was an association of increasing waist circumference and body mass index with increasing score in the meat dietary pattern. The multivariate-adjusted prevalence ratio of metabolic syndrome for the highest quartile of the meat pattern in comparison with the lowest quartile was 1.47 (95% CI: 1.00–2.15, p for trend = 0.016). A positive association between the prevalence of metabolic syndrome and the dietary pattern score was found only for men with the meat dietary pattern (2.15, 95% CI: 1.10–4.21, p for trend = 0.005). The traditional pattern and the snack pattern were not associated with an increased prevalence of metabolic syndrome. The meat dietary pattern was associated with a higher prevalence of metabolic syndrome in Korean male adults.

Editor: Vineet Gupta, University of Pittsburgh Medical Center, United States of America

Funding: This study was supported by a grant from the National Cancer Center, Korea (no. 1210141). The funders had no role in study design, data collection and analysis, decision to publish, or preparation of the manuscript.

Competing Interests: The authors have declared that no competing interests exist.

* Email: jskim@ncc.re.kr

Introduction

Metabolic syndrome is associated with an increased risk of developing type 2 diabetes [1] and cardiovascular disease, as well as general mortality [2,3]. According to the National Health and Nutrition Examination Survey (NHANES), using the revised National Cholesterol Education Program/Adult Treatment Panel III (ATP III) definition, the age-adjusted prevalence of metabolic syndrome in the US adult Americans significantly increased from 29.2% between 1988 and 1994 to 34.2% between 1999 and 2006 [4]. The age-adjusted prevalence of metabolic syndrome was 13.7% between 2000 and 2001, and prevalence of metabolic syndrome was 26.7% between 2007 and 2008, using ATP III criteria modified for the Asia-Pacific subjects in China [5,6]. Based on the Korean National Health and Nutrition Examination Survey (KNHANES), using the ATP III criteria from the Asia-Pacific region for central obesity, the age-adjusted prevalence of metabolic syndrome in the Korean population increased from 24.9% in 1998 to 31.3% in 2007 [7]. Trends in prevalence of diabetes in Asian countries have been increased considerably [8], although no significant change has been observed in recent years [9]. Metabolic syndrome risk factors might be closely related to

diabetes and cardiovascular disease. Thus their potential causative factors need to be explored.

The risk of metabolic syndrome is known to be associated with dietary intake [10–12]. Analysis of dietary patterns could account for the inter-related dietary factors that are potentially important for the development of metabolic syndrome [13]. The dietary patterns of people in developing countries have changed as a result of modernization, which might contribute to an increased risk of obesity and metabolic syndrome [14,15]. Moreover, migration studies have shown that western dietary patterns lead to an accumulation of fat [16,17], which may contribute to the development of metabolic syndrome.

In previous studies, Western pattern characterized by high intakes of protein, processed foods, and refined grains was positively associated with metabolic syndrome, whereas healthy dietary pattern characterized by high intakes of fruits, vegetables and dairy was inversely associated with metabolic syndrome [11,12,18–21]. Recently, several studies were conducted to determine the association between dietary patterns and the prevalence of metabolic syndrome among Koreans [22–24]. The traditional Korean meal, which is low in fat and contains a large portion of vegetables, has been considered a healthy diet [14,25]. However, the traditional Korean dietary pattern was not associated with a lower prevalence of metabolic syndrome, and findings regarding the relationship between dietary pattern and metabolic syndrome have been inconsistent [22–24]. The association between dietary patterns and the metabolic syndrome has not been fully identified in the Korean population. Thus, the purpose of this study is to determine the association between various dietary patterns and the prevalence of metabolic syndrome in Korea.

Methods

Study population

We performed a cross-sectional study of participants who underwent health screening examinations at the Center for Cancer Prevention and Detection at the National Cancer Center in South Korea between October 2007 and December 2009. Visitors are National Health Insurance beneficiaries and those who have all data for survey question including medical history, clinical test result, and dietary consumption data were recruited (n = 2146). No one was excluded due to other diseases such as diabetes, coronary heart disease, stroke, or cancer. We excluded 862 subjects whose medical records lacked data regarding metabolic syndrome components. There was no significant difference in BMI between participants with missing metabolic syndrome components and participants with complete data. Participants with implausible energy intake values (<500 or ≥ 5000 kcal, n = 27) were excluded. The remaining 1257 adults (486 men and 771 women), ranging between 31 and 70 years old, were used in our analysis (Figure 1). Each participant was provided with an informed consent form according to the procedures approved by the institutional review board of the National Cancer Center. Written informed consent was obtained from all participants.

Data collection

The participants completed a self-administered questionnaire, which asked about each participant's demographics, lifestyle, medical history, and diet. Self-reported physical activity was evaluated using an International Physical Activity Questionnaire (IPAQ) short form [26]. The total metabolic equivalent (MET-minutes/week) was a combined score that was calculated by multiplying the frequency, duration, and intensity of physical activity. The dietary intake was assessed using a validated food frequency questionnaire (FFQ) [27]. The participants were asked about their average frequency of intake and portion size of specific foods during the previous year of 103 types of food. Three portion sizes and 9 categories of frequency were specified on the FFQ. The average daily nutrient intake was measured by summing up the intake of associated nutrient content per 100 g for each of the 103 foods. The foods listed in the FFQ were categorized into 37 different food groups, each of which was determined according to the food's nutrient profile and its culinary use (Table S1).

Metabolic syndrome definition

Metabolic syndrome was defined according to the National Cholesterol Education Program Adult Treatment Panel III (NCEP-ATP III) [28]. The abdominal obesity criterion was modified using Asian guidelines [29]. Under these definitions, a person has metabolic syndrome if that person exhibits three or more of the following conditions: 1) triglycerides ≥150 mg/dL; 2) HDL cholesterol <40 mg/dL in men or <50 mg/dL in women; 3) systolic blood pressure (BP) ≥130 mmHg, diastolic blood pressure (BP) ≥85 mmHg, or drug treatment for hypertension; 4) fasting glucose ≥110 mg/dL or drug treatment for elevated glucose levels; and 5) waist circumference ≥90 cm in men or ≥ 80 cm in women.

Statistical analysis

The general characteristics in the group with metabolic syndrome and the group without metabolic syndrome were compared using a Student t-test for continuous variables and a chi-square test for categorical variables. Principal-components analysis (PROC FACTOR) was used to extract the participants' dietary patterns using 37 predefined food groups. We used a varimax rotation to enhance the interpretability of the factors that were analyzed. We determined how many factors to retain after evaluating the eigenvalue, scree test, and interpretability. The dietary patterns were named according to the highest factor driving the food groups for each dietary factor. Each dietary pattern's factor score was categorized by quartile for further analysis. The trend test was performed to analyze the associations between each of the dietary patterns and each of the components of metabolic syndrome using a general linear model with adjustments for confounding factors. Regression analysis in log-log scale of each dietary pattern score and the components of metabolic syndrome was performed to compare with trends across quartiles. Prevalence ratios (PRs) and 95% confidence intervals (CIs) for the metabolic syndrome were calculated across the quartiles of dietary pattern scores using log binomial regression models. The lowest quartile of each dietary pattern was used as the reference. The trend test was performed to analyze the associations between each of dietary pattern score (continuous) and the prevalence of metabolic syndrome using Wald test. PRs and 95% CIs for the metabolic syndrome were also calculated stratified for sex. Model 1 was adjusted according to age, sex (for total), and total energy intake. Model 2 was further adjusted for age, sex (for total), total energy intake, tobacco intake, alcohol consumption, and physical activity. We performed the statistical analysis using SAS version 9.3 (SAS Institute Inc, Cary, NC). All P values were two-tailed (α = 0.05).

Results

The general characteristics of the study participants are reported in Table 1. The overall prevalence of metabolic syndrome was 16.3%, and the prevalence was significantly higher in men (19.8%) than in women (14.1%). Compared to people who

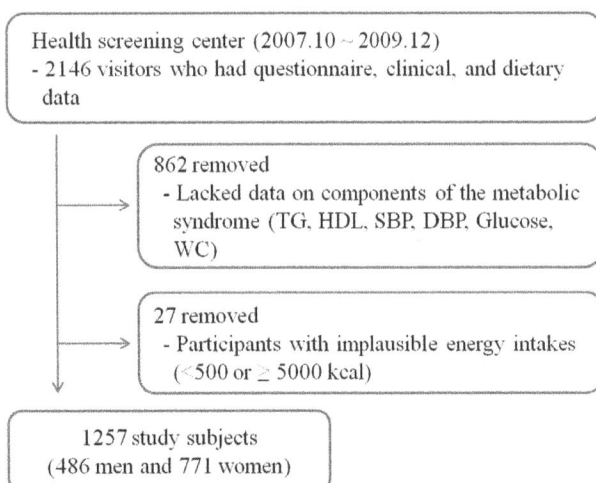

Health screening center (2007.10 ~ 2009.12)
- 2146 visitors who had questionnaire, clinical, and dietary data

862 removed
- Lacked data on components of the metabolic syndrome (TG, HDL, SBP, DBP, Glucose, WC)

27 removed
- Participants with implausible energy intakes (<500 or ≥ 5000 kcal)

1257 study subjects
(486 men and 771 women)

Figure 1. Flow chart of study selection process.

did not have metabolic syndrome, patients who had metabolic syndrome were older (p<0.001) and had a higher BMI (p<0.001). The separate components of metabolic syndrome were significantly different (metabolic syndrome positive vs. negative, Mean (SD): 137.2 (13.1) vs. 124.0 (13.6) for systolic BP, 83.7 (9.5) vs. 75.5 (10.0) for diastolic BP, 105.4 (26.1) vs. 90.7 (16.5) for fasting glucose, 46.9 (8.9) vs. 60.6 (13.9) for HDL cholesterol, 201.6 (81.8) vs. 102.4 (61.2) for triglycerides), depending on whether metabolic syndrome was present or absent (p<0.001).

The PCA identified three major dietary patterns, and their factor-loading scores are shown in Table 2. The 'traditional' dietary pattern included high intakes of condiments, green/yellow vegetables, light-colored vegetables, tubers, clams, tofu/soymilk, and seaweed; the 'meat' dietary pattern included high intakes of red meat, red meat byproducts, other seafood, and high-fat red meat; and the 'snack' pattern included high intakes of cake/pizza, snacks, and bread. Three patterns explained 31.9% of the total variance.

Table 3 shows the associations between each dietary pattern and the components of metabolic syndrome, including BMI, and general characteristics according to quartiles of each dietary pattern score. The traditional and snack patterns showed no relationships with any of the components of metabolic syndrome. Increasing scores in the meat dietary pattern were associated with elevated waist circumference, BMI, triglycerides, blood pressure, and low concentrations of HDL cholesterol (p<0.05). The traditional pattern score increased with an increment of age and physical activity of each quartile, but with decrease in percentage of men and current drinker. The meat pattern score increased with both increment of age and percentage of men and current drinker, but no difference was observed in physical activity. The snack pattern score increased with a decrement of age, but no differences were observed in percentage of men and current drinker, and physical activity.

The association between the PR of metabolic syndrome and the dietary pattern score variables are shown in Table 4. The score variable of the meat pattern was associated with the prevalence of metabolic syndrome in both models (p for trend = 0.006 and 0.016, respectively). The multivariate-adjusted PR of metabolic syndrome for the highest quartile of the meat pattern in comparison with the lowest quartile was 1.47 (95% CI: 1.00–2.15, p for trend = 0.016). We found no association between the

Table 1. General characteristics of the study population, and comparison of individuals with and without metabolic syndrome.

	Total	MS (+)	MS (−)	p value
N (%)	1257	205 (16.3)	1052 (83.7)	
Age (y)	51.7 (9.2)	55.9 (9.2)[†]	50.8 (9.0)	<0.001
Sex, n (%)				
Male	486 (38.7)	96 (19.8)	390 (80.3)	0.009
Female	771 (61.3)	109 (14.1)	662 (85.9)	
Smoking status, n (%)				
Never	774 (63.6)	114 (57.9)	660 (64.7)	0.164
Former	288 (23.7)	52 (26.4)	236 (23.1)	
Current	155 (12.7)	31 (15.7)	124 (12.2)	
Alcohol consumption, n (%)				
Never	472 (38.9)	78 (39.8)	394 (38.8)	0.943
Former	102 (8.4)	17 (8.7)	85 (8.4)	
Current	638 (52.6)	101 (51.5)	537 (52.9)	
Total energy intake (kcal)	2175.7 (857.3)	2282.6 (915.4)	2154.9 (844.4)	0.051
Saturated fatty acid (g/d)	7.9 (4.8)	8.2 (5.0)	7.9 (4.7)	0.987[‡]
Carbohydrates (g/d)	387.1 (163.1)	401.2 (165.3)	384.4 (162.6)	<0.001[‡]
Fiber (g/d)	34.3 (32.6)	37.4 (35.7)	33.7 (32.0)	0.563[‡]
Sodium (mg/d)	3207.0 (1675.9)	3623.8 (2198.5)	3125.8 (1542.0)	<0.001[‡]
Total vegetables (g/d)	177.5 (130.1)	195.9 (149.2)	173.9 (125.8)	0.114[‡]
Fruits (g/d)	221.5 (249.7)	210.3 (219.7)	223.6 (255.2)	0.193[‡]
BMI (Kg/m^2)	23.8 (3.0)	26.8 (2.6)	23.2 (2.7)	<0.001
Waist circumference (cm)	79.0 (8.8)	87.5 (6.7)	77.4 (8.2)	<0.001
Systolic BP (mmHg)	126.2 (14.4)	137.2 (13.1)	124.0 (13.6)	<0.001
Diastolic BP (mmHg)	76.8 (10.4)	83.7 (9.5)	75.5 (10.0)	<0.001
Fasting glucose (mg/dL)	93.1 (19.2)	105.4 (26.1)	90.7 (16.5)	<0.001
HDL cholesterol (mg/dL)	58.3 (14.1)	46.9 (8.9)	60.6 (13.9)	<0.001
Triglycerides (mg/dL)	118.6 (74.6)	201.6 (81.8)	102.4 (61.2)	<0.001
Physical activity (MET-min/wk)	2934.2 (2946.5)	2885.2 (2810.7)	2943.7 (2973.5)	0.795

MS (−), absence of metabolic syndrome; MS (+), presence of metabolic syndrome; MET, metabolic equivalent.
[†]Numbers are Mean (SD), unless otherwise stated.
[‡]Adjusted for total energy intake.

Table 2. Factor loadings of the dietary patterns derived from principal components analysis with orthogonal rotation.

	Traditional pattern	Meat pattern	Snack pattern
Condiments	0.78		0.26
Green/yellow vegetables	0.74		
Light colored vegetables	0.71	0.40	
Tubers	0.67		0.32
Clams	0.63	0.22	0.26
Tofu, soymilk	0.61		0.22
Seaweeds	0.60		
Bonefish	0.54		
Kimchi	0.49		
Lean fish	0.46	0.37	
Mushrooms	0.42	0.36	
Fruits	0.40		
Nuts	0.37		
Legumes	0.29		
Yogurt	0.27		
Eggs	0.27		0.28
Pickled vegetables	0.24		
Milk	0.20		
Red meat	0.23	0.79	
Red meat by-products		0.74	
Other seafood	0.25	0.67	
High-fat red meat		0.60	
Oil		0.50	0.20
Salted fermented seafood		0.44	
Noodles		0.43	
Poultry		0.43	
Fatty fish		0.37	0.29
Carbonated beverages		0.36	0.27
Dairy products		0.30	0.25
Cakes, pizza			0.81
Snacks			0.68
Bread			0.60
Processed meats		0.29	0.50
Sweets		0.28	0.36
Rice cake			0.23
Coffee, tea		0.20	
Grains			
Variance explained (%)	18.8	7.5	5.6

Factor loadings with absolute values <0.2 are not presented.

prevalence of metabolic syndrome and either the snack pattern score variables or the traditional pattern score variables. The association between the dietary pattern scores and the prevalence of metabolic syndrome was analyzed after stratifying by sex. A positive association between the prevalence of metabolic syndrome and the dietary pattern score was found only for men with the meat dietary pattern (multivariate-adjusted PR of the highest group compared with the lowest group: 2.15, 95% CI: 1.10–4.21, p for trend = 0.005).

Discussion

The present study derived three dietary patterns in the Korean adult population: the 'traditional' pattern, the 'meat' pattern, and the 'snack' pattern. We found that the meat dietary pattern score was positively associated with the prevalence of metabolic syndrome especially in men, whereas the traditional pattern and the snack pattern were not associated with metabolic syndrome.

Because Korean meals often include mixed soups or multiple side dishes comprising various vegetables, tubers, tofu, and seaweed with condiments, Factor 1 was labeled as the traditional

Table 3. The association between dietary patterns and the components of metabolic syndrome and BMI*.

| | Quartile of dietary pattern score | | | | p for trend[†] | p value[‡] |
	Q 1	Q 2	Q 3	Q 4		
Traditional pattern (n)	285	263	275	260		
Waist circumference (cm)	78.1 (9.3)	78.6 (8.9)	78.3 (8.8)	78.3 (7.8)	0.790	0.841
Triglyceride (mg/dL)	115.0 (78.3)	116.8 (75.2)	113.5 (64.2)	117.9 (78.3)	0.450	0.063
HDL cholesterol (mg/dL)	58.7 (14.6)	58.7 (14.5)	58.9 (13.9)	59.4 (13. 7)	0.410	0.582
Diastolic BP (mmHg)	76.1 (10.6)	76. 7 (10.5)	76.1 (10.2)	76.7 (10.4)	0.407	0.134
Systolic BP (mmHg)	124.7 (14.4)	126.1 (14.1)	124.5 (13.8)	126.3 (15.0)	0.189	0.593
Fasting glucose (mg/dL)	89.2 (12.9)	91.5 (14.0)	91.0 (15.7)	93.3 (25.9)	0.538	0.177
Body Mass Index (kg/m^2)	23.4 (3.0)	23.6 (3.0)	23.8 (3.0)	23.6 (2.7)	0.960	0.976
Age (yr)	48.9 (8.7)	51.6 (9.4)	51.9 (9.3)	54.2 (8.9)	<0.001	
Male, n (%)	138 (28.4)	135 (27.8)	107 (22.0)	106 (21.8)	0.001	
Current drinker, n (%)	174 (27.3)	173 (27.1)	153 (24.0)	138 (21.6)	<0.001	
Physical activity (MET-min/wk)	2636.5 (2812.3)	2895.4 (3047.1)	3065.5 (3040.0)	3138.7 (2868.2)	<0.001	
Total energy intake	1691.3 (707.7)	2052.4 (753.0)	2303.0 (786.8)	2654.5 (873.1)	<0.001	<0.001
Carbohydrates[§]	397.0 (2.4)	395.1 (2.3)	387.2 (2.3)	369.2 (2.4)	<0.001	<0.001
Fat[§]	31.3 (0.8)	31.4 (0.8)	33.6 (0.8)	39.2 (0.8)	<0.001	<0.001
Protein[§]	71.2 (1.1)	73.4 (1.0)	79.1 (1.0)	89.8 (1.1)	<0.001	<0.001
Fiber[§]	31.3 (1.6)	30.4 (1.6)	35.6 (1.6)	40.0 (1.6)	<0.001	<0.001
Sodium[§]	2226.3 (69.5)	2779.8 (66.6)	3236.7 (66.6)	4580.9 (69.3)	<0.001	<0.001
Saturated fat[§]	7.2 (0.2)	7.6 (0.2)	7.9 (0.2)	9.0 (0.2)	<0.001	<0.001
Meat pattern (n)	273	267	266	277		
Waist circumference (cm)	76.8 (7.7)	77.8 (8.9)	78.0 (8.9)	80.6 (8.9)	<0.001	<0.001
Triglyceride (mg/dL)	108.6 (67.7)	109.9 (69.2)	110.2 (63.8)	133.7 (89.7)	0.018	0.005
HDL cholesterol (mg/dL)	60.5 (14.7)	59.3 (13.8)	58.9 (13.7)	56.9 (14.4)	0.032	0.067
Diastolic BP (mmHg)	75.7 (10.3)	75.2 (10.2)	76.7 (11.0)	77.8 (10.0)	0.034	0.854
Systolic BP (mmHg)	124.4 (13.8)	124.9 (14.4)	125.4 (15.1)	126.7 (14.1)	0.018	0.073
Fasting glucose (mg/dL)	91.9 (20.1)	91.7 (12.7)	88.9 (10.2)	92.3 (24.0)	0.954	0.050
Body Mass Index (kg/m^2)	23.1 (2.6)	23.6 (3.0)	23.4 (2.8)	24.3 (3.1)	<0.001	<0.001
Age (yr)	54.7 (8.5)	51.8 (8.9)	50.9 (9.3)	49.3 (9.4)	<0.001	
Male, n (%)	86 (17.7)	111 (22.8)	128 (26.3)	161 (33.1)	<0.001	
Current drinker, n (%)	123 (19.3)	146 (22.9)	166 (26.0)	203 (31.8)	<0.001	
Physical activity (MET-min/wk)	2950.1 (2951.2)	2963.5 (2991.0)	2880.4 (2785.7)	2942.8 (3064.2)	0.466	
Total energy intake	2116.8 (923.5)	2004.6 (805.9)	2059.3 (723.9)	2520.7 (868.2)	<0.001	<0.001
Carbohydrates[§]	408.9 (2.0)	398.6 (2.0)	387.1 (2.0)	353.9 (2.1)	<0.001	<0.001
Fat[§]	26.2 (0.7)	29.4 (0.7)	33.8 (0.7)	46.1 (0.7)	<0.001	<0.001
Protein[§]	71.6 (1.1)	76.2 (1.1)	78.5 (1.1)	87.2 (1.1)	<0.001	<0.001
Fiber[§]	33.4 (1.6)	37.7 (1.6)	34.5 (1.6)	31.7 (1.6)	0.259	0.629
Sodium[§]	2857.5 (77.3)	2908.1 (77.9)	3226.3 (77.6)	3834.2 (78.9)	<0.001	<0.001
Saturated fat[§]	6.5 (0.2)	6.6 (0.2)	7.9 (0.2)	10.6 (0.2)	<0.001	<0.001
Snack pattern (n)	257	273	275	278		
Waist circumference (cm)	79.8 (8.4)	78.0 (8.4)	78.1 (9.0)	77.2 (8.7)	0.051	0.006
Triglyceride (mg/dL)	117.1 (75.0)	120.5 (76.1)	112.8 (71.7)	112.7 (73.8)	0.833	0.413
HDL cholesterol (mg/dL)	58.6 (13.7)	58.8 (13.7)	58.0 (13.7)	60.2 (15.4)	0.827	0.936
Diastolic BP (mmHg)	77.9 (10.4)	76.2 (10.1)	75.9 (11.1)	75.6 (9.8)	0.124	0.134
Systolic BP (mmHg)	127.2 (13.6)	125.1 (14.0)	124.7 (15.1)	124.6 (14.6)	0.353	0.364
Fasting glucose (mg/dL)	93.2 (14.5)	90.6 (11.8)	90.7 (22.0)	90.4 (20.5)	0.410	0.343
Body Mass Index (kg/m^2)	24.0 (2.8)	23.4 (2.7)	23.6 (3.1)	23.5 (3.0)	0.506	0.115
Age (yr)	53.9 (8.8)	52.5 (8.9)	51.3 (9.4)	48.9 (9.2)	<0.001	
Male, n (%)	132 (27.2)	124 (25.5)	116 (23.9)	114 (23.5)	0.111	

Table 3. Cont.

| | Quartile of dietary pattern score | | | | | |
	Q 1	Q 2	Q 3	Q 4	p for trend[†]	p value[‡]
Current drinker, n (%)	159 (24.9)	167 (26.2)	148 (23.2)	164 (25.7)	0.965	
Physical activity (MET-min/wk)	2964.9 (2944.4)	2797.5 (2876.0)	3095.5 (2946.4)	2879.3 (3023.0)	0.427	
Total energy intake	2087.9 (836.7)	2009.1 (860.9)	2113.8 (761.9)	2491.1 (886.0)	<0.001	<0.001
Carbohydrates[§]	395.6 (2.2)	397.8 (2.2)	387.1 (2.2)	368.1 (2.3)	<0.001	<0.001
Fat[§]	29.4 (0.8)	29.4 (0.8)	34.1 (0.8)	42.5 (0.8)	<0.001	<0.001
Protein[§]	78.5 (1.1)	76.6 (1.1)	77.8 (1.1)	80.6 (1.1)	0.134	0.229
Fiber[§]	34.4 (1.6)	37.1 (1.6)	33.9 (1.6)	31.9 (1.6)	0.141	0.093
Sodium[§]	3340.2 (79.7)	2963.4 (80.1)	3123.5 (79.9)	3399.6 (81.0)	0.349	0.238
Saturated fat[§]	7.0 (0.2)	7.0 (0.2)	8.2 (0.2)	9.5 (0.2)	<0.001	<0.001

*Participants who were taking medication for hypertension and elevated glucose were excluded for the analysis of the components of metabolic syndrome and BMI.
[†]General linear model with adjustments for age, sex, smoking status, alcohol consumption, total energy intake, and physical activity (log-transformed) for the analysis of the components of metabolic syndrome and BMI, and adjustments for total energy intake for the analysis of nutrients.
[‡]Regression analysis in log-log scale with adjustments for age, sex, smoking status, alcohol consumption, total energy intake, and physical activity (log-transformed) for the analysis of the components of metabolic syndrome and BMI, and adjustments for total energy intake for the analysis of nutrients.
[§]Least squares means (SE) adjusted for total energy intake.
Numbers are Mean (SD), unless otherwise stated.

dietary pattern. The traditional Korean meal is low in fat and contains a large portion of vegetables, which can be considered a healthy diet. However, the traditional pattern was not associated with a lower prevalence of metabolic syndrome in our results or in previous studies [22–24]. HDL cholesterol was inversely associated with the traditional pattern score derived from 16 food groups in women [22]. The traditional pattern score was inversely associated, although not statistically significant, with HDL cholesterol, and positively associated with a prevalence of metabolic syndrome in Hong et al. [23]. Another study that used cluster analysis showed that HDL cholesterol was lower in people with the traditional pattern compared with the those of both meats and alcohols pattern and Korean healthy pattern [24]. Grain, especially refined grain was highly correlated with the traditional pattern in the three above Korean studies. Thus it seems that the negative association between HDL cholesterol and the Korean traditional food pattern was substantially affected by high intakes of carbohydrate. HDL cholesterol was negatively related with carbohydrate [30] and glycemic index [31]. However, grain had very low factor loading for the traditional pattern in our study. Traditional Korean foods are usually cooked with condiments that contain high levels of salt. Highest factor loading in the traditional pattern was condiments, and the pattern score of the traditional pattern was highly correlated with sodium intake in our study (r = 0.72, p<0.001). This may have led to a lack of an association between the traditional Korean dietary pattern and the prevalence of metabolic syndrome in our study. Therefore, the Korean traditional pattern is not generally associated with the prevalence of metabolic syndrome, but high sodium intakes could increase the risk of metabolic syndrome.

The western diet, which has high factor loadings for red meat and processed meat, was positively related to metabolic syndrome [11,12,21,32], whereas meat intake was not associated with metabolic syndrome in French adults [33]. The food groups with high factor loadings in the meat dietary pattern in our study were different from those with high factor loadings in the western dietary pattern and in the meat dietary pattern investigated in the previous studies. Both the western dietary pattern and the meat pattern generally had a high factor loading for processed meat,

which is responsible for many of the adverse effects that are characteristic of the meat dietary pattern. Poultry, which is usually found in healthy dietary patterns, was instead characteristic of the meat pattern in our study. The poultry eaten in Korea was mostly cooked by boiling in 1990. However, the majority of poultry eaten changed to fried chicken in 1998 according to KNHANES. Consequently, poultry has become a major source of saturated fat. Despite discrepancies between each of the food groups, our results suggest that the meat pattern is associated with the prevalence of metabolic syndrome in male adults. This association has been consistently observed in the results from previous studies of the western and meat dietary patterns. The positive association was only found in the male group in our study, suggesting that the meat dietary pattern among Korean adults might increase the prevalence of metabolic syndrome in males to a greater extent than in females.

Several possible mechanisms may explain the detrimental effect of the meat pattern in the human body. First, meat is a major source of total fat intake, particularly saturated fat, and the consumption of saturated fat has been associated with plasma lipoprotein levels [34] and higher blood pressure levels [35]. In a group of individuals of Japanese ancestry, red meat consumption was associated with a higher risk of developing metabolic syndrome among men. However, this association was no longer significant after making adjustments for saturated fatty acids [36]. Thus, the prevalence of metabolic syndrome in the meat pattern was analyzed with further adjustments for saturated fatty acids in our study. The prevalence ratio of metabolic syndrome in the highest quartile compared with the lowest quartile was attenuated (data not shown). Second, meat intake is related to the deposition of iron, particularly heme-iron. Metabolic syndrome subjects had a significantly higher prevalence of iron overload than control subjects [37], and high ferritin concentrations were positively associated with the prevalence of metabolic syndrome and with insulin resistance [38,39]. It was suggested that high iron contents of red meat might be related with higher prevalence of metabolic syndrome [40,41]. Meat intake, especially processed meat, was associated with increased risk of coronary heart disease and diabetes in meta-analysis [42]. Exact mechanism is not explained

Table 4. PRs and 95% CIs of metabolic syndrome by quartiles of dietary patterns.

	Dietary pattern	Quartiles of dietary pattern scores			p for trend*
		Q2	Q3	Q4	
Total	Traditional (n)	314	314	315	
(1257)	Model 1[†]	1.04 (0.71–1.53)[§]	1.01 (0.69–1.50)	1.02 (0.69–1.52)	0.408
	Model 2[‡]	1.01 (0.68–1.51)	1.17 (0.79–1.75)	1.08 (0.71–1.63)	0.330
	Meat (n)	313	314	315	
	Model 1[†]	1.16 (0.81–1.67)	1.23 (0.86–1.76)	1.40 (0.98–1.99)	0.006
	Model 2[‡]	1.23 (0.84–1.81)	1.33 (0.91–1.94)	1.47 (1.00–2.15)	0.016
	Snack (n)	314	313	315	
	Model 1[†]	0.77 (0.55–1.07)	0.82 (0.59–1.15)	0.86 (0.62–1.21)	0.249
	Model 2[‡]	0.79 (0.56–1.13)	0.90 (0.64–1.28)	0.93 (0.65–1.32)	0.421
Men	Traditional (n)	135	107	106	
(n = 486)	Model 1[†]	0.99 (0.59–1.66)	1.05 (0.61–1.83)	1.12 (0.64–1.94)	0.195
	Model 2[‡]	0.96 (0.56–1.64)	1.26 (0.72–2.18)	1.18 (0.66–2.10)	0.129
	Meat (n)	111	128	161	
	Model 1[†]	1.23 (0.67–2.27)	1.27 (0.70–2.30)	1.68 (0.94–2.98)	0.005
	Model 2[‡]	1.64 (0.82–3.27)	1.70 (0.86–3.34)	2.15 (1.10–4.21)	0.005
	Snack (n)	124	116	114	
	Model 1[†]	0.57 (0.34–0.96)	0.87 (0.55–1.37)	0.80 (0.49–1.30)	0.314
	Model 2[‡]	0.53 (0.31–0.94)	0.91 (0.57–1.45)	0.80 (0.49–1.31)	0.335
Women	Traditional (n)	179	207	209	
(n = 771)	Model 1[†]	1.18 (0.67–2.08)	0.99 (0.56–1.76)	0.98 (0.55–1.74)	0.932
	Model 2[‡]	1.16 (0.63–2.12)	1.14 (0.63–2.07)	1.07 (0.58–1.97)	0.978
	Meat (n)	202	186	154	
	Model 1[†]	1.21 (0.76–1.92)	1.29 (0.82–2.02)	1.15 (0.71–1.87)	0.248
	Model 2[‡]	1.18 (0.73–1.91)	1.26 (0.78–2.02)	1.14 (0.68–1.92)	0.455
	Snack (n)	190	197	201	
	Model 1[†]	1.03 (0.66–1.61)	0.82 (0.50–1.33)	1.01 (0.63–1.61)	0.685
	Model 2[‡]	1.11 (0.69–1.80)	0.89 (0.53–1.51)	1.11 (0.66–1.85)	0.830

PR: prevalence ratio, CI: confidence interval.
*Trend test were performed by Wald test using continuous variables of each pattern score (log-transformed).
[†]Adjusted for age, sex (for total) and total energy intake.
[‡]Adjusted for age, sex (for total), total energy intake, smoking status, alcohol consumption, and physical activity (log-transformed).
[§]PR (95% CI), compared with quartile 1 as a reference.

clearly, but iron overload increase oxidative stress due to its catalytic properties [43], resulting insulin resistance and decreased insulin secretion [44,45]. Additionally, the meat dietary pattern may be closely related to a high consumption of alcohol in our study, which may have increased the prevalence of metabolic syndrome in these individuals. Food groups with high factor loadings in the meat pattern were often consumed with alcohol. In previous studies that included alcohol as a food for dietary pattern analysis, a 'meats and alcohols' pattern was derived, suggesting that meat and alcohol consumption are highly correlated in Korean diets [23,24]. Heavy drinking was positively associated with metabolic syndrome and its components [46,47], and both alcohol consumption and a meat dietary pattern were associated with an increased prevalence of metabolic syndrome [21]. The percentage of current drinkers was higher in the highest quartile of meat pattern in our study, especially in men. Although it was adjusted for in the analysis, alcohol consumption may partly affect the prevalence of metabolic syndrome in the meat pattern. Drinking habits might explain the sex difference in the meat

pattern as well, as men are more likely to drink heavily than women. Another explanation for the gender difference is body iron stores. It was suggested that a lower incidence of heart diseases in women, especially in premenopausal women, might be related to lower body iron stores [48,49]. A significant association between iron-related genes and type 2 diabetes was observed in men but not in women [50].

The snack pattern was not associated with the prevalence of metabolic syndrome and its components. Women who had a fiber bread pattern had a lower prevalence of metabolic syndrome and higher insulin sensitivity, while a white bread pattern was positively associated with metabolic syndrome and lowered insulin sensitivity [51,52]. Whole grain consumption was inversely associated with type 2 diabetes [53], a higher waist-to-hip ratio, LDL-cholesterol and fasting insulin [52]. Thus the type of grain consumed by individuals with a snack pattern may affect the prevalence of metabolic syndrome. However, the snack pattern score was only slightly correlated with carbohydrate and fiber in our study. The snack pattern score increased with a decrement of age, and waist circumference, which was highly correlated with age, was inversely associated with the snack pattern score. Thus the trend of decreasing age with increasing the snack pattern score, although age was adjusted for in the analysis, may affect the prevalence of metabolic syndrome in the snack pattern. The meat pattern score also increased with a decrement of age, but it still positively associated with the prevalence of metabolic syndrome. It suggests that the association between the meat pattern and the prevalence of metabolic syndrome is strong.

Our study has several limitations. Because this is a cross-sectional study, there is a chance that dietary intake was affected by an individual's health status, which makes it difficult to find a true association between dietary intake and metabolic syndrome. In addition, we cannot exclude of the possibility of measurement errors of study variables and residual confounding. Thus associations identified should be interpreted in caution. The prevalence of metabolic syndrome in our study was lower than that in previous reports that analyzed the KNHANES data [7]. The study participants may have had a healthier lifestyle, as they

volunteered for the health screening examinations, therefore leading to the lower prevalence of metabolic syndrome. The three dietary patterns derived from PCA analysis explained about 32% of the total variation; thus the derived patterns might not explain all Korean dietary patterns thoroughly.

In conclusion, the meat dietary pattern, which was characterized by a high consumption of red meat, red meat byproducts, and high-fat red meat, was associated with an increased prevalence of metabolic syndrome in Korean male adults.

Author Contributions

Conceived and designed the experiments: HDW AS JK. Analyzed the data: HDW JK. Wrote the paper: HDW JK.

References

1. Lorenzo C, Okoloise M, Williams K, Stern M, Haffner S (2003) The metabolic syndrome as predictor of type 2 diabetes. Diabetes Care 26: 3153–3159.
2. Lakka H, Laaksonen D, Lakka T, Niskanen L, Kumpusalo E, et al. (2002) The metabolic syndrome and total and cardiovascular disease mortality in middle-aged men. JAMA 288: 2709–2716.
3. McNeill A, Rosamond W, Girman C, Golden S, Schmidt M, et al. (2005) The metabolic syndrome and 11-year risk of incident cardiovascular disease in the atherosclerosis risk in communities study. Diabetes Care 28: 385–390.
4. Mozumdar A, Liguori G (2011) Persistent Increase of Prevalence of Metabolic Syndrome Among US Adults: NHANES III to NHANES 1999–2006. Diabetes Care 34: 216–219.
5. Li J, Wang X, Zhang J, Gu P, Zhang X, et al. (2010) Metabolic syndrome: prevalence and risk factors in southern China. J Int Med Res 38: 1142–1148.
6. Gu D, Reynolds K, Wu X, Chen J, Duan X, et al. (2005) Prevalence of the metabolic syndrome and overweight among adults in China. Lancet 365: 1398–1405.
7. Lim S, Shin H, Song JH, Kwak SH, Kang SM, et al. (2011) Increasing Prevalence of Metabolic Syndrome in Korea. Diabetes Care 34: 1323–1328.
8. Ramachandran A, Snehalatha C, Shetty AS, Nanditha A (2012) Trends in prevalence of diabetes in Asian countries. World J Diabetes 3: 110–117.
9. Kim HJ, Kim Y, Cho Y, Jun B, Oh KW (2014) Trends in the prevalence of major cardiovascular disease risk factors among Korean adults: Results from the Korea National Health and Nutrition Examination Survey, 1998–2012. Int J Cardiol 174: 64–72.
10. Azadbakht L, Mirmiran P, Esmaillzadeh A, Azizi T, Azizi F (2005) Beneficial effects of a Dietary Approaches to Stop Hypertension eating plan on features of the metabolic syndrome. Diabetes Care 28: 2823–2831.
11. Esmaillzadeh A, Kimiagar M, Mehrabi Y, Azadbakht L, Hu F, et al. (2007) Dietary patterns, insulin resistance, and prevalence of the metabolic syndrome in women. Am J Clin Nutr 85: 910–918.
12. Lutsey P, Steffen L, Stevens J (2008) Dietary intake and the development of the metabolic syndrome: the Atherosclerosis Risk in Communities study. Circulation 117: 754–762.
13. Newby P, Tucker K (2004) Empirically derived eating patterns using factor or cluster analysis: a review. Nutr Rev 62: 177–203.
14. Lee M, Popkin B, Kim S (2002) The unique aspects of the nutrition transition in South Korea: the retention of healthful elements in their traditional diet. Public Health Nutr 5: 197–203.
15. Denova-Gutierrez E, Castanon S, Talavera J, Gallegos-Carrillo K, Flores M, et al. (2010) Dietary Patterns Are Associated with Metabolic Syndrome in an Urban Mexican Population. J Nutr 140: 1855–1863.
16. Ferreira S, Lerario D, Gimeno S, Sanudo A, Franco L (2002) Obesity and central adiposity in Japanese immigrants: role of the Western dietary pattern. J Epidemiol 12: 431–438.
17. Gimeno S, Ferreira S, Franco L, Hirai A, Matsumura L, et al. (2002) Prevalence and 7-year incidence of type II diabetes mellitus in a Japanese-Brazilian population: an alarming public health problem. Diabetologia 45: 1635–1638.
18. DiBello J, McGarvey S, Kraft P, Goldberg R, Campos H, et al. (2009) Dietary patterns are associated with metabolic syndrome in adult Samoans. J Nutr 139: 1933–1943.
19. Deshmukh-Taskar PR, O'Neil CE, Nicklas TA, Yang S-J, Liu Y, et al. (2009) Dietary patterns associated with metabolic syndrome, sociodemographic and lifestyle factors in young adults: the Bogalusa Heart Study. Public Health Nutr 12: 2493–2503.

20. Amini M, Esmaillzadeh A, Shafaeizadeh S, Behrooz J, Zare M (2010) Relationship between major dietary patterns and metabolic syndrome among individuals with impaired glucose tolerance. Nutrition 26: 986–992.
21. Panagiotakos DB, Pitsavos C, Skoumas Y, Stefanadis C (2007) The association between food patterns and the metabolic syndrome using principal components analysis: The ATTICA Study. J Am Diet Assoc 107: 979–987.
22. Cho YA, Kim J, Cho ER, Shin A (2011) Dietary patterns and the prevalence of metabolic syndrome in Korean women. Nutr Metab Cardiovas 21: 893–900.
23. Hong S, Song Y, Lee KH, Lee HS, Lee M, et al. (2012) A fruit and dairy dietary pattern is associated with a reduced risk of metabolic syndrome. Metab Clin Exp 61: 883–890.
24. Song Y, Joung H (2012) A traditional Korean dietary pattern and metabolic syndrome abnormalities. Nutr Metab Cardiovas 22: 456–462.
25. Kim S, Oh S (1996) Cultural and nutritional aspects of traditional Korean diet. World Rev Nutr Diet 79: 109–132.
26. Guidelines for data processing and analysis of the international physical activity questionnaire, 2005. Available: http://www.ipaq.ki.se. Accessed: 2010 Oct 12.
27. Ahn Y, Kwon E, Shim J, Park M, Joo Y, et al. (2007) Validation and reproducibility of food frequency questionnaire for Korean genome epidemiologic study. Eur J Clin Nutr 61: 1435–1441.
28. Antonopoulos S (2002) Third report of the National Cholesterol Education Program (NCEP) expert panel on detection, evaluation, and treatment of high blood cholesterol in adults (Adult Treatment Panel III) final report. Circulation 106: 3143–3421.
29. WHO West Pacific Region. The Asia-Pacific Perspective: Redefining obesity and its treatment. International Obesity Task Force 2000: 15–21.
30. Merchant AT, Anand SS, Kelemen LE, Vuksan V, Jacobs R, et al. (2007) Carbohydrate intake and HDL in a multiethnic population. Am J Clin Nutr 85: 225–230.
31. Frost G, Leeds A, Dore C, Madeiros S, Brading S, et al. (1999) Glycaemic index as a determinant of serum HDL-cholesterol concentration. Lancet 353: 1045–1048.
32. van Dam R, Rimm E, Willett W, Stampfer M, Hu F (2002) Dietary patterns and risk for type 2 diabetes mellitus in US men. Ann Intern Med 136: 201–209.
33. Mennen L, Lafay L, Feskens E, Novak M, Lepinay P, et al. (2000) Possible protective effect of bread and dairy products on the risk of the metabolic syndrome. Nutr Res 20: 335–347.
34. Riccardi G, Giacco R, Rivellese A (2004) Dietary fat, insulin sensitivity and the metabolic syndrome. Clin Nutr 23: 447–456.
35. Trevisan M, Krogh V, Freudenheim J, Blake A, Muti P, et al. (1990) Consumption of olive oil, butter, and vegetable oils and coronary heart disease risk factors. JAMA 263: 688–692.
36. Damiao R, Castro T, Cardoso M, Gimeno S, Ferreira S (2006) Dietary intakes associated with metabolic syndrome in a cohort of Japanese ancestry. Br J Nutr 96: 532–538.
37. Bozzini C, Girelli D, Olivieri O, Martinelli N, Bassi A, et al. (2005) Prevalence of body iron excess in the metabolic syndrome. Diabetes Care 28: 2061–2063.
38. Sun L, Franco O, Hu F, Cai L, Yu Z, et al. (2008) Ferritin concentrations, metabolic syndrome, and type 2 diabetes in middle-aged and elderly Chinese. J Clin Endocrinol Metab 93: 4690–4696.
39. Jehn M, Clark JM, Guallar E (2004) Serum ferritin and risk of the metabolic syndrome in US adults. Diabetes Care 27: 2422–2428.
40. Azadbakht L, Esmaillzadeh A (2009) Red meat intake is associated with metabolic syndrome and the plasma C-reactive protein concentration in women. J Nutr 139: 335–339.

41. Tappel A (2007) Heme of consumed red meat can act as a catalyst of oxidative damage and could initiate colon, breast and prostate cancers, heart disease and other diseases. Med Hypotheses 68: 562–564.

42. Micha R, Wallace SK, Mozaffarian D (2010) Red and processed meat consumption and risk of incident coronary heart disease, stroke, and diabetes mellitus a systematic review and meta-analysis. Circulation 121: 2271–2283.

43. De Valk B, Marx J (1999) Iron, atherosclerosis, and ischemic heart disease. Arch Intern Med 159: 1542–1548.

44. Ford ES, Cogswell ME (1999) Diabetes and serum ferritin concentration among US adults. Diabetes Care 22: 1978–1983.

45. Jiang R, Manson JE, Meigs JB, Ma J, Rifai N, et al. (2004) Body iron stores in relation to risk of type 2 diabetes in apparently healthy women. JAMA 291: 711–717.

46. Baik I, Shin C (2008) Prospective study of alcohol consumption and metabolic syndrome. Am J Clin Nutr 87: 1455–1463.

47. Athyros VG, Liberopoulos EN, Mikhailidis DP, Papageorgiou AA, Ganotakis ES, et al. (2008) Association of drinking pattern and alcohol beverage type with the prevalence of metabolic syndrome, diabetes, coronary heart disease, stroke, and peripheral arterial disease in a Mediterranean cohort. Angiology 58: 689–697.

48. Sullivan J (1981) Iron and the sex difference in heart disease risk. Lancet 317: 1293–1294.

49. Mascitelli L, Goldstein MR, Pezzetta F (2011) Explaining sex difference in coronary heart disease: is it time to shift from the oestrogen hypothesis to the iron hypothesis? J Cardiovasc Med 12: 64–65.

50. He M, Workalemahu T, Manson JE, Hu FB, Qi L (2012) Genetic determinants for body iron store and type 2 diabetes risk in US men and women. PLoS ONE 7: e40919.

51. Wirfalt E, Hedblad B, Gullberg B, Mattisson I, Andren C, et al. (2001) Food patterns and components of the metabolic syndrome in men and women: a cross-sectional study within the Malmo Diet and Cancer cohort. Am J Epidemiol 154: 1150–1159.

52. McKeown NM, Meigs JB, Liu S, Wilson PWF, Jacques PF (2002) Whole-grain intake is favorably associated with metabolic risk factors for type 2 diabetes and cardiovascular disease in the Framingham Offspring Study. Am J Clin Nutr 76: 390–398.

53. Fung T, Schulze M, Manson J, Willett W, Hu F (2004) Dietary patterns, meat intake, and the risk of type 2 diabetes in women. Arch Intern Med 164: 2235–2240.

6

Demographic Model of the Swiss Cattle Population for the Years 2009-2011 Stratified by Gender, Age and Production Type

Sara Schärrer[1]*, Patrick Presi[1], Jan Hattendorf[2], Nakul Chitnis[2,3], Martin Reist[4], Jakob Zinsstag[2]

1 Veterinary Public Health Institute/University of Berne, Berne, Switzerland, 2 Swiss Tropical and Public Health Institute/University of Basel, Basel, Switzerland, 3 Fogarty International Center, National Institutes of Health, Bethesda, Maryland, United States of America, 4 Federal Food Safety and Veterinary Office, Bern, Switzerland

Abstract

Demographic composition and dynamics of animal and human populations are important determinants for the transmission dynamics of infectious disease and for the effect of infectious disease or environmental disasters on productivity. In many circumstances, demographic data are not available or of poor quality. Since 1999 Switzerland has been recording cattle movements, births, deaths and slaughter in an animal movement database (AMD). The data present in the AMD offers the opportunity for analysing and understanding the dynamic of the Swiss cattle population. A dynamic population model can serve as a building block for future disease transmission models and help policy makers in developing strategies regarding animal health, animal welfare, livestock management and productivity. The Swiss cattle population was therefore modelled using a system of ordinary differential equations. The model was stratified by production type (dairy or beef), age and gender (male and female calves: 0–1 year, heifers and young bulls: 1–2 years, cows and bulls: older than 2 years). The simulation of the Swiss cattle population reflects the observed pattern accurately. Parameters were optimized on the basis of the goodness-of-fit (using the Powell algorithm). The fitted rates were compared with calculated rates from the AMD and differed only marginally. This gives confidence in the fitted rates of parameters that are not directly deductible from the AMD (e.g. the proportion of calves that are moved from the dairy system to fattening plants).

Editor: Edna Hillmann, ETH Zurich, Switzerland

Funding: The study was funded as part of a PhD project by the Swiss Federal Veterinary Office. The funders had no role in study design, data collection and analysis, decision to publish, or preparation of the manuscript.

Competing Interests: The authors have declared that no competing interests exist.

* Email: sara.schaerrer@vetsuisse.unibe.ch

Introduction

Switzerland has been collecting data about cattle including date of birth, date of slaughter, date of death (other than slaughter for consumption) and information regarding movements on a mandatory basis since 1999. The purpose of a national database of animal movements was originally to restore consumer trust during the BSE crisis by assuring traceability and therefore a better food safety of beef products and to provide a tool for epizootic disease surveillance and control [1,2]. The AMD contains detailed and complete datasets about the Swiss cattle population for several years offering the opportunity to get an insight into the population dynamics. Understanding the demographic of the livestock population in turn provides accurate parameters needed to develop models of disease transmission and helps policy makers in developing strategies regarding animal health, animal welfare and livestock management [3].

Early detection of disease, monitoring of present agents and substantiation of freedom from disease are described as key tasks of modern public veterinary services in order to allow international trade with agricultural goods and to document a good sanitary status of domestic livestock [4–6].

To monitor the health status of the cattle population, the Swiss veterinary authorities invest substantial resources in yearly surveillance programmes that have to meet international standards. One way to maintain the standards while reducing the costs is the application of risk based targeted approaches (e.g. [7]). Other approaches comprise logistical improvements such as better exploiting infrastructures where already a lot of potential information carriers are available e.g. slaughterhouse or milk quality testing laboratories [8]. With the implementation of bulk milk testing in 2010 [9,10] the production type became an important criterion for shaping the sampling strategy of national surveillance programs. As beef and fattening cattle, correspond to one third of the population, they have to be handled separately. The two production types (dairy and beef) do not only differ with respect to purpose but also with respect to management practices. The resulting differences in age distribution and slaughter rates in the two sub populations are of interest for the planning of stratified surveillance programmes to assure the representativeness of the sample (e.g. for sampling at the slaughterhouse level).

The objective of this study was therefore to create an AMD data driven demographic model that simulates the age and gender specific dynamics of the Swiss cattle population according to the production type. The derived rates describing population dynamics can be used for livestock development planning and associated economic analyses, as a backbone for disease transmission models

Table 1. The Swiss cattle population 2009–2011.

Year	No of farms	No of cattle (January 1th)	No of dairy cows (January 1th)	No of slaughtered animals	No of births
2009	42'966	1'608'062	675'285	647'715	721'810
2010	42'233	1'610'277	671'874	648'313	719'004
2011	41'465	1'612'230	676'253	653'754	718'697

Numbers are extracted from the Swiss animal movement database (AMD).

or for the design of cost-effective disease control and monitoring programmes.

Here we present the first dynamic demographic model of the Swiss cattle population. It is based on over 30 million data points collected in the Swiss animal movement database (AMD) between 2009 and 2011.

Materials and Methods

2.1 The Swiss cattle population

The major livestock species in Switzerland is cattle. Although the number of farms decreases, for the years 2009–2011 the number of cattle in Switzerland is stable at roughly 1.6 million animals (table 1). Two thirds of the Swiss cattle industry is dedicated to dairy production. As a consequence, adult dairy cows (older than two years) make the largest demographic segment (figure 1). The average lifespan of a dairy cow in Switzerland is 6.2 years and the average number of calves in a lifetime is 3.7. The oldest cow that died between 2009 and 2011 was 25 years old.

Due to subsidies for ecological and behaviourally sound husbandry and strict animal protection legislation, small holdings with less than hundred animals are still the most common farm type. Over the summer month (May–October) one fourth of the livestock is moved to alpine pastures.

2.2 Data management

The Swiss animal movement database (AMD) contains information on farm level (e.g. location, production type), animal level (e.g. birthdate, gender, and breed), movement records (date, movement type) and stays (i.e. for every animal the start and end date of a stay on any holding is recorded). The data used for the models was an extract from the AMD, containing all recorded movements (25.5 million entries) and stays (15.8 million entries) from January 1999 until January 2012.

Birthdate, date of death (slaughter or natural) and gender are recorded on individual animal level, while the production type is available on farm level. The production type for each animal was consequently determined by the farm it stayed on at the given time step. Calf mortality consisted of notified stillbirths and mortality. As stays on alpine pastures are recorded only since 2008 and the quality of those recordings improved notably in 2009, only data from 2009 to 2011 was used for fitting of the population model.

2.3 The model

The Swiss cattle population was simulated using a system dynamic software [11]. The model is composed of a series of coupled difference equations. Compartments were defined by production type (dairy or beef), age class and gender. Calves were defined as animal being less than one year old, heifer and young bulls as one to two years old and cows and bulls as older than two years. We assumed that cows calve for the first time at the age of two and therefore the category "heifer" doesn't contribute to births. The beef and dairy system are connected through the transfer of calves from dairy farms to fattening plants, which is represented in the model as "fattening". The model is represented in figure 2.

The dynamic of the cattle population is simulated by month as time unit. Equations (1)–(12) show the number of animals per compartment (for parameter notation see table 2 and 3).

To represent the seasonal fluctuations in the number of births and death calves, we used a sinusodial-function with amplitude (a), phase (φ) and average (μ) as parameters to fit (equations (13)–(20)). The frequency (ω) was set to $\frac{2\pi}{12}$.

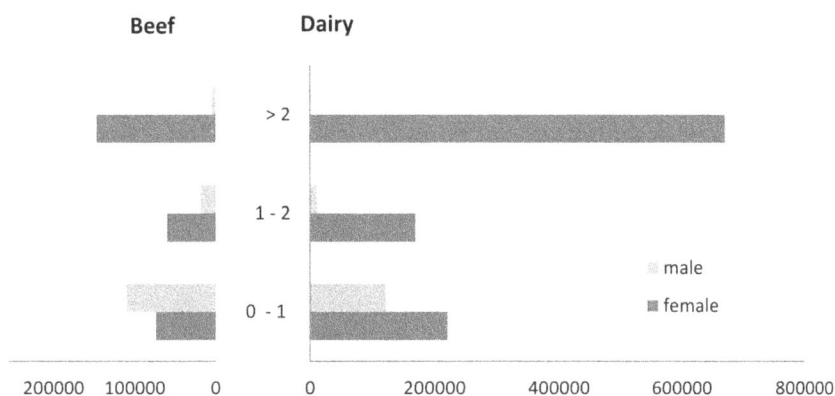

Figure 1. Demographic of the Swiss cattle population per age class and sex in number of animals.

$$\frac{dX_{DF}(t)}{dt} = b_{XDF}(t) * Z_{DF} - (m_{XDF}(t) \\ + s_{XDF} + f_{XDF} + tr_{XDF}) * X_{DF} \qquad (1)$$

$$\frac{dY_{DM}(t)}{dt} = tr_{XDM} * X_{DM} - (m_{YDM} + s_{YDM} + tr_{YDM}) * Y_{DM} \qquad (6)$$

$$\frac{dX_{DM}(t)}{dt} = b_{XDM}(t) * Z_{DF} - (m_{XDM}(t) \\ + s_{XDM} + f_{XDM} + tr_{XDM}) * X_{DM} \qquad (2)$$

$$\frac{dY_{BF}(t)}{dt} = tr_{XBF} * X_{BF} - (m_{YBF} + s_{YBF} + tr_{YBF}) * Y_{BF} \qquad (7)$$

$$\frac{dX_{BF}(t)}{dt} = b_{XBF}(t) * Z_{BF} + f_{XDF} * X_{DF} \\ - (m_{XBF}(t) + s_{XBF} + tr_{XBF}) * X_{BF} \qquad (3)$$

$$\frac{dY_{BM}(t)}{dt} = tr_{XBM} * X_{DF} - (m_{YBM} + s_{YBM} + tr_{YBM}) * Y_{BM} \qquad (8)$$

$$\frac{dX_{BM}(t)}{dt} = b_{XBM}(t) * Z_{BF} + f_{XDM} * X_{DM} \\ - (m_{XBM}(t) + s_{XBM} + tr_{XBM}) * X_{BM} \qquad (4)$$

$$\frac{dZ_{DF}(t)}{dt} = tr_{YDF} * Y_{DF} - (m_{ZDF} + s_{ZDF}) * Z_{DF} \qquad (9)$$

$$\frac{dZ_{DM}(t)}{dt} = tr_{YDM} * Y_{DM} - (m_{ZDM} + s_{ZDM}) * Z_{DM} \qquad (10)$$

$$\frac{dY_{DF}(t)}{dt} = tr_{XDF} * X_{DF} - (m_{YDF} + s_{YDF} + tr_{YDF}) * Y_{DF} \qquad (5)$$

$$\frac{dZ_{BF}(t)}{dt} = tr_{YBF} * Y_{BF} - (m_{ZBF} + s_{ZBF}) * Z_{BF} \qquad (11)$$

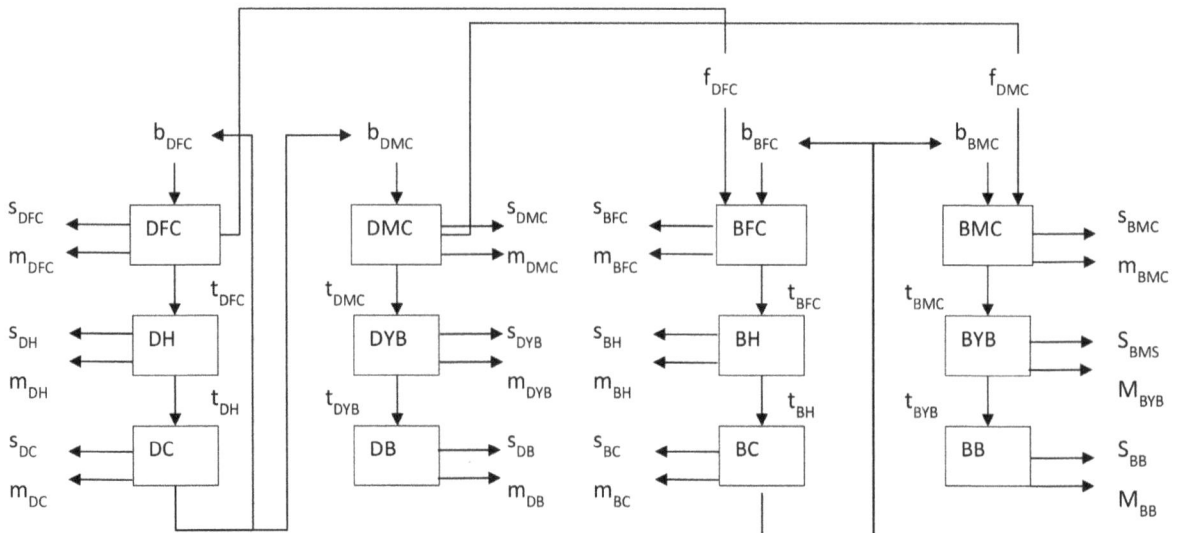

Figure 2. Schematic representation of the Vensim model. Arrows represent flows of animals into or out of a compound, boxes represents numbers of animals at a given time point in a category. s: slaughter; m: mortality; b: birth; tr: transition; f: fattening; D: dairy; B: beef; F: female; M: male; X: calves; Y: subadults; Z: adults.

Table 2. Nomenclature for subscripts in Equations 1–12.

	Description	type
X	Calves	age class
Y	Subadults	age class
Z	Adults	age class
D	Dairy	production type
B	Beef	production type
F	Female	gender
M	Male	gender

$$\frac{dZ_{BM}(t)}{dt}=tr_{YBM}*Y_{BM}-(m_{ZBM}+s_{ZBM})*Z_{BM} \quad (12)$$

$$m_{XDM}(t)=\mu_{2XDM}+a_{2XDM}*\sin(t*\omega+\varphi_{2XDM}) \quad (16)$$

$$b_{XDF}(t)=\mu_{1XDF}+a_{1XDF}*\sin(t*\omega+\varphi_{1XDF}) \quad (13)$$

$$b_{XBF}(t)=\mu_{1XBF}+a_{1XBF}*\sin(t*\omega+\varphi_{1XBF}) \quad (17)$$

$$m_{XDF}(t)=\mu_{2XDF}+a_{2XDF}*\sin(t*\omega+\varphi_{2XDF}) \quad (14)$$

$$m_{XBF}(t)=\mu_{2XBF}+a_{2XBF}*\sin(t*\omega+\varphi_{2XBF}) \quad (18)$$

$$b_{XDM}(t)=\mu_{1XDM}+a_{1XDM}*\sin(t*\omega+\varphi_{1XDM}) \quad (15)$$

$$b_{XBM}(t)=\mu_{1XBM}+a_{1XBM}*\sin(t*\omega+\varphi_{1XBM}) \quad (19)$$

$$m_{XBM}(t)=\mu_{2XBM}+a_{2XBM}*\sin(t*\omega+\varphi_{2XBM}) \quad (20)$$

Table 3. Compartments and parameters in Equations 1–12.

	Description	Unit
X	No of calves	Animals
Y	No of subadults	Animals
Z	No of adults	Animals
s	slaughter rate	month^{-1}
m	mortality rate	month^{-1}
b	birth rate	month^{-1}
tr	transition rate	month^{-1}
f	fattening rate	month^{-1}
μ	Average	month^{-1}
a	Amplitude	month^{-1}
ω	Frequency	month^{-1}
φ	Phase	Dimensionless

Table 4. Monthly population parameters for the Swiss cattle population.

		Dairy			Beef		
			Month^{-1}	95%-CI		Month^{-1}	95%-CI
slaughter rates	Female calf	s_{XDF}	0.0197	[0.0192, 0.02201]	s_{XBF}	0.0396	[0.0389, 0.0403]
	Heifer	s_{YDF}	0.0065	[0.0062, 0.0069]	s_{YBF}	0.0261	[0.0253, 0.0269]
	Cow	s_{ZDF}	0.0190	[0.0189, 0.0191]	s_{ZBF}	0.0233	[0.0231, 0.0235]
	Male calf	s_{XDM}	0.1123	[0.1103, 0.1144]	s_{XBM}	0.0638	[0.0631, 0.0645]
	Young bull	s_{YDM}	0.1702	[0.1658, 0.1748]	s_{YBM}	0.2834	[0.2768, 0.2902]
	Bull	s_{ZDM}	0.1113	[0.1072, 0.1156]	s_{ZBM}	0.0606	[0.0590, 0.0623]
mortality rates	Female calf	μ_{2XDF}	0.0094	[0.0089, 0.0098]	μ_{2XBF}	0.0059	[0.0055, 0.0062]
	Heifer	m_{YDF}	0.0007	[0.0006, 0.0007]	m_{YBF}	0.0008	[0.0007, 0.0009]
	Cow	m_{ZDF}	0.0013	[0.0012, 0.0013]	m_{ZBF}	0.0013	[0.0013, 0.0014]
	Male calf	μ_{2XDM}	0.0255	[0.0241, 0.0269]	μ_{2XBM}	0.0074	[0.0071, 0.0078]
	Young bull	m_{YDM}	0.0017	[0.0015, 0.0019]	m_{YBM}	0.0017	[0.0015, 0.0019]
	Bull	m_{ZDM}	0.0022	[0.0016, 0.0028]	m_{ZBM}	0.0026	[0.0021, 0.00031]
transition rates	Female calf	tr_{XDF}	0.0684	[0.0678, 0.0689]	tr_{XBF}	0.0718	[0.0710, 0.0725]
	Heifer	tr_{YDF}	0.0804	[0.0797, 0.0812]	tr_{YBF}	0.0615	[0.0607, 0.0624]
	Male calf	tr_{XDM}	0.0207	[0.0203, 0.0212]	tr_{XBM}	0.0511	[0.0505, 0.0518]
	Young bull	tr_{YDM}	0.0234	[0.0226, 0.0243]	tr_{YBM}	0.0161	[0.0157, 0.0165]
fattening rates	Female calf	f_{XDF}	0.0172	[0.0170, 0.0175]			
	Male calf	f_{XDM}	0.0731	[0.0722, 0.0740]			
birth rates	Female calf	μ_{1XDF}	0.0374	[0.0373, 0.0376]	μ_{1XBF}	0.0335	[0.0332, 0.0339]
	Male calf	μ_{1XDM}	0.0392	[0.0389, 0.0396]	μ_{1XBM}	0.0352	[0.0347, 0.0357]

D: dairy; B: beef; F: female; M: male; X: calf; Y: subadult; Z: adult. Small letters indicate rates (s: slaughter, m: mortality, f: fattening, tr: transition to next age class). μ_1: average birth rate; μ_2: average mortality rate;

Table 5. Values for the amplitudes and phases in the trigonometric functions of the presented Swiss cattle population model.

	Dairy			Beef	
		95%-CI			**95%-CI**
a_{1XDF}	0.0031	[0.0022, 0.0041]	a_{1XBF}	0.0009	[0, 0.0023]
a_{1XDM}	0.0091	[0.0073, 0.0109]	a_{1XBM}	0.0040	[0.0024, 0.0056]
a_{2XDF}	0.0029	[0.0020, 0.0038]	a_{2XBF}	0.0013	[0.0008, 0.0018]
a_{2XDM}	0.0063	[0.0037, 0.0088]	a_{2XBM}	0.0016	[0.0010, 0.0022]
φ_{1XDF}	1.6799	[1.4046, 1.9574]	φ_{1XBF}	2.9510	[1.1437, 4.7768]
φ_{1XDM}	1.9245	[1.7428, 2.1096]	φ_{1XBM}	2.4772	[2.0699, 2.8935]
φ_{2XDF}	1.6576	[1.3443, 1.9727]	φ_{2XBF}	1.0713	[0.6834, 1.4582]
φ_{2XDM}	1.7900	[1.3820, 2.1969]	φ_{2XBM}	0.9218	[0.5575, 1.2856]

D: dairy; B: beef; F: female; M: male; X: calf, Y: subadult, Z: adult. a 1: amplitude for birth rate; a 2: amplitude for mortality rate; φ_1: phase for birth rate; φ_2: phase for mortality rate;

2.3.1 Model fitting. The number of living animals was extracted at the beginning of each month, number of birth, slaughter and death from the AMD per month, age class, production type and gender from January 2009 to December 2011. This data-set served to optimize the model parameters on the basis of the goodness-of-fit of the nonlinear maximum-likelihood optimization using the Powell algorithm [12]. Parameters were fitted stepwise, adding a variable at every step to the payoff values, using the outcome rates from the previous step as

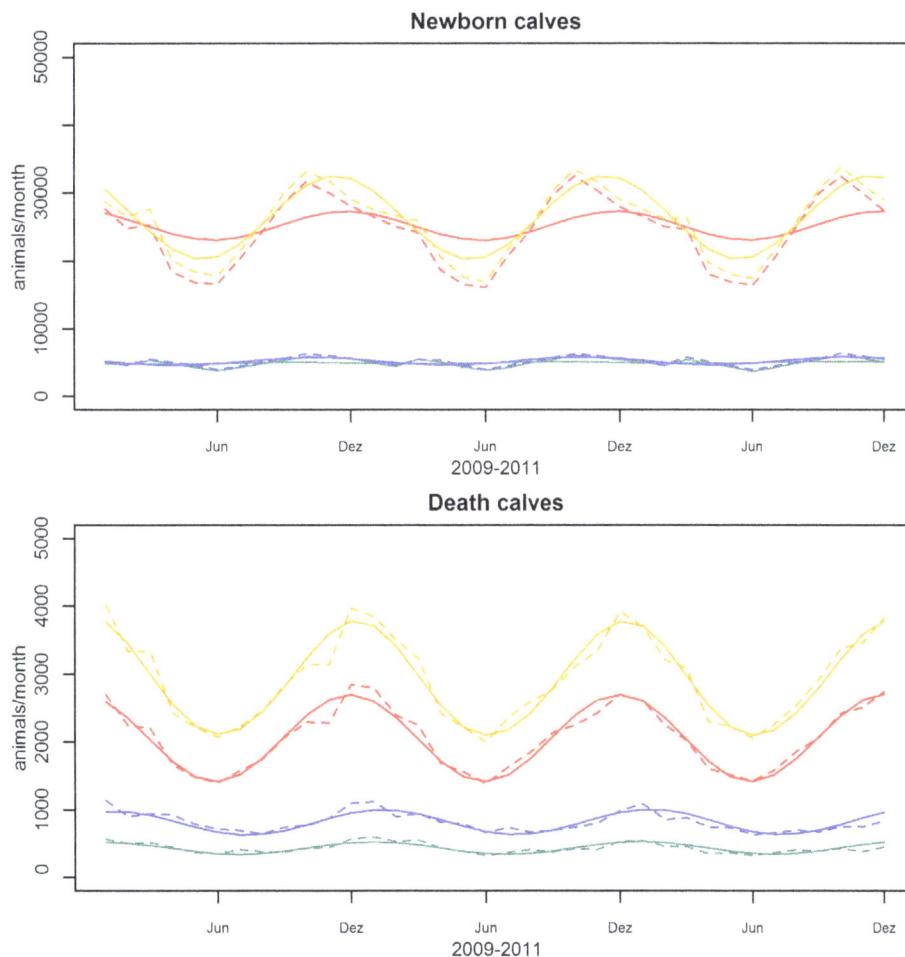

Figure 3. Seasonal pattern of birth and mortality in calves. Solid line: model data, dashed lines: AMD data. Orange: dairy male calf, red: dairy female calf, blue: beef male calf, green: beef female calf.

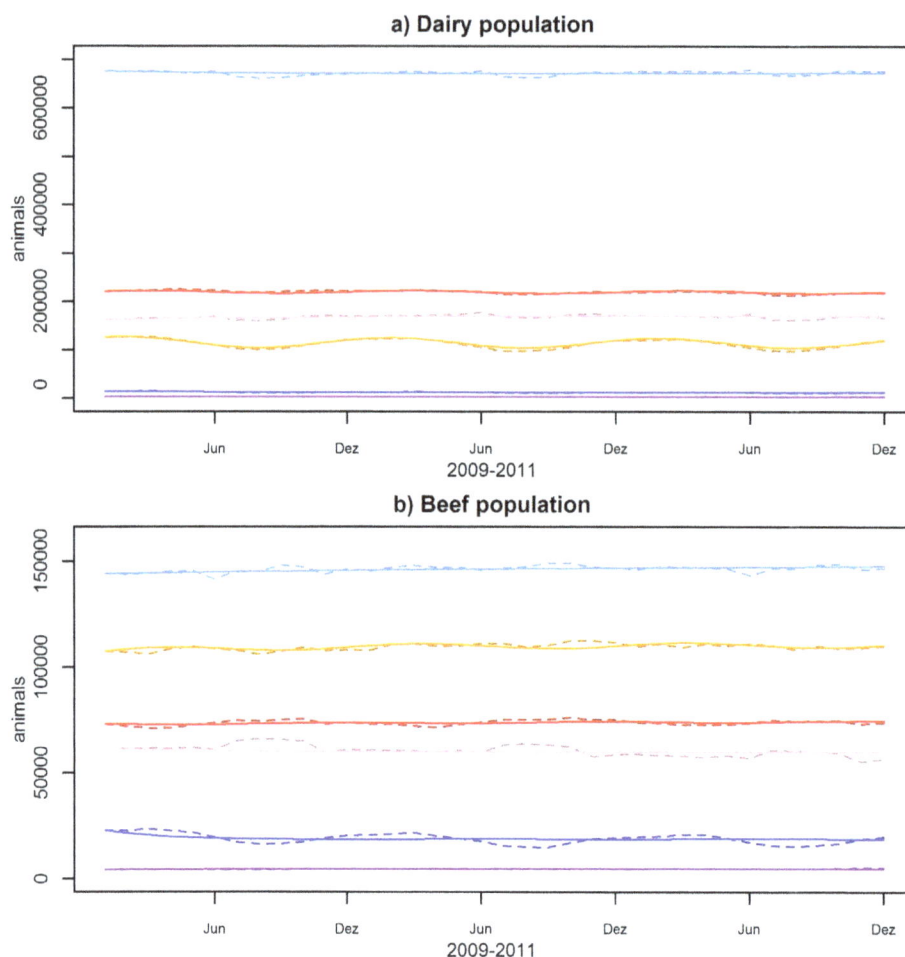

Figure 4. Animal numbers per age category. a) Dairy population. b) Beef population. Solid line: model data, dashed lines: AMD data. Light blue: cow, orange: male calf, red: female calf, pink: heifer, blue: young bull, purple: bull.

initial search point (maximum and minimum values set to +/− 10%).

2.3.2 Comparison of calculated and fitted rates. Birth, slaughter and mortality rates were calculated from the AMD data and compared to the fitted values from the model. Average birth rates were calculated as number of calves per month and category divided by the number of cows on the first of the months of the according production type and averaged over the 3 years period. Mortality and slaughter rates were calculated as number of death or slaughtered animals per month divided by the number of animals of the same age category and production type on the first of the month and averaged over the 3 years period. Model and empirical estimates were correlated in R [13].

2.3.3 Sensitivity analysis. The model was rebuilt with the statistical software R. To assess the sensitivity of the model, each parameter was varied separately using a range from −10% to +10% of the fitted value from the Vensim model (baseline), divided in 100 steps. For each value, the resulting absolute change in total numbers of animals compared to the baseline was represented graphically (Figures S1–S10, supplementary material).

Results

In table 4 the fitted parameter values from the demographic model are shown. The model allowed the calculation of

parameters that are not directly deductible from the AMD (transition rates and fattening rates).

By introducing parameters (amplitude and phase, table 5) to describe calf mortality and birth rates as trigonometric functions, the seasonal dynamic of changes in the population can be described more accurately than with the corresponding linear parameters deducted from the monthly extracts of the AMD (figure 3).

The correlation of the empirical parameters from the AMD and the fitted values gives a correlation coefficient of 0.994. The good fit of the model to the empirical data is also illustrated in figure 4.

As expected, the beef and dairy sector show differences in the demographic composition. While the proportions of young female animals are comparable (18.5% dairy female calves, 17.8% beef female calves 14.2% dairy heifers and 14.6% beef heifers), dairy cows account for around 56.7% of the dairy population while beef cows account for 35.5% of the beef population. For male animals the differences are even more noticeable: beef male calves, young bulls and bulls make 26.5%, 4.6% and 1.1% of the beef population compared to 9.6%, 1.0% and 0.2% for dairy male calves, dairy young bulls and dairy bulls respectively (all proportions are means over the 36 month of data analysis).

As import and export of live cattle are negligible for Switzerland (6'787 imported animals from 2009 to 2011 and 3'318 exported animals over the same period), the beef population is maintained

Calf restocking in the beef sector

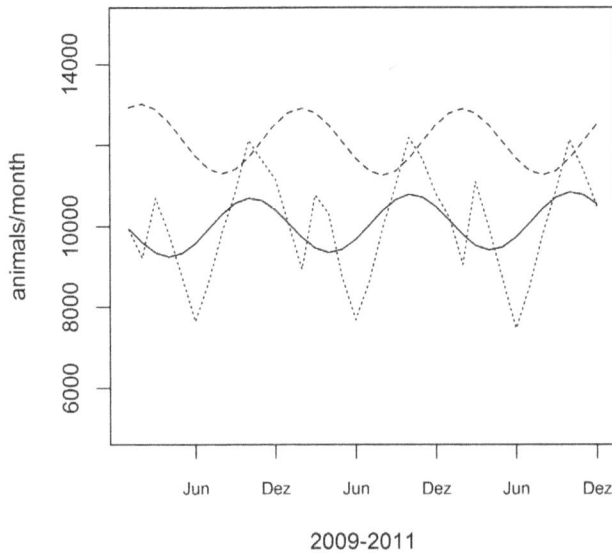

Figure 5. Restocking of calves in the beef sector. Dashed line: dairy calves transferred to fattening plants (VENSIM), solid line:born beef calves (VENSIM), dotted line: born beef calves (AMD).

to a considerable extend by calves from the dairy industry. Almost every month more dairy calves are transferred to fattening plants (i.e. to from the dairy to the beef industry) than were born within the beef industry (figure 5).

The number of slaughtered animals does not show a clear seasonal pattern (AMD data, figure 6) and the slaughter rate in the model is linear.

The sensitivity analysis shows, that the dairy female calf birth average and the dairy cow slaughter rate have the biggest influence on the total population with a change in animal numbers bigger than 50′000 after 3 years of simulation (figures S1–S10, supplementary material).

Discussion

4.1 The Swiss cattle population

The composition of the Swiss cattle population accentuates that the milk industry dominates the domestic production and shapes the population dynamic. Adult dairy cows account for over 40% of all animals (figure 4). The importance of dairy female animals for the total population is reflected in the high sensitivity of the beef population to changes in the dairy cow slaughter rate and the dairy female birth rate (figures S2, S6, supplementary material). The irregular slaughter pattern indicates that the farmers keep the population constant by management decisions.

The higher monthly average mortality of dairy male calves compared to their contemporaries (0.0255 compared to 0.0094 (XDF), 0.0074 (XBM) and 0.0059 (XBF)) is in line with findings of other authors. [14] and [15] found higher mortality rates in dairy breeds than in beef breeds and higher mortality rates in male calves than female calves. As they all defined calves as maximum 180 days of age, the broader categories in our model might explain why dairy male calves differ as much from the others as the effect of early perinatal mortality with higher risk of dystocia for male calves [14] is combined with management decisions, i.e. less care for the economically relatively uninteresting male dairy calves

[15]. As we also determined the production type on farm level and not according to the breed as in the above mentioned studies, effects of management decisions on the calve mortality might be even more manifest.

When deducting yearly rates roughly by multiplying the monthly age transition rates by 12, the difference in the management of beef and dairy animals becomes more obvious: while 82% of female dairy calves reach the next age class, only 25% of dairy male calves live through their first year. For beef calves 86% of the females and 61% of the males reach the next age class which reflects the interest of fattening beef breeds for more than 12 month. The most valued group of animals, dairy heifers, reach adulthood in 96% of the cases while more beef heifers are slaughtered and only 74% get two years old.

4.2 Model assumptions

In high productive agriculture systems of the developed world the population dynamics of livestock is controlled by the farmer and depends on policy and economics rather than on resource limitation or other external factors e.g. [16]. Bleul [14] states, that 80% of Swiss cows are inseminated artificially. For this reason we did not consider a resource constraint i.e. a carrying capacity in our model. The results may be of use for countries in similar economic situation but with less complete records but are to be applied carefully to cattle population that live under more resource dependent natural conditions.

The difference in the birth rates of dairy female and male calves in the model is an artefact presumably due to the difference in the dynamic of the two compartments. Dairy female calves are the most important segment to maintain the population which makes the model sensitive to any change in dairy female calf births. A conservative simulation gives a more stable overall result.

As alpine pastures usually use the gained milk directly for cheese production and it enters therefore not in commerce or they have young stock not yet lactating, they are mostly in the beef category regardless the provenience of the cattle. Therefore the data was corrected over the summer months, using the production type of the farm of origin from the movement records to alpine pastures. The visible seasonal bumps in beef heifers in figure 4 show, that the correction is imperfect due to an incomplete registration of the movements from and to alpine pastures. Since 2012 these are mandatory and improvement of the data quality can be expected.

To integrate the seasonality of birth and mortality in calves, we assumed a sinusoidal pattern and did not investigate other functions.

4.3 Future applications of the model

This is the first dynamic population model for Swiss cattle. As the data source is the complete record of the cattle population, a very good fit could be expected. Nonetheless the fitted population parameters allow a close to reality simulation of the population for future development planning scenario analysis, serve as a backbone to disease transmission models and for the simulation of disease surveillance and control (e.g. [17]).

The fitted population parameters allow building age and sex structured transmission models to simulate disease dynamics with different prevalences in different age classes (e.g. infectious bovine rhinotracheitis IBR, Brucellosis).

Furthermore the transmission rates of different age and production type categories to the slaughterhouse give precise information, which proportions of populations and subpopulations would be basically available for testing at the slaughterhouse in which time period. The slaughterhouse is a very convenient spot for sampling, because it allows taking samples from many animals

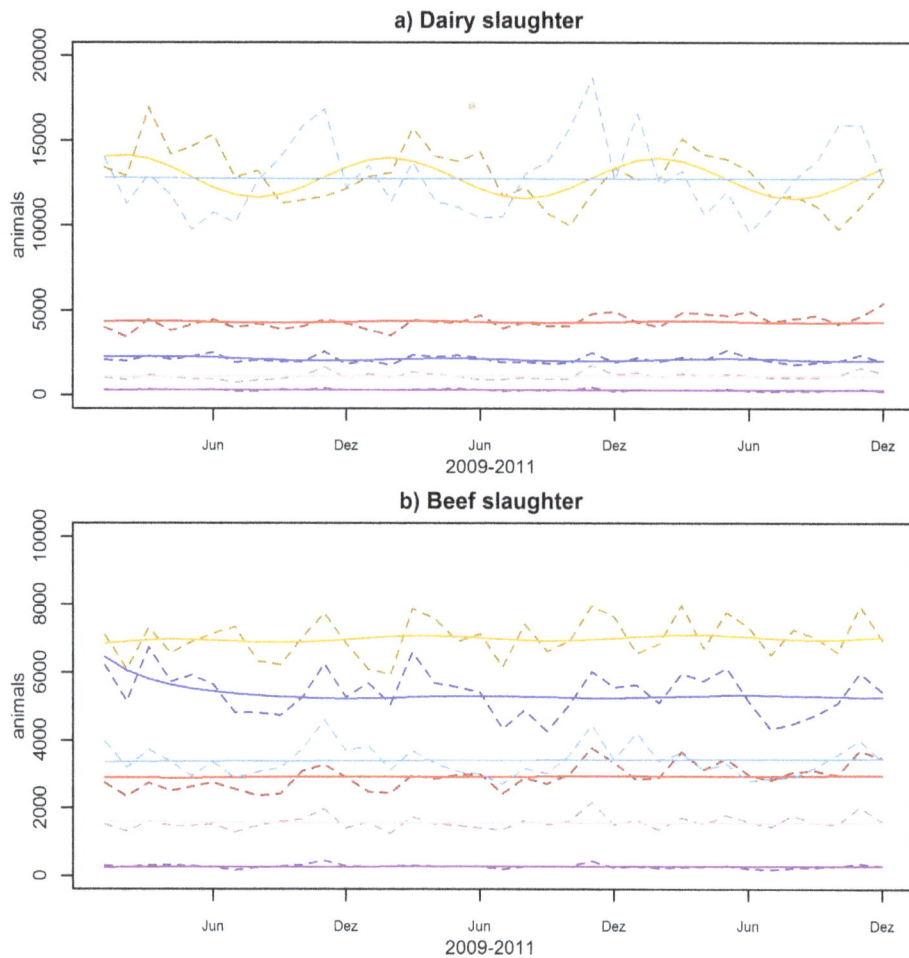

Figure 6. Slaughter numbers per age category. a) Dairy population. b) Beef population. Solid line: model data, dashed lines: AMD data. Light blue: cow, orange: male calf, red: female calf, pink: heifer, blue: young bull, purple: bull.

from different farms of origin within a short time period. Furthermore, there are diseases such as bovine spongiform encephalopathy (BSE) that can only be diagnosed in tissue matrices accessible at slaughter, e.g. brainstem.

As the outcome parameters in the model are calculated for the dairy and beef sector separately, surveillance systems with different components for the different production types can be simulated (e.g. IBR, Brucellosis). For example the efficacy of combining bulk tank milk sampling with slaughterhouse or on farm sampling can be evaluated. As the transfer from calves from the dairy sector to the beef sector is included, the model allows a realistic simulation of disease transmission in the overall population and of the effect of different surveillance strategies on the system sensitivity for different production types.

The fitted population parameters can also be interpreted as baseline parameters for the healthy Swiss cattle population. As seasonal effects are included in the parameter fitting, they can be used to search for aberrations in present data (e.g. increased mortality) to detect health events in an early stage.

In the healthy population most female calves are kept to restock the dairy population, as can be inferred from the relatively low transmission rates of female dairy calves to slaughter. If that segment is affected by an epidemic leading to increased abortions, calf mortality or decreased fertility, consequences on population structure and management are to be expected. Achievement of

breeding objectives might be delayed or even out of reach. Impacts on the milk and meat markets are to be expected. The impact on population structure such as decrease of adult dairy cows in the slaughter population can be estimated by model derived transmission factors.

Conclusions

The Swiss animal movement database is a reliable source of information about the Swiss cattle population and can provide stakeholders and decision makers with important knowledge without expensive and laborious field work. The presented demographic model allows a simulation of Swiss cattle production and economics under different policy scenarios and can be used as the demographic backbone for disease transmission models.

Supporting Information

Figure S1 Influence of varying slaughter rates on the number of animals in the dairy population.

Figure S2 Influence of varying slaughter rates on the number of animals in the beef population.

Figure S3 Influence of varying mortality rates on the number of animals in the dairy population.

Figure S4 Influence of varying mortality rates on the number of animals in the beef population.

Figure S5 Influence of varying average birth rates on the number of animals in the dairy population.

Figure S6 Influence of varying average birth rates on the number of animals in the beef population.

Figure S7 Influence of varying fattening rates (calves transferring from the dairy to the beef sector) on the number of animals in the dairy population.

Figure S8 Influence of varying fattening rates (calves transferring from the dairy to the beef sector) on the number of animals in the beef population.

Figure S9 legends for the colour scales for the dairy population.

Figure S10 legends for the colour scales for the beef population.

Author Contributions

Analyzed the data: SS JZ JH NC. Wrote the paper: SS JH NC MR JZ PP. Model design: SS JZ JH NC PP MR.

References

1. Golan E, Krissoff B, Kuchler F, Calvin L, Nelson K, et al. (2004) Traceability in the U. S. Food Supply: Economic Theory and Industry Studies. Agric Econ Rep 830.
2. Lüdi F (2004) Demographie, räumliche Verteilung und Dynamik der schweizerischen Rindviehpopulation in den Jahren 2002 und 2003.
3. O'Connor J, More S, Griffin J, O'Leary E (2009) Modelling the demographics of the Irish cattle population. Prev Vet Med 89: 249–254. Available: http://www.ncbi.nlm.nih.gov/pubmed/19327855. Accessed 13 June 2012.
4. Anonymous (2002) Agreement between the European Community and the Swiss Confederation on trade in agricultural products.
5. Anonymous (1995) Tierseuchenverordnung, SR 916.401. Bern. Available: http://www.admin.ch/opc/de/classified-compilation/19950206/index.html.
6. WTO (n.d.) The WTO Agreement on the Application of Sanitary and Phytosanitary Measures (SPS Agreement). Available: http://www.wto.org/english/tratop_e/sps_e/spsagr_e.htm.
7. Stärk K, Regula G, Hernandez J, Knopf L, Fuchs K, et al. (2006) Concepts for risk-based surveillance in the field of veterinary medicine and veterinary public health: Review of current approaches. BMC Health Serv Res 6: 20. doi:10.1186/1472-6963-6-20.
8. Hadorn D, Racloz V, Schwermer H, Stärk K (2009) Establishing a cost-effective national surveillance system for Bluetongue using scenario tree modelling. Vet Res 40: 57. doi:10.1051/vetres/2009040.
9. Reber A, Reist M, Schwermer H (2012) Cost-effectiveness of bulk-tank milk testing for surveys to demonstrate freedom from infectious bovine rhinotracheitis and bovine enzootic leucosis in Switzerland. SAT 154: 189–197. doi:10.1024/0036-7281/.
10. Reist M, Jemmi T, Stärk K (2012) Policy-driven development of cost-effective, risk-based surveillance strategies. Prev Vet Med 105: 176–184. Available: http://www.ncbi.nlm.nih.gov/pubmed/22265642. Accessed 4 April 2013.
11. Ventana Systems I (2008) Vensim: Simulator for Business.
12. Press W, Flannery B, Teukolsky S, Viatour P (1991) Numerical Recipies in C. Cambridge.
13. R Development Core Team (2008) R: A language and environment for statistical computing. Available: http://www.r-project.org.
14. Bleul U (2011) Risk factors and rates of perinatal and postnatal mortality in cattle in Switzerland. Livest Sci 135: 257–264. Available: http://linkinghub.elsevier.com/retrieve/pii/S1871141310004142. Accessed 21 August 2014.
15. Perrin J, Ducrot C, Vinard J, Hendrikx P, Calavas D (2011) Analyse de la mortalité bovine en France de 2003 à 2009. INRA Prod Anim 24: 235–244.
16. Rosen S, Murphy K, Scheinkman J (2012) Cattle Cycles. J Polit Econ 102: 468–492.
17. Riley S (2007) Large-scale spatial-transmission models of infectious disease. Science 316: 1298–1301. Available: http://www.ncbi.nlm.nih.gov/pubmed/17540894. Accessed 11 July 2014.

7

Impact of Chronic Neonicotinoid Exposure on Honeybee Colony Performance and Queen Supersedure

Christoph Sandrock[1]*, **Matteo Tanadini**[2], **Lorenzo G. Tanadini**[3], **Aline Fauser-Misslin**[1,4], **Simon G. Potts**[5], **Peter Neumann**[1,4]

1 Agroscope, Swiss Bee Research Centre, Berne, Switzerland, 2 Seminar for Statistics, ETH Zurich, Zurich, Switzerland, 3 Division of Biostatistics, Institute for Social and Preventive Medicine, University of Zurich, Zurich, Switzerland, 4 Institute of Bee Health, Vetsuisse Faculty, University of Berne, Berne, Switzerland, 5 School of Agriculture, Policy and Development, University of Reading, Reading, United Kingdom

Abstract

Background: Honeybees provide economically and ecologically vital pollination services to crops and wild plants. During the last decade elevated colony losses have been documented in Europe and North America. Despite growing consensus on the involvement of multiple causal factors, the underlying interactions impacting on honeybee health and colony failure are not fully resolved. Parasites and pathogens are among the main candidates, but sublethal exposure to widespread agricultural pesticides may also affect bees.

Methodology/Principal Findings: To investigate effects of sublethal dietary neonicotinoid exposure on honeybee colony performance, a fully crossed experimental design was implemented using 24 colonies, including sister-queens from two different strains, and experimental in-hive pollen feeding with or without environmentally relevant concentrations of thiamethoxam and clothianidin. Honeybee colonies chronically exposed to both neonicotinoids over two brood cycles exhibited decreased performance in the short-term resulting in declining numbers of adult bees (−28%) and brood (−13%), as well as a reduction in honey production (−29%) and pollen collections (−19%), but colonies recovered in the medium-term and overwintered successfully. However, significantly decelerated growth of neonicotinoid-exposed colonies during the following spring was associated with queen failure, revealing previously undocumented long-term impacts of neonicotinoids: queen supersedure was observed for 60% of the neonicotinoid-exposed colonies within a one year period, but not for control colonies. Linked to this, neonicotinoid exposure was significantly associated with a reduced propensity to swarm during the next spring. Both short-term and long-term effects of neonicotinoids on colony performance were significantly influenced by the honeybees' genetic background.

Conclusions/Significance: Sublethal neonicotinoid exposure did not provoke increased winter losses. Yet, significant detrimental short and long-term impacts on colony performance and queen fate suggest that neonicotinoids may contribute to colony weakening in a complex manner. Further, we highlight the importance of the genetic basis of neonicotinoid susceptibility in honeybees which can vary substantially.

Editor: Nicolas Desneux, French National Institute for Agricultural Research (INRA), France

Funding: This research received funding from the European Union Framework 7 under grant agreement no. 244090 (CP-FP) STEP (Status and Trends in European Pollinators, www.STEP-project.net). The funders had no role in study design, data collection and analysis, decision to publish, or preparation of the manuscript.

Competing Interests: The authors have declared that no competing interests exist.

* Email: ch.sandrock@gmail.com

Introduction

Ecosystem services provided by pollinating insects are vital for the maintenance of biodiversity [1,2] and food security through agricultural productivity [3,4,5,6]. Recent evidence for globally paralleling declines of various pollinators [7,8,9,10], however, stands in contrast to prospects of continuously increasing demands for pollination services [4,11,12].

The Western honeybee, *Apis mellifera*, is the predominant managed pollinator worldwide [4,8]. Although global stocks of domestic honeybees have increased during the last half century (except in Europe [13] and in the USA [14]), present and predicted agricultural demands for insect pollination far exceed currently available capacities [15]. Repeated, massive declines of managed honeybee colonies during the last decade in North

America, Europe and the middle East [8,16,17,18,19,20,21,22] raised substantial concerns about safeguarding future honeybee pollination services. Despite comprehensive recent research, conclusive evidence for common causal drivers is lacking, which points at multiple interacting factors [8,23,24]. The invasive ectoparasitic mite *Varroa destructor* represents a severe problem for honeybees almost worldwide [25,26,27], in particular due to its vital role as a virus vector [28,29]. Some observations of elevated colony losses were also influenced by the widespread gut parasites *Nosema ceranae* [30,31,32]. Microbes and *V. destructor*-associated viruses are prevalent almost globally and commonly co-occurring, yet there is no uniform evidence so far that even the interactions between some of these pathogenic stressors necessarily result in colony failure [33,34,35,36]. Instead, it appears as if more complex interactions with additional environmental stressors and at multiple levels are of key importance [8,24]. Likely candidates, commonly encountered by honeybees, are routinely applied agricultural pesticides [37,38]. For instance, the use of systemic neonicotinoid insecticides has strongly increased on a global scale during the last decade [39,40,41,42]. Mainly acting as specific agonists of the insect acetylcholine receptors, neonicotinoids disrupt neuromuscular signalling pathways and are thus efficiently used for controlling insect pests [39,43]. Systemic compounds like neonicotinoids can be particularly problematic for pollinating insects through exposure to residues in nectar and pollen of treated crops [41,44,45]. Although field-realistic neonicotinoid residue levels in pollen and nectar are generally assumed to result in sublethal dietary exposure [46], sublethal effects on honeybees include various negative impacts, such as impairment of physiology, cognitive abilities like memory and learning, and foraging and homing behaviour [45,47,48,49,50,51,52,53,54,55,56,57,58,59]. In addition, combined exposure to multiple pesticides can have additive or synergistic adverse effects in bees [60,61,62,63] with eventually underestimated consequences [64]. Similarly, multifactorial impact arising through combined chronic exposure to pathogens and pesticides may trigger detrimental feedbacks of supposedly sublethal individual stressors at the colony level that could explain otherwise enigmatic negative impacts [65]. In this regard, an important but yet poorly understood aspect represents the adverse influence of neonicotinoids on the honeybee's immune system [66,67]. There is growing evidence for detrimental interactions and compromised immunity in honeybees upon combined exposure to neonicotinoids and pathogens, including the prevalent gut-parasite *Nosema* spp. [36,68,69,70,71] and near-ubiquitous viruses, such as the typically *V. destructor*-associated deformed wing virus [72] or the black queen cell virus [36]. To date it remains unclear to what extent these findings can be extrapolated to field settings where, for example, immunity-related patterns in honeybees may be strongly influenced by many more environmental factors, such as the overall nutritional status [24,73,74,75,76]. While several unplanned field exposures resulted in massive effects of neonicotinoids under certain circumstances [41,77,78], there is no compelling evidence so far that field-realistic neonicotinoid exposure resulting from standard agronomic implementations of neonicotinoids pose a serious threat to whole colonies [79,80,81,82,83]. Nevertheless, some regulatory bodies reacted recently [84,85] in order to clarify whether the overall contrasting results may have been influenced by: overestimating individual predictors while dynamics at multiple levels were neglected [65]; by the lack of statistical power for individual studies [44]; or by the possibility that especially sensitive endpoints have simply been missed. For instance, the long-term impacts of neonicotinoids, which might only become evident during sensitive phases like overwintering, have received little attention to date. Moreover, information on how neonicotinoids could impact on queens is virtually lacking [64]. It is unknown whether queens are relatively protected from agricultural pesticides through receiving processed food from hive bees only, or whether trophallactic interactions with the usually most long-lived honeybee in a hive could indeed represent a sink for trace residues of such systemic compounds. Sublethal pesticide exposure could have important consequences for colony fate through compromising the queen's cognitive abilities or immune system and thereby reducing her performance. For instance, it is known that replacement of failing queens by the worker bees, i.e. queen supersedure, can be triggered by reduced oviposition of old or insufficiently mated queens [86,87]. Colony fitness is another sensitive but largely neglected aspect of honeybee colony performance. In bumblebees,

for example, it has repeatedly been shown that the negative impact of chronic neonicotinoid exposure on individuals and colony performance was less pronounced compared to queen production [88,89,90]. Similarly alarming fitness effects upon chronic sublethal neonicotinoid exposure are indicated in solitary bees [91]. Reproductive success, however, is vital for inferring long-term population level consequences. Compared to the assessment of much more general traits of colony performance and productivity in honeybees, the quantification of fitness in the true sense is very difficult because of the complex socio-biology of reproduction in honeybees. In managed honeybee colonies, besides male mating success, swarming can be considered as a tangible proxy of fitness [92], which is in practice primarily linked to beekeeping management decisions though.

Here we experimentally assessed the effects of chronic dietary neonicotinoid exposure on honeybee colony performance and fitness on a temporal scale and in relation to the honeybees' genetic background. In a fully crossed experimental design 24 freely flying honeybee colonies, including two groups of sister-queens from different strains (14 and 10 colonies, respectively), were placed at a single apiary and were either exposed to control pollen or pollen that has been spiked with a combination of the two neonicotinoids thiamethoxam and clothianidin (on average 5.31 µg/kg and 2.05 µg/kg, respectively) via in-hive feeding over two brood cycles (see Figure S1). All colonies were then identically maintained and controlled against *V. destructor* throughout a one year monitoring period. During the study four detailed colony assessments, including estimates of numbers of adult bees and the amounts of brood and stores, were conducted to evaluate potential effects on colony performance and productivity in the short- (1.5 months), medium- (3.5 months) and long-term (1 year). We used this experimental design in order to contribute to better understand three currently poorly resolved aspects. First, while the majority of experiments at the colony level applied sublethal chronic neonicotinoid exposure through sucrose solution, we hypothesized that neonicotinoid-contaminated pollen provides stronger exposure of larval stages and nurse bees that may express sublethal effects when performing more complex tasks in later life cycle stages [53,93], thereby potentially resulting in delayed effects on colony performance. Second, contaminated pollen could result in sublethal exposure of honeybee queens that may affect their performance and subsequently cause failure. Therefore, we assessed the fate of queens one year subsequent to experimental feeding of neonicotinoid-spiked pollen, as well as the colonies' propensity to swarm as an indicator for colony fitness. Third, we addressed the question of whether neonicotinoid susceptibility at the honeybee colony level has a genetic basis.

Materials and Methods

Experimental setup and colony maintenance

Twenty-four honeybee colonies were established using artificial swarms (1.5 kg of bees) in summer 2010. Two groups of sister-queens originating from different, locally adapted breeding populations were introduced in order to control for the honeybees' genetic background and maternal effects: one group of 14 queens from a region in eastern Germany that is characterized by intense agriculture, and one group of 10 queens from an alpine region in central Switzerland. All queens were freely mated at apiaries in corresponding geographic regions during early summer 2010 and then individually tagged and clipped one forewing. The former group of colonies represented *A. m. carnica*, whereas the latter represented predominantly *A. m. mellifera*, and in the following they are referred to as strain A and B, respectively.

As permitted by the veterinary agency of the canton Zurich, we established an apiary on the land of the research station Agroscope Reckenholz-Tänniken in a rural area near the city of Zurich, Switzerland (47°25′38N 8°31′11E). Two groups of each 12 hives were placed in a single row, with all hive-entrances pointing in the same direction. To meet a fully-crossed experimental design, artificial swarms containing queens of the different strains were randomly allocated to the two experimental groups (N = 7 from strain A and N = 5 from strain B each). Hives containing queens of the two different strains A and B were ordered identically in each experimental group (see Figure S1 for details of the setup). Hives within groups were separated by 1 m, and both groups of hives were separated by approximately 20 m distance, including shrubs as landmarks to minimise forager drift between groups [94]. All colonies were equipped with brand new hive material, including polystyrene hive bodies, wooden frames (200×350 mm in size) and organically certified, pesticide-free wax foundations. Commercial pollen traps (Wienold, Lauterbach, Germany), painted in different colours, were installed at all hive entrances (identical colour sequence in each group), but only activated during the exposure phase (see below). All colonies were identically treated against *V. destructor* using oxalic acid (40 ml sucrose solution containing 3.5% w/w oxalic acid per colony) five days after establishment of artificial swarms (in the absence of capped brood), and fed with commercial sugar syrup (sugar beet based, 73% sugar content, containing equal proportions of glucose, fructose and sucrose; Hostettler's, Zurich, Switzerland) during summer 2010 to promote colony growth. Pollen was assumed to be available in sufficient quantities. All colonies overwintered on eleven combs, and oxalic acid treatment against *V. destructor* was repeated in December 2010.

During spring and summer 2011 colonies were not fed but left to freely collect nectar and pollen. All colonies were simultaneously provided with a second and third hive body containing 11 frames with wax-foundations for comb building in early April 2011 and in mid-May 2011, respectively. The upper hive body provided last was separated by a queen excluder to ensure honey storage only, and on the same day it was provided the experimental treatment was initiated and lasted until end of June 2011 (see below). In mid-July 2011 colonies were taken off their honey combs and subsequently maintained on 22 combs. They received 12.5 kg of sugar syrup during late July and late August 2011 (25 kg in total). After each feeding phase, colonies were simultaneously treated against *V. destructor* using 130 ml of formic acid (70% w/w) evaporating from commercial dispensers (Andermatt Biocontrol AG, Grossdietwil, Switzerland) for about one week each during early August and early September 2011. Colonies were then overwintered and treated with oxalic acid in December 2011 (see above). At no point during the study synthetic acaricides were used for varroa mite management. Subsequent to overwintering colonies were monitored until June 2012 without further intervention. No honey supers were provided in 2012 in order to increase the propensity for swarming, which served as a proxy of fitness one year after the treatment (see below).

Treatment procedures and residue analyses

In mid-May 2011 pollen traps were activated to vastly prevent pollen inflow, and in-hive pollen feeding was initiated. Pollen patties consisted of 55% honeybee pollen (common stock of commercial pollen with mixed floral content of at least 19 plants; Sonnentracht Imkerei, Bremen, Germany), 5% brewer's yeast and approximately 40% sucrose (two thirds 73% sugar syrup and one third powder sugar). Three times per week (each Monday, Wednesday and Friday) all colonies were provided with two

200 g pollen patties loosely packed in cellophane paper and placed between the two lower hive bodies (i.e. within the brood nest). Disturbance of the colonies was thus reduced to a minimum. The bees easily corroded the cellophane paper to access the content. Two pollen patties of 200 g were generally consumed completely within 48 hours by each colony. Prior to feeding, pollen was gamma ray irradiated (Leoni Studer Hard AG, Däniken, Switzerland) to prevent putative pathogen spill over (e.g., see [95]). One group of 12 colonies (see above) received plain pollen while the other received patties containing environmentally relevant residues of the two neonicotinoids thiamethoxam and clothianidin (see below). Chronic neonicotinoid exposure through in-hive pollen feeding was performed for 46 days (1.5 months) in order to cover two brood cycles, thereby resulting in total provisions of 8 kg of pollen patties per colony. Pollen traps were then deactivated to no longer prevent colonies from storing pollen collected outside.

For both neonicotinoids pure analytical standards (PESTANAL, Fluka; with purities of 99.7 and 99.9% for thiamethoxam and clothianidin, respectively) were purchased (Sigma-Aldrich, Seelze, Germany), dissolved in distilled water (1 mg/L) and then stored at room temperature and protected from light. Aliquots of a single stock solution for each parent compound were added to the sucrose solution, which was then thoroughly mixed into the pollen and yeast. A commercial kneader was used to produce a homogenous paste to be portioned in cellophane paper (200 g) and kept frozen until usage. In total, 20 mixtures of plain and neonicotinoid-spiked pollen were prepared and fed batch-wise over time. A subsample of each of these batches of pollen patty preparations was subjected to residue analyses performed by the United States Department of Agriculture, Agricultural Marketing Service, Science and Technology Laboratory Approval and Testing Division of the National Science Laboratories in Gastonia, North Carolina. All samples were extracted for analysis of agrochemicals using a refined methodology for the determination of pesticides using an approach of the official pesticide extraction method (AOAC OMA 2007.01) using an acetonitrile:water solution and analysed by liquid chromatography coupled with tandem mass spectrometry detection (LC/MS-MS) utilising the parent and confirmatory ions of thiamethoxam and clothianidin. Samples were analysed using certified standard reference materials for the presence of both compounds with a limit of detection of 4.0 ppb for thiamethoxam and 1.0 ppb for clothianidin. Our target concentrations were 5.0 and 2.0 ppb for thiamethoxam and clothianidin, respectively. Since clothianidin is the major metabolite of thiamethoxam [96,97], both bioactive compounds will co-occur in the pollen and nectar of thiamethoxam-treated crops and were therefore applied in combination. The concentrations used here match field-realistic levels of both compounds previously found in pollen of treated crops [41,42,46,80,82,98]. In order to confirm the absence of unexpected additional exposure, we also subjected six random samples of the sugar syrup used for late season feeding (2010 and 2011), six samples of the original pollen stock, as well as four pollen patty samples of the control group to residue analyses. Moreover, pollen trap contents collected during the experimental feeding were pooled across colonies of each experimental group and samples from five collection dates throughout the treatment were taken for residue analyses. Additional matrix endpoints at the end of the experimental pollen feeding phase for both treatment groups were forager bees (1 sample pooled across all colonies) and pupae close to adult emergence (3 samples pooled across 4 colonies each). Furthermore, in both treatments we sampled wax (2 samples pooled across 6 colonies each) and bee bread (4 samples pooled across 3 colonies each) 3 weeks after the pollen treatment when honey supers were removed, as well as one sample of honey from each colony of the neonicotinoid-exposed group (pooled from at least 3 different combs). All samples for residue analyses were stored frozen and shipped on dry ice.

Population estimates and data collection

Estimates of colony strength were performed using the 'Liebefelder Method' [99,100]. Specifically, we visually estimated the number of adult bees, the amount of capped brood (pupae) and un-capped brood (eggs and larvae), and the amount of honey and pollen stores in dm^2. Colony assessments were consistently performed by the same person and alternated between treatment groups during the day. During each colony assessment the presence of the original queen was checked. Successful queen replacement was counted as the presence of a non-tagged, egg-laying queen which possessed two intact forewings, either in the presence or absence of the originally tagged mother-queen. Based on 1 dm^2 comb containing 400 cells on average, the estimated proportion of comb area comprising of brood was converted into numbers of individuals in order to treat corresponding response variables as counts, whereas estimates of honey stores were converted into total weight (based on an average weight of 2 kg for fully filled honey combs of the used frame format).

In spring 2011 (mid-May) the first colony assessment was performed and three days later the experimental treatment was initiated. During the pollen feeding phase, contents of all pollen traps were collected and weighed each time when new pollen patties were provided, resulting in 20 pollen collection records per colony in total. In summer 2011 (beginning of July), two days after the last pollen patties were fed, the second colony assessment was conducted to evaluate short-term effects on colony performance. After the exposure phase the control and the neonicotinoid-treated colonies were maintained identically and the third colony assessment was performed 3.5 months after the exposure in autumn 2011 (mid-October), to evaluate medium-term effects colony performance before overwintering. Overwintering success was assessed end of March 2012 and thereafter surviving colonies were inspected on a weekly basis. Finally, the fourth colony assessment for all colonies that survived winter was performed one year after the treatment in spring 2012 (late April) to evaluate long-term effects on colony growth. Afterwards colonies were maintained until June 2012 and inspected for queen cells on a weekly basis and for swarming events at least every second day. Since original queens had one clipped forewing, swarms remained nearby the hive and could thus be easily recognized.

Statistical analyses

To investigate the effects of exposure to thiamethoxam and clothianidin on honeybee colony performance over time, we analyzed a set of sensitive endpoints within the framework of mixed models.

Model formulation and selection. The response variables (endpoints) were numbers of adult honeybees, pupae and eggs and larvae. These were modelled including the explanatory variables (factors) treatment (control and neonicotinoids), honeybee strain (A and B), and assessment date (spring 2011, summer 2011, autumn 2011 and spring 2012) as fixed effects, and colony as a random effect. Residual analysis of all response variables indicated the need for variance stabilization and variables were transformed accordingly. The data for the endpoints number of adult honeybees and pupae were square-root transformed. Residuals for eggs and larvae displayed a more complex pattern of a bow-shaped variance being largest for medium fitted values (~8000)

and decreasing for larger and smaller fitted values. This pattern resembles that of a binomially distributed variable, which conforms well to the upper bounded egg-laying rate of a honeybee queen. In such cases effective counts of a given response variable can be divided by the expected maximum number to obtain binomially distributed variables. Based on Khoury et al. [101], we set the limit to a daily egg-laying rate of 2000 and, according to the honeybee life-cycle, 16000 eggs and larvae present at any time. Ratios of actually present and maximum possible numbers of eggs and larvae were arcsine square-root transformed to stabilize variances for further analyses, as commonly performed for binomially distributed variables.

Numbers of adult honeybees exhibited an increased variance at the fourth assessment date in spring 2012, which prompted us to use a weighted linear mixed model for this response variable. Weights were set as the inverse of the residual variances of the two groups (for the assessments between spring and autumn 2011 and the assessment in spring 2012, respectively).

For each endpoint, complete models were fitted based on the threefold interaction term of the explanatory variables (fixed effects) plus the random effect. Model simplification was evaluated by hierarchically removing interaction terms based on likelihood ratio tests. The goodness of corresponding Chi-squared approximations was confirmed by model based parametric bootstrapping. During all steps of the model selection, model assumptions and serial correlations of the residuals of colonies were inspected.

Hypotheses testing with contrasts. To test for effects of neonicotinoid exposure on colony performance and the influence of the honeybees' genetic background on responses to the neonicotinoid-treatment, one-sided contrasts and corresponding P-values (adjusted for multiple testing) were computed for the overall treatment effect at each individual assessment date and, when the threefold interaction significantly contributed to model a given response, also for treatment nested within honeybee strain. Since seasonal effects were expected, contrasts including assessment date were not performed.

Further statistical analyses. To investigate effects of chronic exposure to thiamethoxam and clothianidin on honey production, the difference of the log transformed total weights before and after the experimental pollen feeding was analysed using linear regression. The full model was fitted with neonico-tinoid exposure and honeybee strain as fixed factors and the interaction between them. In the same way we compared comb areas in dm^2 comprising of pollen stores (bee bread) in the hives before and after the experimental pollen feeding in order to evaluate pollen consumptions during the treatment.

The time series of twenty pollen collections per colony sampled during the 1.5 months of experimental in-hive pollen feeding were converted into respective Areas Under the Curve (AUC) for further analysis. The AUC represents the overall colony-specific pollen foraging activity, with higher AUC values corresponding to higher collection performances. We analyzed AUC with a one-sided Mann-Whitney test for a difference between neonicotinoid-exposed and control colonies.

Supersedure of original queens (at any time) and swarming events in spring 2012 yield a yes/no value for each colony. The associations of these variables with neonicotinoid treatment were investigated using Fisher's exact tests for contingency tables of small sample sizes.

All statistical analysis were performed using R [102]. Mixed models were fitted using the lmer function of the *lme4* package [103], and contrasts were performed using the glht function of the *multcomp* package [104].

Results

Residue analyses

Original stocks of honeybee pollen and sugar syrup used to prepare pollen patties in our experiment did not contain traceable amounts of the two neonicotinoids thiamethoxam and clothianidin, even when the limit of detection was reduced to 0.1 ppb for both compounds. Thus, thiamethoxam and clothianidin are considered being absent in the pollen patties fed to the control colonies. In contrast, in all samples from the pollen patties that have been spiked with the two neonicotinoids both parent compounds were detected in the range of the target concentrations: The effective mean concentrations ($\pm$ SD across different pollen patty batches) used during the neonicotinoid treatment were determined to be 5.31 ± 0.75 ppb for thiamethoxam and 2.05 ± 1.18 ppb for clothianidin. The residue analyses document constant chronic exposure to thiamethoxam and clothianidin at field-realistic levels in pollen over a period of two honeybee brood cycles. There was no unexpected additional exposure to thiamethoxam or clothianidin from the outside, as indicated by the lack of traceable residues of both compounds in the pollen collected from pollen traps of both treatment groups during the in-hive pollen feeding phase. In none of the forager bee and pupae samples collected directly after the treatment and in none of the hive samples collected 3 weeks after the treatment (i.e. bee bread, honey and wax) residues of either compound above respective limits of detection could be found in both the control and neonicotinoid treatment. The absence of residues in bees is not surprising given the low concentrations used here, yet the absence of in-hive residues 3 weeks after feeding contaminated pollen over two brood cycles indicates that low level residues may quickly disappear in the hive matrix (see also [80,82]) through consumption, dilution or degradation.

Colony growth

In the control group and in the group exposed to thiamethoxam and clothianidin each 1 colony of strain A lost their queen after the pollen feeding phase during the formic acid treatment against varroa mites in 2011. Moreover, in each experimental group 1 colony became queenless during winter (originating from strain B and A in the control and in the neonicotinoid-exposed group, respectively). Thus, mixed model analyses are based on 12 colonies per experimental group for the first and the second colony assessment (in both treatment groups 7 colonies originating from strain A and 5 from strain B), while 11 colonies per group were available for the third colony assessment (in both treatment groups 6 colonies originating from strain A and 5 from strain B), and 10 colonies were available for the fourth colony assessment (in the control 6 and 4 from strain A and B, respectively, and for the neonicotinoid treatment each 5 from strain A and B). Queenless colonies were removed immediately after queen loss was recognised.

The data for the three endpoints adult bees, pupae and eggs and larvae across assessment dates are summarized in Fig. 1A–C. Dynamics of colony strength and brood curves displayed the expected general pattern of seasonal variation. However, we detected strong effects of neonicotinoid exposure, as well as interactions of the honeybees' genetic background with neonicotinoid exposure. The number of adult bees and the number of eggs and larvae were each best explained by the threefold interaction term model, i.e. neonicotinoid exposure, honeybee strain and assessment date, while for the number of pupae the retained model included the twofold interactions of assessment date with honeybee strain and neonicotinoid exposure, respectively, but no threefold

interaction. Model-based estimates of contrasts and corresponding significance levels are summarized in Table 1. After 1.5 months of experimental pollen feeding, there was a significantly negative influence of the exposure to thiamethoxam and clothianidin on the number of adult bees for both honeybee strains, whereas this effect was much stronger for honeybee strain B than for strain A (Fig. 1A). Overall, average worker populations were 28% smaller in the neonicotinoid treatment compared to the control. There was also a significant overall decrease of the number of eggs and larvae in the neonicotinoid treatment, yet, this effect was not significant when tested within either honeybee strain A or B (Fig. 1B). However, there was no significant effect of nenicotinoid treatment or honeybee strain on the amount of pupae after the experimental pollen feeding (Fig. 1C). Compared to the control, the average amount of total brood had declined by 13% in the colonies exposed to thiamethoxam and clothianidin. No effects of the previous neonicotinoid exposure on the amount of adult honeybees or honeybee brood were detected 3.5 months after the experimental pollen feeding (Fig. 1A–C). Interestingly though, one year after the experimental pollen feeding, the negative impact of the previous neonicotinoid treatment on the number of adult bees was even stronger than directly after exposure to thiamethoxam and clothianidin (Fig. 1A). These effects were significant within honeybee strains A and B, but again much more pronounced in strain B. Similarly, when the overall significant decrease of the amount of eggs and larvae one year after the neonicotinoid treatment was tested within strains, a significant effect was only detected for honeybee strain B (Fig. 1B). Moreover, contrasting to the finding directly after exposure to thiamethoxam and clothianidin, there was also a significantly detrimental effect of the neonicotinoid treatment in the previous year on the amount of pupae (independent of honeybee strain), see Fig. 1C and Table 1.

Honey production

At treatment initiation all colonies already harboured considerable honey stores due to comparatively early spring flowering in 2011. Strain A and B, respectively, had on average 24.8 ± 2.9 and 27.0 ± 2.9 kg in the control group, and 25.4 ± 2.8 and 23.6 ± 0.7 kg in the group subsequently exposed to thiamethoxam and clothianidin. During the experimental pollen feeding, honey stores in the control group increased by 7.7 ± 4.4 kg per colony on average (10.3 ± 3.6 and 4.1 ± 2.2 kg for colonies of strain A and B, respectively). During the same period honey stores in the neonicotinoid-exposed group also slightly increased for colonies of strain A (1.8 ± 2.1 kg on average), but decreased for neonicotinoid-exposed colonies of strain B (-4.9 ± 1.0 kg on average), resulting in an overall decrease of -1.0 ± 3.8 kg per colony on average. Honey production during the treatment was significantly influenced by both neonicotinoid exposure and honeybee strain: While strain B was significantly less productive than strain A independent of the treatment ($F_{1,21} = 40.40$, $P<0.001$), neonicotinoid exposure negatively affected honey production in both strains ($F_{1,21} = 68.18$, $P<0.001$). The interaction between both predictors was not significant ($F_{1,20} = 1.45$, $P = 0.24$) and was thus removed prior to testing the main effects. Overall, the mean honey production over the entire season (including honey production during the pre-treatment phase) remained 29% lower in the neonicotinoid-exposed colonies (23.7 ± 2.5 kg) compared to the control (33.4 ± 5.1 kg).

Pollen consumption

At treatment initiation, pollen stores (bee bread) in the control group comprised of 17.9 ± 10.7 dm^2 and those in the group subsequently exposed to thiamethoam and clothianidin comprised

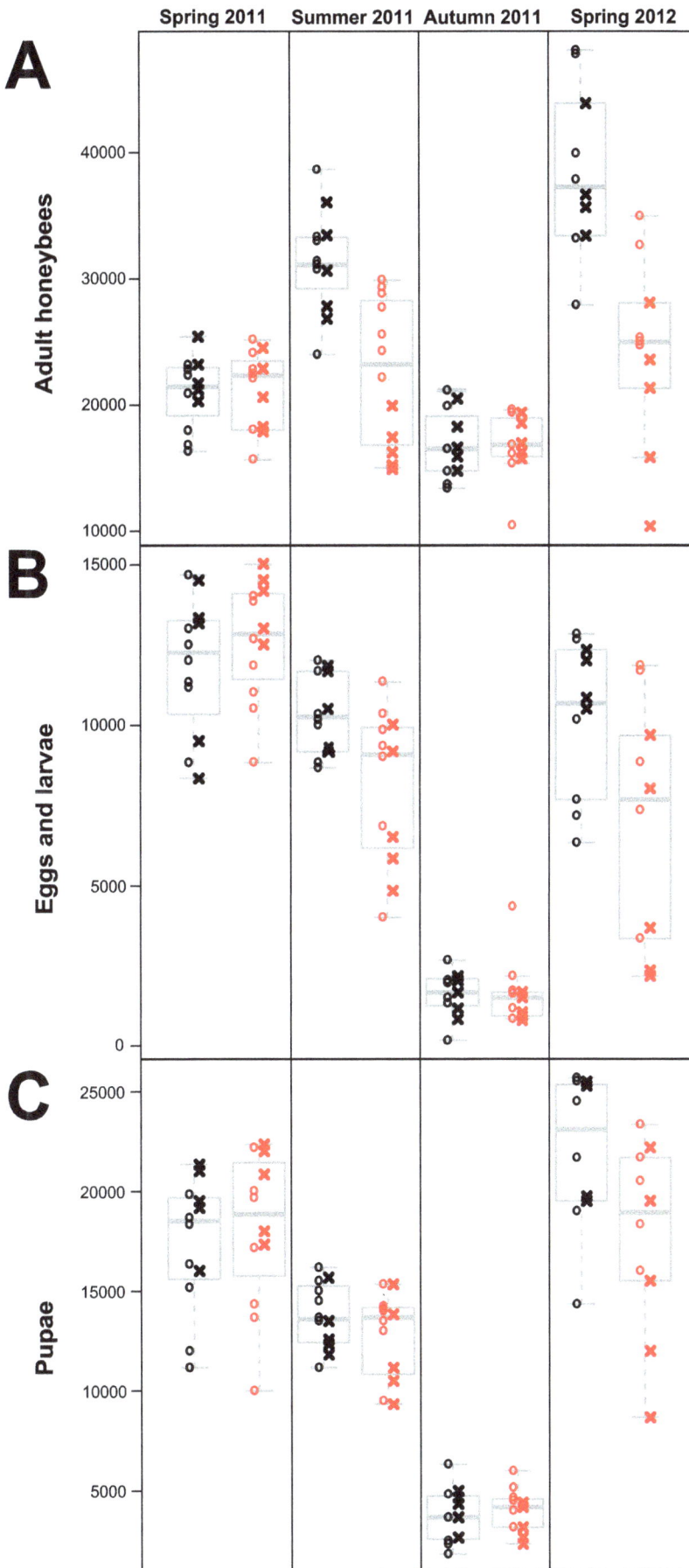

Figure 1. Dynamics of honeybee colony performance. Data of all three endpoints number of adult bees (A), eggs and larvae (B) and pupae (C) for the different pollen feeding treatments (black = control; red = neonicotinoids) and honeybee strains (circles = strain A; crosses = strain B). The data were obtained at four successive colony assessment dates (X-axis subpanels within figures) performed before (Spring 2011) and directly after the 1.5 months of experimental pollen feeding (Summer 2011), 3.5 months after the treatment (Autumn 2011) and one year later (Spring 2012). Estimated numbers on the Y-axes are truncated for adult bees and pupae for better overview.

of 20.8 ± 3.8 dm^2. After the experimental pollen feeding pollen stores within hives comprised of 29.8 ± 8.9 dm^2 and 24.9 ± 11.7 dm^2 in the control and neonicotinoid-exposed group, respectively. There was no indication that pollen storing and pollen consumption, respectively, during the experimental pollen feeding was influenced by neonicotinoid exposure ($F_{1,21} = 2.63$, $P = 0.12$) or honeybee strain ($F_{1,21} = 2.52$, $P = 0.13$). The interaction between the two predictors was not significant ($F_{1,21} = 0.74$, $P = 0.40$) and was thus removed prior to testing the main effects.

Pollen collections

Total mean pollen harvests ($\pm$SD) per colony, as inferred from pollen trap contents, were 4.4 ± 0.47 kg and 3.58 ± 0.43 kg for the control and neonicotinoid-exposed group, respectively. The colonies exposed to thiamethoxam and clothianidin had 19% lower total pollen collections on average. Pollen collections measured as AUCs were found to be significantly lower in neonicotinoid-exposed colonies ($P<0.001$). While both treatment groups collected similar amounts of pollen during the first 3 weeks of the experimental pollen feeding, colonies exposed to thiamethoxam and clothianidin consistently collected less pollen later on, with mean pollen collections barely reaching more than 50% of the control group during the last week of exposure (Fig. 2).

Supersedure of queens and tendency to swarm

We found a significant association of neonicotinoid exposure and queen supersedure (in the absence of swarming) ($P = 0.01$): while all ten queens of the control group survived until the end of the experiment (~2 years or swarmed, see below), 6 out of 10 queens of the colonies experimentally exposed to thiamethoxam and clothianidin over 1.5 months were replaced within one year after treatment. The result remained significant when overall queen loss was assessed, i.e. also including the two colonies (one per treatment group) that lost their queen during winter ($P = 0.02$). A negative association of neonicotinoid exposure and swarming events during spring following experimental treatment was found ($P = 0.005$): in the control group 9 out of 10 colonies swarmed until end of May 2012 (5 out of 6 colonies of strain A and all 4 colonies of strain B), while only 2 colonies (one of strain A and B each) of the group that was exposed to thiamethoxam and clothianidin in the previous season swarmed.

Discussion

The major findings of this study using sublethal chronic exposure of honeybee colonies to thiamethoxam and clothianidin through feeding contaminated pollen were: (*i*) significant short-term (1.5 months) impacts at the colony level resulting in decreased colony performance and productivity; (*ii*) no negative influence in the medium-term (3.5 months) and on colony overwintering; (*iii*) significantly decelerated colony growth in the long-term (1 year) that was associated with higher queen supersedure rates and a reduced tendency to swarm; and (*iv*) significant interactions of the honeybee genetics with the observed effects of neonicotinoids on most parameters. In the following, these findings are discussed in context.

Short-term impact

At the colony level, honeybee foraging efficiency was negatively influenced during chronic exposure to pollen containing environmentally relevant concentrations of the two neonicotinoids thiamethoxam and clothianidin over two brood cycles (1.5 months). The detected decrease in pollen collection and honey production upon sublethal neonicotinoid exposure are in line with earlier findings of impaired honeybee foraging through impacts on neurophysiological traits and cognitive abilities, including sucrose responsiveness, foraging rates, waggle dancing and memory and learning [47,48,49,50,51,52,53,54,55,56,57,58,59]. In addition, increased forager losses resulting from decreased homing success [48] through compromised navigation memory [58] may have contributed to both reduced foraging efficiency and significantly smaller worker populations of the colonies exposed to thiamethoxam and clothianidin compared to controls (Fig. 1A, Table 1). Indeed, there were no significant effects on the numbers of pupae present at the end of the exposure phase (Fig. 1C, Table 1), which points at higher forager losses rather than decreased worker production of neonicotinoid-exposed colonies during the treatment. As opposed to known effects of neonicotinoid ingestion of foragers through nectar-substitutes [47,48,49,50,51,56,57,58], the overall reduced numbers of adult bees through pollen exposure is intriguing. Pollen is generally not stored extensively within hives but consumed quickly. There was no evidence for differences in pollen storing between treatment groups directly after the experimental feeding. Therefore, it can be assumed that control pollen and pollen spiked with thiamethoxam and clothianidin were similarly consumed within experimental colonies. Contaminated pollen may affect various life stages of honeybees [105], yet pollen is predominantly consumed by nurse bees and larvae, but not foragers. In this regard, the recent finding that sublethal neonicotinoid exposure of honeybee larvae impacts post-emergence olfactory associative behaviour of adults [53] is relevant, and might be one of the largely unresolved effects arising from altered larval gene expression profiles upon sublethal neonicotinoid exposure [93] that may help to explain the strong impediment of colony growth documented here. Delayed sublethal effects extending to later life-cycle stages could similarly apply to larvae and nurse bees consuming contaminated pollen [64] and, depending on colony exposure duration, be able to reinforce decreased foraging efficiency. Our observed pattern of increasing divergence of pollen collections between treatment groups after 2–3 weeks, which roughly corresponds to the adult in-hive phase after emergence [106], may be indicative of the recruitment of less efficient foragers, which have encountered pollen contaminated with thiamethoxam and clothianidin during their nursing phase. The successively greater decline of pollen collections of neonicotinoid-exposed colonies after 5 weeks, in turn coincides with the time frame during which foragers could be expected to be exposed to contaminated pollen from the young larva stage onwards and throughout their entire development. Further research is needed to evaluate the potential for delayed impact on implementing complex foraging tasks through altered metabolic networks caused by sublethal dietary neonicotinoid exposure to honeybee brood [53,93] and eventually also nurse bees [64], the developmental

Table 1. Model-based estimates of contrasts and corresponding significance levels of the treatment effect (neonicotinoid *versus* control) and honeybee genetics (strain A *vs.* strain B).

	Adult bees			Eggs and larvae			Pupae		
	Summer 2011	Autumn 2011	Spring 2012	Summer 2011	Autumn 2011	Spring 2012	Summer 2011	Autumn 2011	Spring 2012
Neonicotinoids *vs* Control	−60.56***	−0.73	−82.96***	−0.31*	−0.01	−0.49***	−4.36	1.84	−15.31**
Strain A *vs* strain B							5.31	2.65	7.91
Treatment within strain A	−14.07*	−1.33	−28.59*	−0.10	0.03	−0.06			
Treatment within strain B	−46.49***	0.59	−54.37***	−0.21·	−0.05	−0.42***			

Results are shown in the transformed scale for the three response variables adult bees, eggs and larvae and pupae assessed directly after the 1.5 months of treatment (Summer 2011), 3.5 months later (Autumn 2011) and 1 year later (Spring 2012). For adult bees and eggs and larvae (the models that included a significant threefold interaction between treatment, honeybee strain and assessment date) contrasts for treatment effects were also computed within individual honeybee strains at each assessment date. *P* values are adjusted for multiple testing. ***P<0.001; **P<0.01; *P<0.05; · 0.05<*P*<0.1.

stages during which fundamental physiological processes required for adult olfaction and learning performance are settled [107].

The observation of strongly reduced pollen collection of the colonies exposed to thiamethoxam and clothianidin during the last week of the treatment could also be interpreted as a response to significantly reduced numbers of larvae (Fig. 1B, Table 1). However, a general decrease in foraging efficiency is also indicated by significantly reduced honey production. Because experimental pollen was provided constantly in sufficient amounts, significantly reduced investment in rearing larvae after 1.5 months of combined exposure to thiamethoxam and clothianidin is best explained by the declining numbers of workers: higher forager losses trigger premature forager recruitment, which results in fewer nurse bees available for brood rearing [101,108]. Although physiological changes in nurse bees may have played an additional role [54,55,93,109], reduced investment in brood rearing was probably reinforced by the reallocation of worker resources at the colony level towards the end of the exposure period rather than by adverse effects on the larvae themselves. Although there is some indication for negative effects of neonicotinoids and other pesticides on honeybee larvae [62,93,110], there is so far no evidence that field-realistic neonicotinoid exposure in the absence of additional stressors results in increased larval mortality [36,64]. The latter would also be difficult to reconcile with the observation that the numbers of pupae did not differ between treatments, even after 1.5 months of neonicotinoid exposure. Interestingly, short-term effects of the combined exposure to thiamethoxam and clothianidin on the number of adult bees were influenced by the honeybees' genetic background. While effects on colonies of strain A were only marginally significant, those on colonies of strain B were highly significant (Fig. 1, Table 1).

Medium-term impact

The colony assessment during autumn 2011, 3.5 months after the experimental pollen feeding, revealed that there were neither effects of the combined exposure to thiamethoxam and clothianidin nor of the honeybees' genetic background on colony strength and the amount of brood (Fig. 1A–C, Table 1). All colonies were strong and well-fed, and thus overwintered successfully, except one colony of each treatment which lost their queens during winter. This reiterates the general view that a sustainable varroa mite management is a major aspect of honeybee colony health, thereby limiting colony losses [25,26]. Our comprehensive varroa mite management, comprising of a combination of integrated actions using organic acids [27] but avoiding potentially detrimental synthetic acaricides [47,52,111,112], was apparently sufficient to limit *V. destructor*-associated damage during the sensitive overwintering period, and throughout the entire study. A thorough varroa mite control pre-supposed, the overcoming of the previously observed short-term effects of neonicotinoids on colony strength and brood rearing 4–5 brood cycles later shows that chronic sublethal neonicotinoid-exposure alone does not trigger elevated honeybee colony winter losses [17,20], although contrasting results have been found in other studies [113,114].

There were no obvious clinical symptoms of infections with widespread honeybee pathogens such as the ectoparasitic mite *V. destructor*, microsporidian gut parasites *Nosema* ssp., chalkbrood (the fungal pathogen *Ascophaera apis*) or European and American foulbrood (bacterial diseases caused by *Melissococcus plutonius* and *Paenibacillus larvae*, respectively) at any time during the study. Nevertheless, it cannot be excluded that some of the above described short-term effects on colony performance may have been influenced by detrimental neonicotinoid-pathogen interactions at the level of individual bees [36,68,69,70,71,72]. Higher

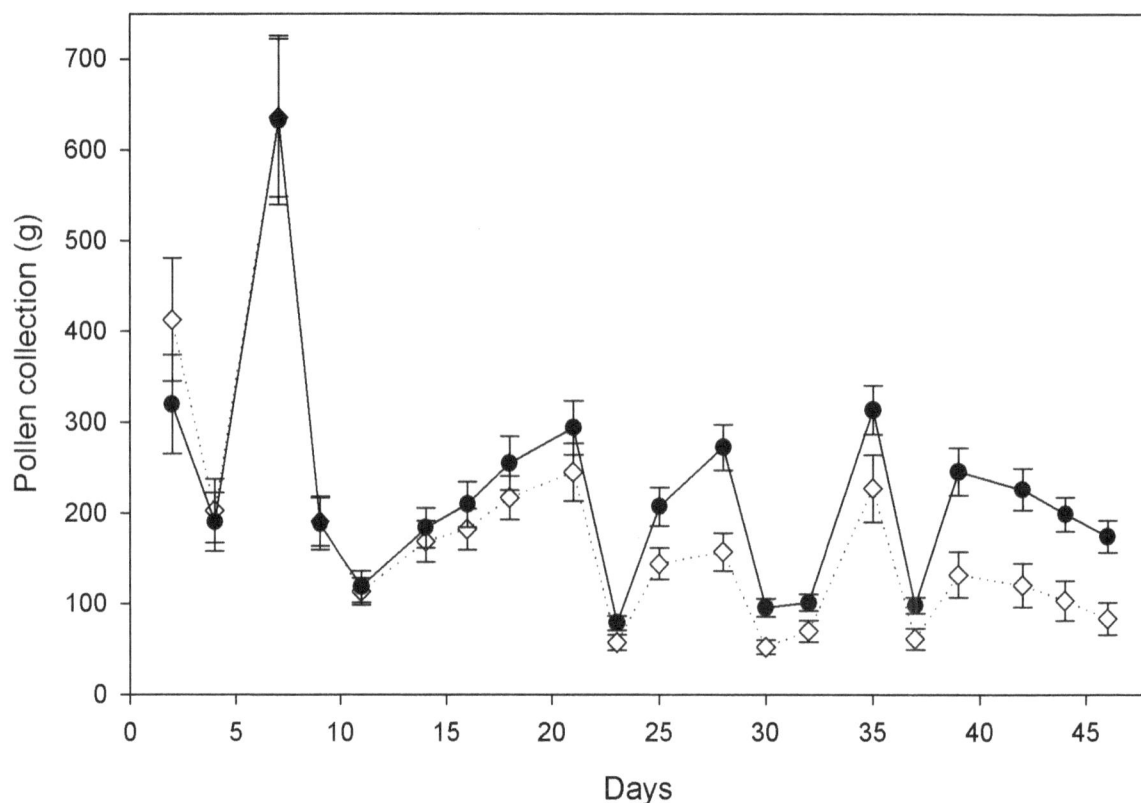

Figure 2. Pollen collections. Mean (±SD) fresh weights of pollen collections for control (black) and neonicotinoid-exposed (white) colonies over the course of the treatment period (pollen-trap contents were weighed in 2-2-3 days intervals throughout the study).

forager losses associated with overall shifts in forager age structure may also cause generally decreased immunological competence at the colony level [115]. However, these differentiations between individual drivers and their interactions are only of limited practical importance because free-flying honeybees will almost inevitably carry ubiquitous pathogens and thus encounter combined pressures. In fact, it is very likely that latent infections with the most common *V. destructor*-associated honeybee viruses and *Nosema* spp. have also been present in the experimental apiary. The common garden approach to maintain all colonies at the same isolated field site for almost one year prior to the experiments actually aimed at 'synchronizing' viral and microbial landscapes across colonies, thereby resulting in potential errors of systematic conservative nature. If multiple pressures were indeed present in our experiment, the observed recovery from decreased performance of previously neonicotinoid-exposed colonies after several brood cycles conforms to model predictions, showing that even detrimental interactions presumably can be tolerated to some extent and do not necessarily result in disrupting colony function [65].

Long-term impact

Given that significantly negative short-term impacts of combined exposure to thiamethoxam and clothianidin during summer 2011 had faded with respect to colony strength during autumn (medium-term) and overwintering success, the patterns observed during the subsequent spring were unexpected. Colonies that received pollen spiked with thiamethoxam and clothianidin during the previous spring exhibited significantly lower numbers of adult bees compared to controls one year later, whereas the effects were

much stronger for strain B than for strain A (Fig. 1A, Table 1). In addition, there were significantly negative overall effects of last year's neonicotinoid exposure on the amount of pupae (Fig. 1C, Table 1) and eggs and larvae (Fig. 1B, Table 1). For the latter endpoint, however, significant effects were only found for strain B (Table 1). These patterns of decelerated colony growth are intriguing, because all hive samples collected 3 weeks after the experimental feeding of neonicotinoid-spiked pollen patties did not contain traceable residues of thiamethoxam and clothianidin. Although trace residues below corresponding limits of detection could have still been present within hives one year after the treatment, it is highly unlikely that this explains detrimental effects that far exceeded those observed directly after the treatment (Table 1). Instead, reduced performance of previously neonicotinoid-exposed queens appears to be a more plausible explanation for long-term effects. While all original, individually tagged and wing-clipped queens were recognized in autumn 2011, the assessment in spring 2012 revealed that there were non-tagged egg-laying queens with intact wings in some colonies previously exposed to thiamethoxam and clothianidin. This clearly shows queen supersedure during the previous autumn as a response to poor queen performance [86,87], which could pose a risk for successful overwintering. Yet, supersedure during autumn bears a considerable risk for colony failure as well, because virgin queens may not be sufficiently mated during their mating flights due to a comparatively lower number of available drones. Queen replacement in the neonicotinoid treatment group in the presence of original queens, and without swarming, was also observed during spring 2012. This indication of impaired performance of previously neonicotinoid-exposed queens, together with a likely reduced mating success of replacement queens from the previous autumn, is thus considered to be responsible for the overall decelerated colony growth of the neonicotinoid treatment group one year after exposure. The causal patterns of queen failure could be manifold, including compromised immunity that has been shown to result in increased pathogen loads in honeybee workers [36,68,69,72]. Taking into account that honeybee colony fate essentially depends on the queen, the long-term impact of neonicotinoids on queens we observed, resulting in 60% supersedure within one year, represents a novel finding of major importance that clearly deserves further research. Increased queen supersedure rates were shown upon sublethal pyrethroid exposure [116] and detrimental effects on queen development and performance have also been reported from exposure to commonly encountered in-hive acaricides [117,118,119,120], some of which are known to have similarly negative influence like neonicotinoids on worker bees [47,52,108,121]. General effects of neurotoxic insecticides on queens may thus not be fundamentally surprising. Queen failure has repeatedly been indicated to be an important aspect shaping the present enigma of widespread colony losses [22,25,122], which may also be influenced by neonicotinoids. While direct contact exposure to agrochemicals can be generally considered much more relaxed for queens compared to workers, this might be different regarding oral exposure to systemic compounds, such as neonicotinoids. Here, ingestion of contaminated food represents a principal risk for all developmental stages and different castes of honeybees. Moreover, the queen's life span as well as her level of trophallactic interactions far exceed that of workers [123], which may predestine her as a potential sink of sublethal exposure to systemic insecticides through pollen-based queen nutrition. Our results suggest that sublethal neonicotinoid effects on queens can jeopardize colony fate in the long-term, and should thus not be ignored.

Compared to the need for maintaining productive queens within colonies, swarming is less relevant for practical beekeeping, especially commercial operations. Nevertheless, apart from drone mating success, colony splitting through swarming represents a true fitness estimate in honeybees. According to overall reduced colony strength, colonies that were exposed to thiamethoxam and clothianidin in the previous season exhibited a significantly lower swarming success.

The role of honeybee genetics

The detection of significant interactions of the honeybees' genetic background with the effects of chronic neonicotinoid exposure on both colony performance in the short-term and queen performance in the long-term was a major insight of our study. Honeybee strain B, *A. m. mellifera* sister queens originating from a Swiss alpine region, was much more susceptible to the combination of thiamethoxam and clothianidin compared to strain A, *A. m. carnica* sister queens originating from a German region characterized by intense agriculture. For example, neonicotinoid-exposed colonies of strain B had the lowest numbers of adult bees directly after exposure (Fig. 1A), and none of their queens survived throughout the one year post-exposure period. This indicates that honeybee susceptibility to the here applied neonicotinoids may also include a genetic or epigenetic component, as has been previously suggested for neonicotinoids and other pesticides [17,93,111,124,125,126]. It remains unresolved to what extent the outcome of our study was influenced by the usage of different honeybee ecotypes and/or by their distinct breeding histories linked to different environments. It could be possible that *A. m. mellifera* tends to be more susceptible to oral neonicotinoid exposure than *A. m. carnica* bees, as similarly found in another

recent study [126]. Yet, compared to strain B, the particular breeding regime of strain A prior to our study may have simply included an unintended selection for higher neonicotinoid tolerance as a non-lineage-specific trait. Further research is required to explore the potential genetic basis underlying variable responses to sublethal neonicotinoid exposure in honeybees. Causal patterns could include detoxification genes and corresponding expression profiles, e.g., cytochrome P450 monooxygenases [93], yet, compared to many other insects, honeybees are equipped with a limited set of detoxification genes [127] and specifically the pathways for nitroguanidine neonicotinoid metabolism in honeybees remain less resolved compared to other insecticidal compounds [61,111,128,129,130,131].

Conclusions

In line with a recent meta-analysis [44], our results clearly indicate that neonicotinoids negatively impact on honeybee colony performance after chronic sublethal exposure throughout two brood cycles. Virtually all tested contrasts (18 out of 21) produced negative estimates for the effects of exposure to thiamethoxam and clothianidin (Table 1). It is supposed that sublethal neonicotinoid exposure through pollen has a stronger impact at the honeybee colony level compared to nectar-substitute feeding (e.g., [132]). Therefore more studies focussing on effects of sublethal exposure of larvae or nurse bees extending to the performance of complex tasks of adult forager honeybees are needed. Similarly, sublethal effects of neonicotinoids on honeybee queens clearly deserve in depth investigation.

It remains uncertain whether the observed colony level responses were stronger influenced by either thiamethoxam or clothianidin, or by the possibility for interactive effects of the combination of these neonicotinoids, both being ranked as having high risk potential to honeybees [64]. Yet, since clothianidin is the major metabolite of thiamethoxam, both bioactive compounds will be present in pollen and nectar of crops treated with thiamethoxam [80,82,96,97,98]. Thus, the combined exposure as well as the residue levels administered in pollen patties in this study represent a biologically relevant exposure scenario.

Exposure to almost any sublethal dosage of these highly potent insecticides could trigger adverse chronic effects in a time-dependant context [133,134]. Our experimental exposure to contaminated pollen over 1.5 months can be considered as a worst-case scenario for agricultural settings, where honeybee colonies may encounter neonicotinoid-treated crops repeatedly throughout the season, but likely for shorter individual exposure periods. In bumblebees, it has recently been shown, that microcolonies can recover after short periods of sublethal exposure [135] to imidacloprid, which may well be similar, or even more pronounced in honeybee colonies. In a similar way, the supposedly exclusive exposure to pollen contaminated with neonicotinoids in our setup can be regarded as a worst case because in the field honeybee colonies generally exploit multiple available pollen resources at any time, some of which may only occasionally be contaminated with agricultural pesticides [38]. Nonetheless, it remains unclear whether repeated pulsed exposures over the entire season, either through consecutively available neonicotinoid-treated honeybee-attractive crops providing pollen and nectar (e.g., oilseed rape, sunflowers and maize that may each flower for 2–3 weeks) or through the general accumulation of neonicotinoid residues in the environment [38,41,42,82], could result in detrimental effects as found here after 1.5 months of continuous chronic exposure. Moreover, although distinguishing between continuous and shorter but repeated periods of neonicotinoid exposure could represent different risk potentials for the honeybee worker population succumbed to generally high turn-over rates [101], the suspected buffering capacities at the colony level may deplete in the long-term, if the queen was still affected through pulsed exposures. Thus, a worst case scenario that applies realistic exposure levels while settling at the upper possible range of overall exposure duration deserves consideration. Currently there is no evidence that field-realistic neonicotinoid exposure could be directly involved in colony losses [79,80,81,82,83]. However, available studies do not fully allow to address the question of whether neonicotinoids are really contributing to colony weakening, because several aspects in corresponding experimental designs remain unsettled, such as lacking sufficient statistical power [44], the putative influence of honeybee genetics (see above), too small distances between test fields and other factors that challenge the prerequisite of adequate control fields, such as no pesticide treatment at all [41,79,80,81,82], queen effects that remain elusive [25] or queen rotation practices that impede uncovering potential effects on queen performance [80]. Interestingly, this long-term impact detected in our study, notably supersedure of failing queens, compounds recent criticism that the presence of a given predictor may not necessarily correlate with impaired colony function [65] by showing that delayed effects can emerge also in the absence of causal stressors, in this case the lack of traceable amounts of the applied neonicotinoids within hives several weeks after exposure. There is an urgent need for more thoroughly designed studies to clarify the threats of neonicotinoids to honeybees, and pollinators in general [42]. The growing body of scientific awareness on sublethal side-effects of pesticides on non-target pollinators, ranging from gene expression profiles of individual developmental stages to entire life-time fitness performances of different species of bees, represents a unique opportu-nity to benefit the current framework of pesticide risk assessment [64,84,136,137].

Finally, the here detected interactions of honeybee genetics and impacts of chronic neonicotinoid exposure on colony performance suggest that there is genetic variability for neonicotinoid suscep-tibility, and thus potential to partly counteract negative effects through selective breeding of more tolerant honeybee strains.

Supporting Information

Figure S1 Colony set up at the experimental apiary. Two groups of honeybee colonies each comprising of 7 colonies originating from strain A (blue squares) and 5 colonies originating from strain B (yellow squares) were placed in a row in front of a small forest with entrances pointing in the same direction. All colonies were maintained approximately one year before and one year after the experimental feeding with either control or neonicotinoid-spiked pollen patties over two brood cycles.

Acknowledgments

We thank Jennifer Brandenburg, Rebecca Basile and Mario Waldburger for assistance and Nicolas Desneux, Werner Stahel and three anonymous reviewers for valuable comments. The infrastructure was courtesy provided by Agroscope Reckenholz-Tänniken ART, Switzerland.

Author Contributions

Conceived and designed the experiments: CS. Performed the experiments: CS AFM. Analyzed the data: MT LGT CS. Contributed reagents/materials/analysis tools: PN CS. Wrote the paper: CS PN SGP.

References

1. Bascompte J, Jordano P, Olesen JM (2006) Asymmetric coevolutionary networks facilitate biodiversity maintenance. Science 312: 431–433.
2. Fontaine C, Dajoz I, Meriguet J, Loreau M (2006) Functional diversity of plant-pollinator interaction webs enhances the persistence of plant communities. PLOS Biol 4: e1.
3. Garibaldi LA, Aizen MA, Klein AM, Cunningham SA, Harder LD (2011) Global growth and stability of agricultural yield decrease with pollinator dependence. Proc Natl Acad Sci USA 108: 5909–5914.
4. Klein AM, Vaissière BE, Cane JH, Steffan-Dewenter I, Cunningham SA, et al. (2007) Importance of pollinators in changing landscapes for world crops. Proc R Soc Lond B Biol Sci 274: 303–313.
5. Gallai N, Salles JM, Settele J, Vaissière BE (2009) Economic valuation of the vulnerability of world agriculture confronted with pollinator decline. Ecol Econ 68: 810–821.
6. Lautenbach S, Seppelt R, Liebscher J, Dormann CF (2012) Spatial and temporal trends of global pollination benefit. PLOS ONE 7: e35954.
7. Biesmeijer JC, Roberts SPM, Reemer M, Ohlemuller R, Edwards M, et al. (2006) Parallel declines in pollinators and insect-pollinated plants in Britain and the Netherlands. Science 313: 351–354.
8. Potts SG, Biesmeijer JC, Kremen C, Neumann P, Schweiger O, et al. (2010) Global pollinator declines: trends, impacts and drivers. Trends Ecol Evol 25: 345–353.
9. Cameron SA, Lozier JD, Strange JP, Koch JB, Cordes N, et al. (2011) Patterns of widespread decline in North American bumble bees. Proc Natl Acad Sci USA 108: 662–667.
10. Goulson D, Lye GC, Darvill B (2008) Decline and conservation of bumble bees. Annu Rev Entomol 53: 191–208.
11. Aizen MA, Garibaldi LA, Cunningham SA, Klein AM (2008) Long-term global trends in crop yield and production reveal no current pollination shortage but increasing pollinator dependency. Curr Biol 18: 1572–1575.
12. Aizen MA, Garibaldi LA, Cunningham SA, Klein AM (2009) How much does agriculture depend on pollinators? Lessons from long-term trends in crop production. Ann Bot 103: 1579–1588.
13. Potts SG, Roberts SPM, Dean R, Marris G, Brown MA, et al. (2010) Declines of managed honey bees and beekeepers in Europe. J Apic Res 49: 15–22.
14. Ellis JD, Evans JD, Pettis JS (2010) Colony losses, managed colony population decline and Colony Collapse Disorder in the United States. J Apic Res 49: 134–136.

15. Aizen MA, Harder LD (2009) The global stock of domesticated honey bees is growing slower than agricultural demand for pollination. Curr Biol 19: 915–918.
16. vanEngelsdorp D, Hayes J Jr., Underwood RM, Pettis J (2008) A survey of honey bee colony losses in the U.S., fall 2007 to spring 2008. PLOS ONE 3: e4071.
17. vanEngelsdorp D, Evans JD, Saegerman C, Mullin C, Haubruge E, et al. (2009) Colony collapse disorder: A descriptive study. PLOS ONE 4: e6481.
18. Cox-Foster DL, Conlan S, Holmes EC, Palacios G, Evans JD, et al. (2007) A metagenomic survey of microbes in honey bee colony collapse disorder. Science 318: 283–287.
19. Anderson D, East IJ (2008) The latest buzz about colony collapse disorder. Science 319: 724–725.
20. van der Zee R, Pisa L, Andronov S, Brodschneider R, Charrière J-D, et al. (2012) Managed honey bee colony losses in Canada, China, Europe, Israel and Turkey, for the winters of 2008–9 and 2009–10. J Apic Res 51: 100–114.
21. Neumann P, Carreck N (2010) Honey bee colony losses. J Apic Res 49: 1–6.
22. Brodschneider R, Moosbeckhofer R, Crailsheim K (2010) Surveys as a tool to record winter losses of honey bee colonies: a two-year case study in Austria and South Tyrol. J Apic Res 49: 23–30.
23. Johnson RM, Evans JD, Robinson GE, Berenbaum MR (2009) Changes in transcript abundance relating to colony collapse disorder in honey bees (*Apis mellifera*). Proc Natl Acad Sci USA 106: 14790–14795.
24. Vanbergen AJ, the Insect Pollinators Initiative (2013) Threats to an ecosystem service: pressures on pollinators. Front Ecol Environ 11: 251–259.
25. Genersch E, von der Ohe W, Kaatz H, Schroeder A, Otten C, et al. (2010) The German bee monitoring project: a long term study to understand periodically high winter losses of honey bee colonies. Apidologie 41: 332–352.
26. Le Conte Y, Ellis M, Ritter W (2010) *Varroa* mites and honey bee health: can Varroa explain part of the colony losses? Apidologie 41: 353–363.
27. Rosenkranz P, Aumeier P, Ziegelmann B (2010) Biology and control of *Varroa destructor*. J Invertebr Pathol 103 Suppl 1: S96–119.
28. Nazzi F, Brown SP, Annoscia D, Del Piccolo F, Di Prisco G, et al. (2012) Synergistic parasite-pathogen interactions mediated by host immunity can drive the collapse of honeybee colonies. PLOS Pathog 8: e1002735.
29. Martin SJ, Highfield AC, Brettell L, Villalobos EM, Budge GE, et al. (2012) Global honey bee viral landscape altered by a parasitic mite. Science 336: 1304–1306.

30. Higes M, Martin-Hernandez R, Garrido-Bailon E, Gonzalez-Porto AV, Garcia-Palencia P, et al. (2009) Honeybee colony collapse due to *Nosema ceranae* in professional apiaries. Environ Microbiol Rep 1: 110–113.

31. Antunez K, Martin-Hernandez R, Prieto L, Meana A, Zunino P, et al. (2009) Immune suppression in the honey bee (*Apis mellifera*) following infection by *Nosema ceranae* (Microsporidia). Environ Microbiol 11: 2284–2290.

32. Higes M, Meana A, Bartolome C, Botias C, Martin-Hernandez R (2013) *Nosema ceranae* (Microsporidia), a controversial 21st century honey bee pathogen. Environ Microbiol Rep 5: 17–29.

33. Evans JD, Schwarz RS (2011) Bees brought to their knees: microbes affecting honey bee health. Trends Microbiol 19: 614–620.

34. Bromenshenk JJ, Henderson CB, Wick CH, Stanford MF, Zulich AW, et al. (2010) Iridovirus and microsporidian linked to honey bee colony decline. PLOS ONE 5: e13181.

35. Cornman RS, Tarpy DR, Chen Y, Jeffreys L, Lopez D, et al. (2012) Pathogen webs in collapsing honey bee colonies. PLOS ONE 7: e43562.

36. Doublet V, Labarussias M, De Miranda JR, Moritz RFA, Paxton JR (2014) Bees under stress: sublethal doses of a neonicotinoid pesticide and pathogens interact to elevate honey bee mortality across the life cycle. Environ Microbiol. (in press) DOI:10.1111/1462-2920.12426.

37. Mullin CA, Frazier M, Frazier JL, Ashcraft S, Simonds R, et al. (2010) High levels of miticides and agrochemicals in North American apiaries: implications for honey bee health. PLOS ONE 5: e9754.

38. Krupke CH, Hunt GJ, Eitzer BD, Andino G, Given K (2012) Multiple routes of pesticide exposure for honey bees living near agricultural fields. PLOS ONE 7: e29268.

39. Elbert A, Haas M, Springer B, Thielert W, Nauen R (2008) Applied aspects of neonicotinoid uses in crop protection. Pest Manag Sci 64: 1099–1105.

40. Jeschke P, Nauen R, Schindler M, Elbert A (2011) Overview of the status and global strategy for neonicotinoids. J Agric Food Chem 59: 2897–2908.

41. van der Sluijs JP, Simon-Delso N, Goulson D, Maxim L, Bonmatin JM, et al. (2013) Neonicotinoids, bee disorders and the sustainability of pollinator services. Curr Opin Environ Sustain 5: 293–305.

42. Goulson D (2013) An overview of the environmental risks posed by neonicotinoid insecticides. J Appl Ecol 50: 977–987.

43. Matsuda K, Buckingham SD, Kleier D, Rauh JJ, Grauso M, et al. (2001) Neonicotinoids: insecticides acting on insect nicotinic acetylcholine receptors. Trends Pharmacol Sci 22: 573–580.

44. Cresswell JE (2011) A meta-analysis of experiments testing the effects of a neonicotinoid insecticide (imidacloprid) on honey bees. Ecotoxicology 20: 149–157.

45. Desneux N, Decourtye A, Delpuech JM (2007) The sublethal effects of pesticides on beneficial arthropods. Annu Rev Entomol 52: 81–106.

46. Blacquière T, Smagghe G, van Gestel CA, Mommaerts V (2012) Neonicotinoids in bees: a review on concentrations, side-effects and risk assessment. Ecotoxicology 21: 973–992.

47. Williamson SM, Wright GA (2013) Exposure to multiple cholinergic pesticides impairs olfactory learning and memory in honeybees. J Exp Biol 216: 1799–1807.

48. Henry M, Beguin M, Requier F, Rollin O, Odoux JF, et al. (2012) A common pesticide decreases foraging success and survival in honey bees. Science 336: 348–350.

49. Belzunces LP, Tchamitchan S, Brunet J (2012) Neural effects of insecticides in the honey bee. Apidologie 43: 348–370.

50. Decourtye A, Devillers J (2010) Ecotoxicity of neonicotinoid insecticides to bees. In: Thany SH, editor. Insect nicotinic acetylcholine receptors. New York: Springer. 85–95.

51. Schneider CW, Tautz J, Grünewald B, Fuchs S (2012) RFID tracking of sublethal effects of two neonicotinoid insecticides on the foraging behavior of *Apis mellifera*. PLOS ONE 7: e30023.

52. Palmer MJ, Moffat C, Saranzewa N, Harvey J, Wright GA, et al. (2013) Cholinergic pesticides cause mushroom body neuronal inactivation in honeybees. Nat Commun 4: 1634.

53. Yang EC, Chang HC, Wu WY, Chen YW (2012) Impaired olfactory associative behavior of honeybee workers due to contamination of imidacloprid in the larval stage. PLOS ONE 7: e49472.

54. Hatjina F, Papaefthimiou C, Charistos L, Dogaroglu T, Bouga M, et al. (2013) Sublethal doses of imidacloprid decreased size of hypopharyngeal glands and respiratory rhythm of honeybees in vivo. Apidologie 44: 467–480.

55. Oliveira RA, Roat TC, Carvalho SM, Malaspina O (2013) Side-effects of thiamethoxam on the brain and midgut of the africanized honeybee *Apis mellifera* (Hymenoptera: Apidae). Environ Toxicol. (in press) DOI:10.1002/tox.21842.

56. Eiri DM, Nieh JC (2012) A nicotinic acetylcholine receptor agonist affects honey bee sucrose responsiveness and decreases waggle dancing. J Exp Biol 215: 2022–2029.

57. Ramirez-Romero R, Chaufaux J, Pham-Delègue MH (2005) Effects of Cry1Ab protoxin, deltamethrin and imidacloprid on the foraging activity and the learning performances of the honeybee *Apis mellifera*, a comparative approach. Apidologie 36: 601–611.

58. Fischer J, Müller T, Spatz AK, Greggers U, Grünewald B, et al. (2014) Neonicotinoids interfere with specific components of navigation in honeybees. PLOS ONE 9: e91364.

59. Han P, Niu CY, Lei CL, Cui JJ, Desneux N (2010) Use of an innovative T-tube maze assay and the proboscis extension response assay to assess sublethal effects of GM products and pesticides on learning capacity of the honey bee *Apis mellifera* L. Ecotoxicology 19: 1612–1619.

60. Gill RJ, Ramos-Rodriguez O, Raine NE (2012) Combined pesticide exposure severely affects individual- and colony-level traits in bees. Nature 491: 105–108.

61. Iwasa T, Motoyama N, Ambrose JT, Roe RM (2004) Mechanism for the differential toxicity of neonicotinoid insecticides in the honey bee, *Apis mellifera*. Crop Prot 23: 371–378.

62. Zhu W, Schmehl DR, Mullin CA, Frazier JL (2014) Four common pesticides, their mixtures and a formulation solvent in the hive environment have high oral toxicity to honey bee larvae. PLOS ONE 9: e77547.

63. Thompson HM, Fryday SL, Harkin S, Milner S (2014) Potential impacts of synergism in honeybees (*Apis mellifera*) of exposure to neonicotinoids and sprayed fungicides in crops. Apidologie (in press): DOI:10.1007/s13592-014-0273-6.

64. Sànchez-Bayo F, Goka K (2014) Pesticide residues and bees - a risk assessment. PLOS ONE 9: e94482.

65. Bryden J, Gill RJ, Mitton RA, Raine NE, Jansen VA (2013) Chronic sublethal stress causes bee colony failure. Ecol Lett. 16: 1463–1469.

66. James RR, Xu J (2012) Mechanisms by which pesticides affect insect immunity. J Invertebr Pathol 109: 175–182.

67. Mason R, Tennekes H, Sànchez-Bayo F, Jepsen PU (2013) Immune suppression by neonicotinoid insecticides at the root of global wildlife declines. J Environ Immunol Toxicol 1: 3–12.

68. Vidau C, Diogon M, Aufauvre J, Fontbonne R, Vigues B, et al. (2011) Exposure to sublethal doses of fipronil and thiacloprid highly increases mortality of honeybees previously infected by *Nosema ceranae*. PLOS ONE 6: e21550.

69. Pettis JS, vanEngelsdorp D, Johnson J, Dively G (2012) Pesticide exposure in honey bees results in increased levels of the gut pathogen *Nosema*. Naturwissenschaften 99: 153–158.

70. Alaux C, Brunet JL, Dussaubat C, Mondet F, Tchamitchan S, et al. (2010) Interactions between *Nosema* microspores and a neonicotinoid weaken honeybees (*Apis mellifera*). Environ Microbiol 12: 774–782.

71. Aufauvre J, Misme-Aucouturier B, Viguès B, Texier C, Delbac F, et al. (2014) Transcriptome analyses of the honeybee response to *Nosema ceranae* and insecticides. PLOS ONE 9: e91686.

72. Di Prisco G, Cavaliere V, Annoscia D, Varricchio P, Caprio E, et al. (2013) Neonicotinoid clothianidin adversely affects insect immunity and promotes replication of a viral pathogen in honey bees. Proc Natl Acad Sci USA 110: 18466–18471.

73. Naug D (2009) Nutritional stress due to habitat loss may explain recent honeybee colony collapses. Biol Conserv 142: 2369–2372.

74. Alaux C, Dantec C, Parrinello H, Le Conte Y (2011) Nutrigenomics in honey bees: digital gene expression analysis of pollen's nutritive effects on healthy and varroa-parasitized bees. BMC Genomics 12: 496.

75. Alaux C, Ducloz F, Crauser D, Le Conte Y (2010) Diet effects on honeybee immunocompetence. Biol Lett 6: 562–565.

76. Mao W, Schuler MA, Berenbaum MR (2013) Honey constituents up-regulate detoxification and immunity genes in the western honey bee *Apis mellifera*. Proc Natl Acad Sci USA 110: 8842–8846.

77. Chauzat MP, Martel AC, Blanchard P, Clément MC, Schurr F, et al. (2010) A case report of a honey bee colony poisoning incident in France. J Apic Res 49: 113–115.

78. Tremolada P, Mazzoleni M, Saliu F, Colombo M, Vighi M (2010) Field trial for evaluating the effects on honeybees of corn sown using Cruiser and Celest xl treated seeds. Bull Environ Contam Toxicol 85: 229–234.

79. Cutler GC, Scott-Dupree CD (2007) Exposure to clothianidin seed-treated canola has no long-term impact on honey bees. J Econ Entomol 100: 765–772.

80. Pilling E, Campbell P, Coulson M, Ruddle N, Tornier I (2013) A four-year field program investigating long-term effects of repeated exposure of honey bee colonies to flowering crops treated with thiamethoxam. PLOS ONE 8: e77193.

81. Nguyen BK, Saegerman C, Pirard C, Mignon J, Widart J, et al. (2009) Does imidacloprid seed-treated maize have an impact on honey bee mortality? J Econ Entomol 102: 616–623.

82. Pohorecka K, Skubida P, Miszczak A, Semkiw P, Sikorski P, et al. (2012) Residues of neonicotinoid insecticides in bee collected plant materials from oilseed rape crops and their effect on bee colonies. Journal of Apicultural Science 56: 115–134.

83. Cresswell JE, Desneux N, vanEngelsdorp D (2012) Dietary traces of neonicotinoid pesticides as a cause of population declines in honey bees: an evaluation by Hill's epidemiological criteria. Pest Manag Sci 68: 819–827.

84. Cressey D (2013) Europe debates risk to bees. Nature 496: 408.

85. Stokstad E (2013) Pesticides under fire for risks to pollinators. Science 340: 674–676.

86. Butler CG (1957) The process of queen supersedure in colonies of honeybees (*Apis mellifera* Linn.). Insectes Soc 4: 211–223.

87. Pettis JS, Higo H, Winston ML, Pankiw T (1997) Queen rearing suppression in the honey bee - evidence for a fecundity signal. Insectes Soc 44: 311–322.

88. Whitehorn PR, O'Connor S, Wackers FL, Goulson D (2012) Neonicotinoid pesticide reduces bumble bee colony growth and queen production. Science 336: 351–352.

89. Larson JL, Redmond CT, Potter DA (2013) Assessing insecticide hazard to bumble bees foraging on flowering weeds in treated lawns. PLOS ONE 8: e66375.

90. Fauser-Misslin A, Sadd BM, Neumann P, Sandrock C (2014) Influence of combined pesticide and parasite exposure on bumblebee colony traits in the laboratory. J Appl Ecol 51: 450–459.

91. Sandrock C, Tanadini LG, Pettis J, Biesmeijer JC, Potts SG, et al. (2014) Sublethal neonicotinoid insecticide exposure reduces solitary bee reproductive success. Agric For Entomol 16: 119–128.

92. Moritz RFA, Southwick EE (1992) Bees as superorganisms: an evolutionary reality. Berlin: Springer-Verlag. 395p.

93. Derecka K, Blythe MJ, Malla S, Genereux DP, Guffanti A, et al. (2013) Transient exposure to low levels of insecticide affects metabolic networks of honeybee larvae. PLOS ONE 8: e68191.

94. Neumann P, Moritz RFA, Mautz D (2000) Colony evaluation is not affected by drifting of drone and worker honeybees (Apis mellifera L.) at a performance testing apiary. Apidologie 31: 67–79.

95. Higes M, Martin-Hernandez R, Garrido-Bailon E, Garcia-Palencia P, Meana A (2008) Detection of infective Nosema ceranae (Microsporidia) spores in corbicular pollen of forager honeybees. J Invertebr Pathol 97: 76–78.

96. Maienfisch P, Angst M, Brandl F, Fischer W, Hofer D, et al. (2001) Chemistry and biology of thiamethoxam: a second generation neonicotinoid. Pest Manag Sci 57: 906–913.

97. Nauen R, Ebbinghaus-Kintscher U, Salgado VL, Kaussmann M (2003) Thiamethoxam is a neonicotinoid precursor converted to clothianidin in insects and plants. Pestic Biochem Phys 76: 55–69.

98. Dively GP, Kamel A (2012) Insecticide residues in pollen and nectar of a cucurbit crop and their potential exposure to pollinators. J Agric Food Chem 60: 4449–4456.

99. Imdorf A, Buehlmann G, Gerig L, Kilchenmann V, Wille H (1987) A test of the method of estimation of brood areas and number of worker bees in free-flying colonies. Apidologie 18: 137–146.

100. Delaplane KS, van der Steen J, Guzman E (2013) Standard methods for estimating strength parameters of Apis mellifera colonies. In: Dietemann V, Ellis JD, Neumann P, editors. The COLOSS BEEBOOK, Volume 1: standard methods for Apis mellifera research.

101. Khoury DS, Myerscough MR, Barron AB (2011) A quantitative model of honey bee colony population dynamics. PLOS ONE 6: e18491.

102. R Core Development_Team (2012) R: A language and environment for statistical computing. R Foundation for Statistical computing, Vienna, Austria. Available: http://cran.r-project.org. Accessed 27 April 2014.

103. Bates D, Maechler M, Bolker B, Walker S (2013) lme4: Linear mixed-effects models using Eigen and S4 classes. R package version 10-4. Available: http://CRANR-projectorg/package = lme4.

104. Hothorn T, Bretz F, Westfall P (2008) Simultaneous inference in general parametric models. Biometrical Journal 50: 346–363.

105. Rortais A, Arnold G, Halm MP, Touffet-Briens F (2005) Modes of honeybees exposure to systemic insecticides: estimated amounts of contaminated pollen and nectar consumed by different categories of bees. Apidologie 36: 71–83.

106. Seeley TD (1982) Adaptive significance of the age polyethism schedule in honeybee colonies. Behav Ecol Sociobiol 11: 287–293.

107. Masson C, Pham-Delègue MH, Fonta C, Gascuel J, Arnold G, et al. (1993) Recent advances in the concept of adaptation to natural odour signals in the honeybee, Apis mellifera L. Apidologie 24: 169–194.

108. Wu JY, Anelli CM, Sheppard WS (2011) Sub-lethal effects of pesticide residues in brood comb on worker honey bee (Apis mellifera) development and longevity. PLOS ONE 6: e14720.

109. Smodis Skerl MI, Gregorc A (2010) Heat shock proteins and cell death in situ localisation in hypopharyngeal glands of honeybee (Apis mellifera carnica) workers after imidacloprid or coumaphos treatment. Apidologie 41: 73–86.

110. Gregorc A, Ellis JD (2011) Cell death localization in situ in laboratory reared honey bee (Apis mellifera L.) larvae treated with pesticides. Pesticide Biochemistry and Physiology 99: 200–207.

111. Johnson RM, Dahlgren L, Siegfried BD, Ellis MD (2013) Acaricide, fungicide and drug interactions in honey bees (Apis mellifera). PLOS ONE 8: e54092.

112. Hawthorne DJ, Dively GP (2011) Killing them with kindness? In-hive medications may inhibit xenobiotic efflux transporters and endanger honey bees. PLOS ONE 6: e26796.

113. Lu C, Warchol KM, Callahan RA (2012) In situ replication of honey bee colony collapse disorder. Bull Insectology 65: 99–106.

114. Lu C, Warchol KM, Callahan RA (2014) Sub-lethal exposure to neonicotinoids impaired honey bees winterization before proceeding to colony collapse disorder. Bull Insectology 67: 125–130.

115. Wilson-Rich N, Dres ST, Starks PT (2008) The ontogeny of immunity: development of innate immune strength in the honey bee (Apis mellifera). J Insect Physiol 54: 1392–1399.

116. Bendahou N, Fleche C, Bounias M (1999) Biological and biochemical effects of chronic exposure to very low levels of dietary cypermethrin (Cymbush) on honeybee colonies (Hymenoptera: Apidae). Ecotoxicol Environ Saf 44: 147–153.

117. Pettis JS, Collins AM, Wilbanks R, Feldlaufer M (2004) Effects of coumaphos on queen rearing in the honey bee, Apis mellifera. Apidologie 35: 605–610.

118. Haarmann T, Spivak M, Weaver D, Weaver B, Glenn T (2002) Effects of fluvalinate and coumaphos on queen honey bees (Hymenoptera: Apidae) in two commercial queen rearing operations. J Econ Entomol 95: 28–35.

119. DeGrandi-Hoffman G, Chen Y, Simonds R (2013) The effects of pesticides on queen rearing and virus titers in honey bees (Apis mellifera L.). Insects 4: 71–89.

120. Collins AM, Pettis JS, Wilbanks R, Feldlaufer M (2004) Performance of honey bee (Apis mellifera) queens reared in beeswax cells impregnated with coumaphos. J Apic Res 43: 128–134.

121. Garrido PM, Antunez K, Martin M, Porrini MP, Zunino P, et al. (2013) Immune-related gene expression in nurse honey bees (Apis mellifera) exposed to synthetic acaricides. J Insect Physiol 59: 113–119.

122. vanEngelsdorp D, Tarpy DR, Lengerich EJ, Pettis JS (2013) Idiopathic brood disease syndrome and queen events as precursors of colony mortality in migratory beekeeping operations in the eastern United States. Prev Vet Med 108: 225–233.

123. Crailsheim K (1998) Trophallactic interactions in the adult honeybee (Apis mellifera L.). Apidologie 29: 97–112.

124. Elzen PJ, Elzen GW, Rubink W (2003) Comparative susceptibility of European and Africanized honey bee (Hymenoptera: Apidae) ecotypes to several insecticide classes. Southwest Entomol 28: 255–260.

125. Suchail S, Guez D, Belzunces LP (2000) Characteristics of imidacloprid toxicity in two Apis mellifera subspecies. Environ Toxicol Chem 19: 1901–1905.

126. Laurino D, Manino A, Patetta A, Porporato M (2013) Toxicity of neonicotinoid insecticides on different honey bee genotypes. Bull Insectology 66: 119–126.

127. Claudianos C, Ranson H, Johnson RM, Biswas S, Schuler MA, et al. (2006) A deficit of detoxification enzymes: pesticide sensitivity and environmental response in the honeybee. Insect Mol Biol 15: 615–636.

128. Johnson RM, Mao W, Pollock HS, Niu G, Schuler MA, et al. (2012) Ecologically appropriate xenobiotics induce cytochrome P450s in Apis mellifera. PLOS ONE 7: e31051.

129. Mao W, Schuler MA, Berenbaum MR (2011) CYP9Q-mediated detoxification of acaricides in the honey bee (Apis mellifera). Proc Natl Acad Sci USA 108: 12657–12662.

130. Suchail S, Debrauwer L, Belzunces LP (2004) Metabolism of imidacloprid in Apis mellifera. Pest Manag Sci 60: 291–296.

131. Johnson RM, Pollock HS, Berenbaum MR (2009) Synergistic interactions between in-hive miticides in Apis mellifera. J Econ Entomol 102: 474–479.

132. Faucon JP, Aurieres C, Drajnudel P, Mathieu L, Ribière M, et al. (2005) Experimental study on the toxicity of imidacloprid given in syrup to honey bee (Apis mellifera) colonies. Pest Manag Sci 61: 111–125.

133. Tennekes H, Sànchez-Bayo F (2012) Time-dependent toxicity of neonicotinoids and other toxicants: implications for a new approach to risk assessment. J Environ Anal Toxicol S4: S4-001.

134. Tennekes HA, Sànchez-Bayo F (2013) The molecular basis of simple relationships between exposure concentration and toxic effects with time. Toxicology 309: 39–51.

135. Laycock I, Cresswell JE (2013) Repression and recuperation of brood production in Bombus terrestris bumble bees exposed to a pulse of the neonicotinoid insecticide imidacloprid. PLOS ONE 8: e79872.

136. Decourtye A, Henry M, Desneux N (2013) Environment: Overhaul pesticide testing on bees. Nature 497: 188.

137. Osborne JL (2012) Ecology: Bumblebees and pesticides. Nature 491: 43–45.

8

Association Mapping and Validation of QTLs for Flour Yield in the Soft Winter Wheat Variety Kitahonami

Goro Ishikawa[1]*, **Kazuhiro Nakamura**[1,2], **Hiroyuki Ito**[1], **Mika Saito**[1], **Mikako Sato**[3,4], **Hironobu Jinno**[3], **Yasuhiro Yoshimura**[3], **Tsutomu Nishimura**[3,5], **Hidekazu Maejima**[6], **Yasushi Uehara**[6], **Fuminori Kobayashi**[7], **Toshiki Nakamura**[1]*

1 NARO Tohoku Agricultural Research Center, Morioka, Iwate, Japan, **2** NARO Kyusyu Okinawa Agricultural Research Center, Chikugo, Fukuoka, Japan, **3** Kitami Agricultural Experiment Station, Hokkaido Research Organization, Tokoro-gun, Hokkaido, Japan, **4** Central Agricultural Experiment Station, Hokkaido Research Organization, Yubari-gun, Hokkaido, Japan, **5** Kamikawa Agricultural Experiment Station, Hokkaido Research Organization, Kamikawa-gun, Hokkaido, Japan, **6** Nagano Agricultural Experiment Station, Suzaka, Nagano, Japan, **7** National Institute of Agrobiological Sciences, Kannondai, Tsukuba, Japan

Abstract

The winter wheat variety Kitahonami shows a superior flour yield in comparison to other Japanese soft wheat varieties. To map the quantitative trait loci (QTL) associated with this trait, association mapping was performed using a panel of lines from Kitahonami's pedigree, along with leading Japanese varieties and advanced breeding lines. Using a mixed linear model corrected for kernel types and familial relatedness, 62 marker-trait associations for flour yield were identified and classified into 21 QTLs. In eighteen of these, Kitahonami alleles showed positive effects. Pedigree analysis demonstrated that a continuous pyramiding of QTLs had occurred throughout the breeding history of Kitahonami. Linkage analyses using three sets of doubled haploid populations from crosses in which Kitahonami was used as a parent were performed, leading to the validation of five of the eight QTLs tested. Among these, QTLs on chromosomes 3B and 7A showed highly significant and consistent effects across the three populations. This study shows that pedigree-based association mapping using breeding materials can be a useful method for QTL identification at the early stages of breeding programs.

Editor: Guangyuan He, Huazhong University of Science and Technology, China

Funding: This work was partially supported by a grant from the Ministry of Agriculture, Forestry and Fisheries of Japan (TRG-1009, NGB-1002 and NGB-2004). The funder had no role in study design, data collection and analysis, decision to publish, or preparation of the manuscript.

Competing Interests: The authors have declared that no competing interests exist.

* Email: goro@affrc.go.jp (GI); tnaka@affrc.go.jp (T. Nakamura)

Introduction

Flour yield, or the percentage of flour from a given quantity of grain, is of great importance to flour milling companies. Flour yield can be increased by the enhancement of techniques in the milling process, or through the development of varieties with higher flour yields. In 2006, the soft winter wheat variety Kitahonami was released in the Hokkaido prefecture of Japan [1]. This variety, which has become a leading variety in Hokkaido, shows the highest flour yield among Japanese soft wheat varieties. Therefore, Kitahonami is now being used as a source of the high flour-yield trait in multiple Japanese wheat breeding programs. Mapping of quantitative trait loci (QTL) associated with this trait and identification of linked markers would accelerate the introgression of the high flour-yield phenotype into other varieties.

However, flour yield is a complex trait that appears to be strongly influenced by genetic background. QTL studies using bi-parental populations have been conducted within hard varieties or within populations of interclass hybridizations between hard and soft varieties. These studies have indicated that QTLs for flour yield are located on 16 out of 21 chromosomes: 1B, 1D, 2A, 2B, 3A, 3B, 4A, 4B, 4D, 5A, 5B, 5D, 6B, 6D, 7A and 7D [2–6]. Interclass hybridization between soft and hard wheat demonstrat-

ed that the hardness locus *Pinb* on 5D chromosome had a strong influence on flour yield [3]. In soft wheat types, only a limited number of studies identifying QTLs associated with flour yield have been reported to date. Using an association mapping approach with 95 soft wheat varieties, Breseghello and Sorrells [7] detected weak QTLs associated with flour yield and break flour yield on 2D and 5B. A study of a bi-parental population derived from two soft wheat cultivars identified QTLs for flour yield on 1B, 2A, 2B, 2D and 3B [8]. Carter et al. [9] found that a large number of QTLs for milling quality and starch functionality were located on 3B and 4D, including QTLs for flour yield. Although the QTLs described in these studies were detected with high confidence, few were consistent between studies, suggesting that they are unlikely to coincide with the high flour yield trait from Kitahonami.

Because developing mapping populations and performing mapping studies are time consuming processes, breeders often have already introgressed target QTL into breeding lines using traditional selection methods before markers are available, especially for highly desirable traits. Thus, the most effective stage for using marker-assisted selection (MAS) to introduce a new trait into breeding programs is often missed. One solution to this could be to take advantage of populations developed within a breeding program to identify QTLs. Jannink et al. [10] proposed an approach applying family-based methods that are generally used within human and animal populations. Family-based QTL mapping for resistance to Fusarium head blight has been reported [11], and Malosetti et al. [12] used pedigree-data in association mapping of resistance to *Phytophthora infestans* in potato. Such association mapping techniques might be useful in rapid marker development for MAS.

The objective of this study was to dissect the genetic factors contributing to the high flour-yield trait of Kitahonami by an association mapping approach. After identification of QTLs related to flour yield, the pedigree record of Kitahonami was used to trace the origin of QTLs which had been inadvertently accumulated through selection for high flour yield. To confirm the utility of this approach, QTLs identified by association mapping were validated using our own bi-parental populations.

Materials and Methods

Plant materials: One hundred eighty-five accessions were used in this study (Table S1 in File S1). Of these, 65 accessions were winter wheat varieties related to Kitahonami and lines from the pedigree of Kitahonami (Fig. S1), along with advanced breeding materials and varieties developed at Kitami Agricultural Experimental Station (KAES), NARO Tohoku Agricultural Research Center (TARC) and Nagano Agricultural Experimental Station (NAES). These lines, which made up the association panel, were subjected to intensive phenotyping and were used in an association analysis. The remaining 120 accessions, which were included in a diversity analysis to investigate the genetic diversity of Japanese breeding materials, consisted of leading varieties and advanced breeding lines from across the country, along with introduced varieties from other countries and experimental lines such as Chinese Spring and *T. spelta* var. *duhamelianum*.

DNA isolation and genome-wide marker analysis: Genomic DNA of each accession was extracted from 100 mg of young leaf tissue using the automated DNA isolation systems PI-50α or PI-

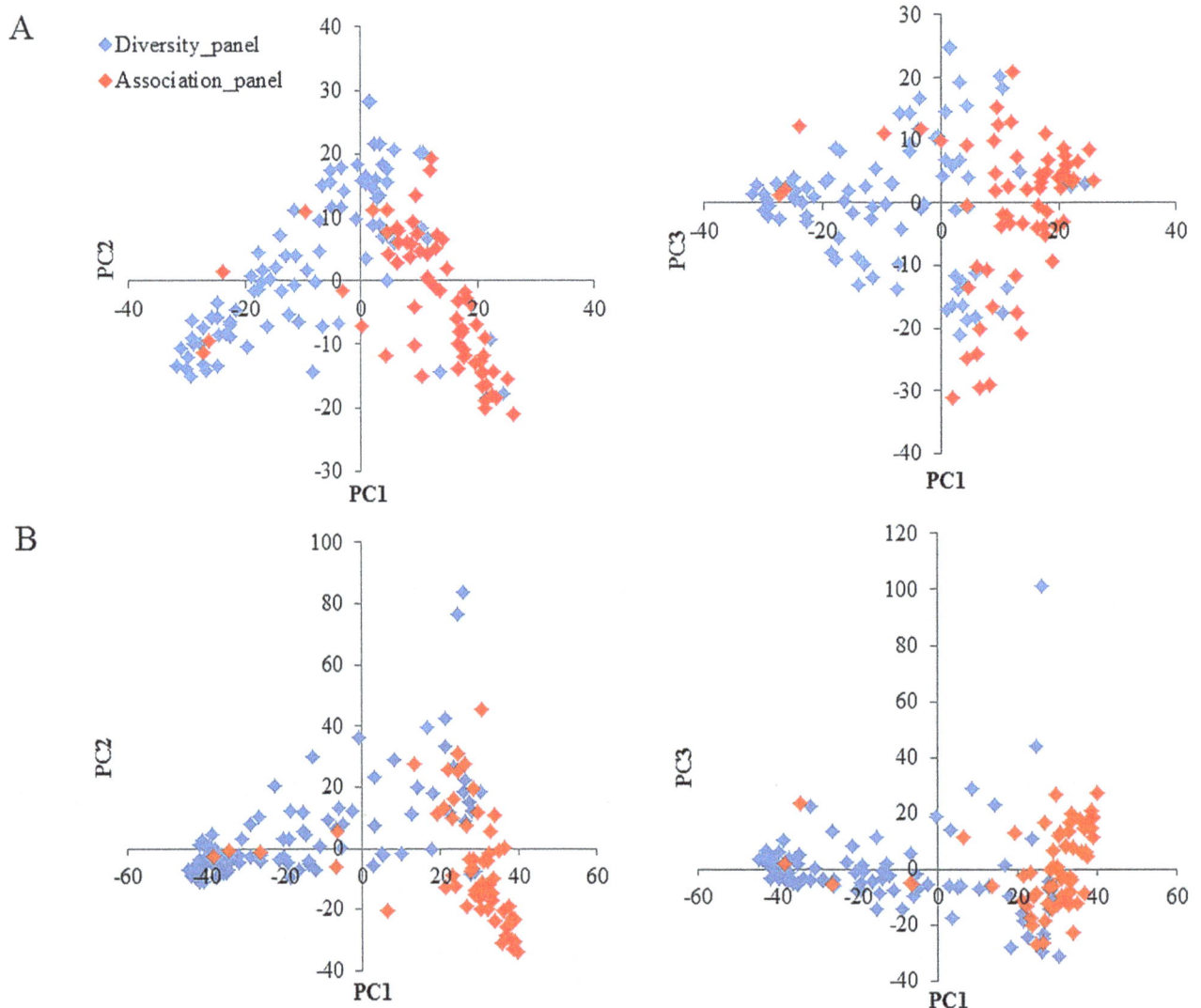

Figure 1. Scatter diagrams of principal component (PC) 1, 2 and 3 values calculated by the PCA function of TASSEL 3.0 using 2,933 DArT (A) and 6,042 SNP (B) markers.

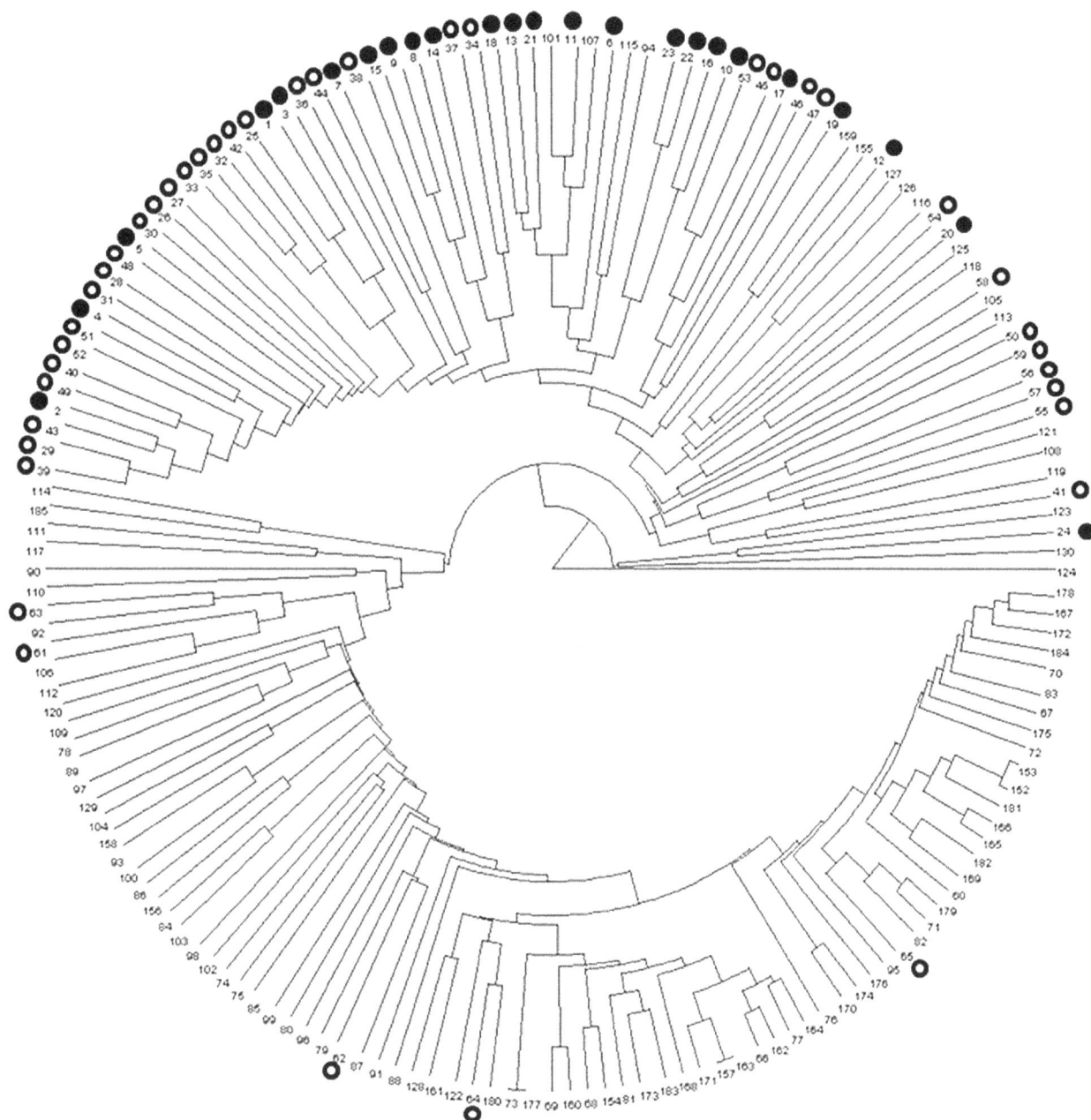

Figure 2. UPGMA dendrogram showing the pattern of genetic diversity among the 164 accessions based on the analysis of 6,042 SNP markers. Open and black circles indicate accessions found in the Association panel. Black circle means accessions in Kitahonami's pedigree. Numbers outside of the dendrogram correspond to accession numbers in Table S1 in File S1.

80X (Kurabo Industries Ltd., Osaka, Japan) according to the manufacturer's instructions. For the diversity analysis, 151 accessions were genotyped by DArT [Wheat *Pst*I (*Taq*I) v.3] (Diversity Arrays Technology Pty Ltd., http://www.diversityarrays.com/) and 164 accessions were genotyped by SNP (PrivKSU_WheatCons_9k) [13] markers. After removing data with minor allele frequencies (MAF) of less than 0.01, genotyping data from 2,933 DArT and 6,042 SNP markers were forwarded for diversity analysis. In addition to DArT and SNP genotyping, materials in the association panel were also genotyped using SSR markers (GrainGenes 2.0, http://wheat.pw.usda.gov/GG2/index.shtml/) and established diagnostic markers, such as

Pina-D1, *Pinb-D1*, *Wx-A1*, *Wx-B1*, *Ppo-A1*, *Ppo-D1*, *Psy-A1* and *Psy-B1* (reviewed in Liu et al. [14]). Genotyping data from SSR markers was recorded in the bi-allelic state: each fragment derived from a SSR marker was recorded as presence (1) or absence (0). All genotyping data from the association panel was merged. Data with MAF of less than 0.1 and redundancies among markers were removed. After these processes, genotyping data from 3,815 selected markers was used for the association analysis. Distribution of these markers across the wheat genome is shown in Table S2 in File S1.

Field experiments: Accessions in the association panel were field-grown in three locations [Kitami (Hokkaido island, 43.7°N,

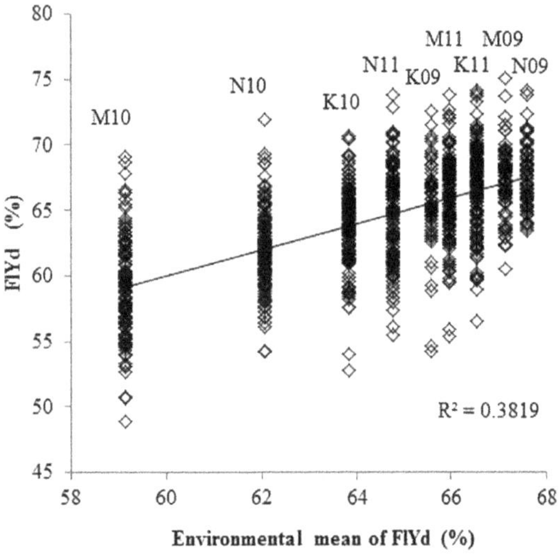

Figure 3. Leverage plots of flour yield (FlYd) values. Abbreviations for environments consist of first letter of location and harvest year. K: Kitami, M: Morioka, N: Nagano. For example, K09 means samples of 2008/2009 cropping season at Kitami.

143.7°E), Morioka (Northern Honsyu island, 39.7°N, 141.1°E) and Nagano (Central of Honsyu island, 36.7°N, 138.3°E)], during the three successive cropping seasons from 2008/2009 to 2010/2011. The plot size was 3.0 m×0.7 m, and each plot consisted of 40–50 plants separated from one another by 10–15 cm. Two replications were conducted in each season except 2008/2009.

Trait analyses: Grain samples were tempered to 14.5% moisture and 100 g of each sample was milled on a Quadrumat Junior mill (Brabender Co., Hackensack, NJ). The mill was preheated to prevent expansion of the rolls during operation. Ground grain samples were sifted with an 8XX silk reel sieve and a 94-mesh (180 μm) screen. Flour yield (FlYd) was expressed as the percentage of total flour weight to initial sample weight. Measurements were also taken for the following 14 traits: flour efficiency (FlEf), median size of flour particles (x50), specific surface area of flour particles (Sv), flour protein content (FPC), flour ash content (Fash), flour color L* (FlL), flour color a* (Fla), flour color b* (Flb), grain protein content (GPC), grain ash content (Gash), test weight (TestW), 1000-kernel weight (TKW), heading date (HD) and maturity date (MD). Detailed explanations for each trait are described in Table S3 in File S1.

Statistical analysis: Principal component analysis (PCA) and cladogram construction were performed with TASSEL 3.0 [15]. Analyses of DArT and SNP data were conducted separately. A correlation-based PCA was performed, and missing data was imputed with following settings: use manhatten distance, use unweighted average, 3 numbers of neighbors, and 0.80 minimum frequency of row data. Statistical analysis of traits was performed with JMP 9 (SAS Institute, Raleigh, NC). The mean value of two replications was used as the environmental value for each accession in the 2009/2010 and 2010/2011 cropping seasons. For two-dimensional analysis of variance, the fit model function was used with standard least squares method. Random effects of the genotypes and environments were applied to estimate the variance components. Heritability in the broad sense was estimated from the results of the variance analysis according to the formula used by Burton and DeVane [16]. Associations between markers and traits were calculated with TASSEL 3.0 [15] using the mixed linear model. The kinship matrix calculated by

Table 1. Correlation coefficients among environments for flour yield (FlYd).

Harvest year	Location	2009			2010			2011			Mean
		Kitami	Morioka	Nagano	Kitami	Morioka	Nagano	Kitami	Morioka	Nagano	
2009	Kitami		0.591	0.492	0.624	0.575	0.517	0.635	0.624	0.419	0.723
	Morioka			0.630	0.800	0.828	0.769	0.817	0.849	0.562	0.893
	Nagano				0.613	0.577	0.664	0.566	0.694	0.522	0.739
2010	Kitami					0.791	0.733	0.892	0.887	0.682	0.921
	Morioka						0.669	0.773	0.829	0.693	0.888
	Nagano							0.745	0.769	0.552	0.832
2011	Kitami								0.892	0.688	0.920
	Morioka									0.688	0.945
	Nagano										0.772

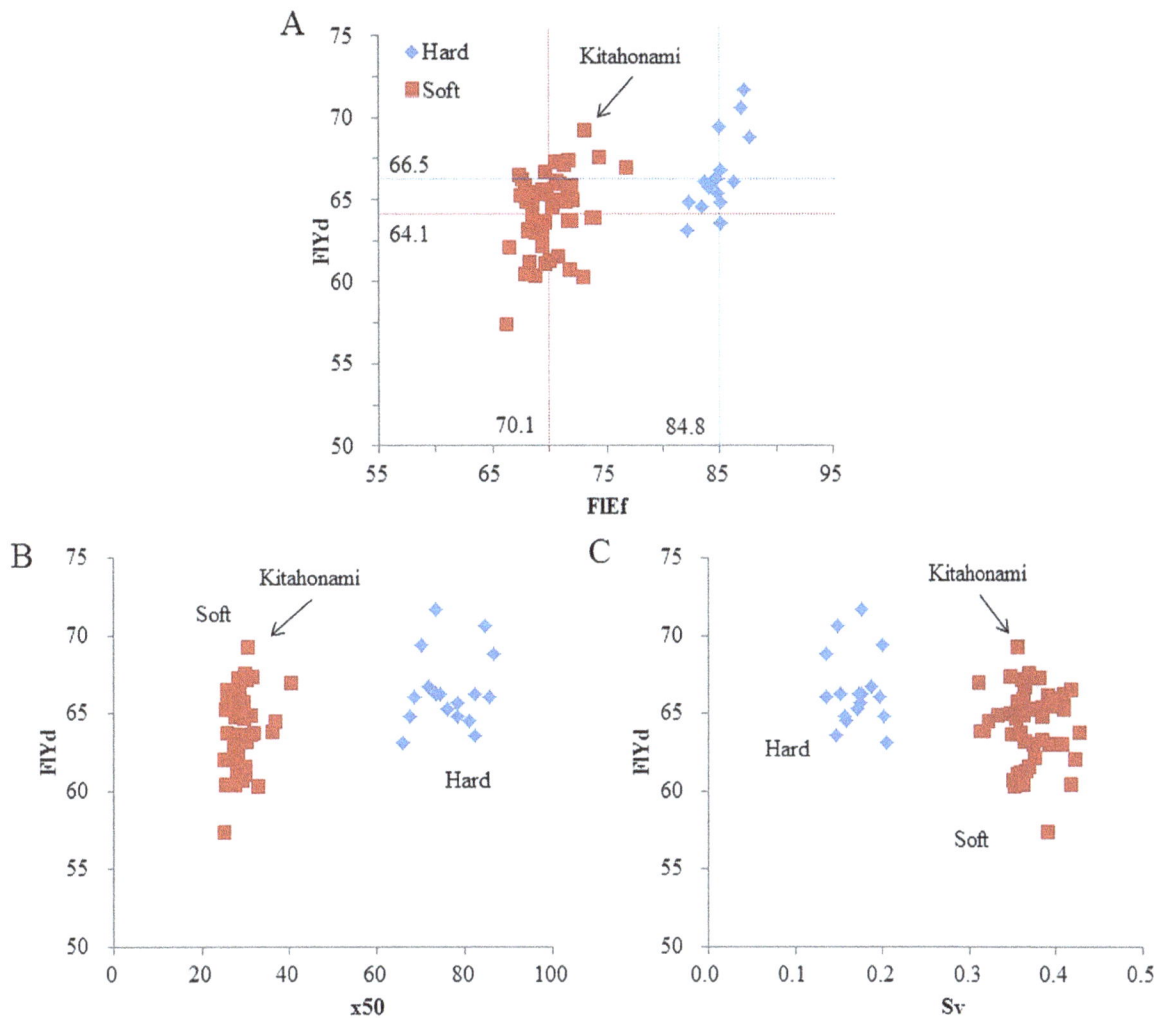

Figure 4. Relationships between flour yield (FlYd) and flour efficiency (FlEf) (A), FlYd and median diameter of particles (x50) (B) and FlYd and specific surface area of particles (Sv) (C). Accessions could be classified into either soft or hard kernel types based on *Pina-D1/Pinb-D1* genotypes. Soft accessions have *Pina-D1a* and *Pinb-D1a*, while hard have *Pina-D1b* or *Pinb-D1b*.

TASSEL was used for considering familial relatedness of accessions. Since a different distribution pattern was observed between soft and hard kernel types (see Results' section), the effect of kernel type was considered as an additional term of fixed effect in the model: 0 and 1 values were rendered to soft and hard kernel type, respectively. To take into account multiple comparisons, significance was tested using a 0.5 false discovery rate implemented in the q value software [17].

QTL validation: QTLs obtained as described above were validated using three doubled haploid (DH) populations, which were developed from F_1 plants from crosses between Kitahonami and three other varieties, namely Kinuhime, Tohoku224 and Shunyou. At least 151 lines from each population were field-grown without replication during the 2010/2011 season and subjected to validation. FlYd values for these lines were obtained as described above. To reduce genotyping costs, markers showing significant association with this trait in the association mapping analysis were converted from array-based SNP markers into PCR-based markers. To do this, probe sequences of the SNPs of interest were identified; since most of these sequences have been mapped, the chromosome number to which they have been assigned is known. These sequences were used as queries in BLASTN searches (E-value<e-40) against the wheat survey sequences (IWGSC, http://www.wheatgenome.org/) [18]. Generally, this allowed the identification of three highly homologous contigs from the relevant A, B and D homoeologous chromosomes, one of which showed 100% match to the probe sequence and originated from the same chromosome to which the probe sequence had been mapped. Contigs were aligned and regions that were polymorphic among the three genomes were used to design genome-specific primers (GSPs) upstream and downstream from the SNP of interest, with the SNP location set at approximately one third of the interval between primers. Two additional allele-specific primers (ASPs) were designing, with the 3′ base of these primers being concurrent with the SNP. To increase allele specificity, an artificial mismatch was integrated at the third nucleotide from the 3′ end of the ASPs, as described by Liu et al. [19]. Each PCR reaction included both GSPs and one ASP, and two reactions, each containing a different ASP, were used for genotyping. For PCR analysis, each 25-µL PCR mixture included 50–100 ng of DNA, 1.5 mM $MgCl_2$, 0.2 mM dNTP (each), 1×Ex Taq buffer, and 0.5 U of TaKaRa Ex Taq (Takara, Osaka,

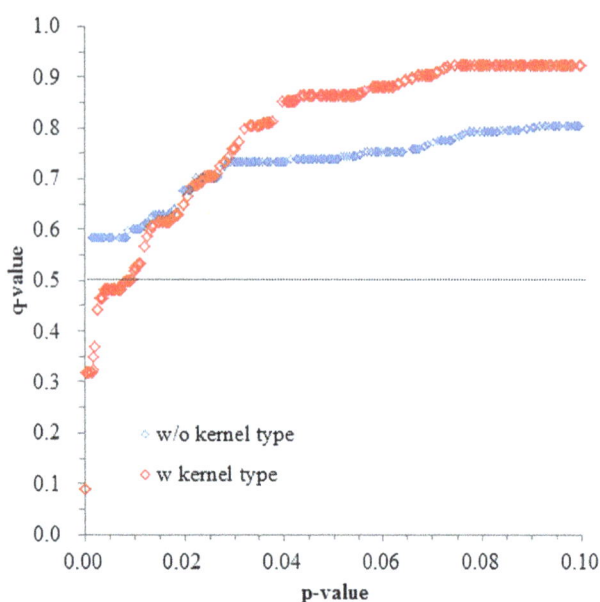

Figure 5. Distributions of p- and q-values and impact of kernel type correction on association mapping results for flour yield (FlYd). The distribution was calculated without (w/o) or with (w) employing kernel type as a covariant.

Japan). The concentrations of GSPs and ASPs in PCR mixtures are provided in the Table S4 in File S1. The PCR cycle consisted of an initial 5 min denaturation at 95°C, followed by 32 cycles of 95°C for 30 s, 55 −62°C for 30 s, and 72°C for 1 min, followed by a final extension at 72°C for 7 min. PCR products were separated by electrophoresis on a QIAxcel system (Qiagen, Hilden, Germany) using a QIAxcel DNA screening kit. Differences between allele mean values were tested by a T-test function implemented in JMP 9 for each combination of QTL and population. As well, the effects of the eight selected markers on FlYd were estimated using a multiple regression model. The fit model function of JMP 9 was used with the standard least squares method. Regression models were constructed based on the three combined DH populations (DH_total) or on each population separately (DH_each). The models were used to predict FlYd values of the individual DH lines. Correlation coefficients between predicted and actual values were determined.

Results

Population structure and familial relationships

Array-based marker analyses allowed the identification of 2,933 polymorphic DArT (Dataset S1 in File S2) and 6,042 polymorphic SNP markers (Dataset S2 in File S2) using 151 and 164 accessions, respectively. To obtain an overview of the genetic diversity of the accessions, a PCA was performed with each marker type. With the DArT markers, principal component (PC) 1, PC2 and PC3 explained 9.0, 4.6 and 3.9% of the total variation, while the first three PCs from the SNP markers explained 15.0, 5.2 and 3.5% of the variation. Scatter plots of PC1 and either PC2 or PC3 showed similar distribution patterns for both marker types (Fig. 1). These plots indicated that the accessions were distributed continuously and did not form any clear clusters. For the association panel, most accessions showed high PC1 values, but no clear tendencies were observed with PC2 and PC3 values. Based on the source of the accessions, PC1 represents the axis of earliness or growth habit

(data not shown). In the scatter plots of SNP markers, two accessions, U24 (Acc. no. 114) and Gabo (124), showed outlier values in PC2 and PC3. This indicates that the SNP markers used in this study have more power to distinguish lines than the DArT markers.

To investigate relationships among accessions, a cladogram was generated based on a distance matrix. Little difference was observed between the marker types, therefore only the cladogram generated from the SNP genotyping was employed here. As shown in Fig. 2, accessions were classified into two main clusters, and most accessions in the association panel into the same cluster. Particularly, clusters within the first four nodes from Kitahonami (Acc. no. 1) displayed relatively short distances and contained 35.4% (23/65) of accessions in the association panel. This indicates that familial relationship should be taken into account for association mapping.

Flour yield and relationship with other traits

The FlYd of lines in each environment and their mean values over the nine environments are provided in Dataset S3 in File S2. The analysis of variance showed significant genetic and environmental variation in FlYd compared to residual errors. Mean squares of accessions and environments were 61.76 (F = 19.65***) and 475.57 (F = 151.26***), respectively. The heritability of FlYd was 38.5%, which was relatively low compared to other traits investigated (Table S5 in File S1). This indicates that environmental factors have strong influence on FlYd. Leverage plots of environment also indicated a significant environmental effect on FlYd (Fig. 3). Samples harvested in 2010 showed a lower mean value for FlYd in all locations, and the lowest mean value was observed in the samples from Morioka in 2010. However, correlations across nine environments ranged from 0.419 to 0.945 (average 0.717), indicating that relative differences among accessions were consistent over the environments (Table 1). Therefore, we considered the mean value from the nine environments as the genotypic value of each accession.

Relationships between FlYd and other quality traits were also investigated (TableS6 in File S1). The FlEf (r = 0.511), x50 (0.436), Sv (−0.430) and Fash (0.349) each showed a significant relationship with FlYd, while no relationship with the other traits investigated was observed. It has been reported that x50, Sv and FlEf, as well as FlYd, have strong correlations with soft and hard kernel types (reviewed in Morris [20]). Therefore, the kernel types of 65 accessions were genotyped by *Pina-D1/Pinb-D1* markers [21,22]. All were classified as either soft (48) or hard (17) type. Taking kernel type into consideration, the relationships between FlYd and the other four traits were reanalyzed. This revealed two clear clusters attributable to kernel type, and no correlation with FlYd was detected in either cluster; only FlEf was correlated with kernel type (Fig. 4; TableS6 in File S1). Kitahonami showed the highest FlYd value among accessions in the soft cluster, although its value was considerably lower than the highest value observed in the hard cluster. These results clearly indicate that kernel type is an important element to consider in the association analysis.

Association analysis for flour yield using mixed-model

Genotype data obtained with the 3,815 selected markers was used for association mapping (Dataset S4 in File S2). Calculations were performed using a mixed linear model, with and without using kernel type as a covariant. To take into account multiple comparisons, a false discovery rate (q value) was adopted in determining significant marker-trait associations (MTAs). Distributions of q values with and without using kernel type as a covariant are shown in Fig. 5. When the kernel type was not used

Table 2. Flour yield (FlYd) QTLs detected by genome-wide association mapping.

| QTL | No of markers | Mean of markers within QTL | | | | Map[c] | Linkage group | (cM) |
		MAF[a]	q-value	R² (%)	Effect[b]			
1B.1	1	0.117	0.319	17.0	−2.572	DArT	1B	17.5
1B.2	1	0.316	0.482	11.5	1.846	DArT	1B	40.3
2B.1	2	0.359	0.482	9.4	1.755	SNP	2B	56.2
2B.2	5	0.379	0.322	14.4	2.031	SNP	2B	210.2–217.2
3B.1	4	0.335	0.437	10.9	1.897	SNP	3B	61.0–71.7
3B.2	1	0.313	0.482	9.3	2.022	SNP	3B	91.1
3B.3	1	0.270	0.482	9.3	1.614	DArT	3B	53.2
3D	1	0.346	0.482	11.0	1.708	DArT	3D	51.0
4B	1	0.238	0.482	10.3	1.851	DArT	4B	20.4
5A	1	0.234	0.443	11.3	2.180	SNP	5A	27.8
5B	1	0.156	0.482	9.7	−2.119	SNP	5B	212.5
5D.1	4	0.359	0.431	11.7	1.910	SNP	5D1cult	42.5–47.6
5D.2	4	0.188	0.482	9.5	2.172	SNP	5D3cult	8.4–10.6
6A.1	1	0.219	0.319	14.5	2.676	SNP	6A	75.5
6A.2	21	0.322	0.385	11.5	1.999	SNP	6A	114.5–117.0
6B	1	0.387	0.482	9.8	1.750	SNP	6B	59.5
7A	2	0.211	0.415	10.6	2.343	SNP	7A	58.6
7B.1	1	0.266	0.466	11.0	1.914	SNP	7B	73.3
7B.2	2	0.198	0.400	11.3	2.167	SNP	7B	115.9
7B.3	1	0.203	0.482	10.3	−2.104	SNP	7B	164.9
7D	1	0.220	0.466	11.4	2.353	DArT	7D	1.6

[a]Minor allele frequency.
[b]Increasing values indicate increasing effect of Kitahonami alleles.
[c]SNP and DArT locations are based on the consensus map of Cavanagh et al. [13] and Huang et al. [23], respectively.

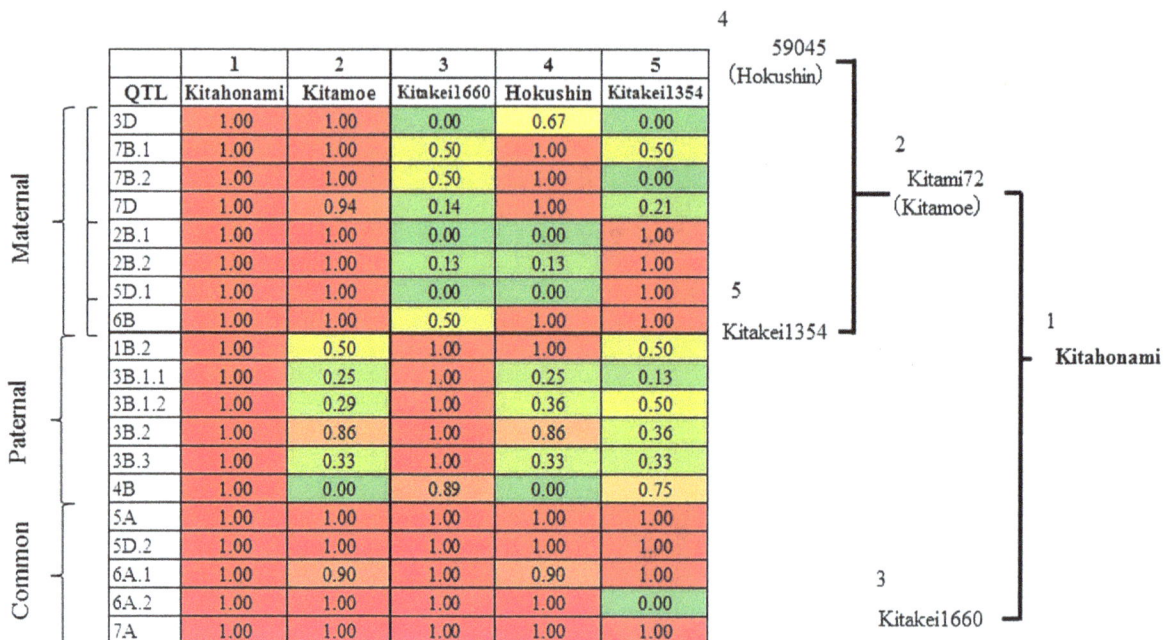

Figure 6. Pedigree analysis of flour yield (FlYd) QTLs. Values indicate frequencies of the same genotype as Kitahonami for each QTL.

in the model, the q value was within 0.55–0.85. When kernel type was accounted for, more accurate MTAs were detected, as indicated by q values ranging from 0.09 to 0.93. Therefore, to select reliable markers, kernel type was used as a covariant and the threshold of the q value was set at 0.5. This led to the identification of a total of 62 markers (Table S7 in File S1). Based on the locations of the markers [13,23], MTAs were classified into 21 QTLs (Table 2), although five MTA locations remained undetermined (Table S7 in File S1). Among the 21 QTLs, 18 had positive effects when the Kitahonami allele was present (Table 2). Since QTLs were classified based on two consensus genetic maps, it is possible that some QTLs overlapped: for example, 3B.3 may represent the same QTL as 3B.1 or 3B.2 (Table 2). Among the 62 MTAs, r^2 ranged from 9.2 to 20.5% and effects on FlYd ranged from 1.51 to 2.68 (Table S7 in File S1).

Pedigree analysis for flour yield QTL

For the 18 QTLs that showed positive effects on FlYd when Kitahonami alleles were present, linkage disequilibrium (LD) analyses of the associated loci were performed. Obvious LD blocks were observed in 11 of 18 QTLs (Table 3; Fig. S2). The 3B.1 QTL consisted of two blocks (3B.1.1 and 3B.1.2). The sizes of the blocks ranged from 0.5 to 23.5 cM, with an average size of 7.5 cM. By referencing Kitahonami's pedigree tree (Fig. S1), we investigated the origin and routes of transfer of QTLs into Kitahonami, based on similarities of genotypes in the LD blocks. Results indicated that the QTLs on 2B.1, 2B.2, 3D, 5D.1, 6B, 7B.1, 7B.2 and 7D were derived from the maternal variety, Kitamoe (Acc. no. 2) and 1B.2, 3B.1.1, 3B.1.2, 3B.2, 3B.3 and 4B from the paternal line, Kitakei1660 (Acc. no. 3) (Fig. 6). Since QTLs on 5A, 5D.2, 6A.1, 6A.2 and 7A existed in both parents, it could not be determined which side was the source of these QTLs in Kitahonami. QTLs originating from the maternal donor were further traced back to either Hokushin (Acc. no. 4) (3D, 7B.1, 7B.2 and 7D) or Kitakei1354 (Acc. no. 5) (2B.1, 2B.2 and 5D.1) (Fig. 6). When we attempted to trace the QTLs further back in Kitahonami's lineage, some showed discrepancies with the

pedigree record (Fig. S1). However, the origins of several QTLs could be attributed to varieties introduced from abroad: it was concluded that 2B.1, 2B.2, 4B and 7D were from Ibis (Acc. no. 19), 3B.1.1 and 3B.1.2 were from Wichita (Acc. no. 12), and 3D and 5D.1 from Newthach (Acc. no. 24). The QTL on 6B seems to have originated from Norman (Acc. no. 9). The full matrix of similarities within LD blocks is shown in Table S8 in File S1.

Validation of the QTLs with newly developed PCR markers

For validation of the QTLs detected by association mapping, we performed linkage analysis using three sets of DH populations (Dataset S5 in File S2). The distribution of FlYd among lines in the three DH populations is shown in Fig. S3. For QTL validation, we wished to convert the SNP markers associated with the 11 QTLs into PCR-based markers to reduce the analysis cost involved with using the 9,000 SNP chip detection system. We succeeded in designing primer sets for eight of the 11 QTLs (Table S4 in File S1). Before use of these markers for QTL validation, their genome and allele specificity were confirmed in all four parents used for the DH populations. Based on the design of the primers, two bands were expected if the sample sequence matched with the sequence of the ASP, and a single band if it did not. All amplified products showed the expected band patterns (Fig. 7), indicating the new PCR-based markers were capable of identifying the eight QTLs.

Using these markers, it was determined that four of the QTLs had significant effects on FlYd in the Kinuhime/Kitahonami population, three had significant effects in the Tohoku224/Kitahonami population, and four QTLs had significant effects in the Shunyou/Kitahonami population (Table 4). Notably, QTLs on 3B.1.1, 3B.2 and 7A showed highly significant and consistent effects across the populations.

Multiple regression analysis showed significant effects in five of the eight markers using the DH_total dataset, which is based on the combination of all three DH populations (Table S9 in File S1). For the DH_each dataset, based on individual DH populations,

Table 3. Linkage disequilibrium blocks around flour yield (FlYd) QTLs detected in this study.

QTL	Map[a]	LD block[b]	Linkage group	No. of markers	LD range (cM) Start	End	Size
1B.2	DArT	–	1B	2			
2B.1	SNP	–	2B	1			
2B.2	SNP	+	2B	15	202.0	217.2	15.2
3B.1.1	SNP	+	3B	16	61.0	66.4	5.4
3B.1.2	SNP	+	3B	14	68.4	71.7	3.3
3B.2	SNP	+	3B	14	89.4	94.2	4.8
3B.3	DArT	–	3B	3			
3D	DArT	–	3D	4			
4B	DArT	+	4B	10	15.0	20.4	5.4
5A	SNP	–	5A	2			
5D.1	SNP	+	5D1cult	5	42.5	47.6	5.1
5D.2	SNP	+	5D3cult	3	8.4	10.6	2.2
6A.1	SNP	+	6A	20	75.5	99.0	23.5
6A.2	SNP	+	6A	12	114.5	126.4	11.9
6B	SNP	–	6B	2			
7A	SNP	+	7A	4	58.6	63.9	5.3
7B.1	SNP	–	7B	2			
7B.2	SNP	–	7B	2			
7D	DArT	+	7D	35	1.6	2.1	0.5

[a]Map information is based on SNP (Cavanagh et al. [13]) and DArT (Huang et al. [23]) markers.
[b]Presence (+) and absence (−) of LD block.

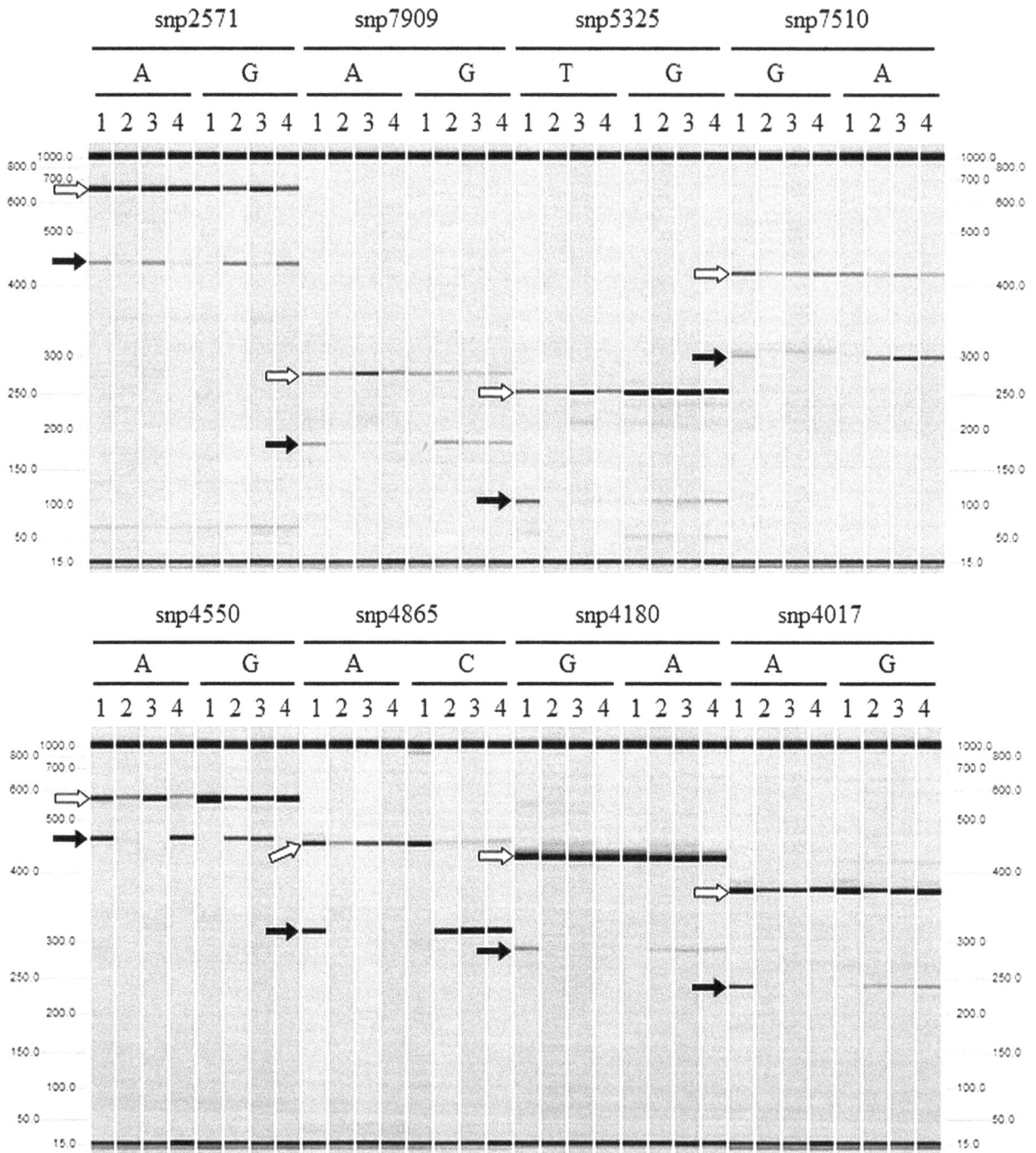

Figure 7. PCR assays to detect of polymorphic SNPs between Kitahonami and the three other varieties used as parents in the DH populations. White and black arrows indicate bands derived from genome-specific and allele-specific amplicons, respectively. 1: Kitahonami, 2: Kinuhime, 3: Tohoku224, 4: Shunyou.

four, two and three markers were significant in Kinuhime/Kitahonami, Tohoku224/Kitahonami and Shunyou/Kitahonami populations, respectively (Table S9 in File S1). The regression models were used to predict FlYd in the individual DH lines (Dataset S5 in File S2). When we considered all DH lines together, the correlation coefficients between predicted and actual values were 0.479 for DH_total and 0.607 for DH_each (Fig. 8). In each

population, prediction accuracies based on the DH_each dataset were consistently higher than those based on the DH_total dataset (Fig. 8).

Discussion

Array-based systems allow genotyping using a large number of markers simultaneously, and the use of these systems has become

Table 4. Validations of flour yield (FlYd) QTLs using three DH populations.

QTL	Marker[a]	Allele[b]	Kinuhime/Kitahonami			Tohoku224/Kitahonami			Shunyou/Kitahonami		
			Number	Mean	p	Number	Mean	p	Number	Mean	p
2B.1	snp2571	A	79	63.01	0.559	160	64.14		75	66.69	0.694
		G	72	63.31					80	66.51	
2B.2	snp7909	A	86	63.75	0.007**	80	64.55	0.091	76	66.82	0.323
		G	65	62.36		80	63.73		79	66.38	
3B.1.1	snp5325	T	70	63.92	0.005**	67	65.24	<.0001***	71	67.66	<.0001***
		G	81	62.49		93	63.35		83	65.69	
3B.2	snp7510	G	72	63.77	0.022*	68	65.05	0.001**	66	67.34	0.004**
		A	79	62.59		92	63.47		89	66.05	
5D.1	snp4550	A	81	63.23	0.755	87	64.19	0.844	155	66.60	0.408
		G	70	63.07		73	64.09				
6A.2	snp4865	A	74	63.56	0.117	75	64.20	0.811	75	66.78	
		C	77	62.76		85	64.09		77	66.40	
7A	snp4180	G	71	64.07	0.001***	76	64.89	0.003**	84	67.12	0.011*
		A	80	62.34		84	63.46		71	65.98	
7B.1	snp4017	A	82	63.43	0.246	75	64.52	0.138	84	67.29	0.001***
		G	69	62.83		85	63.81		70	65.77	

[a]Numbers in marker names correspond to the index of Illumina's 9K Infinium array (Cavanagh et al. [13]).
[b]Kitahonami's allele is shown at the top for each marker.
*, ** and *** indicate significant differences between allele mean values at 5%, 1% and 0.1% level, respectively.

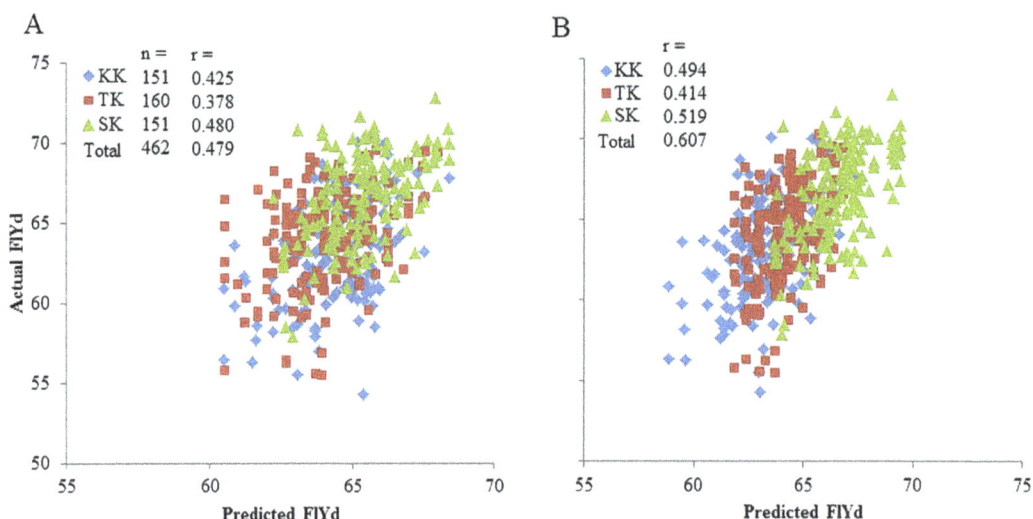

Figure 8. Scatter diagrams of predicted and actual FlYd values in DH lines. Regression models were constructed using the three DH populations together (DH_total) (A) or separately (DH_each) (B). KK: Kinuhime/Kitahonami, TK: Tohoku224/Kitahonami, SK: Shunyou/Kitahonami population.

popular for rapidly determining the genetic diversity and population structure of samples [13,24–26]. The DArT and SNP arrays used in this study contain approximately 7,000 and 9,000 markers, respectively. Among these markers, 42% of the DArT and 67% of the SNP markers showed polymorphisms among the accessions used here. By using these high-density genome-wide markers, we could provide the first precise overview of genetic variation in Japanese wheat varieties (Fig. 1; Fig. 2). The 185 lines used in the diversity analysis included an association panel of 65 lines. As expected, the lines in the association panel showed a higher level of similarity compared to other accessions, which is reasonable given that the association panel consists mainly of lines developed at KAES in the Hokkaido region; most are winter lines that are well-adapted to the northern region of Japan. Such regional adaptation is an important feature in evaluating genetic performances of complex traits such as yield and grain quality. Of the 33 lines in the pedigree record of Kitahonami (Fig. S1), 24 accessions were still available and were employed in this study. The cladograms generated by SNP markers clearly showed that Kitamoe (Acc. no. 2), Kitakei1660 (3), Hokushin (4) and Kitakei1354 (5) were clustered close to Kitahonami (Fig. 2), agreeing with the pedigree record. This indicates that kinship matrix generated with markers can be useful for representing the familial relation of accessions.

The analysis of variance indicated there was significant genetic and environmental variation for the target trait (Fig. 3). Samples collected at Morioka in 2010 showed the lowest mean FlYd values compared to other environments. The meteorological data recorded during the cropping seasons did not indicate any clear reason for this (data not shown). Although the differences among mean values between environments were significant (Fig. 3), the relative differences among accessions were consistent over the environments (Table 1). For example, Kitahonami consistently grouped within the eight accessions showing highest values for FlYd in all environments, indicating that Kitahonami carries alleles affecting this trait that will be useful across environments.

In the diversity analysis, the lines used for association mapping did not fall into distinct groups, but did show high familial relatedness. Therefore, we performed association mapping using

kinship matrix (K) rather than population structure (Q) as a covariant. Based on a plot of expected versus observed p values, this correction achieved a reduction in the false positive rate (data not shown). Kernel type is known to have a great impact on milling yield traits [3], and the *Pin* genes, which encode puroindolines, determine kernel type [27,28]. In this study, a significant difference in mean values of FlYd was observed between soft and hard accessions grouped by *Pina-D1* and *Pinb-D1* genotypes (Fig. 4). Therefore, kernel type was used as a fixed effect term in the statistical model. This treatment resulted in a major improvement in the association mapping results. When kernel type was not considered, no significant MTAs were detected at a q value of 0.5. However, when kernel type was used as a covariate, 62 MTAs were identified at this q value. This implies that the statistical model employed in this study, which used both kinship and kernel type as covariates, was appropriate for identifying genetic factors related to FlYd in Kitahonami.

It was not possible to precisely compare the positions of QTLs detected in this study to those identified in previous reports, since few identical markers were used. However, based on the microsatellite consensus map [29], the QTLs on 2B.1, 2B.2, 3B.2, 6A.2 and 7A observed here may correspond to those reported by Smith et al. [4], Lehmensiek et al. [5], Carter et al. [9], Fox et al. [6] and Lehmensiek et al. [5], respectively. The pedigree analysis showed that ongoing pyramiding of QTLs had occurred during the history of wheat breeding in the Hokkaido region. The high FlYd values of Kitahonami were achieved by combining eight positive QTLs from maternal lines, six from paternal lines, and five from both sides (Fig. 6). This information will be useful in developing effective breeding strategies to improve FlYd, since it allows us to predict the performance of progeny lines from a specific cross based on the genotypes of parental varieties or lines.

The level of LD can be affected by various factors including linkage, selection, and admixture [30]. Although there have been several studies on LD levels in various wheat populations [7,31–35], direct comparisons between studies is difficult, since LD levels are influenced by the type of markers used for genotyping and by sample size. However, generally LD decays to half of the initial value within less than 9 cM. In this study, LD blocks with more than 10 cM were identified in the QTLs on 2B.2, 3B.1+3B.2 (these two QTLs are closely linked), 6A.1 and 6A.2. Since the accessions in this study consist mainly of breeding materials, it is possible that LD blocks detected in the QTL regions result from selection for favorable phenotypes during the history of wheat breeding in KAES.

Segregation analysis confirmed that five of the eight QTLs tested had significant effects on the FlYd (Table 4). Previous studies using bi-parental populations reported that flour yield QTLs were detected on most wheat chromosomes but it was not demonstrated whether these QTLs maintained their favorable effects in materials with different genetic backgrounds. In this study, we used three different populations in which Kitahonami served as pollen donor to confirm positive effects for five out of eight QTLs. The contributions of 3B.1.1, 3B.2 and 7A were significant in all three populations, and the contributions of 2B.2 and 7B.1 were significant in one population. Although the effects of 2B.1, 5D.1 and 6A.2 were not confirmed in the three DH populations, this does not mean these QTLs have no positive effects on FlYd; rather, they have significant effects in a specific genetic background. Generally, it can be expected that during long term breeding programs, positive QTLs will accumulate in most breeding materials. However, the usefulness of these QTLs for crop improvement via breeding will be determined by their robustness, or their ability to predict effects in a range of genetic backgrounds. In this study, the QTLs on 3B and 7A consistently showed highly significant effects across three DH populations (Table 4). QTL analyses using a joint linkage map from these three populations indicated that the 3B and 7A QTLs explained 6.0% and 11.7% of the total variation, respectively (data not shown). Besides being significant in the DH populations, these QTLs were also present in lines with high flour yield originating from three separate breeding programs where Kitahonami was used as a parent (data not shown).

The effects of the markers on FlYd were not identical between single and multiple regression models (Table 4; Table S9 in File S1). These differences may be caused by relationships among the eight markers, since there were significant relationship between snp2571 and snp7909, between snp2571 and snp4550, and between snp5325 and snp7510 (data not shown). The correlation coefficients between actual and predicted values indicated that the model based on the DH_each dataset showed higher prediction accuracy than that based on the DH_total (Fig. 8). This result implies that the model should be constructed based on each cross combination. This is reasonable, because the number and effects of QTLs segregating in a biparental population varied within crosses. Since the three DH populations in this study share Kitahonami as a paternal parent, the differences observed between the two regression models among populations were caused by the differences in genetic backgrounds of maternal parents. In this study, the multiple regression analysis indicated that a relatively high prediction accuracy ($r = 0.607$) was achieved when the eight markers were applied for each DH population at the same time (Fig. 8). Not all markers showed positive effects in each DH population, yet the prediction accuracy was higher when all markers were used concurrently, as opposed to using only those markers that showed a positive effect for a specific DH population. Therefore, in terms of practical breeding, the construction of a regression model using all QTLs identified by GWAS in this study represents an attractive approach for increasing the selection efficiency for FlYd. Notably, the prediction accuracy was higher when the multiple regression model was based on data from each DH population ($r = 0.607$), rather than on data from combining the three DH populations (0.479) or data from the panel (0.370).

In this study, PCR-based markers linked to eight QTLs were developed (Fig. 7; Table S4 in File S1). Recently, large numbers of SNPs have been identified and characterized in wheat [13,24,36–38]. Using these resources, high density array-based markers have been established and used for diversity and LD analyses. Those tools have opened a new gate for understanding the genetic architecture of populations of interest. However, although the cost of genotyping per data point has dramatically declined, array-based systems are still relatively costly to access. This hinders the adoption of array-based systems in crop breeding programs, especially in the public sector. Since PCR-based markers are commonly used in MAS and are well suited to breeding programs, we decided to convert array-based SNP markers to PCR-based markers. In hexaploid wheat, difficulties arise in distinguishing allelic from genomic SNPs [38]. This is especially problematic because most SNP resources originate from exonic sequences [39,40] which tend to maintain higher similarity among A, B and D genomes than intronic sequences. This led us to design not only allele specific but also genome specific primers for precise targeting of SNPs. The chromosome locations of the target SNPs and the corresponding probe sequences were used to sort out the A, B and D homoeologous contigs from chromosome-arm specific survey sequences (IWGSC, http://www.wheatgenome.org/) [18]. By aligning the three contig sequences, we could identify polymorphic positions flanking the target SNP. These regions were used to design genome specific primer sets, allowing the amplification of fragments specific to the genome from which the SNP originated. Only eight markers were developed in this study, but using the same strategy we have succeeded in developing PCR-based markers capable of detecting 54 additional SNPs. We estimate that approximately 70% of the publicly available wheat SNP markers can be converted to genome specific PCR-based markers.

Milling tests require a substantial quantity of grain, meaning that selection for flour yield cannot be performed at the early stages of wheat breeding programs. The identification of markers linked to the flour yield trait can circumvent this problem. Using association mapping with a mixed model, we identified 21 QTLs influencing flour yield. The role of these QTLs was supported by pedigree information and results of linkage analysis. Notably, we identified several QTLs which were consistently associated with high flour yield across different genetic backgrounds. The introduction of these QTLs from Kitahonami into other lines by MAS is a promising method of improving flour yield in Japanese soft wheat varieties.

Supporting Information

Figure S1 **Record of Kitahonami's pedigree.**

Figure S2 **LD charts produced by TASSEL 3.0.**

Figure S3 **Distribution of FlYd in the three DH populations.**

File S1 **Supporting tables.** Table S1, List of accessions used in this study. Table S2, Distribution of the markers used for association mapping across the wheat genome. Table S3, Description of traits investigated. Table S4, Details of PCR-based markers developed in this study. Table S5, Variance components and heritabilities of traits investigated. Table S6, Correlation coefficients among traits. Table S7, List of marker-trait associations (MTAs) detected by genome-wide association mapping. Table S8, Full matrix produced by similarity analysis in LD blocks.

Table S9, Results of multiple regression analysis using the three DH populations.

File S2 Datasets. Dataset S1, DArT genotyping data for diversity analyses. Dataset S2, SNP genotyping data for diversity analyses. Dataset S3, Phenotyping data for association mapping. Dataset S4, Genotyping data for association mapping. Dataset S5, Genotype and phenotype data of doubled haploid populations.

Acknowledgments

The authors thank Dr. Patricia Vrinten for her useful comments on the manuscript. We also thank Drs. Shiaoman Chao, Andrzej Kilian, Kazuhiro Sato, Peter Bradbury and Mark Sorrells for their support in the molecular and statistical analyses of our materials. Thanks are due to support staff at NARO Tohoku Agricultural Research Center, Kitami Agricultural Experiment Station and Nagano Agricultural Experiment Station.

Author Contributions

Conceived and designed the experiments: GI KN YY T. Nishimura HM T. Nakamura. Performed the experiments: GI KN HI M. Saito M. Sato HJ YY T. Nishimura HM YU FK. Analyzed the data: GI M. Saito FK. Contributed reagents/materials/analysis tools: KN HI M. Sato HJ YY T. Nishimura HM YU. Contributed to the writing of the manuscript: GI T. Nakamura.

References

1. Yanagisawa A, Yoshimura Y, Amano Y, Kobayashi S, Nishimura T, et al. (2007) A new winter wheat variety Kitahonami'. Bulletin of Hokkaido Prefectural Agricultural Experiment Stations 91: 1–13.
2. Parker GD, Chalmers KJ, Rathjen AJ, Langridge P (1999) Mapping loci associated with milling yield in wheat (*Triticum aestivum* L.). Mol Breed 5: 561–568.
3. Campbell KG, Finney PL, Bergman CJ, Gualberto DG, Anderson JA, et al. (2001) Quantitative trait loci associated with milling and baking quality in a soft x hard wheat cross. Crop Sci 41: 1275–1285.
4. Smith AB, Cullis BR, Appels R, Campbell AW, Cornish GB, et al. (2001) The statistical analysis of quality traits in plant improvement programs with application to the mapping of milling yield in wheat. Aust J Agr Res 52: 1207–1219.
5. Lehmensiek A, Eckermann PJ, Verbyla AP, Appels R, Sutherland MW, et al. (2006) Flour yield QTLs in three Australian doubled haploid wheat populations. Aust J Agric Res 57: 1115–1122.
6. Fox GP, Martin A, Kelly AM, Sutherland MW, Martin D, et al. (2013) QTLs for water absorption and flour yield identified in the doubled haploid wheat population Lang/QT8766. Euphytica 192: 453–462.
7. Breseghello F, Sorrells ME (2006) Association mapping of kernel size and milling quality in wheat (*Triticum aestivum* L.) cultivars. Genetics 172: 1165–1177.
8. Smith N, Guttieri M, Souza E, Shoots J, Sorrells M, et al. (2011) Identification and validation of QTL for grain quality traits in a cross of soft wheat cultivars Pioneer brand 25R26 and Foster. Crop Sci 51: 1424–1436.
9. Carter AH, Garland-Campbell K, Morris CF, Kidwell KK (2012) Chromosomes 3B and 4D are associated with several milling and baking quality traits in a soft white spring wheat (*Triticum aestivum* L.) population. Theor Appl Genet 124: 1079–1096.
10. Jannik JL, Bink MCAM, Jansen RC (2001) Using complex plant pedigrees to map valuable genes. Trends Plant Sci 6: 337–342.
11. Rosyara UR, Gonzalez-Hernandez JL, Glover KD, Gedye KR, Stein JM (2009) Family-based mapping of quantitative trait loci in plant breeding populations with resistance to Fusarium head blight in wheat as an illustration. Theor Appl Genet 118: 1617–1631.
12. Malosetti M, van der Linden CG, Vosman B, van Eeuwijk FA (2007) A mixed-model approach to association mapping using pedigree information with an illustration of resistance to *Phytophthora infestans* in potato. Genetics 175: 879–889.
13. Cavanagh CR, Chao S, Wang S, Huang BE, Stephen S, et al. (2013) Genome-wide comparative diversity uncovers multiple targets of selection for improvement in hexaploid wheat landraces and cultivars. Proc Natl Acad Sci U S A 110: 8057–8062.
14. Liu Y, He Z, Appels R, Xia X (2012) Functional markers in wheat: current status and future prospects. Theor Appl Genet 125: 1–10.
15. Bradbury PJ, Zhang Z, Kroon DE, Casstevens TM, Ramdoss Y, et al. (2007) TASSEL: software for association mapping of complex traits in diverse samples. Bioinformatics 23: 2633–2635.
16. Burton GW, DeVane EH (1953) Estimating heritability in tall fescue (*Festuca Arundinacea*) from replicated clonal material. Agronomy Journal 45: 478–481.
17. Storey JD, Tibshirani R (2003) Statistical significance for genomewide studies. Proc Natl Acad Sci U S A 100: 9440–9445.
18. The International Wheat Genome Sequencing Consortium (2014) A chromosome-based draft sequence of the hexaploid bread wheat (*Triticum aestivum*) genome. Science 345: 1251788.
19. Liu J, Huang S, Sun M, Liu S, Liu Y, et al. (2012) An improved allele-specific PCR primer design method for SNP marker analysis and its application. Plant Methods 8: 34.
20. Morris CF (2002) Puroindolines: the molecular genetic basis of wheat grain hardness. Plant Mol Biol 48: 633–647.
21. Giroux MJ, Morris CF (1997) A glycine to serine change in puroindoline b is associated with wheat grain hardness and low levels of starch-surface friabilin. Theor Appl Genet 95: 857–864.
22. Gautier MF, Aleman ME, Guirao A, Marion D, Joudrier P (1994) *Triticum aestivum* puroindolines, two basic cystine-rich seed proteins: cDNA sequence analysis and developmental gene expression. Plant Mol Biol 25: 43–57.
23. Huang BE, George AW, Forrest KL, Kilian A, Hayden MJ, et al. (2012) A multiparent advanced generation inter-cross population for genetic analysis in wheat. Plant Biotechnol J 10: 826–839.
24. Wang S, Wong D, Forrest K, Allen A, Chao S, et al. (2014) Characterization of polyploid wheat genomic diversity using a high-density 90,000 single nucleotide polymorphism array. Plant Biotechnol J.
25. Stodart BJ, Mackay MC, Raman H (2007) Assessment of molecular diversity in landraces of bread wheat (*Triticum aestivum* L.) held in an *ex situ* collection with Diversity Arrays Technology (DArT™). Aust J Agric Res 58: 1174–1182.
26. White J, Law JR, Mackay I, Chalmers KJ, Smith JS, et al. (2008) The genetic diversity of UK, US and Australian cultivars of *Triticum aestivum* measured by DArT markers and considered by genome. Theor Appl Genet 116: 439–453.
27. Bhave M, Morris CF (2008) Molecular genetics of puroindolines and related genes: allelic diversity in wheat and other grasses. Plant Mol Biol 66: 205–219.
28. Bhave M, Morris CF (2008) Molecular genetics of puroindolines and related genes: regulation of expression, membrane binding properties and applications. Plant Mol Biol 66: 221–231.
29. Somers DJ, Isaac P, Edwards K (2004) A high-density microsatellite consensus map for bread wheat (*Triticum aestivum* L.). Theor Appl Genet 109: 1105–1114.
30. Flint-Garcia SA, Thornsberry JM, Buckler ES (2003) Structure of linkage disequilibrium in plants. Annu Rev Plant Biol 54: 357–374.
31. Maccaferri M, Sanguineti MC, Noli E, Tuberosa R (2005) Population structure and long-range linkage disequilibrium in a durum wheat elite collection. Mol Breed 15: 271–289.
32. Chao SM, Zhang WJ, Dubcovsky J, Sorrells M (2007) Evaluation of genetic diversity and genome-wide linkage disequilibrium among US wheat (*Triticum aestivum* L.) germplasm representing different market classes. Crop Sci 47: 1018–1030.
33. Chao SM, Dubcovsky J, Dvorak J, Luo MC, Baenziger SP, et al. (2010) Population- and genome-specific patterns of linkage disequilibrium and SNP variation in spring and winter wheat (*Triticum aestivum* L.). BMC Genomics 11.
34. Hao C, Wang L, Ge H, Dong Y, Zhang X (2011) Genetic diversity and linkage disequilibrium in Chinese bread wheat (*Triticum aestivum* L.) revealed by SSR markers. PLoS One 6: e17279.
35. Zhang K, Wang J, Zhang L, Rong C, Zhao F, et al. (2013) Association analysis of genomic loci important for grain weight control in elite common wheat varieties cultivated with variable water and fertiliser supply. PLoS One 8: e57853.
36. Akhunov E, Nicolet C, Dvorak J (2009) Single nucleotide polymorphism genotyping in polyploid wheat with the Illumina GoldenGate assay. Theor Appl Genet 119: 507–517.
37. Wilkinson PA, Winfield MO, Barker GL, Allen AM, Burridge A, et al. (2012) CerealsDB 2.0: an integrated resource for plant breeders and scientists. BMC Bioinformatics 13: 219.
38. Allen AM, Barker GL, Wilkinson P, Burridge A, Winfield M, et al. (2013) Discovery and development of exome-based, co-dominant single nucleotide polymorphism markers in hexaploid wheat (*Triticum aestivum* L.). Plant Biotechnol J 11: 279–295.
39. Akhunov ED, Akhunova AR, Anderson OD, Anderson JA, Blake N, et al. (2010) Nucleotide diversity maps reveal variation in diversity among wheat genomes and chromosomes. BMC Genomics 11: 702.
40. Allen AM, Barker GL, Berry ST, Coghill JA, Gwilliam R, et al. (2011) Transcript-specific, single-nucleotide polymorphism discovery and linkage analysis in hexaploid bread wheat (*Triticum aestivum* L.). Plant Biotechnol J 9: 1086–1099.

9

Minerals in the Foods Eaten by Mountain Gorillas (*Gorilla beringei*)

Emma C. Cancelliere[1,2], Nicole DeAngelis[3], John Bosco Nkurunungi[4], David Raubenheimer[5], Jessica M. Rothman[1,2,6]*

1 Department of Anthropology, Graduate Center of the City University of New York, New York, New York, United States of America, **2** New York Consortium in Evolutionary Primatology, New York, New York, United States of America, **3** Department of Animal Science, Cornell University, Ithaca, New York, United States of America, **4** Department of Biology, Mbarara University of Science and Technology, Mbarara, Uganda, **5** Charles Perkins Centre, Faculty of Veterinary Sciences, School of Biological Sciences, The University of Sydney, Sydney, NSW, Australia, **6** Department of Anthropology, Hunter College of the City University of New York, New York City, New York, United States of America

Abstract

Minerals are critical to an individual's health and fitness, and yet little is known about mineral nutrition and requirements in free-ranging primates. We estimated the mineral content of foods consumed by mountain gorillas (*Gorilla beringei beringei*) in the Bwindi Impenetrable National Park, Uganda. Mountain gorillas acquire the majority of their minerals from herbaceous leaves, which constitute the bulk of their diet. However, less commonly eaten foods were sometimes found to be higher in specific minerals, suggesting their potential importance. A principal component analysis demonstrated little correlation among minerals in food items, which further suggests that mountain gorillas might increase dietary diversity to obtain a full complement of minerals in their diet. Future work is needed to examine the bioavailability of minerals to mountain gorillas in order to better understand their intake in relation to estimated needs and the consequences of suboptimal mineral balance in gorilla foods.

Editor: Andrea B. Taylor, Duke University School of Medicine, United States of America

Funding: Cornell University and Hunter College funded this research within larger research projects (no specific funding was received for this work). The funders had no role in study design, data collection and analysis, decision to publish, or preparation of the manuscript.

Competing Interests: The authors have declared that no competing interests exist.

* Email: jessica.rothman@hunter.cuny.edu

Introduction

Minerals play a vital role in the growth and maintenance of animal tissues [1], including their involvement in maintaining structural components (e.g. magnesium [Mg], manganese [Mn], and phosphorus [P]), mediating enzymatic reactions (e.g. calcium [Ca], potassium [K], Mg, and zinc [Zn]), and maintaining acid-base balance (e.g. Ca) in the body [2,3]. Mineral deficiency has both short- and long-term health costs, including compromised neuromuscular, gastrointestinal, cardiovascular, cognitive, or immune functioning [2]. This compromised functioning can impact fitness, with detrimental effects on fertility, growth, and mortality [3]. For example, short-term deficiencies in Ca can affect muscle function, nerve transmission, and blood clotting [4]. Prolonged Ca deficiency can cause chronic conditions including rickets and osteomalacia/osteoporosis in humans [5] as well as retard growth and cause abnormalities to bone and teeth [6]. Despite its importance, our understanding of the mineral intake and requirements of wild primates is limited [2,7]. Few studies have investigated the dietary minerals of primates [7,8,9,10,11,12,13,14], and particularly of apes [15,16,17,18,19].

The environmental availability of minerals in primate habitats has been suggested as a potential limiting factor to population growth in redtail monkeys (*Cercopithecus ascanius*) [20], and movement patterns of black and white colobus (*Colobus guereza*)

are dictated to some extent by the sodium found in *Eucalyptus* trees [21]. Gorillas (*Gorilla gorilla*) in swampy areas select foods that are rich in Ca and Na [15]. Mineral acquisition strategies also vary based on ability to utilize consumed minerals, and the bioavailability of minerals within food items; minerals first have to be found in suitable foods, and then must be available for digestion and absorption. Given the variety of dietary types and digestive systems within the primate order, taxa may differ greatly in their mineral requirements and strategies to acquire mineral nutrients. For example, howler monkeys rely heavily on figs in their diets, a fruit that is high in Ca [9], and thus may not seek to otherwise supplement their diet with Ca, and colobines that host foregut microbes may have a lower need for certain minerals (similar to ruminants) [22,23]. When staple foods do not provide sufficient minerals, primates can meet their mineral needs by supplementing their typical diet. Several distinctive and unusual feeding behaviors have been suggested to serve a mineral acquiring function, including geophagy (e.g. in chimpanzees [24], *Pithecia* [25], and *Macaca* [26]), consumption of wood (e.g. *Ateles* [8], mountain gorillas [17], chimpanzees [18]), insectivory [27], or consumption of liquids like urine and swampy waters (e.g. *Procolobus* monkeys [7,28]).

Bwindi mountain gorillas (*Gorilla beringei beringei*) live in montane forests characterized by high-protein herbaceous plants, with seasonal availability of fruit [29,30], while the mountain gorillas in the neighboring Virunga Volcanoes are almost exclusively folivorous as fruiting trees are not available in their high altitude habitat [31,32]. In comparison, western lowland gorillas (*Gorilla gorilla gorilla*) are more frugivorous than the Bwindi population, consuming fruit almost daily [32]. Their respective food availability may have implications for the mineral compositions of their diets.

The diet of the Bwindi gorilla has been previously described [30,33,34,35] with some reference to mineral nutrition. Their diet is relatively diverse, comprising 148 food items from 107 species of plant [30]. Nevertheless, over 90% of the Bwindi mountain gorilla diet consists of 17 staple foods, while the remaining food items each contribute less than 1% to the total diet [30]. Some of the less commonly eaten foods may be important sources of minerals; in particular decaying wood has been shown to provide the majority of sodium in Bwindi gorilla diets [17]. A study in Bwindi gorillas addressing nutrition across age/sex classes found that mineral intake varied [35]. However, in all age/sex classes, mineral intake was consistent with or exceeded adequate daily concentrations recommended for Hominidae by the National Research Council (NRC) [2], with exceptions being Zn, Na, and P [35].

We characterized the mineral compositions of mountain gorilla (*G. b. beringei*) food items to better understand potential mineral acquisition strategies. We predicted that mountain gorillas would gain most of their minerals from the herbaceous leaves in their diet, and fruits would be a relatively poor source of minerals compared to leaves [11].

We also examined the ratios of minerals in food items known to have interactive effects in the diets of mammals [3]. Interactions between co-occurring minerals can profoundly impact their bioavailability, such that excesses or deficits in one mineral can inhibit the absorption of another [36]. For example, when P is excessively high in relation to Ca, the body will stop absorbing Ca and the mineral may be actively removed from the blood plasma [37]. If minerals are not properly balanced (i.e., consumed in specific proportion relative to other minerals, in order to be used optimally by biological tissues [3]), mineral deficiency may occur at a cellular or tissue level, despite the consumption of a sufficient amount of each mineral in isolation.

Finally, we compared the minerals in mountain gorilla foods to minerals in the diets eaten by the more frugivorous western lowland gorilla (*Gorilla gorilla*) [15,16], and the minerals in of leaves eaten by a diversity of primates.

Methods

Study site and animals

Bwindi Impenetrable National Park (BINP) is located between 0°53′ and 1°08′S, 29°35′ and 29°50′E in southwestern Uganda, and our research was conducted at the Institute of Tropical Forest Conservation in Ruhija sector. The landscape is characterized by rugged mountainous rainforest, with steep hills and narrow valleys. BINP contains one group of mountain gorillas specifically habituated for research, the Kyagurilo group [35]. Details of the Kyagurilo group are outlined in previous publications [33,34,35].

Researchers are permitted to carry out observations for a maximum of four hours a day, in order to minimize both disturbance and disease risk to the gorillas. Typically, these four hours occurred between 0830 to 1500 hours for this study, but they varied throughout the day.

Plant collection and nutritional analysis

As outlined previously [33,34,35], food items consumed by the gorillas during observation were collected within the same week they were consumed. When possible, samples were taken from the exact plant consumed, or from directly adjacent plants of the same species. Food items were processed in a manner similar to how the gorillas processed the food (i.e. if only certain parts of the plant were eaten, only those plant parts were processed for analysis). For mineral analysis, 103 plants were analyzed and one rock seen to be ingested by the gorillas.

Plants were weighed immediately after collection using a portable balance and then the samples were dried at ≤22°C at the field station until a constant weight was achieved. Dried samples were ground at Makerere University in Uganda, using a Wiley Mill with a 1-mm screen. Mineral content (sodium [Na], Ca, P, Mg, K, iron [Fe], Zn, copper [Cu], and Mn) was determined using a Thermo Jarrell Ash IRIS Advantage Inductively Coupled Plasma Radial Spectrometer at Dairy One Forage Laboratory, Ithaca, New York, USA. We present mineral content on a dry matter basis.

Statistical analysis

Samples were grouped into one of six plant part categories: bark, fruit, herbaceous leaves, tree leaves, pith/stem (including both pith, the outer green peel on herbs, and stem material), and root. Bark was defined as the outer bark of trees and twigs (woody material). Mineral compositions across plant parts were compared using nonparametric Kruskal-Wallis tests, with multiple comparisons conducted based on Dwass, Steel, & Critchlow-Fliger pairwise rankings, with an *a priori* alpha level of 0.05 [38]. These analyses were conducted using StatsDirect. We also conducted a principal component analysis (PCA) in R to assess potential underlying trends in plant part categories and mineral content of individual samples [39].

The ratios of minerals in foods were calculated for select mineral pairs (Ca:P, Ca:Na, Ca:K, Ca:Mg, Na:Mg, Zn:Cu, and Fe:Cu) as per the NRC's nutritional guidelines for non-human primates [2], to enable comparisons to recommended ratios. Ratios for food items were calculated by weighting mineral contents for the most commonly consumed gorilla food items (accounting for 80% of the total diet) [40] by the percent intake of each food item [40]. Mineral ratios then were presented as averages based on total dietary intake [40]. The mineral content of decaying wood was previously reported [17], and was therefore not included in summary figures.

The NRC non-human primate guidelines for Hominidae, based on recommended human values (Table 11-1 of NRC [2]), were used as a standard for comparison to observed mineral intake in mountain gorillas [2].

Results

Mineral content of food items

Plant parts differed in concentrations of Ca ($H = 36.56$, $P < 0.001$), P ($H = 14.99$, $P = 0.01$), Mg ($H = 31.85$, $P < 0.001$), K ($H = 27.15$, $P < 0.001$), Na ($H = 15.26$, $P = 0.01$), Zn ($H = 12.24$, $P = 0.03$), Fe ($H = 29.15$, $P < 0.001$), and Mn ($H = 23.64$, $P < 0.001$), but not in concentrations of Cu (*Figure 1*). Roots were higher than all other plant parts in Fe (*Table 1*), and a single ingested rock sample analyzed from the site was also very high in Fe (2,520 PPM). Pith/stem was higher than bark, fruit, herbaceous leaves, and tree leaves in K, and had the highest mean concentrations for P, Zn, and Cu. Herbaceous leaves had the highest mean concentrations of Ca, Mg, and Mn, and there were differences between herbaceous leaves and fruit for Ca, herbaceous leaves and fruit for Mn, and herbaceous leaves, bark, and fruit in Mg (*Table 1*).

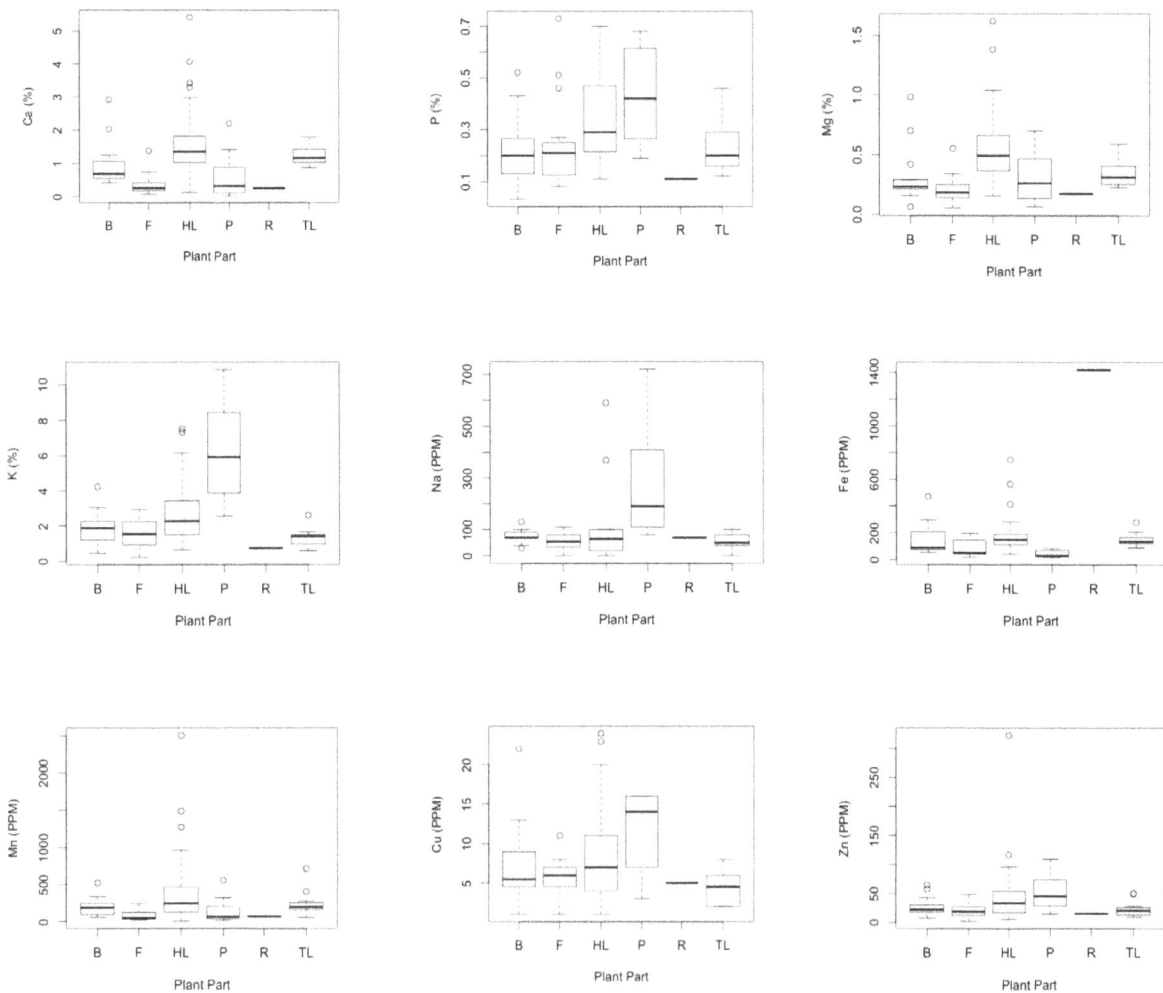

Figure 1. Comparisons of mean mineral composition in gorilla food items at Bwindi Impenetrable National Park.

The mineral ratios of gorilla food items analyzed rarely met ratios suggested by the NRC [2] (*Table 2*). Na:K ratios, Na:Mg ratios, and Fe:Cu ratios were consistently outside of the recommended range.

Mineral diversity and associations between minerals

The PCA showed similar patterns of association between minerals, with Ca, Mg, and Zn grouping most closely together according to the first and second principal components (*Figure 2*). Food part categories did not cluster in multivariate space, with the first and second principal components together explaining only 50.8% of the variation in mineral quantities between plant parts (34.8% and 16.1% respectively) (*Table 3*). A plot of the principal components indicates that subsequent components each explain a fairly even, consistent proportion of variation in data. Thus, no single underlying association seems to have greatly influenced mineral presence or concentrations in plant parts or samples; rather, individual foods and groups of food items are highly variable in their mineral profiles and scatter relatively randomly in multivariate space.

Comparison to mineral nutrition in western lowland gorilla foods

Mineral compositions of Bwindi gorilla food items are generally similar to those in western lowland gorilla food items (*Table 4*).

However, Cu in western lowland gorilla fruits, leaves, and shoots was higher than in comparable foods eaten by Bwindi gorillas while leaves at Bwindi were lower in Na, and higher in P and Mg.

Discussion

Mineral Composition

Our study suggests that food items consumed by Bwindi mountain gorillas differ substantially in their mineral profiles both between and within plant part categories. Roots were higher in Fe compared to all other plant parts, pith/stem was higher in K compared to bark, fruit, herbaceous leaves, and tree leaves, and herbaceous leaves had the highest mean concentrations of Ca, Mg, and Mn compared to all other plant/plant parts tested.

Conversely, certain food items were found to be very low in their mineral concentrations when compared to other food items. The fruits analyzed in this study were low in their mineral content. The low mineral quality of fruits is well documented [11] (an exception being figs, which act as an important source of Ca for many primate species [9,41]). The mineral content of mountain gorilla food items was similar to the foods consumed by western lowland gorillas in Cameroon ([16]; *Table 4*). Herbaceous leaves, an important food item for mountain gorillas [30], were equal or higher in Ca, K, Mn, and P than tree and herbaceous leaves

Table 1. Mean mineral concentrations in food items eaten by mountain gorillas (Gorilla beringei) in Bwindi Impenetrable National Park.

Part	N	Ca (%)	P (%)	Mg (%)	K (%)	Na (PPM)	Fe (PPM)	Zn (PPM)	Cu (PPM)	Mn (PPM)
Bark	13	**0.96a**	**0.22a**	**0.31ab**	**1.89ab**	**75a**	**149abc**	**26a**	**7a**	**194a**
SD		0.68	0.14	0.23	0.95	25	115	16	5	122
Range		0.41–2.92	0.03–0.52	0.06–0.98	0.47–4.23	30–130	52–472	7–64	1–22	58–522
Fruit	16	**0.36b**	**0.25a**	**0.21a**	**1.59ab**	**55a**	**81ac**	**20a**	**5a**	**79b**
SD		0.33	0.18	0.12	0.79	29	58	11	2	63
Range		0.07–1.38	0.08–0.73	0.05–0.55	0.25–2.94	20–110	19–195	2–48	1–11	22–240
Herb leaves	27	**1.62c**	**0.34a**	**0.56c**	**2.71a**	**81a**	**181b**	**44a**	**8a**	**411a**
SD		1.08	0.17	0.31	1.72	106	136	54	6	504
Range		0.12–5.41	0.11–0.7	0.15–1.62	0.65–7.51	20–590	39–748	5–323	1–24	7–2511
Tree leaves	20	**1.25ac**	**0.24a**	**0.35bc**	**1.35b**	**56a**	**152ab**	**23a**	**4a**	**246a**
SD		0.29	0.11	0.12	0.53	32	52	14	2	181
Range		0.87–1.8	0.12–0.46	0.25–1.67	0.6–2.6	40–100	90–280	9–50	2–8	55–718
Pith	7	**0.65abc**	**0.44a**	**0.32abc**	**6.28c**	**290b**	**45c**	**53a**	**11a**	**159ab**
SD		0.83	0.20	0.23	3.25	236	29	36	5	205
Range		0.02–2.2	0.19–0.68	0.06–0.48	2.56–10.85	80–720	19–86	14–109	3–16	17–560
Root	1	**0.25**	**0.17**	**0.11**	**0.75**	**1420**	**15**	**5**	**70**	**70**
SD		-	-	-	-	-	-	-	-	-
Range		-	-	-	-	-	-	-	-	-
Silverback male daily mineral intake (mg per unit M) [35]		392	67	116	612	0.05	2.42	0.64	0.18	8.30
Female daily mineral intake (mg per unit M) [35]		733	131	225	1013	0.06	4.34	1.34	0.33	16.2
Juvenile daily mineral intake (mg per unit M) [35]		931	192	292	1597	0.08	7.52	2.09	0.50	21.5

Differences in mineral concentrations between plant parts (P<0.005) are indicated by letter differences (per column). Shared letters indicate no significant differences in mineral concentration (per column).[1]
[1]Food items include bark, fruit, herbaceous leaves, tree leaves, pith/stem (including both pith, the outer green peel on herbs, and stem material), and root. Bark was defined as the outer bark of trees and twigs (woody material). Ca = calcium, P = phosphorus, Mg = magnesium, K = potassium, Na = sodium, Fe = iron, Zn = zinc, Cu = copper, Mn = manganese. PPM = Parts per million, % = Percentage on a dry matter basis. All pairwise comparisons are based on Dwass, Steel, & Critchlow-Fliger pairwise rankings.

Table 2. Mean ratios of minerals in staple foods eaten by Bwindi Mountain gorillas, weighted by daily intake (measured in g) [40].

	% Daily Intake [40]	Ca:P	Ca:Na	Ca:K	Ca:Mg	Na:Mg	Zn:Cu	Fe:Cu
Herbaceous leaves	61%	4.36	703.46	0.66	2.14	<0.001	3.24	9.81
Fruit	13%	0.16	2.40	0.02	0.19	<0.001	0.21	1.07
Pith	6%	0.19	1.37	0.01	0.14	0.01	0.23	0.14
Ratio for all staple foods [40]		2.81	427.01	0.41	1.43	0.01	2.17	6.19
Recommended ratio [2,61]	–	1.57	0.67	0.21	2.97	10.50	8.89	1.57

Staple foods considered accounted for 80% of total diet [40].

consumed by other free-ranging primates at sites across Africa, Asia, and the Americas (*Table 5*).

Dietary diversity, food selection, and mineral composition

To obtain sufficient quantities of all minerals, it is important for gorillas to consume a wide range of different food items. As suggested by Milton [42], this strategy of selecting a diversity of food items, each item high in particular minerals, may allow primates to achieve optimal micronutrient nutrition in habitats that are typically mineral-poor [7,8,9,10]. The relationship between dietary diversity and likelihood of obtaining an adequate complement of nutrients has been observed across animals in general [43,43,44,45]. In humans, increases in dietary diversity can contribute to longer life expectancy and lower infant mortality [46,47], and dietary diversity is often used as an indicator of nutritional adequacy [48].

The PCA results support the idea that high dietary diversity allows for the acquisition of a full complement of minerals. The first two principal components were driven by weak or moderate associations between minerals [39], indicating that minerals do not associate strongly along an underlying gradient or set of parameters. The overlap in plant parts and the high variation within groups together indicate that plant part does not indicate any generality to the mineral composition of a food item.

Much debate exists in the current literature as to the importance of minerals in driving food selection [7,9,10,15,17,18,28,42,49]. The consumption of certain foods that are low in macronutrients, like wood and roots, is likely explainable by their mineral composition. Wood consumption in mountain gorillas has been previously related to its high Na content [17] and gorillas select stumps that are high in sodium, a behavior observed in other primates as well [8,18]. Nevertheless, it remains unclear as to whether mountain gorillas are selecting specifically for mineral content in their food. Future studies should investigate mineral temporal and spatial availability in relation to consumption.

Although pith might be selected for water content, or its high level of easily digestible sugars or hemicellulose [50], it is possible that pith consumption is at least in part driven by its high K composition. The piths consumed by Bwindi mountain gorillas contain large percentages of water (up to 96% water content) [33], high levels of fiber, and low levels of crude protein [34]. Although K deficiencies are rare due to its abundance in plants [2], selection for K has been noted in folivorous mammals. For example, the folivorous Brazilian rodent *Kerodon rupestris* has been shown to select low quality foods (low macronutrient content) in order to meet daily minimum K requirements, even during periods of food resource limitation [51]. However, given that mountain gorillas consume a much higher level of K than minimally required [35], alternative explanations are likely required to account for this behavior in mountain gorillas.

Mineral Ratios

The ratios of minerals found in individual gorilla food items rarely met acceptable targets as per the NRC's guidelines for nonhuman primates [2]. Minerals consumed in excess or deficit relative to the proportion of other minerals may have compromised bioavailability, as a result of mineral interactions within the body [3,36]. Such unbalanced mineral ratios can have adverse health effects. The relationships among Ca, P, and Mg in particular have strong implications for health; for example, captive primates fed diets unbalanced in these minerals can develop a series of skeletal deformities throughout their lifespan [37]. Within this group of minerals, the proportion of P in relation to Ca is

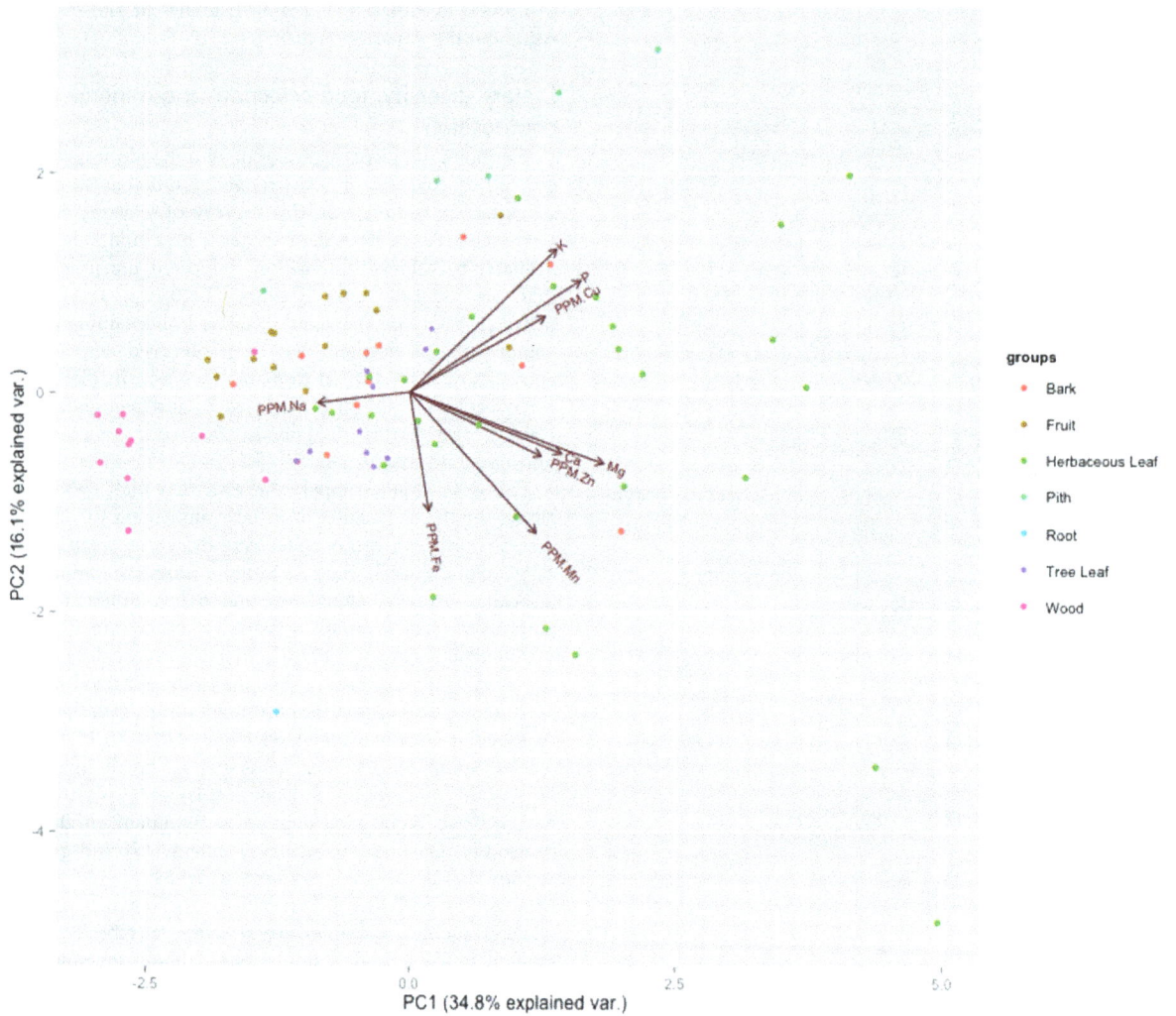

Figure 2. Biplot showing the first two loadings of the principal component analysis of all mineral values in samples analyzed. Food items grouped by color.

especially tightly interwoven [52]. Calcium:Phosphorus ratios in gorilla food items were generally higher than the NRC-recommended ratio, even when considered in the context of dietary intake [35,40].

Unbalanced ratios occur in other dietary minerals, as well. When considered within the context of dietary intake per unit body mass, the Ca:Mg ratio for silverback males is lower than the ratio recommended for good health in primates, whereas the Ca:K

Table 3. Loadings of the first two components of a principal component analysis of associations between minerals.

	Component 1	Component 2
Ca	0.37	−0.21
P	0.41	0.39
Mg	0.47	−0.24
K	0.35	0.49
Na	−0.22	−0.04
Fe	0.05	−0.41
Zn	0.32	−0.22
Cu	0.33	0.26
Mn	0.31	−0.48

Table 4. Comparison of mineral composition of food items between *Gorilla beringei* and *Gorilla gorilla* [16][1].

Part	n	n	Ca (%) G. beringei	Ca (%) G. gorilla	P (%) G. beringei	P (%) G. gorilla	Mg (%) G. beringei	Mg (%) G. gorilla
Leaf	47	8	1.53±0.96	0.18±0.07	*0.32±0.16 (p=0.022)	0.18±0.07	*0.51±0.29 (p=0.038)	0.29±0.153
Bark	13	2	1.02±0.72	0.07±0.02	0.22±0.14	0.07±0.02	0.34±0.23	0.12±0.113
Fruit	16	7	0.36±0.33	0.18±0.11	0.25±0.17	0.18±0.11	0.21±0.12	0.16±0.069
Shoot	2	5	0.07±0.07	0.22±0.08	0.42±0.29	0.22±0.08	0.16±0.14	0.26±0.089
Root	1	1	0.25	0.11	0.11	0.07	0.17	0.04

Part	n	n	K (%) G. beringei	K (%) G. gorilla	Fe (PPM) G. beringei	Fe (PPM) G. gorilla	Zn (PPM) G. beringei	Zn (PPM) G. gorilla
Leaf	47	8	2.39±1.62	1.68±0.770	175.02±122.31	274±217.881	39.87±48.46	34.88±26.292
Bark	13	2	2.01±0.96	1.65±0.919	162.23±119.29	105.5±77.074	27.92±16.81	10.5±3.535
Fruit	16	7	1.59±0.79	1.81±1.009	81.38±58.63	206±404.691	20.13±11.49	52.75±101.66
Shoot	2	5	3.34±1.09	3.02±1.207	43.5±20.50	93.40±16.890	38±16.97	48±16.093
Root	1	1	0.75	1	1420	37	15	14

Part	n	n	Cu (PPM) G. beringei	Cu (PPM) G. gorilla	Na (PPM) G. beringei	Na (PPM) G. gorilla	Mn (PPM) G. beringei	Mn (PPM) G. gorilla
Leaf	47	8	7.74±5.97	*14.13±6.010 (p=0.012)	75.75±95.04	*178.75±194.546 (p=0.021)	372.6±453.32	284±136.32
Bark	13	2	8.2±5.69	6±0	76.15±26.94	85±49.497	208.39±125.41	155±49.497
Fruit	16	7	5.53±2.47	*13.43±9.288 (p=0.008)	55±29.44	*119.38±160.14 (p=0.013)	79.75±63.29	140±158.19
Shoot	2	5	5±2.82	14.00±5.612	85±7.071	155±133.32	47±42.43	552±371.44
Root	1	1	5	4	70	70	70	0

[1]*, denotes significantly higher values ($P<0.05$) as determined by Mann-Whitney U tests of significance. Samples with an $n<2$ were excluded from analysis.

Table 5. Mineral content of leaves consumed by gorillas at BINP compared to leaves consumed by wild primates at other research sites.

Species	Location	n	% Ca	% P	% Mg	% K	% Na	PPM Fe	PPM Zn	PPM Cu	PPM Mn
Gorilla beringei (this study)	BINP	47	1.53	0.32	0.513	2.387	0.007	175.021	39.87	7.743	372.6
Nasalis larvatus [14]	Indonesia	17	0.634	0.069	0.312	0.726	0.003	35.2	13.9	–	122
Alouatta nigra [13]	Belize	33	1.305	0.27	0.52	1.83	0.032	95.56	30.19	15.025	–
Gorilla gorilla [15]	Congo	–	2.63	0.77	0.58	0.307	0.218	–	–	–	–
Macaca sylvanus, Lemur catta [10]	Georgia, USA	37	0.73	0.15	0.32	–	0.04	84.2	24.1	5.6	114.4
Gorilla gorilla [16]	Cameroon	8	1.341	0.181	0.286	1.675	0.018	274	34.88	14.13	284
Pan troglodytes [18]	Budongo, Uganda	4	0.597	0.132	0.268	1.624	–	84.25	103.5	–	93.25
Colobus guereza, Procolobus tephrosceles [7]	Kibale, Uganda	106	1.02	–	0.28	1.64	–	146.45	25.85	9.95	139.6

ratio of total dietary intake lies above the recommended ratio [35,40]. In dairy cattle it has been suggested that high levels of K interfere with Ca absorption and lead to higher incidence of milk fever (a hypocalcemic state that leads to appetite loss, weakness, and heart failure), though this interaction may be unique to foregut fermenters [53]. Both mountain gorilla foods and their diets overall have a higher Ca:K ratio than recommended, but the direct implications of this ratio are unknown. It should also be noted that published mineral requirements and ratios might be conservative because primates vary in body size, physiology and digestion, and foods vary in bioavailability [54]. While we compared gorilla foods to recommended Hominidae requirements, the recommendations for adequate concentrations of minerals for nonhuman primates may be higher than what primates actually require. For example, the NRC suggests that primates need to consume a diet of 0.25% Na on a dry matter basis, but the diets of most wild primates are much lower [7,35], indicating that primates are able to survive on lower dietary concentrations of Na.

Given that the mineral ratios do not always meet recommended target ratios when considered within the context of overall diet, the bioavailability of minerals in gorilla foods becomes important in understanding the implication of mineral ratios in mountain gorilla diets. The ratio of Ca to P ratio in gorilla feces is 4.04 (Rothman, *unpublished data*), higher than the ratio averaged across all staple foods suggested by this study (2.81) but lower than that of the major dietary component, herbaceous leaves (4.36). The Ca:P ratio in feces, however, is lower than that observed in the daily diets of gorilla females (5.59), silverback males (5.85), and juveniles (4.85) [35], which is considerably higher than the recommended ideal ratio for humans (1.57, *Table 2*). This suggests that the high levels of Ca in the diet might inhibit the absorption of P, which occurs when Ca is consumed in excessive amounts [33,52].

In addition to mineral interactions, the bioavailability of minerals can also be affected by plant physiology. Roots, for example, carry high percentages of the minerals abundant in the surrounding soil [55], and the single rock sample ingested by an individual at the site was found to be exceedingly high in Fe (2,520 PPM). The Fe in plant tissues tends to be predominantly unavailable for digestion, as it is usually bound to organic compounds in the plant structure that may past through animal

digestive tracts [56]. While little is known about the use and uptake of Fe in roots [56], the availability of Fe has been tested in legumes, where levels of Fe-binding polyphenols and the presence of phytate (an inhibitor of Fe absorption) render most Fe unusable to animals, despite the high overall content of Fe found in these plant structures [57]. Understanding patterns of bioavailability in Fe is especially crucial in primates, as captive primates have been shown to be highly vulnerable to hemosiderosis (iron overload) as a result of overconsumption of Fe [11,58].

Future Directions

Moving forward, it is critical to better understand the biological availability of minerals to gorillas, and how the mineral composition of plants relates to dietary selection and mineral nutrient acquisition. Information on bioavailability in primates is scarce [41,59], but non-invasive methods to estimate bioavailability are available [60]. Employing these methods would allow researchers to make more accurate statements pertaining to mineral ratios, potential mineral targets, and the importance of minerals as a deciding factor in dietary choices. Lastly, understanding mineral composition within the context of dietary contribution would allow us to further explore the hypothesis that increasing dietary diversity and supplementation with low-macronutrient, high-mineral foods optimizes mineral intakes in mountain gorillas.

Acknowledgments

We are grateful for helpful discussions with Ellen Dierenfeld, Alice Pell, Peter Van Soest and Harold Hintz. We are very appreciative of the comments of Andrea Taylor (Academic Editor), Erin Vogel, and one anonymous reviewer. Thank you to the assistants at the Institute of Tropical Forest Conservation, John Makombo, Alastair McNeilage, Robert Bitariho, Dennis Babaasa and Aventino Kasangaki for logistical support. Permission to conduct research in Bwindi Impenetrable National Park was granted by the Uganda Wildlife Authority and the Uganda National Council of Science and Technology.

Author Contributions

Conceived and designed the experiments: ECC JMR. Performed the experiments: JBN ND. Analyzed the data: ECC ND. Contributed reagents/materials/analysis tools: JMR. Contributed to the writing of the manuscript: ECC JMR ND DR.

References

1. Barboza P, Parker K, Hume I (2008) Integrative Wildlife Nutrition. Springer, USA.

2. National Research Council (2003) Nutrient Requirements of Nonhuman Primates, 2nd edn. The National Academies Press, Washington, DC.

3. Robbins C (1993) Wildlife Feeding and Nutrition. Academic Press, San Diego, California.

4. Weaver CM, Heaney RP (1999) Calcium. In: Modern Nutrition in Health and Disease, 9th edition. Shils ME, Olson JA, Shike M, and Ross CA, editors. Philadelphia: Lippincott Williams & Wilkins. pp. 141–155.

5. Power ML, Heaney RP, Kalkwarf HJ, Pitkin RM, Repke JT, et al. (1999) The role of calcium in health and disease. Am J Obstet Gynecol 181: 1560–9.

6. Underwood EJ, Suttle NF (1999) The Mineral Nutrition of Livestock, 3rd edition. CABI Publishing, New York.

7. Rode KD, Chapman CA, Chapman LJ, Mcdowell LR (2003) Mineral Resource Availability and Consumption by Colobus in Kibale National Park, Uganda. Int J Primatol 24: 541–573.

8. Chaves OM, Stoner KE, Angeles-Campos S, Arroyo-Rodríguez V (2011) Wood consumption by Geoffroyi's spider monkeys and its role in mineral supplementation. PloS One, 6: e25070.

9. Behie AM, Pavelka MSM (2012) The role of minerals in food selection in a black howler monkey (Alouatta pigra) population in Belize following a major hurricane. Am J Primatol 74: 1054–63.

10. Blake JG, Guerra J, Mosquera D, Torres R, Loiselle B, et al. (2010) Use of mineral licks by white-bellied spider monkeys (Ateles belzebuth) and red howler monkeys (Alouatta seniculus) in Eastern Ecuador. Int J Primatol, 31: 471–483.

11. Dierenfeld ES, McCann CM (1999) Nutrient composition of selected plant species consumed by semi free-ranging lion-tailed macaques (Macaca silenus) and ring-tailed lemurs (Lemur catta) on St. Catherines Island, Georgia, U.S.A. Zoo Biol, 18: 481–494.

12. Isbell LA, Rothman JM, Young PJ, Rudolph K (2013) Nutritional benefits of Crematogaster mimosae ants and Acacia drepanolobium gum for patas monkeys and vervets in Laikipia, Kenya. Am J Phys Anthropol, 150: 286–300.

13. Silver SC, Ostro LET, Yeager CP, Dierenfeld ES (2000) Phytochemical and mineral components of foods consumed by black howler monkeys (Alouatta nigra) at two sites in Belize. Zoo Biol, 19: 95–109.

14. Yeager CP, Silver SC, Dierenfeld ES (1997) Mineral and phytochemical influences on foliage selection by the proboscis monkey (Nasalis larvatus). Am J Primatol, 41: 117–28.

15. Magliocca F, Gautier-Hion A (2002) Mineral content as a basis for food selection by western lowland gorillas in a forest clearing. Am J Primatol, 57: 67–77.

16. Calvert JJ (1985) Food selection by western gorillas (Gorilla gorilla) in relation to food chemistry. Oecologia, 65: 236–246.

17. Rothman JM, Van Soest PJ, Pell AN (2006a) Decaying wood is a sodium source for mountain gorillas. Biol Letters, 2: 321–4.

18. Reynolds V, Lloyd AW, Babweteera F, English CJ (2009) Decaying Raphia farinifera palm trees provide a source of sodium for wild chimpanzees in the Budongo Forest, Uganda. PloS One, 4: e6194.

19. O'Malley RC, Power ML (2014) The energetic and nutritional yields from insectivory for Kasekela chimpanzees. J Hum Evol, 71: 46–58.

20. Rode KD, Chapman CA, McDowell LR, Stickler C (2006) Nutritional Correlates of Population Density Across Habitats and Logging Intensities in Redtail Monkeys (*Cercopithecus ascanius*). Biotropica, 38: 625–634.

21. Harris TR, Chapman CA (2007) Variation in diet and ranging of black and white colobus monkeys in Kibale National Park, Uganda. Primates, 48: 208–221.

22. van Soest P (1994) Nutritional Ecology of the Ruminant. Cornell University Press, Ithica, NY.

23. Kay R, Davies A (1994) Digestive physiology. In: Colobine Monkeys: Their ecology, behavior, and evolution. Davies A & Oates J, editors. Cambridge University Press, Cambridge, UK. pp. 229–259.

24. Mahaney WC, Zippin J, Milner MW, Sanmugadas K, Hancock RGV, et al. (1999) Chemistry, mineralogy and microbiology of termite mound soil eaten by the chimpanzees of the Mahale Mountains, Western Tanzania. J Trop Ecol, 15: 565–588.

25. Setz EZF, Enzweiler J, Solferini VN, Amendola MP, Berton RS (1999) Geophagy in the golden-faced saki monkey (*Pithecia pithecia chrysocephala*) in the Central Amazon. J Zool, 247: 91–103.

26. Mahaney W, Hancock R, Inoue M (1993) Geochemistry and clay mineralogy of soils eaten by Japanese macaques. Primates, 34: 85–91.

27. Rothman JM, Raubenheimer D, Bryer MAH, Takahashi M, Gilbert CC (2014) Nutritional contributions of insects primate diets: Implications for primate evolution. J Hum Evol, 71: 59–69.

28. Oates JF (1978) Water–plant and soil consumption by guereza monkeys (*Colobus guereza*): A relationships with minerals and toxins in the diet. Biotropica, 10: 241–253.

29. Stanford CB, Nkurunungi JB (2003) Behavioral ecology of sympatric chimpanzees and gorillas in Bwindi Impenetrable National Park, Uganda: Diet. Int J Primatol, 24: 901–918.

30. Rothman JM, Nkurunungi JB, Shannon BF, Bryer MAH (2013) High altitude diets: implications for the feeding and nutritional ecology of mountain gorillas. In: High Altitude Primates. Grow N, Gursky-Doyen S, & Krzton A, editors. Springer: USA.

31. Watts DP (1984) Composition and variability of mountain gorilla diets in the Central Virungas. Am J Primatol, 7: 323–356.

32. Doran D, McNeilage A (2001) Subspecific variation in gorilla behavior: the influence of ecological and social factors. In: Mountain Gorillas: Three decades of research at Karisoke. Robbins MM, Sicotte P, & Stewart KJ, editors. Cambridge: Cambridge University Press. 123–149.

33. Rothman JM, Dierenfeld ES, Molina D, Shaw A, Hintz H, et al. (2006) Nutritional chemistry of foods eaten by gorillas in Bwindi Impenetrable National Park, Uganda. Am J Primatol 68: 675–691.

34. Rothma JM, Plumptre AJ, Dierenfeld ES, Pell AN (2007) Nutritional composition of the diet of the gorilla (*Gorilla beringei*): A comparison between two montane habitats. J Trop Ecol, 23: 673–682.

35. Rothman JM, Dierenfeld ES, Hintz HF, Pell AN (2008) Nutritional quality of gorilla diets: consequences of age, sex, and season. Oecologia, 155: 111–122.

36. Mills CF (1985) Dietary interactions involving the trace elements. Annu Rev Nutr, 5: 173–193.

37. Frye FF (1997) The importance of calcium in relation to phosphorus, especially in folivorous reptiles. P Nutr Soc, 56: 1105–1117.

38. Hollander M, Wolfe ND (1999) Nonparametric statistical methods. New York, NY: Wiley.

39. McGarigal K, Cushman S, Stafford SG (2000) Multivariate Statistics for Wildlife and Ecology Research. Springer-Verlag, New York.

40. Reiner W, Petzinger C, Power ML, Hyeroba D, Rothman JM (2014) Fatty acids in mountain gorilla diets: Implications for primate nutrition and health. Am J Primatol, doi:10.1002/ajp.22232.

41. O'Brien TG, Kinnaird MF, Dierenfeld ES, Conklin-Brittain NL, Wrangham RW, et al. (1998) What's so special about figs? Nature, 392: 668–668.

42. Milton K (2003) Micronutrient intakes of wild primates: are humans different? Comp Biochem Phys A, 136: 47–59.

43. Cruz-Rivera E, Hay ME (2000) The effects of diet mixing on consumer fitness: Macroalgae, epiphytes, and animal matter as food for marine amphipods. Oceologia, 123: 252–264.

44. Kleppel GS, Burkart CA (1995) Egg production and the nutritional environment of *Acartia tonsa:* the role of food quality in copepod nutrition. J Mar Sci, 52: 297–304.

45. Provenza FD, Villalba J, Dziba LE, Atwood SB, Banner RE (2003) Linking herbivore experience, varied diets, and plant biochemical diversity. Small Ruminant Res, 49: 257–274.

46. Westoby M (1978) What are the biological bases of varied diets? Am Nat, 112: 627–631.

47. Hockett B, Haws J (2003) Nutritional ecology & diachronic trends in Paleolithic diet and health. Evol Anthropol, 12: 211–216.

48. Hatløy A, Torheim LE, Oshaug A (1998) Food variety – a good indicator of nutritional adequacy of the diet? A study from an urban area in Mali, West Africa. Eur J Clin Nutr, 52: 891–898.

49. Felton AM, Felton A, Lindenmayer DB, Foley WJ (2009) Nutritional goals of wild primates. Funct Ecol, 23: 70–78.

50. Wrangham RW, Conklin N, Chapman CA, Hunt K, Milton K, et al. (1991) The significance of fibrous foods for Kibale forest chimpanzees. Philos T R Roy Soc B, 334: 171–178.

51. Willig M, Lacher T (1991) Food selection of a tropical mammalian folivore in relation to leaf-nutrient content. J Mammal. 73: 314–321.

52. Heaney RP, Nordin B (2002) Calcium effects on phosphorus absorption: implications for the prevention and co-therapy of osteoporosis. J Am Coll Nutr, 21: 239–244.

53. Goff JP, Horst RL (1997) Effects of the addition of potassium or sodium, but not calcium, to prepartum ratios in milk fever in dairy cows. J Dairy Sci, 80: 176–186.

54. Oftedal OT, Allen ME (1996) The feeding and nutrition of omnivores with emphasis on primates. In: Wild Mammals in Captivity, Kleiman DG, Allen ME, Thompson KV, &Lumpkin S, editors. University Chicago Press, Chicago. pp. 148–157.

55. Chapin FS (1980) Mineral nutrition of wild plants. Annu Rev Ecol Syst, 11: 233–260.

56. Frossard E, Bucher M, Machler F, Mozafar A, Hurrell R (2000) Potential for increasing the content and bioavailability of Fe, Zn, and Ca in plants for human nutrition. J Sci Food Agr, 80: 861–879.

57. Sandberg AS (2002) Bioavailability of minerals in legumes. Brit J Nutr, 88: 281–285.

58. Spelman LH, Osborn KG, Aderson KG (1989) Pathogenesis of hemosiderosis in lemurs: role of dietary iron, tannin, and ascorbic acid. Zoo Biol, 8: 239–251.

59. Schmidt DA, Iambana RB, Britt A, Junge RE, Welch CR, et al. (2010) Nutrient composition of plants consumed by black and white ruffed lemurs, Varecia variegata, in the Betampona Natural Reserve, Madagascar. Zoo Biol, 29: 375–96.

60. Ammerman CB (1995) Methods for estimation of mineral bioavailability. In Bioavailability of Nutrients for Animals. Ammerman CB, Baker D, & Lewis A, editors. Elsevier, pp. 83–94.

61. Hellwig JP, Otten JJ, Meyers LD (2006) *Dietary Reference Intakes: The Essential Guide to Nutrient Requirements.* National Academies Press.

10

Households across All Income Quintiles, Especially the Poorest, Increased Animal Source Food Expenditures Substantially during Recent Peruvian Economic Growth

Debbie L. Humphries[1]*, **Jere R. Behrman**[2], **Benjamin T. Crookston**[3], **Kirk A. Dearden**[4], **Whitney Schott**[2], **Mary E. Penny**[5], **on behalf of the Young Lives Determinants and Consequences of Child Growth Project Team**¶

1 Department of Epidemiology of Microbial Disease, Yale School of Public Health, New Haven, Connecticut, United States of America, 2 Department of Economics, University of Pennsylvania, Philadelphia, Pennsylvania, United States of America, 3 Department of Health Science, Brigham Young University, Provo, Utah, United States of America, 4 Department of Global Health, Boston University School of Public Health, Boston, Massachusetts, United States of America, 5 Instituto de Investigación Nutricional, Lima, Peru

Abstract

Background: Relative to plant-based foods, animal source foods (ASFs) are richer in accessible protein, iron, zinc, calcium, vitamin B-12 and other nutrients. Because of their nutritional value, particularly for childhood growth and nutrition, it is important to identify factors influencing ASF consumption, especially for poorer households that generally consume less ASFs.

Objective: To estimate differential responsiveness of ASF consumption to changes in total household expenditures for households with different expenditures in a middle-income country with substantial recent income increases.

Methods: The Peruvian Young Lives household panel (n = 1750) from 2002, 2006 and 2009 was used to characterize patterns of ASF expenditures. Multivariate models with controls for unobserved household fixed effects and common secular trends were used to examine nonlinear relationships between changes in household expenditures and in ASF expenditures.

Results: Households with lower total expenditures dedicated greater percentages of expenditures to food (58.4% vs.17.9% in 2002 and 24.2% vs. 21.5% in 2009 for lowest and highest quintiles respectively) and lower percentages of food expenditures to ASF (22.8% vs. 33.9% in 2002 and 30.3% vs. 37.6% in 2009 for lowest and highest quintiles respectively). Average percentages of overall expenditures spent on food dropped from 47% to 23.2% between 2002 and 2009. Households in the lowest quintiles of expenditures showed greater increases in ASF expenditures relative to total consumption than households in the highest quintiles. Among ASF components, meat and poultry expenditures increased more than proportionately for households in the lowest quintiles, and eggs and fish expenditures increased less than proportionately for all households.

Conclusions: Increases in household expenditures were associated with substantial increases in consumption of ASFs for households, particularly households with lower total expenditures. Increases in ASF expenditures for all but the top quintile of households were proportionately greater than increases in total food expenditures, and proportionately less than overall expenditures.

Editor: Andrea S. Wiley, Indiana University, United States of America

Funding: The study was conducted in compliance with the Principles of Ethical Practice of Public Health. The analysis in this manuscript was based on research funded by the Bill & Melinda Gates Foundation (Global Health Grant OPP1032713), Eunice Shriver Kennedy National Institute of Child Health and Development (Grant R01 HD070993) and Grand Challenges Canada (Grant 0072-03 to the Grantee, The Trustees of the University of Pennsylvania). The funders had no role in study design, data analysis, decision to publish, or preparation of the manuscript.

Competing Interests: The authors have declared that no competing interests exist.

* Email: debbie.humphries@yale.edu

¶ Membership of the Young Lives Determinants and Consequences of Child Growth Project Team is provided in the Acknowledgments.

Introduction

Over the past two decades, developing countries on average and middle-income countries in particular, have experienced substantial economic growth. As a result, there has been a worldwide convergence in per capita income as developing countries have closed somewhat the gap with high-income countries and the number of people living below international poverty lines has dropped by about a billion [1]. In this global context, a better understanding of the effects of economic growth on consumption of foods that are rich in critical nutrients strengthens understanding of whether policies aimed at improving economic growth lead to improvements in nutritional status.

Relative to plant-based foods, animal source foods (ASFs) are richer sources of accessible protein, iron, zinc, calcium, vitamin B-12 and other nutrients [2]. Their high nutrient density and the bioavailability of the minerals make ASFs particularly important for children in resource-limited settings during critical growth periods. Consuming ASFs is associated with better length- and weight-for-age [3–5] and improved cognitive function among children [3,6]. Studies in Peru have identified micronutrients such as zinc [7], and animal source foods [4] as key influences on improved nutritional status of lower income children. Globally, 25 percent of children under 5 years of age are stunted, and 18 percent are affected by iron deficiency anemia [8]. In Peru, 20 percent of children under 5 years of age are stunted, and 32 percent are anemic [9]. Despite their importance, ASFs on average provide less than 10 percent of total energy intake in most of Sub-Saharan Africa and South Asia and less than 20 percent in most of the rest of the world [10].

Household resources, as reflected in household expenditures as well as income, might be expected to have a major influence on the consumption of ASFs if these foods are highly valued. Because of the importance of ASFs in the diets of children, the question of how responsive the consumption of ASFs is to changes in overall household expenditures is critical for poorer households in the face of income changes. In the economics literature there have been studies of demand systems, where demands for various groups of goods and services that households consume have been shown to depend on household resources (particularly income as reflected in overall expenditures), prices, household demographics, and other factors [11–15]. Ernst Engel (1821–1896) was the first to systematically investigate the relationship between goods expenditure and income, and in 1857 proposed "Engel's Law" stating that the poorer a family is, the larger the budget share it spends on food. A corollary of that law is that as income increases, expenditures on nourishment (food) increase by smaller percentages than the percentage increase in income. The demand for such commodities is said to be inelastic (see discussion of "elasticities" below). In addition, as incomes rise "Bennett's Law" states that food expenditures will favor more nutrient-rich foods, such as animal source foods, and food expenditures on starchy staples will decrease [16]. However, as Abdulai and Aubert [17] note in sub-Saharan Africa, evidence on responsiveness to expenditures for individual food and food groups in developing countries is limited. Most of the existing evidence, such as Bhaumik and Nugent's analysis of competing demands for food and livestock feed in Peru [18] and the recent analysis of price influences on demand for animal products in the BRIIC countries (Brazil, Russia, India, Indonesia, China) [19], are based primarily on cross-sectional associations rather than longitudinal analysis. Estimates resulting from cross-sectional analyses may confound expenditure effects with unobserved household characteristics such as preferences for different types of food. To our knowledge, moreover, there is no published evidence based on longitudinal analyses with control for unobserved household fixed factors during periods of fairly rapid overall economic growth in developing countries.

There are many studies consistent with Engel's Law for total food consumption in different time periods and countries [11–15]. But food is not homogenous. At very low incomes households may consume largely basic staples. With more income they may increase the shares of other foods, such as green leafy vegetables, fruits and animal source foods [20,21]. When demand studies disaggregate food into food groups, they have found variations in income elasticities, with some foods (such as basic staples) responding relatively little to income changes (income inelastic) and others responding proportionally more than income changes (income elastic) [17,21].

Our basic methodological question in this study is whether household fixed-effects estimates are preferred to random-effects estimates. If they are, the cross-sectional estimates that predominate in the previous literature may be confounded if, for example, patterns in unobserved preferences for different types of food are associated with household expenditures.

We analyzed: (1) how ASF expenditures change with changes in household resources in a middle-income country (Peru) during a period of substantial economic growth? and (2) to what extent are these changes dependent on whether a household is poor or relatively better off? If increases in household resources lead to substantial increases in ASF expenditures for low-income households, dietary quality and perhaps quantity improves, with the potential for decreasing malnutrition. Peru experienced rapid overall income growth during the period of the study (2002–2009), with a 40 percent increase in Peruvian Gross National Product per capita measured in constant 2005 international purchasing power parity (PPP) terms and a decrease from 24.2 percent to 14.0 percent of the prevalence of poverty (percentage of the population living below $2 per day per capita in PPP terms) [22]. Improvements between 2002 and 2009 are also apparent among Young Lives (YL) households, as 83.7 percent reported increased total expenditures per adult equivalent (AE) in 2009. Peru has also seen a dramatic decrease in stunting rates, from 29.8 percent in 2005 to 18.1 percent in 2011 [23].

The term "elasticity" is used to describe responsiveness of one variable X to another variable Y. The elasticity of X in response to Y is the percentage change in X given a specific percentage change in Y. For example, the elasticity of ASF expenditures with respect to food expenditures is the percentage change in ASF expenditures for a given percentage change in food expenditures. If this elasticity is 0.80, then ASF expenditures increase by 8 percent for a 10 percent increase in food expenditures. If the elasticity is 1.0, the percentage change in ASF equals the percentage change in food expenditures. A food demand with an elasticity less than 1.0 is called "inelastic" while a food demand with an elasticity greater than 1.0 is called "elastic". Consumer demands for basic commodities such as food staples are generally inelastic and they are likely to account for a considerable share of consumption expenditures at low overall expenditure or income levels, but increase less than proportionately as overall expenditure or income increases. However the demand elasticities for some food items, perhaps including foods rich in micronutrients such as ASF, are likely to be higher than for basic staples and for food in general. If household resources are very low, food consumption tends to concentrate on cheap sources of basic macronutrients that are necessary for survival. But with more household resources, people choose to diversify their food consumption because of their implications for health beyond survival and/or because they prefer diversity in their diet. [21,24].

Methods

Ethics statement

The original protocol and each subsequent round of YL data collection was approved by the Ethics committees of the Instituto de Investigación Nutricional and the University of Oxford. Ethical approval for this analysis was obtained from the University of Pennsylvania Institutional Review Board.

Study design and participants

The Peruvian Young Lives Younger Cohort (YL-YC) study is a part of the YL 4-country study following children and their households. The YL study so far consists of three waves of data collection in 2002, 2006, and 2009, hereafter referred to as rounds 1–3, respectively. Households in the Peruvian cohort were enrolled in 20 districts randomly selected from all but the 5 percent of wealthiest districts in Peru [25]. Details of the YL study have been described [26] and can be found at http://www. younglives.org.uk. For each round, trained enumerators visited each household and completed detailed surveys of household expenditures. The prices and availabilities of a standard list of items from shops in each community were collected. We utilized YL-YC data from all Peruvian households that had complete household expenditure data in all three years (n = 1750). Peru was the only YL country with food expenditure data for all three rounds.

Study indicators

Expenditures. Total expenditures, food expenditures, and ASF expenditures were calculated for all three rounds. Each household was asked about total expenditures and categories of expenditures over the previous two weeks. Detailed information was collected on food expenditures, covering 33 different food groups. Animal source food groups included: red meat and processed meat, poultry, fish, milk, yogurt, cheese, and eggs. Three of the eight categories of ASF (milk, yogurt, cheese) were combined into one group (dairy) because of low levels of consumption of yogurt and cheese, and processed meat was combined with red meat. Locality- and goods-specific consumer price indices from Peru's Instituto Nacional de Estadística e Informática (INEI) were used to adjust the 2002 and 2009 household food expenditures to 2006 food prices in order to control for location- and goods-specific price changes. Total expenditures were deflated to 2006 levels. Household adult equivalents (AE) were calculated [27], and all expenditures were expressed per AE.

Statistical analyses

All data were analyzed using SPSS for Windows (version 19.0, 2011, IBM) and Stata (version 12.0, 2011. Stata Corp).

We compared the cross-sectional relations between overall food expenditures, ASF expenditures, and household consumption at different levels of household expenditures for 2002, 2006 and 2009 using ANOVA. Differences between years were tested with a paired t-test or paired Wilcoxon rank sum test, adjusted for multiple comparisons.

We estimated multivariate relations for each specification among overall food expenditures, ASF expenditures, and household resources using two sets of data: 2002 and 2006, and 2006 and 2009. Our preferred estimates control for unobserved household fixed effects, but to test whether those estimates are preferred with our data and to investigate possible associations with other observed controls such as maternal schooling we also undertook household random effects, with and without observed controls.

Model framework

We used total household consumption expenditures to represent long-term household resources. This was done as income tends to fluctuate from year to year, particularly in poor households, but households can use savings and withdrawals from savings to smooth resources so that total household consumption expenditures better represent longer-run household resources than does annual income [28,29]. We represented the relations between ASF expenditure and total household consumption expenditures as expressed in the conceptual framework presented in Figure 1. Each of the two relations pertains to a division between two categories of expenditures:

Relation 1: division of food expenditures between ASF and non-ASF.

Relation 2: division of total expenditures between food and non-food expenditures.

We estimated these two relations as they pertain to the center, solid boxes in Figure 1. Further details and equations are available in Appendix S1. We were interested in questions such as, if food expenditures increase by 10 percent, what happens to ASF expenditures? And, does it matter if the households have relatively low or relatively high overall expenditures? Because of the dichotomous division of food expenditures into ASF and non-ASF food expenditures, if we learned that ASF expenditures increase by, for example, more than 10 percent when total food expenditures increase by 10 percent, then necessarily non-ASF food expenditures must increase by less than 10 percent. Although we focused our attention on the solid arrows between the center boxes in the figure, we also learned about the dotted arrows pointing to the dotted (left) boxes in the figure. Observed controls included other important determinants of ASF and food consumption expenditures, such as maternal schooling [30,31], paternal schooling and community wealth. Unobserved controls included all other household characteristics, such as food preferences. We were interested in these controls for two reasons. First, some controls, such as women's schooling, have previously been reported to impact household food purchases, so we undertook random effects estimates with and without them to see if their exclusion confounds the estimates of interest [30,31]. Second, we wanted to avoid confounding the estimates of the elasticities for the primary variables of interest as happens in cross-sectional estimates when correlated controls are excluded. Indeed an important contribution of our study is to avoid such confounding with factors such as unobserved preferences that may be correlated with income and expenditures.

We also estimated parallel relations for each of the five types of ASF. Expenditures on each type of ASF were the dependent variables and total ASF expenditures and total expenditures were the key right-side variables. Elasticities for individual ASF component expenditures were constrained to be consistent with the elasticity for total ASF expenditures, using the identity that the elasticity for total ASF expenditures is a weighted-average of the elasticities of the ASF component expenditures with the shares of the components in the total ASF expenditures as the weights.

Multivariate specifications

In the economics literature on commodity demands, two of the most commonly-used functional forms are linear and log-linear. The former assumes constant marginal consumption propensities (and varying elasticities) and the latter assumes constant elasticities. Both can be viewed as first-order Taylor series approximations to more general specifications. We used natural logarithm (ln) functional forms to estimate elasticities because of the straightforward relation between ln functional forms and elasticities. In the ln-linear case, the coefficient estimate of the predictor gives a direct estimate of the elasticity of the dependent variable with respect to that predictor. Since the ln-linear form constrains the

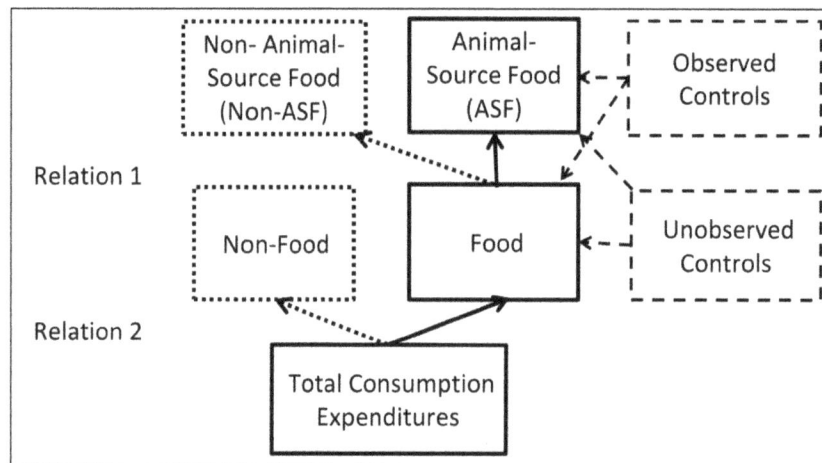

Figure 1. Framework for Analysis. Relation 1 represents the division of food expenditures between ASF and non-ASF, as shown in the top row. Relation 2 represents the division of total expenditures between food and non-food expenditure, as shown in the middle row. We focused our attention on the solid arrows between the center boxes in the figure. Observed controls (dashed boxes on the right) included other important determinants of ASF and food consumption expenditures, such as maternal schooling [31,49], paternal schooling and community wealth. Unobserved controls included other household characteristics, such as food preferences.

elasticity to be the same for all values of the predictor, we added the square of the ln predictor to permit nonlinear elasticities.

We began by estimating the sequence of two relations pertaining to the solid arrows between the center boxes in Figure 1 discussed above, (1) ASF expenditures as a function of food expenditures, and (2) food expenditures as a function of total expenditures (see Appendix S1 for details). While estimates of relations (1) and (2) are of interest to better understand the sequence through which total consumption expenditures may affect ASF expenditures through food expenditures, it also is of interest to know the direct elasticity of ASF expenditures with respect to total consumption expenditures, which can be obtained from estimating (3) ASF expenditures as a function of total expenditures. For each of the three relations we estimated three models: (a) random effects, (b) random effects with observed controls, and (c) fixed effects. Household fixed-effects estimates control for all fixed characteristics of the household, whether observed or not. If there are important unobserved fixed characteristics, such as the fixed component of preferences, these fixed-effect estimates avoid confounding that might occur in the absence of such controls. Household fixed effects also control for any secular trends in time-varying unobservables that are common across observations – but not for time-varying unobservables that differ systematically across households such as ones that affected poorer households differently from better-off households. We report Hausman's p that tests whether the fixed-effects models are preferred over the random-effects models with controls. Based on the estimates of the multivariate relations, we next estimated elasticities of ASF expenditures under multiple scenarios as detailed below.

In preliminary estimates we analyzed the robustness of our estimates to some alternatives: for instance, distinguishing between rural and urban areas, allowing the elasticities to vary by the tertiles of the right-side variables rather than including the squares of those variables, and allowing the elasticities to vary by the tertiles of the wealth asset index. We were not able to identify time-varying changes in elasticities between the different rounds. There were no significant differences between the rural and urban

coefficients, and the tertile estimates were consistent with the estimates with linear and squared ln terms.

Results

Changes in Food and ASF Expenditures between 2002 and 2009

Table 1 provides descriptive statistics for 2002, 2006 and 2009. Between 2002 and 2009, total expenditures per adult equivalent increased by 109 Peruvian soles or USD 31.24 (in constant 2006 prices, September 2006 exchange rate), and total food expenditures per AE increased by 14.1 soles (USD 4.04). Food expenditures as a percentage of total expenditures decreased dramatically between 2002 and 2009 (47% to 23.2%). ASF expenditures increased by 8.6 soles (USD 2.46) from 2002 to 2009. Changes between 2002 and 2009 were significant for all food components except the proportion of the food budget devoted to meat and the proportion of the food budget used for dairy. Dairy, poultry, and meat together accounted for most of the ASF expenditures.

Households in the lower quintiles of total expenditures spent a higher percentage on food (Figure 2a). In 2002, these proportions for all quintiles were different (p<0.05) except for the lowest and second quintiles, and the second and third quintiles. In 2009 all quintiles were significantly different from the highest, and the lowest quintile was also significantly different from the fourth quintile (p<0.05). Households from the lower quintiles of total expenditures devoted smaller percentages of their overall food expenditures to ASFs (Figure 2b). In 2002 all quintiles were significantly different except for the third and fourth quintiles and the fourth and highest quintiles. In 2009 all were significantly different except the second and third, second and fourth, and the third and fourth quintiles (p<0.05).

Figure S1a–c shows the percentage of ASF expenditures devoted to each type of ASF. Spending was significantly different (p<0.05) across quintiles for poultry, meat and eggs in 2002, and for eggs in 2009.

Table 1. Average Household Expenditures on Food and Groups of ASF per Adult Equivalent (AE).

	2002	2006	2009
	Mean 2006 soles (SD)	**or Mean % (SD)***	
Total 15-day expenditures (2006 soles) per AE	326.2 (548.2)[a]	356.0 (238.7)[a]	453.5 (240.2)[b]
Total 15-day food expenditure per AE	82.5 (65.3)[a]	85.6 (47.1)[a]	96.6 (43.2)[b]
Food expenditure as % of total expenditure	46.6% (21.0)[a]	25.9% (6.1)[b]	23.0% (6.4)[c]
ASF 15-day expenditure per AE	26.0 (24.5)[a]	26.1 (19.4)[a]	34.6 (21.3)[b]
ASF expenditures as % of total food expenditures	29.7% (13.6)[a]	28.9% (12.1)[b]	34.1% (11.3)[c]
15-day meat expenditure per AE	7.0 (16.9)[a]	6.4 (8.6)[a]	9.1 (10.0)[b]
15-day meat % of ASF	23.2% (23.2)[a]	21.5% (21.2)[b]	23.9% (19.9)[a]
15-day egg expenditure per AE	2.2 (2.1)[a]	2.2 (2.0)[a]	2.57 (2.1)[b]
15-day egg % of ASF	12.7% (17.2)[a]	11.4% (12.9)[a]	9.3% (9.9)[b]
15-day dairy expenditure per AE	8.2 (9.0)[a]	8.3 (7.9)[a]	10.4 (8.5)[b]
15-day dairy % of ASF	30.8% (23.4)[a]	32.3% (21.6)[b]	30.3% (17.4)[a]
15-day poultry expenditure per AE	6.3 (7.2)[a]	5.8 (6.2)[a]	8.60 (7.8)[b]
15-day poultry % of ASF	24.3% (21.6)[a]	21.8% (18.5)[b]	24.7% (17.1)[c]
15-day fish expenditure per AE	2.3 (3.7)[a]	3.3 (4.5)[b]	3.9 (4.3)[c]
15-day fish % of ASF	9.0% (13.3)[a]	13.0% (14.1)[b]	11.9% (11.4)[b]

*2002, 2006 and 2009 columns represent the mean in soles or mean percent of expenditures (SD).
[a,b,c]Columns with a different letter are significantly different (p<0.05) in a matched comparison of means. Mean expenditures were compared with a pairwise t-test, and percentages were compared with the Wilcoxin signed rank test for non-parametric data.

Elasticities for Total ASF Expenditures

Table 2 gives the basic estimates for the three relations estimated (relations (1)–(3) in Appendix S1), in each case with random effects and no controls in the first column (Model 1), random effects with controls in the second column (Model 2) and fixed effects in the third column (Model 3). In addition to the patterns in elasticities that are of primary interest, two basic features of these estimates emerge. First, the estimates for the random effects with controls are very similar to the estimates of random effects without controls. Therefore, in terms of the estimates of the elasticities of interest, parental schooling and community wealth do not add much to the explanatory power of the relation nor change much the estimated coefficients of the right-side food expenditure and total consumption expenditure variables of primary interest. Second, in each case, the fixed-effects estimates are preferred over the random-effects estimates at the < 0.01 level based on the Hausman test comparing fixed-effects and random-effects models.

The preferred fixed-effects estimates (as well as the random-effects alternatives) suggest for all three relations higher elasticities for poorer households (as reflected in the larger coefficient for the ln linear term) and larger declines in the elasticities as expenditures increase (as reflected in the absolute magnitudes of the coefficients for the squared ln terms). The right-side of Table 2 gives the elasticities for the three fixed-effects relations at the 10th, 25th, 50th, 75th and 90th percentiles, respectively, of the distributions for the right-side expenditure variables (food expenditures for relation 1; total expenditures for relations 2 and 3). Figure 3 graphs the three sets of elasticities against the percentiles of the relevant expenditures.

The elasticities of ASF expenditures with respect to food expenditures (relation 1) are similar in the two time periods: above 1 at lower percentiles of food expenditures, and not significantly different from 1 at the higher income percentiles. Thus those at the low end of the food expenditure distribution increase their

expenditures on ASF more than the increase in food expenditures overall, and those high in the distribution change their ASF expenditures with the same percentage as their food expenditures. In 2002–2006 the elasticities of food expenditures with respect to total consumption expenditures are always significantly different from 1.0, and in 2006–2009 the elasticities of food expenditures with respect to total consumption expenditures are significantly different from 1.0 at all expenditure percentiles except below the 5th percentile. Relation (3) combines the response of ASF expenditures to food expenditures in relation (1) and the response of food expenditures to overall consumption expenditures in relation (2) to give the elasticities of ASF expenditures with respect to overall consumption expenditures. These elasticities are below 1 across all percentiles of expenditures in 2002–2006, and in 2006–2009 are above 1 for the lower expenditure percentiles, not significantly different from 1 between the 35th and 45th percentiles, and less than 1 above the 45th percentile.

Therefore, between 2002 and 2006 ASF expenditures increase 5.0 percent at the 10th percentile in response to a 10 percent increase in total consumption expenditures, due primarily to a relatively large increase in ASF expenditures with respect to food expenditures and a smaller proportional increase in total food expenditures in response to total consumption expenditure. In contrast, at the 90th percentile, the increase in ASF expenditure is only 0.6 percent with a 10 percent increase in total consumption expenditure, due to a relatively small increase in ASF expenditure relative to total food expenditure and a relatively smaller increase in total food expenditure relative to overall consumption expenditure. Similar patterns are observed between 2006 and 2009.

Despite the limited explanatory power of the observed controls and the fact that their inclusion does not change the expenditure elasticities of primary interest, the effects of these controls are of interest in themselves. At levels of maternal schooling above primary school, more schooling is associated with a greater

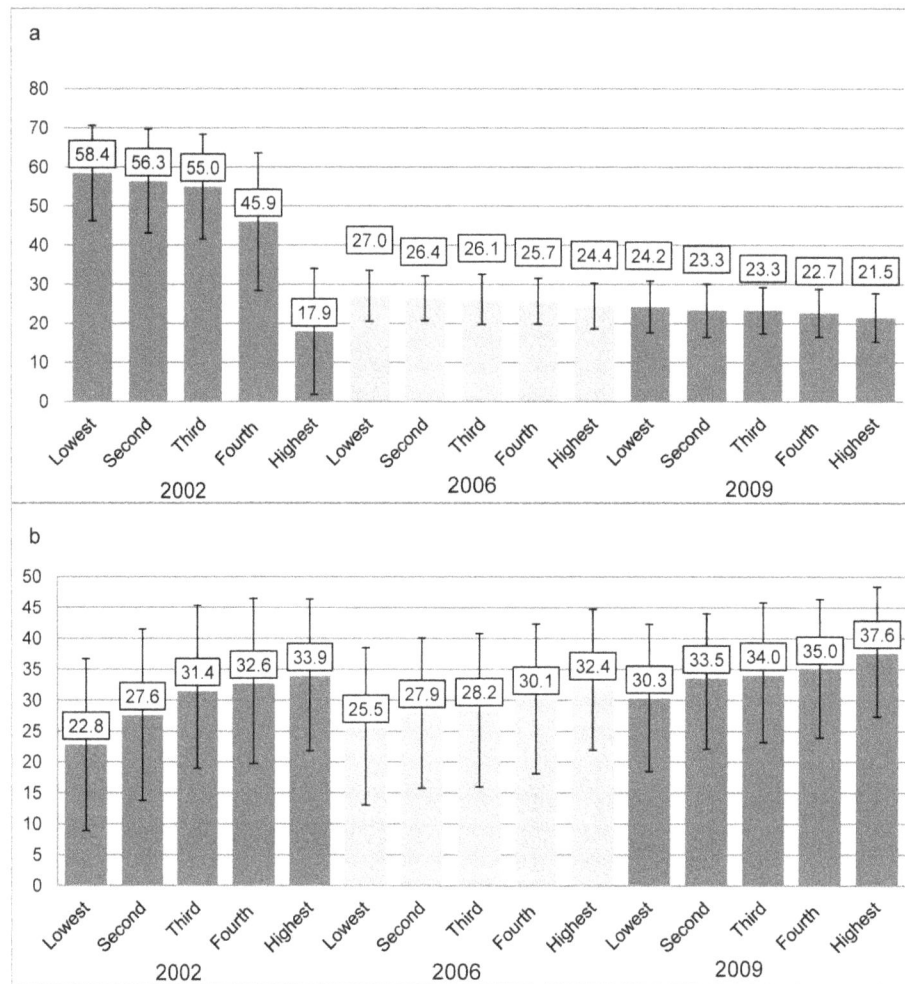

Figure 2. (a) Percent total expenditures devoted to food in 2002, 2006 and 2009, and (b) percent food expenditures devoted to ASF in 2002, 2006 and 2009, by 2002 quintiles of total expenditures. (a) In 2002 each quintile is significantly different from all other quintiles in that year (p<0.05) except for the first and second quintiles, and the second and third. In 2006 each quintile is significantly different from most other quintiles in that year (p<0.05). In 2009 the lowest quintile is significantly different from all other quintiles, and the second, third and fourth quintiles are significantly different from the highest quintile (p<0.05). All quintiles are based on 2002 total expenditures. **(b)** In 2002 each quintile is significantly different from all other quintiles in that year except for the third and fourth quintiles, and the fourth and highest quintiles (p<0.05). In 2006 each quintile is significantly different from all other quintiles for that year except for the second and third quintiles. In 2009 most quintiles are significantly different from all other quintiles in that year (p<0.05). All quintiles are based on 2002 total expenditures.

proportional increase in ASF expenditures than food expenditures, and total consumption expenditures (Table 3). At levels of paternal schooling above primary school in 2002–2006, and secondary school in 2006–2009, more schooling is associated with a greater proportional increase in ASF expenditures with reference to food expenditures and increases in both ASF expenditures and food expenditures with reference to total consumption expenditures. The effect is slightly smaller for paternal schooling than it is for maternal schooling. Community wealth in 2002–2006 is associated with a greater proportional increase in ASF expenditures with reference to total expenditures than ASF expenditures with reference to food expenditures and total expenditures.

Elasticities for Specific ASF Foods

Table 4 and Figure 4 give the elasticities for the types of ASF expenditures with respect to total ASF expenditures. In both 2002–2006 and 2006–2009, meat and fish expenditures have increasing elasticities at higher total ASF expenditures, while poultry and dairy expenditure have decreasing elasticities at higher total ASF expenditures. Eggs show a decrease across the percentiles of total ASF expenditures in 2002–2006 and an increase across the percentiles in 2006–2009. In 2002–2006 the elasticities for the components of ASF expenditures with respect to total ASF expenditures are not significantly different from 1 for meat below the 10th percentile, poultry above the 10th percentile, dairy between the 15th and the 75th percentiles, and for fish above the 70th percentile. In 2006–2009 elasticities are significantly different from 1 for all components except fish at and above the 90th percentile. Aside from these exceptions, the shares of the ASF component expenditures in total ASF expenditures change differentially over the distribution of the total ASF expenditures. Expenditures on eggs and fish over the entire range of total ASF expenditures in both time periods, and dairy in 2006–2009, increase proportionately less than total ASF expenditures as total ASF expenditures increase. Expenditures on meat in both time

Table 2. Multivariate analysis of ASF expenditures with total and food expenditures as predictors.

Dependent Variable	Predictor Variable	Random Effects	Random Effects with Controls	Fixed Effects	Elasticity at Given Percentiles				
					10%	25%	50%	75%	90%
2002-2006									
1 ASF	Food	3.04	3.06	3.13	1.48	1.32	1.15	1.00*	0.85
	(Food)²	−0.20	−0.21	−0.23					
	R²	0.64	0.65	0.64					
	Hausman p			<0.001					
2 Food	Consumption	1.64	1.66	1.16					
	(Consumption)²	−0.11	−0.11	−0.08					
	R²	0.49	0.52	0.49					
	Hausman p			<0.001					
3 ASF	Consumption	2.20	2.28	1.48	0.45	0.38	0.29	0.21	0.12
	(Consumption)²	−0.15	−0.16	−0.11					
	R²	0.32	0.38	0.32					
	Hausman p			<0.001	0.50	0.41	0.29	0.17	0.06
2006-2009					10%	25%	50%	75%	90%
1 ASF	Food	2.69	2.75	2.64	1.37	1.25	1.13	1.03	0.94
	(Food)²	−0.16	−0.17	−0.17					
	R²	0.67	0.69	0.67					
	Hausman p			<0.001					
2 Food	Consumption	2.33	2.32	2.17					
	(Consumption)²	−0.14	−0.14	−0.13					
	R²	0.72	0.72	0.72					
	Hausman p			<0.01					
3 ASF	Consumption	3.89	3.93	3.52	0.84	0.75	0.65	0.55	0.46
	(Consumption)²	−0.24	−0.25	−0.22					
	R²	0.56	0.57	0.56					
	Hausman p			<0.001	1.27	1.11	0.95	0.78	0.54

All elasticities are significantly different from 0 (p<0.001); *Elasticity is not significantly different from 1 (p>0.01).

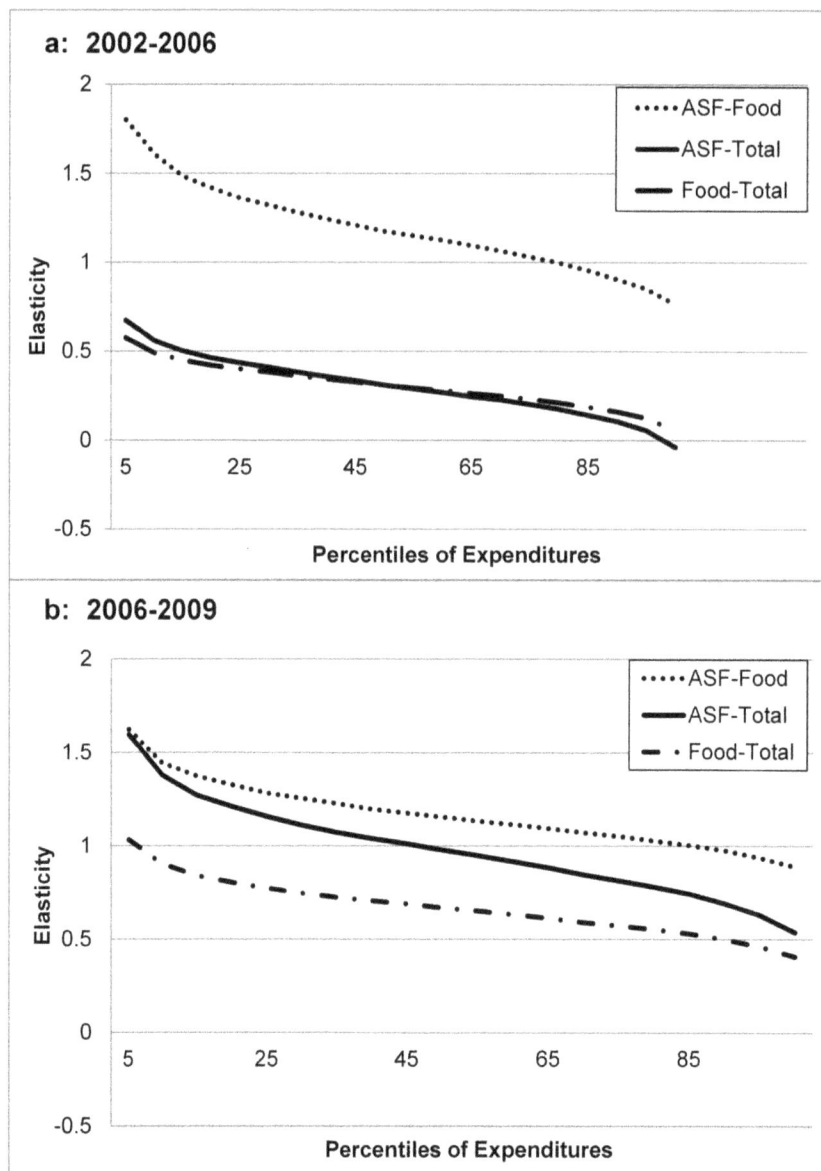

Figure 3. Elasticities of ASF and Food Expenditures with respect to Food or Total Expenditures. (a) From 2002–2006 elasticities are significantly different from 0 (p<0.05) except for ASF-Total and Food-Total >95th percentile. All are significantly different from 1.0 (p<0.05) except for ASF-Food between percentiles 65 and 80. (b) From 2006–2009 all elasticities are significantly different from 0 (p<0.05). All are significantly different from 1.0 (p<0.05) except for ASF-Food between percentiles 75 and 98, Food-Total below the 5th percentile, and ASF-Total between percentiles 35 and 45.

periods increase by larger percentages as total ASF expenditures increase.

While the overall patterns are similar in the two time periods, there are some interesting differences. In 2002–2006 dairy and poultry elasticities with respect to total ASF expenditures are not significantly different from 1 across much of the expenditure percentiles, in contrast to 2006–2009 when the elasticities are significantly above 1 for poultry and significantly below 1 for dairy. In addition, eggs showed an increasing elasticity with ASF expenditure percentiles in 2002–2006, and a decreasing elasticity with ASF expenditures in 2006–2009. Dairy elasticities decreased with ASF expenditures in both time periods, and the elasticities were lower across lower ASF expenditure percentiles in 2006–2009.

Discussion

Based on our preferred household fixed-effects estimates in a fairly rapidly growing economy, we found that 1) relative to households in the highest quintile of expenditures, lowest quintile households dedicated a greater percentage of overall spending to food; 2) households from the lowest expenditure quintiles devoted a smaller percentage of their overall food expenditures to ASFs; 3) households in the lower quintiles of expenditures showed a greater increase in ASF expenditures with an increase in food expenditures (elasticity >1); and 4) households in the higher quintiles of consumption showed a less than proportional increase in ASF expenditures (elasticity <1) with increases in total consumption expenditures.

Table 3. Controls for Random Effects Models.

	ASF-Food	Food-Total	ASF-Total
	2002–2006 Estimates		
Community Wealth	0.015*	0.016**	0.035**
Maternal Schooling			
6 grades	0.064*	0.010	0.086*
7–12 grades	0.158**	0.087**	0.254**
>12 grades	0.213*	0.180**	0.389**
Paternal Schooling			
6 grades	0.012	0.040	0.058
7–12 grades	0.068*	0.064*	0.135*
>12 grades	0.146*	0.165**	0.324**
	2006–2009 Estimates		
Community Wealth	0.012*	0.001	0.010*
Maternal Schooling			
6 grades	0.05	0.004	0.06
7–12 grades	0.139**	−0.019	0.113**
>12 grades	0.179*	−0.030	0.122
Paternal Schooling			
6 grades	−0.042	0.023	−0.012
7–12 grades	0.051	0.011	0.064*
>12 grades	0.116*	−0.002	0.089

*$p < 0.05$; **$p < 0.001$.

Similarly, Abdulai and Aubert [17] found that in Tanzania the diets of high-income households were richer in all micro- and macronutrients. Though they did not specify whether micro- and macronutrients were from animal sources, they did report that the commodity groups most responsive to expenditure fluctuations were animal source foods including meat, fish and eggs, and milk products, and that elasticities were lower for those micronutrients that are consumed through staple foods and higher for micronutrients obtained mainly through animal products. Likewise, Ecker and Qaim [32] generally found that in Malawi, higher household incomes were associated with a more diversified diet as measured by the number of different food items consumed.

Our findings from Peru indicate that increases in total expenditures are associated with greater increases in consumption of ASFs for households in the lowest quintiles of expenditures, a finding also reported by Ecker and Qaim [32] in urban Malawi. In addition to increasing substantially the demand for ASFs, increases in household resources for poorer households allow such families to purchase greater quantities of or higher quality healthcare, education, water and sanitation, and so on [32–34]. However, several studies [33,35] show that growth in income is more effective in improving nutritional status when coupled with nutritional education than when there is an increase in income alone.

Increases in ASF consumption have been shown to contribute to improved performance on school tests [36], improved cognitive functioning for undernourished children [3], and improved anthropometric indices [3]. However, a recent intervention that provided 6–18 month old children with 30–45g of meat daily for twelve months did not find a treatment effect on linear growth [37]. The authors suggest the lack of effect may be due to the high

rates of undernutrition in the population (mean −1.4 LAZ at 6 months) [37]. This recent finding highlights the challenge of determining what levels of dietary ASF are needed to have public health impact, and in interpreting the implications of changes in ASF spending.

There is some concern that the health and nutritional benefits of increasing ASF consumption might have negative environmental costs due to the resources needed for animal production as well as animal waste and other negative environmental impacts [38]. Other reports have suggested that this is a very complex question, with potential changes in natural ecosystems, zoonoses associated with livestock farming, as well as climate and greenhouse gas emissions [36]. A recent analysis highlights the heterogeneity of livestock production systems, with eight different ruminant systems identified, and a range in environmental impacts across the systems [39]. Hence, the impact of increasing consumption of animal source foods on environment depends on the livestock production systems in use.

We found that household fixed-effects estimates of demand for ASF are preferable to random-effects estimates. This means that there are unobserved factors such as preferences that significantly affect household demands for ASF and for total food that may confound the cross-sectional estimates that dominate in most of the literature.

This study had several limitations. The YL-YC study data is limited to expenditures and so we are not able to comment on actual dietary intakes, although substantial detail was collected on different food groups. It should also be recognized that increases in expenditures on ASFs may reflect in part, purchases of foods in these categories that are higher priced due to greater quality or more convenience, and not just increases in quantities of ASFs

Table 4. Elasticities of ASF components with reference to ASF expenditures and total expenditures at specified percentiles of total expenditures.[a]

Percentile	Meat	Poultry	Dairy	Eggs	Fish
2002–2006					
ASF component-Total ASF Expenditures					
10	1.184	1.131	1.127	0.363	0.634
25	1.310	1.081**	1.017**	0.427	0.732
50	1.407	1.042**	0.931**	0.477	0.809
75	1.474	1.015**	0.873**	0.511	0.861**
90	1.524	0.995**	0.829	0.536	0.900**
ASF component-Total Consumption					
10	0.828	0.658	0.781	0.444	1.095**
25	0.684	0.551	0.625	0.364	0.890**
50	0.499	0.413	0.425	0.262	0.627
75	0.323	0.281	0.235	0.165	0.376
90	0.139***	0.144***	0.035***	0.063***	0.115***
2006–2009					
ASF component-Total ASF Expenditures					
10	1.311	1.208	0.921	0.555	0.725
25	1.368	1.185	0.888	0.510	0.789
50	1.410	1.168	0.864	0.476	0.837
75	1.440	1.157	0.847	0.452	0.870
90	1.460	1.148	0.835	0.436	0.893**
ASF component-Total Consumption					
10	1.524	1.555	1.226	0.855**	1.097**
25	1.369	1.342	1.119**	0.738	0.966**
50	1.211	1.125**	1.010**	0.620	0.832
75	1.050**	0.903**	0.899**	0.499	0.695
90	0.902**	0.699	0.230**	0.388	0.569

[a]Elasticities for individual ASF component expenditures are constrained to be consistent with elasticity for total ASF expenditures.
**Elasticity is not different from 1, p value>0.05.
***Elasticity is not different from 0, p value>0.05.

consumed [24,37–39]. The nutritional impact of increase in ASF consumption expenditures is not possible to predict without knowing the food quantity in grams. This study focused on the household as a unit of ASF consumption, and the economic influences on ASF availability. This study did not address individual access: intra-household distribution may vary due to influences such as maternal knowledge [40], and cultural norms or labor and marriage market incentives that may discourage women and children from consuming ASFs [41–47]. While data on prepared foods were collected, they were not included in food expenditures as they could not be allocated to a food expenditure category. However, prepared food expenditures are included in the measure of total consumption expenditures.

Policy makers should consider the growing evidence base that suggests that efforts to increase income may be effective than food price regulations in achieving nutritional goals such as increasing demand for more nutrient-dense foods, and enhancing macro- and micronutrient consumption across a wide variety of nutrients [17,32].

While the purpose of this study was not to evaluate the impact of various efforts to increase income, microcredit, cash transfers and employment-generating interventions are three promising approaches that can play an important part in increasing income, thereby improving nutrition [32,48]. However, evidence demonstrating the impact of these efforts, along with nutritional education, is mixed. Regardless of the policies and programs governments and NGOs implement, identifying the most cost-effective means to improving consumption of ASFs requires detailed monitoring and evaluation as well as rigorous research, all of which has been too limited to date.

a: 2002-2006

b: 2006-2009

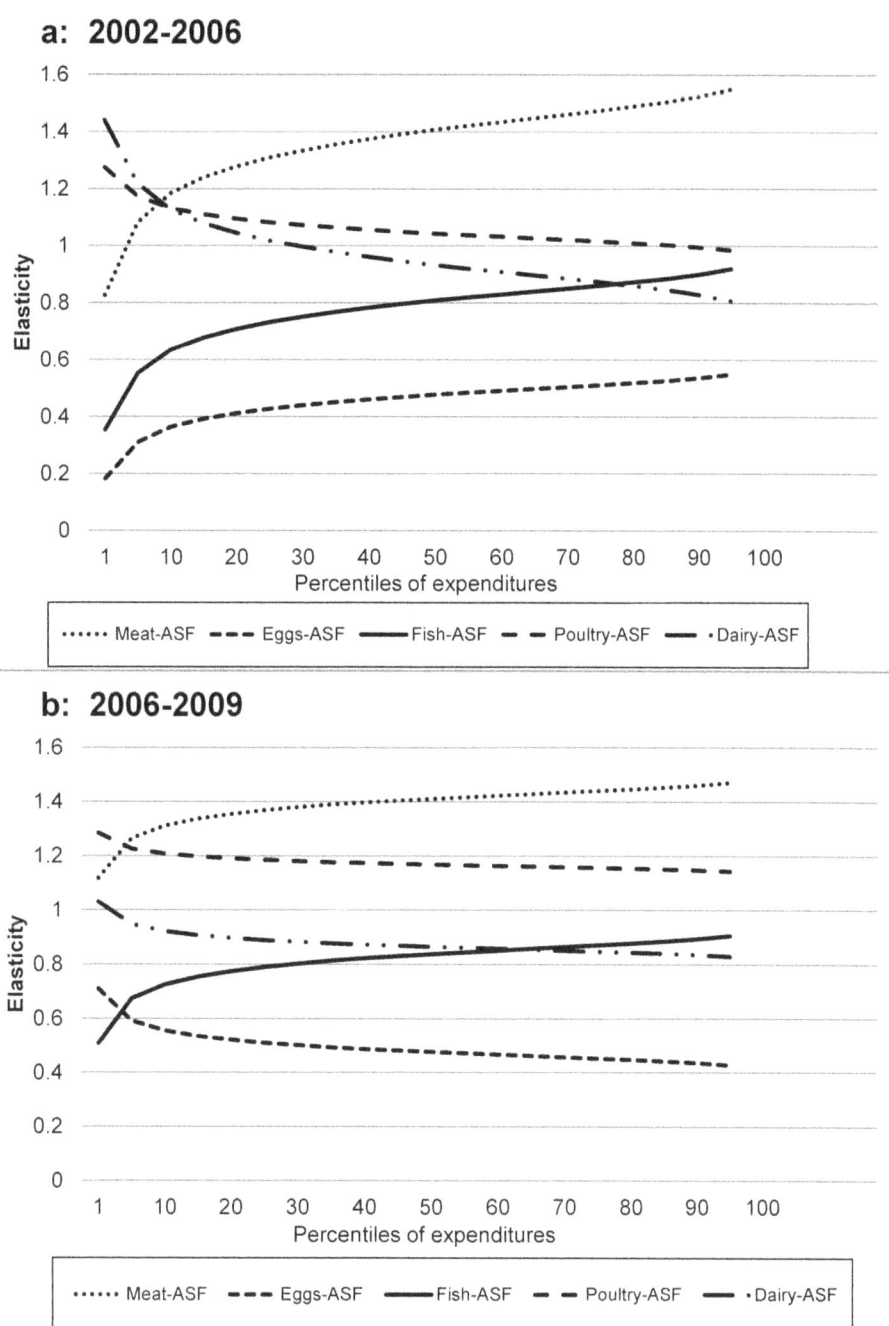

Figure 4. Elasticities of ASF components with reference to total ASF expenditures. Elasticities for individual ASF component expenditures are constrained to be consistent with elasticity for total ASF expenditures. (**a**) From 2002–2006 elasticities are not significantly different from 1 for meat below the 10th percentile, for poultry above the 15th percentile, for dairy between the 15th and 75th percentiles, and for fish >70th percentile. Elasticities with respect to total ASF expenditures are always significantly different from 1 for eggs. (**b**) From 2006–2009 elasticities are not significantly different from 1 for fish between the 10th and 25th percentiles. Elasticities with respect to total ASF expenditures are always significantly different from 1 for meat, poultry, dairy and eggs.

Supporting Information

Figure S1　Mean Expenditures on ASF subgroups as a Percent of ASF Expenditures by 2002 Total Consumption Quintiles in (a) 2002, (b) 2006 and (c) 2009. In 2002 the percent of ASF expenditures on poultry, eggs and meat varied significantly (p<0.05) across 2002 total expenditure quintiles. For meat, the lowest quintile was significantly different from all others; the first quintile was different from the 2nd, and the third and fourth quintiles were different from the highest. For eggs, the lowest and second quintiles were both significantly different from all other quintiles. For poultry, the lowest quintile was different from the highest, the second quintile was different from the fourth and highest, and the third quintile was different from the highest. For dairy, the second and fourth quintiles were different and the fourth quintile was different from the highest. In 2006 percent of ASF expenditures on fish and eggs varied significantly by 2002 total expenditure quintiles. For fish, the lowest quintile was significantly different from the middle quintile and the second and third quintiles were significantly different. For eggs, the lowest quintile was significantly different from the fourth and highest quintiles, and the second and third quintiles were significantly different from the highest quintile. In 2009 the only significant differences between quintiles were in eggs (lowest different from fourth and highest quintiles) and poultry (lowest different from highest quintile).

Acknowledgments

The data used in this study come from Young Lives, a 15-year survey investigating the changing nature of childhood poverty in Ethiopia, India (Andhra Pradesh), Peru and Vietnam (www.younglives.org.uk). We are grateful to the Instituto Nacional de Estadistica e Informatica for providing information on regional food prices. We are grateful to the editor, an anonymous reviewer and William Masters who provided useful comments on a previous version of the manuscript. The Young Lives Determinants and Consequences of Child Growth Project team includes, in addition to the co-authors of this paper, Aryeh Stein, Andreas Georgiadis, Santiago Cueto, Elizabeth Lundeen, Le Thuc Duc, Javier Escobal, Lia Fernald, Shaik Galab, Priscila Hermida, Subha Mani, and Tassew Woldehanna.

Author Contributions

Conceived and designed the experiments: MP JB DH KD BC WS. Analyzed the data: DH BC WS. Contributed reagents/materials/analysis tools: JB. Wrote the paper: DH JB BC KD. Implemented field study: MP.

References

1. Rodrik D (2014) The Past, Present, and Future of Economic Growth. In: Behrman JR, Fardoust S, editors. Towards a Better Global Economy: Policy Implications for Global Citizens in the 21st Century. Oxford, UK: Oxford University Press.
2. Murphy SP, Allen LH (2003) Nutritional importance of animal source foods. J Nutr 133: 3932S–3935S.
3. Dror DK, Allen LH (2011) The importance of milk and other animal-source foods for children in low-income countries. Food Nutr Bull 32: 227–243.
4. Marquis GS, Habicht JP, Lanata CF, Black RE, Rasmussen KM (1997) Breast milk or animal-product foods improve linear growth of Peruvian toddlers consuming marginal diets. Am J Clin Nutr 66: 1102–1109.
5. Krebs NF, Mazariegos M, Tshefu A, Bose C, Sami N, et al. (2011) Meat consumption is associated with less stunting among toddlers in four diverse low-income settings. Food Nutr Bull 32: 185–191.
6. Neumann CG, Murphy SP, Gewa C, Grillenberger M, Bwibo NO (2007) Meat supplementation improves growth, cognitive, and behavioral outcomes in Kenyan children. J Nutr 137: 1119–1123.
7. Iannotti LL, Zavaleta N, Leon Z, Shankar AH, Caulfield LE (2008) Maternal zinc supplementation and growth in Peruvian infants. Am J Clin Nutr 88: 154–160.
8. Black RE, Victora CG, Walker SP, Grp MCNS (2013) Maternal and Child Nutrition Study Group. Maternal and child undernutrition and overweight in low-income and middle-income countries (vol 382, pg 427, 2013). Lancet 382: 396–396.
9. Loret de Mola C, Quispe R, Valle GA, Poterico JA (2014) Nutritional Transition in Children under Five Years and Women of Reproductive Age: A 15-Years Trend Analysis in Peru. PLoS One 9: e92550.
10. Allen LH (2006) Causes of nutrition-related public health problems of preschool children: available diet. J Pediatr Gastroenterol Nutr 43 Suppl 3: S8–12.
11. Banks J, Blundell R, Lewbel A (1997) Quadratic Engel Curves and Consumer Demand. The Review of Economics and statistics 79: 527–539.
12. Chai A, Moneta A (2010) Retrospectives: Engel Curves. Journal of Economic Perspectives 24: 225–240.
13. Deaton A, Muellbauer J (1980) Economics and Consumer Behavior. Cambridge: Cambridge University Press.
14. Gorman WM (1981) Some Engel Curves. In: Deaton A, editor. Essays on the Theory and Measurements of Demand in Honour of Sir Richard Stone. Cambridge: Cambridge University Press.
15. Lewbel A (2007) Engel Curves. The New Palgrave Dictionary of Economics.
16. Bennett MK (1941) Wheat in National Diets. Food Research Institute Studies 18: 37–76.
17. Abdulai A, Aubert D (2004) A cross-section analysis of household demand for food and nutrients in Tanzania. Agricultural Economics 31: 67–79.
18. Bhaumik SK, Nugent JB (1999) Analysis of Food Demand in Peru: Implications for Food-Feed Competition. Review of Development Economics 3: 242–257.
19. Chen D, Abler D (2014) Demand Growth for Animal Products in the BRIIC Countries. Agribusiness 30: 85–97.
20. Block SA, Kiess L, Webb P, Kosen S, Moench-Pfanner R, et al. (2004) Macro shocks and micro outcomes: child nutrition during Indonesia's crisis. Econ Hum Biol 2: 21–44.
21. Bouis HE, Eozenou P, Rahman A (2011) Food prices, household income, and resource allocation: socioeconomic perspectives on their effects on dietary quality and nutritional status. Food Nutr Bull 32: S14–23.
22. World Bank. International Comparison Program database. Available: http://data.worldbank.org/country/peru. Accessed 2013 December 31.
23. Acosta AM, Haddad L (2014) The politics of success in the fight against malnutrition in Peru. Food Policy 44: 26–35.
24. Behrman JR, Deolalikar AB (1989) Is Variety the Spice of Life? Implications for Calorie Intake. Review of Economics and Statistics 71: 4: 666–672.
25. Wilson I, Huttly S, Fenn B (2006) A case study of sample design for longitudinal research. International Journal of Social Research Methodology 9: 351–356.
26. Barnett I, Ariana P, Petrou S, Penny ME, Duc LT, et al. (2012) Cohort Profile: The Young Lives Study. Int J Epidemiol.
27. Glewwe P, Twum-Baah K (1991) The Distribution of Welfare in Ghana, 1987–88. Washington, D.C.: The World Bank.
28. Deaton A (1997) The Analysis of Household Surveys: A Microeconometric Approach to Development Policy. Baltimore and London: The Johns Hopkins University Press for the World Bank.
29. Behrman JR, Knowles JC (1999) Household income and child schooling in Vietnam. World Bank Economic Review 13: 211–256.
30. Behrman JR, Wolfe BL (1987) How Does Mothers Schooling Affect Family Health, Nutrition, Medical-Care Usage, and Household Sanitation. Journal of Econometrics 36: 185–204.
31. Wolfe BL, Behrman JR (1983) Is Income Overrated in Determining Adequate Nutrition. Economic Development and Cultural Change 31: 525–549.
32. Ecker O, Qaim M (2011) Analyzing nutritional impacts of policies: an empirical study for Malawi. World Development 39: 412–428.
33. Alderman H, Hoogeveen H, Rossi M (2006) Reducing child malnutrition in Tanzania. Combined effects of income growth and program interventions. Econ Hum Biol 4: 1–23.
34. Anand S, Ravallion M (1993) Human development in poor countries: on the role of private incomes and public services. Journal of Economic Perspective 7: 133–150.
35. Agee MD (2010) Reducing child malnutrition in Nigeria: combined effects of income growth and provision of information about mothers' access to health care services. Soc Sci Med 71: 1973–1980.
36. Dury S, Alpha A, Bichard A (2014) What risks do agricultural interventions entail for nutrition? Montpelier, France: Unite mixte de recherche: Marches, Organisations, Institutions et Strategies d'Acteurs.
37. Behrman JR, Deolalikar AB (1987) Will Developing-Country Nutrition Improve with Income - a Case-Study for Rural South-India. Journal of Political Economy 95: 492–507.
38. Bouis HE (1996) A food demand system based on demand for characteristics: If there is curvature' in the Slutsky matrix, what do the curves look like and why? Journal of Development Economics 51: 239–266.
39. Bouis HE, Haddad LJ (1992) Are Estimates of Calorie Income Elasticities Too High - a Recalibration of the Plausible Range. Journal of Development Economics 39: 333–364.
40. Block SA (2007) Maternal nutrition knowledge versus schooling as determinants of child micronutrient status. Oxford Economic Papers 59: 330–353.
41. Gittelsohn J, Vastine AE (2003) Sociocultural and household factors impacting on the selection, allocation and consumption of animal source foods: current knowledge and application. J Nutr 133: 4036S–4041S.
42. Behrman JR (1988) Nutrition, Health, Birth-Order and Seasonality - Intrahousehold Allocation among Children in Rural India. Journal of Development Economics 28: 43–62.
43. Behrman JR (1988) Intrahousehold Allocation of Nutrients in Rural India - Are Boys Favored - Do Parents Exhibit Inequality Aversion. Oxford Economic Papers-New Series 40: 32–54.
44. Behrman JR, Deolalikar AB (1990) The Intrahousehold Demand for Nutrients in Rural South-India - Individual Estimates, Fixed Effects, and Permanent Income. Journal of Human Resources 25: 665–696.
45. Pitt MM, Rosenzweig MR (1985) Health and Nutrient Consumption across and within Farm Households. Review of Economics and Statistics 67: 212–223.

46. Pitt MM, Rosenzweig MR, Hassan MN (1990) Productivity, Health, and Inequality in the Intrahousehold Distribution of Food in Low-Income Countries. American Economic Review 80: 1139–1156.

47. Rosenzweig MR, Schultz TP (1984) Market Opportunities, Genetic Endowments, and Intrafamily Resource Distribution - Reply. American Economic Review 74: 521–522.

48. DeLoach S, Lamanna E (2011) Measuring the impact of microfinance on child health outcomes in Indonesia. World Development 39: 1808–1819.

49. Behrman JR, Wolfe BL (1987) How Does Mother's Schooling Affect the Family's Health, Nutrition, Medical Care Usage, and Household Sanitation? Journal of Econometrics 36: 185–204.

11

A Survey of Chinese Citizens' Perceptions on Farm Animal Welfare

Xiaolin You¹, Yibo Li², Min Zhang³, Huoqi Yan⁴*, Ruqian Zhao⁵*

1 History of Science, Nanjing Agricultural University, Nanjing, P.R. China, 2 Department of Sociology, Nanjing Agricultural University, Nanjing, P.R. China, 3 Department of Laws, Nanjing Agricultural University, Nanjing, P.R. China, 4 Philosophy of Science and Technology, Nanjing Agricultural University, Nanjing, P.R. China, 5 Key Laboratory of Animal Physiology and Biochemistry, Nanjing Agricultural University, Nanjing, P.R. China

Abstract

Farm animal welfare has been gradually recognized as an important issue in most parts of the world. In China, domestic animals were traditionally raised in backyard and treated as an important component of family wealth. Industrialization of animal production brings forth the farm animal welfare concerns recently in China, yet the modern concept of animal welfare has not been publicized and a comprehensive recognition on how consumers and farmers perceive animal welfare is lacking. Therefore, we conducted a survey on public opinions toward farm animal welfare in China, based on pigs (including sows, piglets, and fattening pigs), domestic fowls (including layers and broilers) and their products. From 6,006 effective questionnaires approximately two thirds of the respondents had never heard of 'animal welfare'; 72.9% of the respondents claimed that, for the sake of animal derived food safety, human beings should improve the rearing conditions for pigs and domestic fowls; 65.8% of the respondents totally or partly agreed on establishing laws to improve animal welfare; more than half of the respondents were willing, or to some extent willing, to pay more for high-welfare animal products, whereas 45.5% of the respondents were not willing or reluctant to pay more. In summary, farm animal welfare is still in its early stage of development and more efforts are needed to improve the public conception to animal welfare in the process of establishing farm animal welfare standards and legislations in China.

Editor: Georges Chapouthier, Université Pierre et Marie Curie, France

Funding: These authors have no support or funding to report.

Competing Interests: The authors have declared that no competing interests exist.

* Email: yhq@njau.edu.cn (HY); zhao.ruqian@gmail.com (RZ)

Introduction

Concerns for the well-being of animals have long been attached importance in history of human beings. Nowadays, over 100 countries have enacted variety of laws on animal welfare, which not only demonstrates people's respect for animals but also guarantees the safety of animal derived food. Farm animal welfare is also crucial in issues such as international trade, human health, and the environment protection. In China, animal well-being concerns can be traced back to ancient times in some literatures, such as "kindness to humans and other creatures" and "loving human and every creature", which have become prevalent quotations among Chinese people for generations.

Recently, researchers carried on surveys on societal attitudes to animal welfare. A working paper of "Animal Welfare Project" presents a description of some major findings from surveys carried out in seven countries (France, the United Kingdom, Hungary, Italy, the Netherlands, Norway, and Sweden) in September 2005. The paper concentrated on describing national variations in public views on farm animal welfare in general and related shopping practices. The analyses indicate some common features in public opinions about farm animal welfare across Europe, and draw a conclusion that farm animal welfare is clearly an important issue for ordinary people across Europe. It is also found that, even though to a more limited extent, many Europeans still think about animal friendliness when shopping for food and making responsible purchases depends on what people mean by animal friendliness [1].

In a survey on United States households, researchers found sharp differences between direct and indirect questions related to farm animal welfare. This finding, coupled with the extant literature on indirect questioning, suggests that people's concerns for farm animal welfare are actually much lower than what they say they are. It is suggested that responses from indirect questions provide a very different picture of the importance of farm animal welfare to the public than what might be suggested both by direct questioning [2].

In Belgium, a research has been conducted to develop a conception of farm animal welfare that starts from the public's perception and integrates the opinion of different stakeholder representatives. The resulting conception revealed seven dimensions grouped in two different levels. Three dimensions were animal-based: "Suffering and Stress", "Ability to Engage in Natural Behavior" and "Animal Health". Four dimensions were resource-based: "Housing and Barn climate", "Transport and Slaughter", "Feed and Water" and "Human-Animal Relationship" [3].

Opinion surveys indicate that concerns about animal welfare resonate with the general public. In 2005, the European Commission's Health and Consumer Protection Directorate

General commissioned a comprehensive survey of public attitudes towards animal welfare, involving 24,708 citizens in 25 Member States of the European Union. Only 32% of respondents had a positive view about the welfare of laying hens and 22% had a very negative view of their welfare. More than 40% of respondents chose laying hens and broilers among the top three species needing improvements in their welfare. However, there are regional differences in the level of concern for animal welfare, and only 52% of respondents reported that they consider animal welfare when they are making their food purchases. Similarly, in an American Farm Bureau sponsored survey, more than 60% of respondents felt that the government should take an active role in promoting farm animal welfare, and 69–88% of respondents agreed with the statement "I would vote for a law in my state that would require farmers to treat their animals more humanely". Fifty-six percent of respondents in this study felt that decisions about animal welfare should be made by the "experts" rather than the public. Interestingly, a survey of animal science faculty at US universities revealed support for general principles of animal welfare, and greatest concerns were directed at the welfare of poultry relative to other food producing species [4].

However, in China, animal welfare is still at the early stage of development. It didn't draw attention from Chinese public until 2003. In effect, there is a long way to go for China to promote animal welfare. As China has its distinctive history and mode of economic and social development, it is necessary to investigate the public attitude towards animal welfare before establishing strategies of how to promote it more efficiently. This survey, as a part of the "Project of Research and Demonstration on Key Technological System for Farm Animals and Fowls", is conducted to investigate the societal attitudes of the Chinese public towards animal welfare.

Methods

Problem and Strategy

Currently, the subjects of welfare of animals are diverse, including farm animals, experiment animals, and working animals, etc. This study exclusively focuses on the public attitudes to farm animals, with pigs and domestic fowls selected as question topics in that the scope of study can be narrowed and the reality and tradition of China be considered. In modern farming system of China, pigs and domestic fowls and their products are two main sources of meat consumption. They are the two animals of the 'Six Farm Animals' described in Chinese historical literatures. There-fore, the study on animals with which the public are most familiar and keep most relations will probably be the most appropriate initiation.

In accordance with the general interpretation of scholars worldwide, the scope of animal welfare covers the following five aspects that came to be known as the Five Freedoms, proposed by Farm Animal Welfare Council (now the Farm Animal Welfare Committee) and were pivotal in the advancement of animal welfare worldwide:

1. Freedom from thirst, hunger or malnutrition by ready access to fresh water and a diet to maintain full health and vigor.

2. Freedom from discomfort by providing a suitable environment including shelter and a comfortable resting area.

3. Freedom from pain, injury, and disease by prevention or rapid diagnosis and treatment.

4. Freedom to express normal behavior by providing sufficient space, proper facilities, and company of the animal's own kind.

5. Freedom from fear and distress by ensuring conditions and treatment which avoid mental suffering [5].

This comprehensive survey, based on the 5 aspects mentioned above, designs a series of questions to get answers from respondents to know the public attitudes to animal welfare. The backgrounds of respondents such as gender, age, level of education, career, income, birthplace, and working place are considered as independent variables based on the hypothesis that these variables might have impact on the China's public attitudes to animal welfare. Their relations to public attitudes are analyzed through Bi-category Logistic Model.

Data

The data used in this study are collected from a questionnaire survey in January 2011 when the researchers allocated the questionnaires to a number of undergraduate students who brought them back to their hometowns to do the investigation. The survey was planned to cover every provinces and autonomous regions in China, with 4–5 cities or counties selected from each of them. For each city or county, 50 questionnaires have been distributed. Adding up, there have been totally about 8,000 questionnaires collected from all the areas surveyed. However, due to the restriction of admission quota of the researcher's university, few students from Tibet, Hainan, Taiwan, Hong Kong, and Macao are available for this survey. (Nanjing Agricultural University annually admits only 4 undergraduate students from Tibet and only about 10 from Hainan province). With the exception of those five areas, this survey covers the remaining 29 provinces in China.

Before the survey, all students to do the survey got through necessary training, especially the ethical requirements of conduct-ing survey faithfully. In the questionnaire, the item 'Telephone Number of Respondent' is designed to verify the effectiveness of the survey to ensure that every questionnaire has been completed properly. Finally, 6,006 effective questionnaires were received, accounting for 75.1% of total questionnaires.

During the survey, the students are allowed to use non-random approach to select respondents with the gender ratio being kept as approximately 1:1. In all respondents of this survey, the male interviewees account for 51.5%, and female ones 48.5%, which approximates the gender ratio of China. As for the age proportions of samples, 17.3% are below 20 years old; 44.9% between 21 and 34 years old; 27.6% between 35 to 49 years old; 7.0% between 50 to 59 years old; and 2.9% above 60 years old. With regard to the level of education, 5.5% are below the level of elementary school; 15.1% are at the level of junior high school; 24.5% reach the level of senior high school; and 47.0% receive higher education. These data show that the respondents reflect relatively low age and high education level, not precisely representing the real percentages of China. The reason probably lies in the fact that the students conducting the survey most likely to send questionnaires to their peers. At least, the survey results can reflect a trend on public perception to this issue.

Variables

(1) **Variable Declaration and Value Assignment.** The main focus of this survey will be put on four issues: first, public awareness of the concept and connotation of 'animal welfare'; second, public opinions on current intensive factory rearing; third, the public's level of satisfaction on legislation of animal welfare; fourth, the public's level of contentment on the market supply of pork and egg. Practical questions have been designed for each

Table 1. Variables and Assignment.

Independent variables	Gender	male = 1, female = 2
	Age	below 34 = 1, 35–59 = 2, 60 = 3
	Education	below elementary = 1, junior high = 2, senior high = 3, college and above = 4
	Career category	farmer = 1, urban and rural non-agriculture employee = 2, professional = 3, government and NGO employee = 4
	Annual household income	below 40,000 Yuan = 1, 40,000–80,000 = 2, 80,000–150,000 = 3, above 150,000 = 4
	Birthplace (urban or rural)	urban = 1, rural = 2
	Birthplace (eastern, central or western)	eastern = 1, central = 2, western = 3
	Working place (urban or rural)	urban = 1, rural = 2
	Working place (eastern, central or western)	eastern = 1, central = 2, western = 3
Dependent variables	Awareness of concept and connotation of animal welfare	
	V1 being aware of animal welfare or not	Y = 1, N = 0
	V2 being appropriate or not to use cement floor for raising pig	Y = 1, N = 0
	V3 being appropriate or not to kill fowls near cages in which they are kept	Y = 1, N = 0
	Attitude and response to factory rearing	
	V4 evaluation on factory rearing	positive = 1, negative = 0
	V5 willing or not willing to pay more for meat products for the sake of animal welfare	Y = 1, N = 0
	Legislative issues on animal welfare	
	V6 agreeing or not agreeing on establishing legislation for animal welfare	Y = 1, N = 0
	V7 agreeing or not agreeing on introducing foreign legislations of animal welfare into China	Y = 1, N = 0
	The satisfaction degree of public to pork and egg supply	
	V8 satisfaction degree on pork supply	satisfy = 1, dissatisfy = 0
	V9 satisfaction degree on egg supply	satisfy = 1, dissatisfy = 0

category and all variables are bi-category variables expressed as $V_i(i = 1, 2, ..., 9)$ as in Table 1.

In this study, the elements of gender, age, education, career, income, and locality have been designed as independent variables, some of which are partly adjusted in order to be more conveniently analyzed by Logistic Model. The scope of age variables has been narrowed down from original 5 levels to 3 levels of 'youth, middle age, and senior'. The scope of income variables has been reduced from original 6 levels to 4 levels, with the family income of below 40,000 Yuan merged into only one level. The career variables, initially defined as constant category variables, have been divided into 4 levels based on the technical characteristics of industry or career attributes as well as its social power. The two variables of birthplace and working place initially had 9 levels, but in analysis they have sub-variables of 'urban or rural' and 'eastern, central or western'. Table 1 shows the assignment on independent variables.

(2) The Mutual Impacts between Independent Variables and Verification. There may exit some mutual impacts between independent variables. Hence, it is critical to measure such impacts. Table 2 shows the Kendall's tau-b coefficient results between independent variables.

In Table 2, there are 17 corelativities among all 21 independent variable pairs. In the dual-trail verification on condition of $P < 0.01$, the corelativity between 'birthplace' and 'working place' demonstrates the maximum of '0.736', indicating that there is fairly large overlap between the place of birth and work. The second maximum is −0.418 between 'age' and 'education', implying that the lower the age is, the higher education will be received. Other corelativities are all less than $|0.3|$. Although Table 2 reflects the corelativities between independent variable pairs, the massive number of 6006 samples itself can overcome the

Table 2. Kendall's tau-b Coefficient Results of Independent Variable Pairs.

	Gender	Age	Birthplace	Education	Working place	Career
Age	−0.055**					
Birthplace	−0.013	0.027*				
Education	0.027	−0.418**	−0.147**			
Working place	−0.029*	0.182**	0.736**	−0.284**		
Career	0.033*	−0.155**	−0.054**	0.227**	−0.126**	
Income	−0.024	−0.068**	−0.235**	0.262**	−0.195**	0.027

Note: **P<0.01 and *P<0.05 were through dual-trail verification.

extreme correlativity (for instance, r>0.95) between variable pairs and thus achieve a high standard capacity of statistics. This advantage can avoid any interpretation problem caused by multicollinearity between independent variable pairs. Therefore, the bi-category Logistic regression model can be utilized to analyze and verify the significance of model fitting variation.

This study has been approved by Ethics Committee of Scientific Research of Nanjing Agricultural University, and the respondent information is anonymized.

A Descriptive Analysis of the Public Attitudes to 'Animal Welfare'

The Public Cognition of Concept and Connotation of Animal Welfare

As revealed in the survey with 5,982 respondents, 2,187 of them (36.6%), a little more than one third, has ever heard of 'animal welfare'. In other words, the majority of the public did not ever hear of this concept. But it's undeniable that ideas of treating animals with love, which can be found in Chinese traditional culture, are similar to the concept of animal welfare in the western culture. Such ideas include "kindness to humans and other creatures" and "loving human and every creature". To measure China's public awareness about animal welfare, two sets of questions have been designed by the researchers.

The first set of questions includes three statements with focus shifting from human beings to animals. The first statement is human-centered: "Pigs and domestic fowls are only beast, and people can treat them as they wish". The second one sees animals as tools: "Humans should improve the rearing conditions for pigs and domestic fowls to ensure the quality and safety of animal products". The third one says that animals should have some basic rights: "Pigs and domestic fowls should enjoy happy life and be free from troubles as humans do". The results show that, among 5,916 respondents, 4,314 of them (72.9%) choose an "instrumental reason" to decide how humans should treat animals; 1,135 of them (19.2%) agree that animals themselves should enjoy some basic rights; and 468 people (7.9%) support anthropocentrism. So it can be inferred that the majority of Chinese public treat animals as instruments and part of the public think that animals themselves should enjoy some basic rights, the number of whom is 1.43 times larger than those who assert that "Pigs and domestic fowls are only beast, and people can treat them as they wish".

The second set of questions involves two common situations about pigs and fowls in the daily lives of Chinese people. The public attitude is sought by investigating the public opinions on these matters. Table 3 shows the statistical results as follows.

Table 3 emphasizes two points: first, behavioral welfare to give animals freedom to live in a natural way; second, psychological welfare to avoid anxiety and fear. As shown in the data about animal behavioral welfare (take pig rearing as an example), 20.5% of respondents think it is "extremely inappropriate" to rear pigs on cement floor, 49.2% of respondents choose "somewhat inappropriate", these two groups totaling 69.7%. However, 15.0% consider this acceptable and 15.4% don't care. In the meantime, as for animal psychological welfare by the example of fowl killing, 30.8% of respondents consider killing fowls near cage as "extremely inappropriate" and 43.5% view it as "somewhat inappropriate", with a total percentage of 74.3%. Only 10.4% support this and 15.2% choose "unimportant". The result reveals that most of the answers correspond to the ideas of animal welfare.

A Description of Public Attitudes to Factory Rearing

To find out public attitude to factory rearing in China, four choices are given for respondents to select: factory rearing is "a very good way of production", "a scientific way of production", "a way limiting the freedom of pigs and domestic fowls", or "a cruel way of production". Among the 5,705 respondents, 1,228 of them (21.5%) select "a very good way of production"; 1,970 of them (34.5%) believe it is a scientific way; 1,357 people (23.8%) think this way limits the freedom of pigs and domestic fowls; 1,150 respondents (20.2%) dismiss this as a cruel way. The above data show serious discrepancies among people's evaluations of the current mode of factory farming in China: a little more than half (56%) show positive attitude and a little less than half (44%) express negative opinion.

When giving comprehensive evaluation of current factory rearing modes, the respondents are also asked to further evaluate the details of factory rearing. The findings are listed in Table 4.

Table 4 shows that quite a number of people make negative comments on factory rearing of pigs and fowls–71.9% of respondents are worried about the overuse of additives; 49.9% are concerned about the overuse of antibiotics and 48.3% complain about "bad taste". In positive comments, 38.0% select "fast growth for slaughter"; 33.3% choose "high productivity" and 22.4% think it has a satisfying commercial return. Generally, on the topic of factory rearing, negative comments by the public outnumber positive ones.

Given the fact that there were more respondents express negative views, are people willing to pay more for animal-friendly production with a higher cost? Take an instance of pork purchasing, among the 5,974 respondents, 564 of them (9.4%) are "gladly willing to" spend more; 2,693 of them (45.1%) are willing just "to some degree"; 2,029 people (34.0%) show

Table 3. China's Public Attitudes to the Way of Treating Pigs and Domestic Fowls in their Daily Life.

	Though pigs naturally like to nose the earth, most of the piggeries use cement floor.		The venders kill fowls near the cages in which fowls are kept.	
	n	%	n	%
Appropriate	886	15.0	618	10.4
Somewhat inappropriate	2913	49.2	2581	43.5
Extremely inappropriate	1215	20.5	1827	30.8
Unimportant	911	15.4	900	15.2
N	5925	100.0	5926	100.0

reluctance; and 688 of them (11.5%) say no to it. Shown in the above data, a little more than half of the public are willing to pay more for pork reaching the standards of animal welfare.

The Public Attitudes to Animal Welfare Legislation

Generally, there is a global agreement that animal welfare is not only a moral issue but also a legal one. So far, more than one hundred states have established laws on animal welfare, but China is not one of them. Do Chinese also need such laws? Among 5,772 respondents, 4,712 of them think it is necessary; whereas 1,060 of them don't think so, with the proportion being 81.6% and 18.4% respectively. Judging from this, the necessity of establishing animal welfare laws is widely recognized by the public in China.

Nevertheless, when it comes to specific animals and certain human behavior, what will happen to public opinion? To the question "Do you agree on establishing mandatory laws of animal welfare to compel producers to provide better living conditions for farm animals such as pigs and fowls to help them grow and survive?", 1,249 people, 20.8% of the 5,996 respondents, express complete approval; 2,699 of them (45%) approve "to some degree"; 188 of them (3.1%) disapprove completely; 856 of them (14.4%) have never thought about it; and the answer from 1,004 people (16.7%) is "not completely approving". Compared to previous data, the proportion of people who approve (including both "completely approving" and "to some degree approving") has a drop of 16 percent from 81.6% to 65.99%, while the proportion of those who disapprove (including "not completely approving" and "completely disapproving") has slightly risen to 19.8% from 18.4%, though there are some respondents "never having thought about it".

Analysis of Satisfaction on the Market Supply of Pigs and Fowls

As for the situation of pork supply in China's market, among the 5,976 respondents, 1,176 of them (19.7%) feel "satisfactory"; 1,967 of them (32.9%) choose "to some degree satisfactory"; 2,056 of them (34.4%) are "not very satisfactory"; and 777 people (12.9%) regard it as unsatisfactory. In general, more people, 52.6% of the whole sample quantity, feel satisfied (their answers include "satisfactory" or "to some degree satisfactory"). Concerning the situation of egg supply in China' market, among the 5,967 respondents, 1,412 of them (23.7%) feel "satisfactory" and 1,952 people (32.7%) make the choice of "to some degree satisfactory"; while 2,033 of them (34.1%) feel "not very satisfactory" and 570 of them (9.6%) are "unsatisfactory". In general, there are more people, 56.4% of the whole sample quantity, who express

satisfaction (including "satisfactory" or "to some degree satisfactory").

But what lead to people's dissatisfaction on the supply of pork and egg? Table 5 probes into this question, showing that pork receives lower level of satisfaction than egg in the market.

Then the researchers measure the levels of satisfaction on pork and egg respectively. In Table 5, under the category of "the most unsatisfactory" of pork, the reason chosen by the largest number of respondents (almost half) is "higher price", followed by "uncertainty of food safety", then "taste worse than before". Less than 5 percent are dissatisfied because of "deficiency of supply" or "weak market supervision". While for egg, high price leads to the greatest level of dissatisfaction, though less than pork. The factors leading to discontentment that are ranked second and third are "uncertainty of food safety" and "taste worse than before". Just a very small percent are disappointed by "deficiency of supply" and "weak market supervision". Obviously, when people purchase pork and egg, the top three main reasons for dissatisfaction are price, food safety and taste.

Factors that May Influence Public Attitudes to Animal Welfare

As this survey is conducted among citizens nationwide, it should be noted that people from different regions tend to take different attitudes on the same subject due to the gap of economic and social development. In order to obtain a more objective understanding of the influence of different factors on people's attitudes to animal welfare, individual characteristics, including gender, age, education, career, income, birthplace and working place which are related to social attitude to animal welfare, should be taken into consideration. Table 6 shows the results of relevant data analysis.

In Table 6, all the dependent variables are forced into model through the approach of "Enter" statistics. It reflects influence on each dependent variable from each independent variable. For instance, the four independent variables of education, birthplace (divided by eastern, central and western), age and working place (divided by eastern, central and western) have notable impact on whether the respondents are aware of concept of animal welfare (V1) or not. Accordingly, the respective influential factors of the dependent variables V1–V9 can be deduced.

It is necessary to carry out further analysis because there are other determining factors, such as general approach and mechanism of the spread of new concepts or ideas, level of acceptance of legislation by different social groups, and degree of sensibility to price. Hence, manual screening is adopted to remove

Table 4. China's Public Attitudes to the Process and Final Products of Factory Rearing of Pigs and Domestic Fowls (Multiple Choices Can Be Made).

	Number of people who approve	Number of effective sample	Proportion		Number of people who approve	Number of effective sample	Proportion
Bad taste	2894	5987	48.3	high productivity	1991	5986	33.3
Overusing additives	4299	5980	71.9	fast growth for slaughter	2272	5985	38.0
Overusing antibiotics	2987	5987	49.9	high profits	1340	5985	22.4

the variables which do not markedly affect the dependent variables in Table 6. The findings are reported in Table 7.

In Table 7, variables of birthplace (divided as eastern, central and western) no more markedly affect V1 and V3; variables of birthplace (divided as urban and rural) and working place (divided as urban and rural, and divided as eastern, central and western) no longer have significant influence on V7.

According to Table 7, the citizens with higher educational backgrounds, lower age, and working in eastern regions are more likely to be aware of animal welfare (V1); those at older ages, born in rural areas, with lower educational backgrounds, male citizens and those engaging in relatively simple career (such as farmers) are more inclined to support the use of cement flooring for pig rearing (V2); those with lower educational backgrounds, male citizens, those at older age and engaging in relatively simple career are more likely to consider killing fowls near their cage (V3) as appropriate; citizens at lower age, female citizens and those engaging in relatively complicated career (such as in government and NGO) are more likely to support factory rearing (V4); those with higher annual household income, working in rural areas, engaging in relatively complicated career, born in urban regions and those of higher educational backgrounds are more willing to pay more for the more expensive animal products living with better animal welfare (V5); those born or working in urban regions, engaging in relatively complicated career, female citizens and those generating higher annual household income are more likely to agree on making mandatory laws of animal welfare (V6); those having higher educational backgrounds, female citizens and engaging in relatively complicated career are more likely to accept the idea of learning from abroad to establish laws of animal welfare (V7); those born in urban regions and those with higher annual household income are more likely to feel satisfied with the pork supply in China (V8); female citizens, those with higher annual household income and those engaging in relatively complicated career are more inclined to be satisfied with the egg supply in China (V9).

Discussion

The survey conducted by the researchers reveals that only about one third of Chinese public have ever heard of animal welfare. In other words, most Chinese have never heard of it. Moreover, considering relatively high educational level and young age of the respondents, it is possible that the real proportion of Chinese who have ever heard of animal welfare would be even lower. Consequently, though the past decade saw the spread of the concept of animal welfare after its introduction to China from the west, it has not been truly popularized in China partly due to the lack of introduction via mainstream media. Public awareness for animal welfare, on both legislation and farming system, still need to be enhanced thus there is a long way to go for China to promote the concept of animal welfare.

While most Chinese have never heard of animal welfare, this does not mean that Chinese people do not care about the well-being of animals. As a developing country, China needs to adopt factory rearing so as to meet the growing demand for products of livestock and fowls. The government and media should carry on massive and extensive campaign to improve public awareness on such issue. However, the survey has found that 44% of respondents make negative comments on current factory rearing; 23.8% think it limits the freedom of pigs and domestic fowls; 20.2% regard it as a cruel way of production of pigs and fowls. The opposition to current rearing system reflected from this survey matches, to some extent, to a conclusion in Welfare Quality

Table 5. Chinese Public's Satisfaction Degrees on Market Supply of Pork and Egg.

	The first three most unsatisfactory items of pork supply in China's market						The first three most unsatisfactory items of egg supply in China's market					
	The most unsatisfactory		The second most unsatisfactory		The third most unsatisfactory		The most unsatisfactory		The second most unsatisfactory		The third most unsatisfactory	
	n	%	n	%	n	%	n	%	n	%	n	%
Deficiency of supply	203	4.9	126	3.7	138	5.1	237	6.1	140	4.7	153	6.6
Higher price	2049	49.7	610	18.1	324	12.0	1725	44.3	538	18.0	316	13.6
Taste worse than before	662	16.1	990	29.4	475	17.6	931	23.9	1037	34.7	435	18.7
Uncertainty of food safety	974	23.6	1071	31.8	693	25.7	811	20.8	868	29.0	639	27.4
Weak market supervision	189	4.6	514	15.3	772	28.6	141	3.6	355	11.9	561	24.1
Others	47	1.1	51	1.5	292	10.8	52	1.3	50	1.7	228	9.8
N	4124	100.0	3362	100.0	2695	100.0	3898	100.0	2988	100.0	2332	100.0
Proportion	68.7		56.0		44.9		64.9		49.8		38.8	

Table 6. The Logistic Regression Model I of China's Public Attitudes to Animal Welfare and its Influential Factors.

Independent Variables		V1	V2	V3	V4	V5	V6	V7	V8
Gender	B	-0.091	-0.265**	-0.235**	0.231**	0.092	0.280**	0.123	-0.091
	S.E	0.07	0.074	0.077	0.068	0.067	0.075	0.086	0.066
	Exp (B)	0.913	0.767	0.791	1.259	1.097	1.323	1.131	0.913
Age sector	B	-0.311**	0.292**	0.215**	-0.343**	0.024	0.095	-0.06	0.083
	S.E	0.074	0.073	0.076	0.072	0.069	0.077	0.087	0.068
	Exp (B)	0.733	1.34	1.24	0.712	1.024	1.1	0.942	1.087
Birthplace (urban or rural)	B	-0.006	0.312**	0.038	-0.042	-0.211*	-0.311**	-0.223*	-0.279**
	S.E	0.088	0.094	0.1	0.087	0.085	0.093	0.109	0.084
	Exp (B)	0.994	1.366	1.039	0.959	0.81	0.733	0.9	0.756
Birthplace (eastern, central or western)	B	0.147*	0.028	0.229**	0.078	-0.04	-0.104	-0.167	-0.092
	S.E	0.072	0.081	0.083	0.072	0.072	0.079	0.092	0.07
	Exp (B)	1.159	1.028	1.257	1.082	0.961	0.902	0.846	0.912
Education	B	0.368**	-0.248**	-0.229**	0.08	0.111*	0.17	0.242**	0.004
	S.E	0.055	0.053	0.055	0.052	0.05	0.055	0.061	0.05
	Exp (B)	1.445	0.781	0.742	1.083	1.118	1.186	1.274	1.004
Career category	B	-0.093	-0.179**	-0.134*	0.115*	0.178*	0.143*	0.232**	0.031
	S.E	0.061	0.061	0.064	0.059	0.058	0.064	0.074	0.056
	Exp (B)	0.912	0.836	0.875	1.122	1.195	1.154	1.262	1.031
Working place (urban or rural)	B	-0.141	-0.028	0.053	-0.015	0.231*	0.393**	0.285*	0.072
	S.E	0.117	0.116	0.123	0.113	0.11	0.12	0.136	0.103
	Exp (B)	0.869	0.973	1.054	0.985	1.26	1.482	1.33	1.075
Working place (eastern, central or western)	B	-0.227**	-0.041	-0.114	-0.065	-0.03	0.07	0.206*	-0.076
	S.E	0.075	0.083	0.084	0.074	0.073	0.081	0.094	0.072
	Exp (B)	0.797	0.96	0.892	0.937	0.97	1.072	1.229	0.927
Annual Household income	B	0.11	0.041	0.076	0.018	0.319**	0.100*	0.113	0.092*
	S.E	0.043	0.047	0.049	0.043	0.045	0.05	0.059	0.042
	Exp (B)	1.116	1.042	1.079	1.018	1.375	1.105	1.12	1.096
Effective sample quantity		3755	3746	3746	3605	3750	3766	3673	3753
Constant		-0.827	-0.102	-0.214	-0.662	-1.1	-0.563	-0.191	0.49
Likelihood logarithm		4751.473	4389.244	4119.78	4838.019	5020.266	4317.292	3452.232	5151.484
chi-square value		185.055	211.159	158.935	94.664	135.59	76.172	102.478	39.858

Note: **$P<0.01$ and *$P<0.05$ were through dual-trail verification.

Table 7. The Logistic Regression Model II of China's Public Attitudes to Animal Welfare and its Influential Factors.

Independent variables		V1	V2	V3	V4	V5	V6	V7	V8	V9
Gender	B		-0.237**	-0.248**	0.200**		0.227**			-0.241**
	S.E		0.069	0.072	0.060		0.073			0.060
	Exp (B)		0.789	0.780	1.222		1.255			0.786
Age sector	B	-0.270**	0.339**	0.225**	-0.425**					
	S.E	0.066	0.069	0.071	0.060					
	Exp (B)	0.763	1.403	1.253	0.654					
Birthplace (urban or rural)	B		0.272**			-0.202*	-0.323**	-0.185	-0.151**	
	S.E		0.073			0.081	0.090	0.103	0.058	
	Exp (B)		1.313			0.817	0.724	0.831	0.859	
Birthplace (eastern, central or western)	B	0.044		0.090						
	S.E	0.064		0.047						
	Exp (B)	1.045		1.094						
Education	B	0.349**	-0.239**	-0.282**		0.096*		0.252**		
	S.E	0.040	0.049	0.050		0.046		0.055		
	Exp (B)	1.418	0.788	0.754		1.100		1.287		
Career category	B		-0.211**	-0.156**	0.145**	0.213**	0.234**	0.231**		0.087*
	S.E		0.057	0.059	0.041	0.054	0.051	0.068		0.038
	Exp (B)		0.810	0.856	1.156	1.238	1.264	1.260		1.091
Working place (urban or rural)	B					0.231*	-0.365**	0.182		
	S.E					0.103	0.115	0.127		
	Exp (B)					1.260	1.441	1.200		
Working place (eastern, central or western)	B	-0.152*						0.014		
	S.E	0.067						0.056		
	Exp (B)	0.859						1.015		
Annual family income	B					0.308**	0.130**		0.128**	0.130**
	S.E					0.042	0.047		0.035	0.037
	Exp (B)					1.360	1.139		1.137	1.139
Effective sample quantity		4588	4357	4360	4673	4094	3963	4212	5103	4542
Constant		-1.148	-0.108	0.080	-0.339	-1.093	-0.098	0.146	0.130	0.194
Likelihood logarithm		5845.928	5025.725	4744.709	6288.666	5493.409	4550.549	3937.298	7033.893	6187.647
Chi-square value		183.860	269.779	180.704	110.939	139.737	62.362	99.193	24.664	36.675

Note: **P<0.01 and *P<0.05 were through dual-trail verification.

Projects that, significantly, consumers believe that intensive systems are unnatural and, therefore, unsafe. [6] On several detailed welfare aspects, Chinese public shows a high level of approval. For instance, when asked about the statement "Though pigs naturally like to nose the earth, most of the piggeries use cement floor.", 69.7% of respondents consider it to be not very appropriate or extremely inappropriate. When asked about whether it is appropriate to kill fowls near their cages when they are sold in market, as high as 74.3% of respondents oppose it by choosing "not very appropriate or extremely appropriate". For Chinese people, even though they have never heard of the concept of animal welfare, they are very likely caring for animals in mind, influenced by traditional ethics in such as Confucianism, Taoism, and Buddhism. Therefore, Chinese cultural legacy might make contribution to the promotion of welfare farming as a psychological driving force.

In this study, among the answers to the question "Do you agree on establishing mandatory laws for animal welfare to compel producers to provide better living conditions for farm animals such as pigs and fowls to help them grow and survive?", 65.8% of respondents completely or partly agree, and 19.8% completely disapprove or not very much approve of it, with 14.4% "never having thought about it". The data shows that there do exist positive social background for promoting animal welfare legislation in China. However, this background may vary in different social stratifications. Based on data in Table 6 and Table 7, the idea of animal welfare is more likely to be held by people with relatively sophisticated professions (related to higher educational background and career prestige), higher educational background (related to better jobs with higher pay), people working in eastern regions (where there are more developed social and economic situation and the related better living conditions), and people at a younger age (more open and sensitive to new ideas). In addition, the young and the highly educated tend to accept more easily the concept of animal welfare. These findings match, to some extent, a survey conducted by Welfare Quality Project on the points of view of citizens in several European countries which showed that "generally everybody cares about animal welfare, both in general and in relation to food production" [7].

With the application of factory farming system, improved living conditions of pigs means not only improving the quality of pork, but also resulting in higher cost of production. Among all the respondents, 9.4% of them are quite willing to spend more; 45.1% are "to some degree willing to do so"; 34.0% show reluctance; and 11.5% are unwilling to do so. These findings imply an existing advantage for China to improve animal welfare. Anyway, it is not wise to be too optimistic because people's support is based on the condition that the meat quality will be improved and 45% of the respondents are completely or partly not willing to pay more.

Therefore, it is necessary to realize that welfare rearing in China will still be hindered by many societal barriers.

However, what consumers think do not always influence what they have being doing. A majority of Norwegians (57%) believe that animal welfare should be considered to a greater extent, while 38% are content with the situation today. Although many seem to worry about animal welfare in food production, it doesn't seem to influence their consumption of meat and fish to a great extent. [8] Therefore, instead of the being dependent on voluntary behavior change from consumers, animal welfare legislation could be a necessary way to get a quick and obvious result. Nevertheless, while there has been, to some extent, social supports for animal welfare improvement in which the most practical way might be the animal welfare legislation, it may take long time to achieve this tough object. A comprehensive time schedules supplemented to rigid laws may have to be adopted. Bennett and Blaney empirically showed that the vast majority of UK respondents are concerned about animal welfare and supported proposed legislation to phase out the use of battery cages for egg production within the EU. [9] EU legislation allows for transitional periods of several years in order to facilitate the implementation of structural changes in certain farming systems; however, this approach has not always led to timely conversion. Indeed cultural appreciation of animal welfare aspects plays a fundamental role in enhancing the respect of both the spirit and the actual stipulations of the legislation [10].

In conclusion, it can be found that the majority of the public in China take a stand of weak anthropocentrism–their support for improving rearing conditions of pigs and fowls stems largely from their hope for better food quality and safety of animal products. This standpoint is different from the gist of such animal welfare laws as "Martin's Act" advocating people to respect animal lives, protect animal rights, and not abuse animals, which should be considered in the promotion of animal welfare in China.

Acknowledgments

We appreciate the massive efforts by many undergraduate and graduate students who were conducting the survey, which is the essential foundation for this paper. During the preparation and revision on this paper, many works have been conducted on the questionnaire design and statistics process, which should be deemed as the endeavor of both people listed as authors and ones who have not been listed, but worked on it as a part of their academic practices.

Author Contributions

Conceived and designed the experiments: XLY YBL MZ HQY RQZ. Performed the experiments: XLY YBL MZ HQY RQZ. Analyzed the data: XLY YBL MZ HQY RQZ. Contributed reagents/materials/analysis tools: XLY YBL MZ HQY RQZ. Contributed to the writing of the manuscript: XLY YBL MZ HQY RQZ.

References

1. Kjærnes U, Miele M, Roex J (2007) Attitudes of Consumers, Retailers and Producers to Farm Animal Welfare, Welfare Quality Reports No. 2.
2. Lusk JL, Norwood FB (2009) Direct Versus Indirect Questioning: An Application to the Well-Being of Farm Animals. Soc Indic Res (2010) 96: 551–565. DOI 10.1007/s11205-009-9492-z.
3. Vanhonacker F (2010) The Concept of Farm Animal Welfare: Citizen Perceptions and Stakeholder Opinion in Flanders, Belgium, J Agric Environ Ethics. DOI 10.1007/s10806-010-9299-6.
4. Duncan IJH, Hawkins P (2010) The Welfare of Domestic Fowl and Other Captive Birds, Animal Welfare 9: 279. DOI 10.1007/978-90-481-3650-6_12.
5. Radford M (2001) Animal Welfare Law in Britain: Regulation and Responsibility: 265. "FAWC (1993) Report Priorities for Animal Welfare Research and Development, paragraphs 8 and 9".
6. Roex J, Miele M (2005) Farm Animal Welfare Concerns: Consumers, Retailers and Producers Welfare Quality Reports No. 1: 42.
7. Roex J, Miele M (2005) Farm Animal Welfare Concerns: Consumers, Retailers and Producers Welfare Quality Reports No. 1: 10.
8. Roex J, Miele M (2005) Farm Animal Welfare Concerns: Consumers, Retailers and Producers Welfare Quality Reports No. 1: 13–14.
9. Roex J, Miele M (2005) Farm Animal Welfare Concerns: Consumers, Retailers and Producers Welfare Quality Reports No. 1: 13.
10. Communication from the Commission to the European Parliament, the Council and the European Economic and Social Committee, on the European Union Strategy for the Protection and Welfare of Animals 2012–2015: 4.

12

Serological Investigation of Food Specific Immunoglobulin G Antibodies in Patients with Inflammatory Bowel Diseases

Chenwen Cai, Jun Shen, Di Zhao, Yuqi Qiao, Antao Xu, Shuang Jin, Zhihua Ran, Qing Zheng*

Key Laboratory of Gastroenterology & Hepatology, Ministry of Health, Division of Gastroenterology and Hepatology, Ren Ji Hospital, School of Medicine, Shanghai Jiao Tong University, Shanghai Institute of Digestive Diseases, 145 Middle Shandong Road, Shanghai 200001, China

Abstract

Objective: Dietary factors have been indicated to influence the pathogenesis and nature course of inflammatory bowel diseases (IBD) with their wide variances. The aim of the study was to assess the prevalence and clinical significance of 14 serum food specific immunoglobulin G (sIgG) antibodies in patients with IBD.

Methods: This retrospective study comprised a total of 112 patients with IBD, including 79 with Crohn's disease (CD) and 33 with ulcerative colitis (UC). Medical records, clinical data and laboratory results were collected for analysis. Serum IgG antibodies against 14 unique food allergens were detected by semi-quantitative enzyme linked immunosorbent assay (ELISA).

Results: Food sIgG antibodies were detected in 75.9% (60/79) of CD patients, 63.6% (21/33) of UC patients and 33.1% (88/266) of healthy controls (HC). IBD patients showed the significantly higher antibodies prevalence than healthy controls (CD vs. HC, $P = 0.000$; UC vs. HC, $P = 0.001$). However no marked difference was observed between CD and UC groups ($P = 0.184$). More subjects were found with sensitivity to multiple antigens (≥ 3) in IBD than in HC group (33.9% vs.0.8%, $P = 0.000$). Egg was the most prevalent food allergen. There was a remarkable difference in the levels of general serum IgM ($P = 0.045$) and IgG ($P = 0.041$) between patients with positive and negative sIgG antibodies. Patients with multiple positive allergens (≥ 3) were especially found with significant higher total IgG levels compared with sIgG-negative patients ($P = 0.003$). Age was suggested as a protective factor against the occurrence of sIgG antibodies ($P = 0.002$).

Conclusions: The study demonstrates a high prevalence of serum IgG antibodies to specific food allergens in patients with IBD. sIgG antibodies may potentially indicate disease status in clinical and be utilized to guide diets for patients.

Editor: David L Boone, University of Chicago, United States of America

Funding: The work was supported by grants from the National Key Technology R&D Program of China (No. 2012BAI06B03), http://kjzc.jhgl.org/and National Natural Science Foundations of China (No. 81170362, 81200280 and 81370508), http://isisn.nsfc.gov.cn. The funders had no role in study design, data collection and analysis, decision to publish, or preparation of the manuscript.

Competing Interests: The authors have declared that no competing interests exist.

* Email: qingzheng101@163.com

Introduction

Inflammatory bowel diseases (IBD) include two main types, Crohn's disease (CD) and ulcerative colitis (UC), both characterized by mucosal ulceration in gut. Patients with IBD suffer from abdominal pain, diarrhea, weight loss and fatigue. As diseases progress, they can cause perianal lesion, abdominal abscess, intestinal perforation or even canceration as diverse complications. Extraintestinal manifestations including oral ulcer, arthritis, iritis and cholangitis may also occur simultaneously [1]. The tendentiousness of the constant alternation of relapses and remissions seriously affects the quality of life in IBD patients [2]. For the moment the mechanisms involved in the pathogenesis can be summarized that the influence of environmental factors to genetically susceptible individuals triggers abnormal mucosal immune reaction in gut with the involvement of intestinal flora and eventually leads to the bowel inflammation and ulceration [3].

IBD has long been regarded as a problem threatening public health in the Western world. However in recent decades the rising incidence of it in developing nations has made IBD a global issue. In Asia, according to an epidemiological survey, the rapid westernization and modernization may attribute to the condition [4]. Among various environmental factors, diet has been implicated to play a considerable role in the course of IBD [5–8]. A Japanese study suggested the intake of a high fat diet and sweets may associate with CD and UC [9]. Another research also pointed an increased consumption of alcohol and red meat might cause higher incidence of relapse in UC patients [10]. Meanwhile food intolerance, which is defined as a reproducible adverse reaction to specific food or food ingredients, has become more common in recent years [11–12]. Banai J. considered all kinds of

Table 1. Demographic data of all subjects.

Clinicopathological features	CD (N = 79)	UC (N = 33)	HC (N = 266)
Male (n, %)	47(59.5)	18(54.5)	146(54.9)
Female (n, %)	32(40.5)	15(45.5)	120(45.1)
Age (yr) (mean, 95% CI)	36.5(33.6–39.4)	40.7 (35.9–45.7)	46.4(45.2–47.6)
Age range (yr)	18–68	17–73	24–71
Duration of disease (yr) (n, %)			
<1	21 (26.6)	13 (39.4)	/
1–5	38 (48.1)	11 (33.3)	/
5–10	16 (20.3)	7 (21.2)	/
>10	4 (5.1)	2 (6.1)	/
Disease activity (n, %)			
Remission	7 (8.9)	0 (0)	/
Mild	16 (20.3)	12 (36.4)	/
Moderate	33 (41.8)	14 (42.4)	/
Severe	23 (29.1)	7 (21.2)	/
Localization of disease (n, %)			
L1 (terminal ileum)	40 (50.6)	/	/
L2 (colon)	10 (12.7)	/	/
L3 (ileocolon)	29 (36.7)	/	/
E1 (rectum)	/	1 (3.0)	/
E2 (left-sided colon)	/	14 (42.4)	/
E3 (entire colon)	/	18 (54.5)	/
Complications of disease (n, %)			
None	32 (40.5)	31 (93.9)	/
1 item	42 (53.2)	2 (6.1)	/
2 items	5 (6.3)	/	/
Extraintestinal manifestations (n, %)			
None	60 (75.9)	26 (78.8)	/
1 item	18 (22.8)	7 (21.2))	/
2 items	1 (1.3)	/	/
Intestinal surgery	15 (19.0)	0 (0.0)	/

CD: Crohn's disease; UC: ulcerative colitis; HC: healthy controls; yr: year; 95% CI: 95% confidence interval; L1,L2,L3: disease localization of Crohn's disease by Montreal Classification; E1,E2,E3: disease localization of ulcerative colitis by Montreal Classification.

food could be antigenic properties to cause chronic mild inflammation in gut and eventually lead to ulcerative colitis [13]. For example, intolerance to milk has long been believed involving in the pathogenesis of IBD [14,15]. Glassman et al reported during infancy stage the frequency of symptoms compatible with milk protein sensitivity was greater in UC compared with control population (P<0.03) [16]. Furthermore patients who underwent milk intolerance were found to develop UC at an earlier age compared to those without a history of hypersensitivity to milk (P<0.02) [16]. The statistics suggested the early antigenic stimuli might play a role in development of IBD at a later age [17]. However the causality of food intolerance and IBD still remains controversial and needs further researches to figure out.

Studies on food adverse reactions mediated by immunoglobulin G (IgG) in certain intestinal diseases, such as irritable bowel syndrome (IBS), have been increasingly reported [18–19]. Researchers used to mainly focus on food intolerance classically by the presence of serous IgE antibodies but it seemed that the characteristically immediate allergic reactions were quite rare in some conditions [20–21]. By contrast, the circulating IgG antibodies provide a more delayed or even asymptomatic response after the exposure to a unique food antigen [22]. Considering that IBD patients mostly suffer a long course of the chronic disease, we hypothesize IgG may have a stronger relevance with IBD than IgE.

Food is a complicated field to study because of its enormous varieties. In another way this feature also makes it a resource palace for us researchers to explore. In this study we aimed to identify the prevalence and significance of 14 food specific IgG antibodies in IBD patients through serological investigation. We expected the results may provide a more clear and detailed connection between food intolerance and IBD in our patient population as well as a supporting evidence for the antibody test to serve the clinical.

Subjects and Methods

Ethics Statement

This study was approved by Medical Ethics Committee of Ren Ji Hospital of Shanghai Jiao Tong University School of Medicine.

Table 2. Prevalence of food specific IgG antibodies in patients with Crohn's disease (CD), ulcerative colitis (UC) and healthy controls (HC).

Group	N	Seropositive Degree (n)				Antibodies (−)	Antibodies (+)
		0	+1	+2	+3	(n, %)	(n, %)
CD	79	19	10	14	36	19 (24.1)	60 (75.9)
UC	33	12	9	4	8	12 (36.4)	21 (63.6)
HC	266	178	41	24	23	178 (66.9)	88 (33.1)

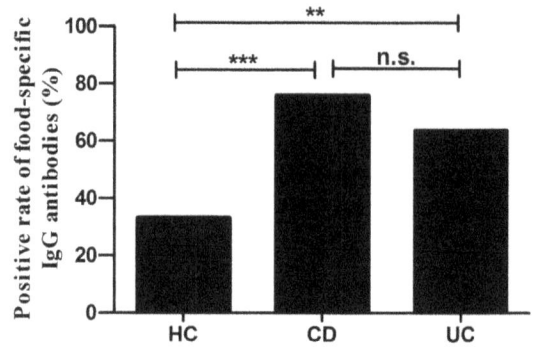

Figure 1. Positive rate of food-specific IgG antibodies in Crohn's disease (CD), ulcerative colitis (UC) and healthy control (HC) groups. Chi-square test, *** $P<0.001$, **$P<0.005$, n.s. not significant.

Informed consent wasn't applied as the medical records and private information of all the subjects were anonymized and de-identified prior to analysis.

Subjects

The study included a total of 112 patients with CD (n = 79) or UC (n = 33) in Ren Ji Hospital from June 2011 to December 2013. All patients met the diagnostic criteria for CD or UC according to the consensus on the diagnosis of IBD drawn up by European Crohn's and Colitis Organization (ECCO) [23,24] and were serologically tested of the food sIgG antibodies during their visit in hospital. In addition another 266 people, who came to our Health Care Centre to do checkups which contained the test as a routine item, were randomly chosen to represent the general population as healthy controls (HC).

Data Collection

Medical records and clinical data of all the IBD patients were comprehensively reviewed and the general demographic data were summarized. The disease severity of CD or UC was based on Harvey-Bradshaw Index [25] or Modified Truelove-Witts Classi-

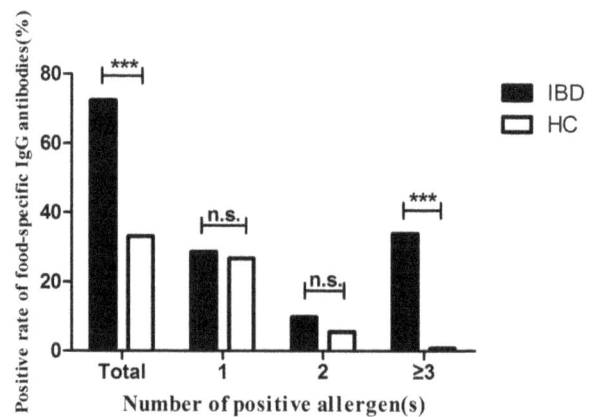

Figure 2. Distribution of the number of positive allergen(s) with positive rate of food-specific IgG antibodies in patients with inflammatory bowel diseases (IBD) and healthy controls (HC). Chi-square test, *** $P<0.001$, n.s. not significant.

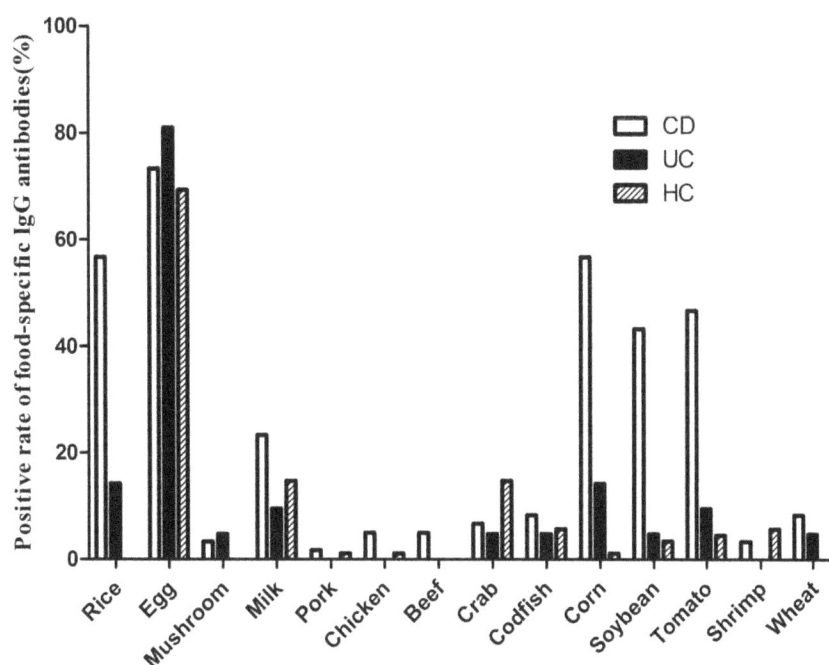

Figure 3. Distribution of positive food allergens in Crohn's disease (CD), ulcerative colitis (UC) and healthy control (HC) groups.

fication [26], respectively. The disease localization of both CD and UC was determined by Montreal Classification [27]. The complications that patients carried consisted of fistula, abdominal abscess, intestinal obstruction, perianal disease, hemorrhage of gastrointestinal tract and acute perforation. The simultaneous extraintestinal manifestations included oral ulcer, sacro-iliitis, rheumatoid arthritis, iritis, primary sclerotic cholangitis, hepatic adipose infiltration and cholelithiasis. Relevant laboratory findings contained peripheral white blood cell (WBC) counts, eosinophile granulocyte (EOS) counts, lymphocyte (LYM) counts, haemoglobin (Hb), serum albumin, erythrocyte sedimentation rate (ESR), high-sensitivity C-reactive protein (hs-CRP), serum total immunoglobulin (IgM, IgA and IgG) and anti double-stranded DNA antibodies (anti-dsDNA).

Enzyme Linked Immunosorbent Assay (ELISA)

ELISA tests for the semi-quantitative analysis of serum IgG antibodies to 14 unique food allergens, including rice, egg, mushroom, milk, pork, chicken, beef, crab, codfish, corn, soybean, tomato, shrimp and wheat, were performed by the detection kit according to the operation manual (Biomerica, Inc. USA). The IgG concentration less than 50U/ml was considered negative

(Grade 0). The values between 50–100U/ml, 100–200U/ml and more than 200U/ml were represented mild sensitivity (Grade +1), moderate sensitivity (Grade +2) and high sensitivity (Grade +3), respectively.

Data Analysis

Statistics were performed with SPSS 19.0 (SPSS Inc. USA). Enumeration data were analyzed using chi-squared test, in which rates of multiple samples were compared by R×C contingency table analysis. Continuous numerical variables were expressed in the form of mean with 95% confidence interval (95% CI) and analyzed by Student's t test. Regression analysis was utilized to identify the correlation/risk factors among variables. Two-tailed P-value<0.05 was regarded statistically significant.

Results

Demographic data of all subjects

The characteristic information of IBD patients and healthy controls were summarized in **Table 1**.

Table 3. Distribution of food specific IgG antibodies in different disease localizations.

Localization	sIgG antibodies (+)	sIgG antibodies (−)
	(n, %)	(n, %)
Only small intestine (N = 40)	33 (82.5)	3 (17.5)
Only large intestine (N = 43)	26 (60.5)	17 (39.5)
Both small & large intestine (N = 29)	22 (75.9)	7 (24.1)
P-value	0.072△	

△ Chi-square test (R×C contingency table analysis), not statistically significant (P>0.05).

Table 4. Distribution of food specific IgG antibodies at different IBD activity status.

Disease status	sIgG antibodies (+) n (%)	Multiple positive (≥2) n (%)	High sensitivity n (%)
Remission (N = 7)	6 (85.7)	5 (71.4)	3 (42.9)
Mild (N = 28)	18 (64.3)	7 (25)	9 (32.1)
Moderate (N = 47)	35 (74.5)	23 (48.9)	18 (38.3)
Severe (N = 30)	22 (73.3)	14 (46.7)	14 (46.7)
P-value	0.647△	0.079△	0.719△

△Chi-square test (R×C contingency table analysis), not statistically significant (P>0.05).

Prevalence of serum IgG antibodies to 14 unique food allergens

Food specific IgG antibodies were detected positive in 75.9% (60/79) of CD patients, 63.6% (21/33) of UC patients and 33.1% (88/266) of HC (**Table 2**). The antibodies showed a significantly higher frequency in both CD and UC groups than in healthy controls (CD vs HC, P = 0.000; UC vs HC, P = 0.001) (**Figure 1**). However, there was no significant difference between CD and UC groups (P = 0.184). In general the total positive rate of all IBD patients was 72.3% (81/112), higher than the control group (P = 0.000). Among them 28.6% (32/112), 9.8% (11/112) and 33.9% (38/112) of subjects were respectively sensitive to one, two and more than two food allergens while the corresponding ratios of healthy controls were 26.7% (71/266), 5.6% (15/266) and 0.8% (2/266). There were more subjects who got intolerant to 3 or more antigens in IBD group than in HC group (P = 0.000) (**Figure 2**).

In the present study, the top five prevalent food allergens which caused positive sIgG antibodies in CD patients were egg (44/60, 73.3%), rice (34/60, 56.7%), corn (34/60, 56.7%), tomato (28/60, 46.7%) and soybean (26/60, 43.3%). Besides, the top five prevalent food allergens in UC group were egg (17/21, 81.0%), rice (3/21, 14.3%), corn (3/21, 14.3%), tomato (2/21, 9.5%) and

milk (2/21, 9.5%). And healthy controls demonstrated the compositions of egg (61/88, 69.3%), milk (13/88, 14.8%), crab (13/88, 14.8%), codfish (5/88, 5.7%) and shrimp (5/88, 5.7%), which were similar to the results of a reported epidemiological survey in general population [28]. **Figure 3** showed the distribution of positive food allergens in CD, UC and HC groups.

Association of food specific IgG antibodies with inflammatory segments

To analyze the association between food sIgG antibodies status and disease extent in gut, all 112 patients were divided into three subgroups according to endoscopic results (**Table 3**). The positive rate was found higher (82.5%) in patients with only small intestine involved but the statistical difference wasn't remarkable (P = 0.072).

Relevance of food specific IgG antibodies with disease activity

We divided all 112 IBD patients into four subgroups according to clinical disease activity and compared the ratios of patients with positive IgG antibodies, multiple positive antibodies (≥2) and high sensitivity to at least one food allergen (**Table 4**). Although the

Table 5. Comparison of laboratory results in IBD patients with positive and negative food specific IgG antibodies.

Laboratory results	sIgG antibodies (+) (N = 81)	sIgG antibodies (−) (N = 31)	P-value
WBC (×10⁹/L)	7.15(6.50–7.86)	6.66(5.79–7.56)	0.437
EOS (×10⁹/L)	0.16(0.12–0.23)	0.15(0.10–0.21)	0.833
LYM (×10⁹/L)	1.44(1.31–1.59)	1.54(1.32–1.81)	0.501
Hb (g/L)	114.31(109.52–119.06)	114.23(104.84–122.58)	0.986
Albumin (g/L)	36.19(34.82–37.72)	35.98(33.67–38.28)	0.889
ESR (mm/h)	28.07(23.01–33.58)	21.35(15.65–27.61)	0.160
hs-CRP (mg/L)	18.84(13.78–23.89)	11.75(6.51–18.51)	0.083
IgM (g/L)	1.22(0.99–1.49)	0.79(0.60–1.03)	0.045*
IgA (g/L)	2.80(2.37–3.27)	2.60(2.13–3.05)	0.597
IgG (g/L)	12.90(11.94–13.94)	10.92(9.50–12.48)	0.041*
anti-dsDNA (IU/mL)	3.31(2.85–3.75)	3.56(2.72–4.61)	0.609

Statistics were expressed as mean with 95% confidence interval.
Normal range: white blood cell (WBC), 3.69–9.16×10⁹/L; eosinophile granulocyte (EOS), 0.02–0.50×10⁹/L; lymphocyte (LYM), 0.8–4.0×10⁹/L; haemoglobin (Hb), 113–172 g/L; albumin, 35–55 g/L; erythrocyte sedimentation rate (ESR) 0–20 mm/h; high-sensitivity C-reactive protein (hs-CRP), 0–3 mg/L; immunoglobulin M (IgM) 0.4–2.3 g/L; IgA, 0.7–4.0 g/L; IgG, 7–16 g/L; anti double-stranded DNA antibodies (anti-dsDNA), 0–7.0 IU/mL
Student's t test, *P<0.05.

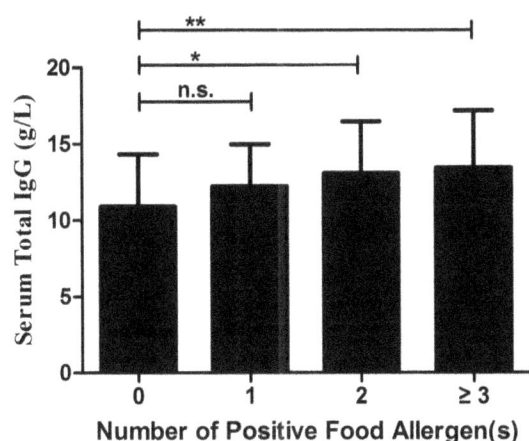

Figure 4. Serum total IgG values in inflammatory bowel disease patients with multiple positive food allergens. Statistics were shown as mean ± standard deviation. Chi-square test, * $P<0.05$; ** $P<0.005$, n.s. not significant.

general antibodies and multi-antibodies were found more common in remission group (85.7% and 71.4%) while the patients who were highly sensitive to specific food allergens seemed to be severer (46.7%), there was no significant differences among the four subgroups.

To further analyze the possible association with several laboratory values related to disease activity, IBD patients were classified into positive and negative IgG antibodies subgroups (**Table 5**). We did find a notable difference in general serum IgM ($P = 0.045$) and IgG ($P = 0.041$) levels between the two groups. Mean values of WBC, ESR and hs-CRP in the positive patients were recognized higher compared with the other group although no marked differences were found. The rest lab findings of the two groups were resembled.

Considering the relative levels of total IgG in relation to antigen specific IgG, we then divided all the IBD patients into 4 groups according to the number of positive allergen(s) (0, 1, 2, ≥3) and compared the difference of serum general IgG values among the groups (**Figure 4**). The mean values of each group were 10.92 (9.50–12.48) g/L, 12.24 (11.31–13.32) g/L, 13.08 (11.31–14.97) g/L and 13.42 (12.24–14.58) g/L, respectively. Patients with multiple positive allergens (≥2) were found with significant higher total IgG levels than sIgG-negative patients ($P = 0.044$, $P = 0.003$, respectively).

Predictors for the occurrence of food specific IgG antibodies

To further identify the predictive factors, we utilized binary logistic regression analyses of several demographic and clinical parameters. The independent variables contained disease type (CD/UC), gender, age, complication, extraintestinal manifestation and intestinal surgery (**Table 6**). Only age was found as a predisposing as well as protective factor against the development of serum food-related IgG antibodies ($P = 0.002$).

Discussion

In the present study, we have demonstrated that IgG antibodies against food specific allergens are distinctly elevated in IBD patients compared with healthy controls. This serologic antibody investigation is served as an aid in the diagnosis and management of food intolerance in clinical activities [29]. Food intolerance can be defined as a series of unpleasant symptoms including eructation, abdominal pain, diarrhea, fatigue, headache and palpitation after the intake of particular food products [29]. It is usually caused by enzyme deficiency as well as pharmacological effects of vasoactive amines present in foods. When certain food can't be fully digested, bodies may produce specific IgG antibodies which would form immune complexes with food particles and lead to excessive protective immune responses [30]. Food intolerance is not so similar to food allergy as the latter mainly occurs through the classically immediate IgE-mediated antibody responses and the obvious symptoms can be detected by most patients [31]. Meanwhile, IgG-mediated reaction characteristically acts as a particular delayed-type hypersensitivity response after the exposure to antigens and sometimes the symptoms are too occult for patients to recognize [22]. Some investigators considered them to be physiological since that the food sIgG antibodies can appear in healthy individuals [32], which was also proved in our data. However, several studies including our results have shown the higher prevalence of sIgG antibodies to food in IBD patients or animal models in contrast to normal controls [19,33], which arouse the attention to the relevance between IBD and food intolerance. Van Den Bogaerde *et al.* used colonoscopic allergen provocation test to determine the gut mucosal response to food antigens in Crohn's disease [34]. This intuitionistic study provided both *in vivo* and *in vitro* evidence that CD patients were more sensitive to exogenous food antigens than healthy people and the reactions were gut specific.

Food sIgG antibodies were discovered frequent in IBD patients with small intestine involved in the present research, which resembled the result of a study on the correlation of lactose malabsorption and disease segments [35]. Mishkin *et al.* demon-

Table 6. Correlation of food specific IgG antibodies with demographic and clinical parameters.

Parameters	Odds ratio	95% CI	P-value
Disease type (CD vs. UC)	1.167	0.389–3.498	0.783
Gender	0.758	0.306–1.879	0.549
Age	0.945	0.912–0.979	0.002**
Complication	2.269	0.719–7.157	0.162
Extraintestinal manifestation	1.302	0.438–3.868	0.635
Intestinal surgery	1.613	0.329–7.905	0.556

95% CI: 95% confidence interval; CD: Crohn's disease; UC: ulcerative colitis.
Binary logistic regression, **$P<0.005$.

strated that CD of the proximal small bowel (duodenum, jejunum), terminal ileum, terminal ileum plus colon and colon alone were related with lactose malabsorption of 100%, 68.1%, 54.5% and 43.5%, respectively. Regression analysis of our data indicated age as a protective factor of food intolerance in IBD patients, which is opposite to the result of a cross sectional epidemiological study in other area of China [28]. We assume the difference might be associated with age structure of onset as inflammatory bowel diseases, especially Crohn's disease, tend to occur in younger people based on our clinical observation.

A survey demonstrated that 15.6% IBD patients believed diet could initiate the disease while 57.8% were convinced certain foods could play a role in causing relapses [36]. In our study, we found food sIgG antibodies were more frequent in patients who were during remission. Meanwhile, the patients who were highly sensitive to food allergens tended to be serious in disease activity and mean levels of ESR and hs-CRP were detected higher in sIgG-positive patients. Although the results didn't show statistical significance, they did provide us with a tendency. We consider the few remission cases (only 7) contributed to the above conflicting results.

The increased serum IgM values tested in sIgG-positive patients may indicate recent infection. Besides, we also found a higher level of serum total IgG antibodies in patients with positive allergens, especially with multiple ones in the study. The general increase on IgG levels may be an important reason for the increased amount of sIgG antibodies against food epitopes. It has been reported that mast cells can respond not only to IgE antibodies but also to IgG antibodies [37]. In addition IgG antibodies in food allergy may even influence allergen-IgE complex formation and bind to B cells, which is quite opposite the traditional concept that IgG antibodies are supposed to inhibit these processes [38]. Therefore IgG antibodies probably serve as mediation effects rather than inhibition in hypersensitivity reaction caused by food intolerance. Serum sIgG antibodies to non-food related antigens are mostly studied within inhaled antigens in respiratory allergic disorders such as asthma. Wang et al. investigated the relevance between asthma morbidity and sIgE/sIgG levels to inhaled allergen exposure [39]. They concluded that sIgE levels could serve as markers of asthma however sIgG was not that important as predictor or modifier. Due to the rich diversity of food and non-food allergens, we believe that the immune reactions can be very distinct for different allergens and the mechanisms of how sIgG functions may be complicated and needs further study.

How IBD and food intolerance interact with each other remains controversial. Disruption of epithelial tight junctions causes hyperpermeability in the gut of IBD patients and allows the antigen presenting cells (dendritic cells) to directly encounter food antigens in lamina propria to activate Th/B cells, thus resulting in high levels of sIgG antibodies [40]. Simultaneously this defect may also exacerbate inflammatory conditions. In another respect, lack of certain enzyme may lead to incomplete digestion of food and the remaining polypeptides then stimulate the secretion of sIgG antibodies as well as inflammatory cytokines, thereby bringing impairments to normal intestines [41]. The definite mechanism still remains unclear.

Although we used several methods to improve our study, there were still some limitations to the present results. The small quantity of cases may bring about the first shortcoming. The sample size needs to be enlarged to see the exact trend. We didn't contain the follow-up data of whether or not the IBD patients eliminated the food based on the presence of sIgG antibodies and whether the subsequent diet elimination had an effect on the diseases, which is also a limitation of our research. However, promising results have been reported in a randomized controlled trial [19]. Bentz et al. designed a double-blind, cross-over nutritional diet intervention according to circulating food sIgG antibodies with 40 CD patients and it did show therapeutic effects with regard to abdominal pain, stool frequency and general well-being. However, the investigators could not elucidate the exact mechanisms of the contribution of sIgG antibodies to disease activity. Anyway, we still consider serologic investigation of IgG antibodies to food antigens may bring benefits to dietary modification for IBD patients.

In conclusion, the prevalence of food specific IgG antibodies is remarkably higher in patients with inflammatory bowel diseases than healthy controls, and age may acts as a protective factor for their occurrence. Although the mechanisms of the interaction between food intolerance and IBD remain obscure, the sIgG antibodies may potentially show the clinical significance to indicate disease status and to ameliorate symptoms by guiding diets for patients.

Acknowledgments

The authors would like to thank the Department of Clinical Laboratory at Ren Ji Hospital for assisting us to conduct the study.

Author Contributions

Conceived and designed the experiments: CC QZ ZR. Performed the experiments: CC QZ. Analyzed the data: CC DZ AX SJ. Contributed reagents/materials/analysis tools: JS YQ DZ SJ. Wrote the paper: CC JS DZ ZR QZ.

References

1. Bernstein CN (2001) Extraintestinal manifestations of inflammatory bowel disease. Curr Gastroenterol Rep 3: 477–483.
2. Vidal A, Gomez-Gil E, Sans M, Portella MJ, Salamero M, et al. (2008) Health-related quality of life in inflammatory bowel disease patients: the role of psychopathology and personality. Inflamm Bowel Dis 14: 977–983.
3. Macdonald TT, Monteleone G (2005) Immunity, inflammation, and allergy in the gut. Science 307: 1920–1925.
4. Goh K, Xiao SD (2009) Inflammatory bowel disease: a survey of the epidemiology in Asia. J Dig Dis 10: 1–6.
5. Magee EA, Edmond LM, Tasker SM, Kong SC, Curno R, et al. (2005) Associations between diet and disease activity in ulcerative colitis patients using a novel method of data analysis. Nutr J 4: 7.
6. Hou JK, Abraham B, El-Serag H (2011) Dietary intake and risk of developing inflammatory bowel disease: a systematic review of the literature. Am J Gastroenterol 106: 563–573.
7. Brunner B, Scheurer U, Seibold F (2007) Differences in yeast intolerance between patients with Crohn's disease and ulcerative colitis. Dis Colon Rectum 50: 83–88.
8. Haboubi NY, Jones S (2005) Influence of dietary factors on the clinical course of inflammatory bowel disease. Gut 54: 567.
9. Sakamoto N, Kono S, Wakai K, Fukuda Y, Satomi M, et al. (2005) Dietary risk factors for inflammatory bowel disease: a multicenter case-control study in Japan. Inflamm Bowel Dis 11: 154–163.
10. Jowett SL, Seal CJ, Pearce MS, Phillips E, Gregory W, et al. (2004) Influence of dietary factors on the clinical course of ulcerative colitis: a prospective cohort study. Gut 53: 1479–1484.
11. Sampson HA (2004) Update on food allergy. J Allergy Clin Immunol 113: 805–819; quiz 820.
12. David TJ (2000) Adverse reactions and intolerance to foods. Br Med Bull 56: 34–50.
13. Banai J (2009) Nutrition in inflammatory bowel disease. Orv Hetil 150: 839–845.
14. Binder JH, Gryboski JD, Thayer WR Jr, Spiro HM (1966) Intolerance to milk in ulcerative colitis. A preliminary report. Am J Dig Dis 11: 858–864.
15. Truelove SC (1961) Ulcerative colitis provoked by milk. Br Med J 1: 154–160.

16. Glassman MS, Newman LJ, Berezin S, Gryboski JD (1990) Cow's milk protein sensitivity during infancy in patients with inflammatory bowel disease. Am J Gastroenterol 85: 838–840.

17. Cashman KD, Shanahan F (2003) Is nutrition an aetiological factor for inflammatory bowel disease? Eur J Gastroenterol Hepatol 15: 607–613.

18. Atkinson W, Sheldon TA, Shaath N, Whorwell PJ (2004) Food elimination based on IgG antibodies in irritable bowel syndrome: a randomised controlled trial. Gut 53: 1459–1464.

19. Bentz S, Hausmann M, Piberger H, Kellermeier S, Paul S, et al. (2010) Clinical relevance of IgG antibodies against food antigens in Crohn's disease: a double-blind cross-over diet intervention study. Digestion 81: 252–264.

20. Zar S, Kumar D, Benson MJ (2001) Food hypersensitivity and irritable bowel syndrome. Aliment Pharmacol Ther 15: 439–449.

21. Mekkel G, Barta Z, Ress Z, Gyimesi E, Sipka S, et al. (2005) Increased IgE-type antibody response to food allergens in irritable bowel syndrome and inflammatory bowel diseases. Orv Hetil 146: 797–802.

22. Crowe SE, Perdue MH (1992) Gastrointestinal food hypersensitivity: basic mechanisms of pathophysiology. Gastroenterology 103: 1075–1095.

23. Stange EF, Travis SP, Vermeire S, Reinisch W, Geboes K, et al. (2008) European evidence-based Consensus on the diagnosis and management of ulcerative colitis: Definitions and diagnosis. J Crohns Colitis 2: 1–23.

24. Van Assche G, Dignass A, Panes J, Beaugerie L, Karagiannis J, et al. (2010) The second European evidence-based Consensus on the diagnosis and management of Crohn's disease: Definitions and diagnosis. J Crohns Colitis 4: 7–27.

25. Harvey RF, Bradshaw JM (1980) A simple index of Crohn's-disease activity. Lancet 1: 514.

26. Truelove SC, Witts LJ (1955) Cortisone in ulcerative colitis; final report on a therapeutic trial. Br Med J 2: 1041–1048.

27. Satsangi J, Silverberg MS, Vermeire S, Colombel JF (2006) The Montreal classification of inflammatory bowel disease: controversies, consensus, and implications. Gut 55: 749–753.

28. Sai XY, Zheng YS, Zhao JM, Zhao W (2011) A cross sectional survey on the prevalence of food intolerance and its determinants in Beijing, China. Chin J Epidemiol 32(3):302–05 (in Chinese).

29. Palmieri B, Esposito A, Capone S, Fistetto G, Iannitti T (2011) Food intolerance: reliability and characteristics of different diagnostic alternative tests. Minerva Gastroenterol Dietol 57: 1–10.

30. Ortolani C, Pastorello EA (2006) Food allergies and food intolerances. Best Pract Res Clin Gastroenterol 20: 467–483.

31. Wuthrich B (2009) Food allergy, food intolerance or functional disorder. Praxis (Bern 1994) 98: 375–387.

32. Husby S, Oxelius VA, Teisner B, Jensenius JC, Svehag SE (1985) Humoral immunity to dietary antigens in healthy adults. Occurrence, isotype and IgG subclass distribution of serum antibodies to protein antigens. Int Arch Allergy Appl Immunol 77: 416–422.

33. Foster AP, Knowles TG, Moore AH, Cousins PD, Day MJ, et al. (2003) Serum IgE and IgG responses to food antigens in normal and atopic dogs, and dogs with gastrointestinal disease. Vet Immunol Immunopathol 92: 113–124.

34. Van Den Bogaerde J, Cahill J, Emmanuel AV, Vaizey CJ, Talbot IC, et al. (2002) Gut mucosal response to food antigens in Crohn's disease. Aliment Pharmacol Ther 16: 1903–1915.

35. Mishkin B, Yalovsky M, Mishkin S (1997) Increased prevalence of lactose malabsorption in Crohn's disease patients at low risk for lactose malabsorption based on ethnic origin. Am J Gastroenterol 92: 1148–1153.

36. Zallot C, Quilliot D, Chevaux JB, Peyrin-Biroulet C, Gueant-Rodriguez RM, et al. (2013) Dietary beliefs and behavior among inflammatory bowel disease patients. Inflamm Bowel Dis 19: 66–72.

37. Malbec O, Daeron M (2007) The mast cell IgG receptors and their roles in tissue inflammation. Immunol Rev 217: 206–221.

38. Meulenbroek LA, de Jong RJ, den Hartog Jager CF, Monsuur HN, Wouters D, et al. (2013) IgG antibodies in food allergy influence allergen-antibody complex formation and binding to B cells: a role for complement receptors. J Immunol 191: 3526–3533.

39. Wang J, Visness CM, Calatroni A, Gergen PJ, Mitchell HE, et al. (2009) Effect of environmental allergen sensitization on asthma morbidity in inner-city asthmatic children. Clin Exp Allergy 39: 1381–1389.

40. Chahine BG, Bahna SL (2010) The role of the gut mucosal immunity in the development of tolerance against allergy to food. Curr Opin Allergy Clin Immunol 10: 220–225.

41. Isolauri E, Rautava S, Kalliomaki M (2004) Food allergy in irritable bowel syndrome: new facts and old fallacies. Gut 53: 1391–1393.

13

Effect of Breadmaking Process on *In Vitro* Gut Microbiota Parameters in Irritable Bowel Syndrome

Adele Costabile[1]*, **Sara Santarelli**[1], **Sandrine P. Claus**[1], **Jeremy Sanderson**[2], **Barry N. Hudspith**[2], **Jonathan Brostoff**[2], **Jane L. Ward**[3], **Alison Lovegrove**[3], **Peter R. Shewry**[3,4], **Hannah E. Jones**[4], **Andrew M. Whitley**[5], **Glenn R. Gibson**[1]

1 Department of Food and Nutritional Sciences, The University of Reading, Reading, United Kingdom, 2 King's College London, Biomedical & Health Sciences, Dept. of Nutrition and Dietetics, London, United Kingdom, 3 Rothamsted Research, Harpenden, Hertfordshire, United Kingdom, 4 School of Agriculture, Policy and Development, Earley Gate, Reading, United Kingdom, 5 Bread Matters Limited, Macbiehill Farmhouse, Lamancha, West Linton, Peeblesshire, Scotland

Abstract

A variety of foods have been implicated in symptoms of patients with Irritable Bowel Syndrome (IBS) but wheat products are most frequently cited by patients as a trigger. Our aim was to investigate the effects of breads, which were fermented for different lengths of time, on the colonic microbiota using *in vitro* batch culture experiments. A set of *in vitro* anaerobic culture systems were run over a period of 24 h using faeces from 3 different IBS donors (Rome Criteria–mainly constipated) and 3 healthy donors. Changes in gut microbiota during a time course were identified by fluorescence *in situ* hybridisation (FISH), whilst the small -molecular weight metabolomic profile was determined by NMR analysis. Gas production was separately investigated in non pH-controlled, 36 h batch culture experiments. Numbers of bifidobacteria were higher in healthy subjects compared to IBS donors. In addition, the healthy donors showed a significant increase in bifidobacteria ($P < 0.005$) after 8 h of fermentation of a bread produced using a sourdough process (type C) compared to breads produced with commercial yeasted dough (type B) and no time fermentation (Chorleywood Breadmaking process) (type A). A significant decrease of δ-*Proteobacteria* and most *Gemmatimonadetes* species was observed after 24 h fermentation of type C bread in both IBS and healthy donors. In general, IBS donors showed higher rates of gas production compared to healthy donors. Rates of gas production for type A and conventional long fermentation (type B) breads were almost identical in IBS and healthy donors. Sourdough bread produced significantly lower cumulative gas after 15 h fermentation as compared to type A and B breads in IBS donors but not in the healthy controls. In conclusion, breads fermented by the traditional long fermentation and sourdough are less likely to lead to IBS symptoms compared to bread made using the Chorleywood Breadmaking Process.

Editor: Hauke Smidt, Wageningen University, Netherlands

Funding: The authors have no support or funding to report.

Competing Interests: The authors have declared that no competing interests exist.

* Email: a.costabile@reading.ac.uk

Introduction

Irritable bowel syndrome (IBS) is a common functional gastrointestinal disorder defined by the coexistence of abdominal discomfort or pain associated with alterations in bowel habits [1]. Several studies have indicated that the aetiology of IBS is most likely multi-factorial, due to abnormalities in intestinal motility, visceral hypersensitivity, altered brain-gut interaction, food intolerance, abnormal gut microbiota and persistence of low-grade inflammatory conditions [2]. Due to effects on modulating the immune function, motility, secretion and gut sensation, probiotics have been suggested to have the potential to exert a beneficial role in managing IBS symptoms [3]. Furthermore, it has been suggested that IBS patients could be characterized by a potential dysregulation in energy homeostasis and liver function, which may be improved through probiotic supplementation [4]. A recent review of clinical trials using lactic acid bacteria (LAB) in patients with IBS [5] showed improvement in abdominal pain, discomfort, abdominal bloating and distension as main endpoints. Dietary

factors are also important in IBS as they are considered major drivers for changes in the compositional and functional relationship between microbiota and the host [6]. In fact, dietary components are substrates for metabolism by the intestinal microbial ecosystem, particularly influencing the growth and metabolic activities of dynamic bacterial populations thriving in the human colon. Studies on the relationships between diet and symptoms in IBS suggest that elimination of potential culprit foods can be helpful [7,8]. A variety of foods are thought to contribute to IBS, but wheat is the dietary ingredient frequently cited by patients as a trigger, with the exclusion of bread and other wheat products often leading to partial or complete resolution of symptoms [7,9]. In particular, changes in the type of bread generally available to consumers and the overall wheat content of the average diet may be significant underlying reasons why problems of gas-related gastrointestinal problems have increased. However, few studies have examined the impact of different types of bread on gastrointestinal symptoms in IBS and this is a topic worthy of further consideration [9]. To date, there is evidence that a diet low in fermentable carbohydrates, particularly fermentable oligosaccharides, disaccharides, monosaccharides and polyols (FODMAPs; also referred to as fermentable short-chain carbohydrate) reduces some symptoms associated with IBS [10–11]. In particular, Gibson and Shepherd suggest that fermentable short-chain carbohydrates can be a 'problem high food source' for those susceptible to IBS when consumed in large amounts (no specific number suggested) [10]. A recent study found that significantly more patients with IBS who followed a low-FODMAP diet (76%) reported satisfaction with their symptom response (decrease in symptoms) compared with patients following a standard diet recommended by the National Institute for Health and Clinical Excellence (54%) [11]. Although interesting, it is not possible to say which particular FODMAPs or sources of these are associated with gastrointestinal symptoms. Therefore, based on these studies, no conclusions about the impact of bread (or specific types of bread) on gastrointestinal symptoms in IBS sufferers can be drawn, although, this would seem to be a topic worthy of consideration.

A component of bread that has been suggested to help relieve IBS symptoms by shortening transit time (mainly in those suffering from constipation) is dietary fibre. However, two systematic reviews found no effect of cereal bran on IBS symptoms [12–13]. In fact, insoluble fibre, the main fibre component of bran, may increase symptoms in some IBS sufferers among whom reducing intakes of insoluble fibre may reduce symptoms. Therefore, eating white rather than whole meal bread may actually help relieve symptoms [14–15]. More specifically, a change in the process of wheat fermentation from the traditional long fermentation process to the shorter, incomplete fermentation of the Chorleywood Breadmaking Process (CBP) may have contributed to intolerance to bread through effects on gut microbiota and fermentation. Furthermore, another component of the CBP that has been suggested to be related to gastrointestinal symptoms is an increased percentage of yeast used in the fermentation process. However, whereas no evidence supporting the role of yeasts in the production of symptoms has been reported from clinical trials, dietary elimination of yeasts and anti-fungal therapy have been shown to be beneficial in IBS subjects [16]. Therefore, it is not possible to confirm or reject claims that the higher amount of yeast added to dough of bread made with CBP may be responsible for gastrointestinal problems.

Recently, there has been a growing interest in investigating the role of an altered gut microbiota in the pathogenesis of IBS [17–22]. "Healthy" gut microbiota have either direct bactericidal effects or can prevent the adherence of pathogenic bacteria to the wall of the gastrointestinal tract [23]. Dysbiosis in the gut may facilitate the adhesion of enteric pathogens in the human gut, which can be associated with IBS symptoms [23]. Alteration in the composition of the healthy microbiota and disturbed colonic fermentation in IBS patients may play an important role in development of IBS symptoms. Intestinal inflammation is generally believed to be associated with a reduced bacterial diversity and, in particular, a lower abundance of, and a reduced complexity in, the *Bacteroidetes* and *Firmicutes* phyla with a specific reduction of abundance in the *Clostridium* coccoides groups [24]. It has also been indicated that while *Firmicutes* are reduced there is an increase in gammaproteobacteria in patients with IBS [25]. In contrast to the general microbial dysbiosis theory, some researchers have suggested the involvement of specific taxa [26]. There have been a number of studies that have also highlighted a lower abundance of *F. prausnitzii* [26].

In the present study, we investigated the impact of breads fermented for different lengths of time on the human intestinal microbiota, using *in vitro* batch culture experiments with faecal donors from IBS patients and healthy control subjects. The main bacterial groups of the faecal microbiota were determined using 16S rRNA-based analyses. Metabolic effects of the breads on the microbial physiology were also studied using high resolution [1]NMR-based metabolic profiling. Finally, the *in vitro* gas production was determined in non-pH-controlled, 36 h faecal static batch cultures. As such, the intention was to assess the influence of bread making process on gut microbial fermentation *in vitro*.

Materials and Methods

Preparation of three selected breads

Grain of wheat (cv Maris Widgeon) was milled commercially to an extraction rate of 85% and was kindly supplied by Mr Andrew Whitley and Bread Matters Limited (Macbiehill Farmhouse, Lamancha, West Linton, Peeblesshire EH46 7AZ). Three types of bread, A) conventional yeasted dough, zero bulk fermentation time; B) conventional yeasted 16-hour sponge-and-dough and C) 30% sourdough, 4-hour refreshment stage, 5-hour final proof were produced. Type A was prepared accordingly to the Chorleywood Breadmaking Process (CBP). Types B and C include the metabolism of endogenous flour components (yeasts, LAB, enzymes, micro- and macro-nutrients) that are present in greater quantity in flours containing more of the germ and bran layers.

All different type of breads (A, B and C) were prepared in the Food Processing Centre (FPC) of the Department of Food and Nutritional Sciences at the University of Reading (UK).

Simulated human digestion of bread (from mouth to small intestine)

Frozen bread samples were thawed and 60 g of each sample was processed by an *in vitro* simulation of upper gut digestion and freeze dried as described by Maccaferri et al. [27]. Dialyses with membrane of 100–200 Daltons cut off (Spectra/por 100–200 Da MWCO dialysis membrane, Spectrum Laboratories Inc., UK) were used to remove monosaccharides from the pre-digested breads.

Compositional analyses of the dough and breadflour samples by ^{1}H NMR

NMR sample preparation was carried out according to a modification of the procedures described [28–29]. Extraction into 80:20 $D_2O:CD_3OD$ (1 mL) containing 0.05% w/v d_4-TSP (1 mL) was performed for three technical replicates, of 30 mg, for each biological sample. ^{1}H NMR data were collected as described below.

After analysis, to minimize variation due to differing sample pH, samples were evaporated and reconstituted in sodium phosphate buffer in D_2O (750 µL, pH = 6, 300 mM) and data collection repeated. ^{1}H NMR spectra were acquired under automation at 300°K on an Avance Spectrometer (Bruker Biospin, Coventry, UK) operating at 600.0528 MHz and equipped with a 5 mm selective inverse probe. Spectra were collected using a water suppression pulse sequence with a relaxation delay of 5 s. Each spectrum was acquired using 128 scans of 64 k data points with a spectral width of 7309.99 Hz. Spectra were automatically Fourier transformed using an exponential window with a line broadening value of 0.5 Hz. Phasing and baseline correction were carried out within the instrument software (Topspin v.2.1 and Amix (Analysis of MIXtures software, v.3.9.11), Bruker Biospin). [1]H chemical shifts were referenced to d_4-TSP at δ0.00. Quantification of individual metabolites was achieved using Chenomx Profiler (Chenomx Inc., Alberta) software against an in-house reference library of metabolite signatures of authentic compounds, with known concentrations, ran under identical conditions.

Collection and stool sample preparation

Faecal samples were obtained from 3 healthy human volunteers (two males, one female; age 30 to 38 years; BMI: 18.5–25) who were free of known metabolic and gastrointestinal diseases (e.g. diabetes, ulcerative colitis, Crohn's disease, irritable bowel syndrome, peptic ulcers and cancer). All healthy faecal donors had the experimental procedure explained to them and were given the opportunity to ask questions. All donors then provided verbal informed consent for the use of their faeces in the study and a standard questionnaire to collect information regarding the health status, drugs use, clinical anamnesis, and lifestyle was administrated before the donor was ask to provide a faecal sample. The University of Reading research Ethics Committee exempted this study from review because no donors were involved in any intervention and waived the need for written consent due to the fact the samples received were not collected by means of intervention. For the IBS donors (Rome criteria - mainly constipated), written informed consent was obtained in each case and the study was approved by the St. Thomas' Hospital Research Ethics Committee (Ref 06/Q0702/74 - A study of mucosal and luminal bacteria microbiota in irritable bowel syndrome). All faecal samples collected from healthy and IBS donors were collected on site, kept in an anaerobic cabinet (10% H_2, 10% CO_2 and 80% N_2) and used within a maximum of 15 minutes after collection. Samples were diluted 1/10 w/v in anaerobic PBS (0.1 mol/L phosphate buffer solution, pH 7.4) and homogenised (Stomacher 400, Seward, West Sussex, UK) for 2 minutes at 460 paddle-beats.

In vitro fermentations

Sterile stirred batch culture fermentation systems (50 ml working volume) were set up and aseptically filled, with 45 ml sterile, pre-reduced, basal medium [peptone water 2 g/L (Oxoid), yeast extract 2 g/L (Oxoid, Basingstoke, UK), NaCl 0.1 g/L, K_2HPO_4 0.04 g/L KH_2PO_4 0.04 g/L, $MgSO_4.7H_2O$ 0.01 g/L, $CaCl_2.6H_2O$ 0.01 g/L, $NaHCO_3$ 2 g/L, Tween 80 2 mL (BDH, Poole, UK), haemin 0.05 g/L, vitamin K1 10 µL, cysteine.HCl 0.5 g/L, bile salts 0.5 g/L, pH 7.0)] and gassed overnight with oxygen free nitrogen (15 mL/min). The different pre-digested breads, 5 g (1/10 w/v) were added to the respective fermentation vessels just prior to the addition of the faecal slurry. The temperature was kept at 37°C and pH was controlled between 6.7 and 6.9 using an automated pH controller (Fermac 260, Electrolab, Tewkesbury, UK). Each vessel was inoculated with 5 ml of fresh faecal slurry (1/10 w/v) for both healthy and IBS donors. The batch cultures (n = 3) were ran over a period of 24 h and 5 mL samples were obtained from each vessel at 0, 4, 8 and 24 h for fluorescence in situ hybridisation (FISH) and ^{1}H NMR analysis.

In vitro enumeration of bacteria population by FISH

Numbers of predominant intestinal bacterial groups, as well as total bacterial populations, were evaluated in samples from in vitro batch culture system by fluorescence in situ hybridization (FISH) analysis, as previously described by Martin-Pelaez and colleagues [30]. The probes used are reported in Table 1. They were commercially synthesised and 5′-labelled with the fluorescent Cy3 dye (Sigma, UK).

Short chain fatty acid analysis

Analysis was performed using ion exclusion high performance liquid chromatography (HPLC) system (LaChrom Merck Hitachi, Poole, Dorset UK) equipped with pump (L-7100), RI detector (L-7490) and autosampler (L-7200). Samples (1 mL) from each fermentation time point (1 mL) were centrifuged at 13,000×g for 10 min to remove bacterial cells and any particulate material. Supernatants were filtered through a 0.22 µm filter unit (Millipore, Cork, Ireland) and 20 µL injected into the HPLC, operating at a flow rate of 0.5 mL/min with heated column at 84.2°C.SCFAs (acetate, propionate, butyrate) and lactate were determined by HPLC on an Aminex HPX-87H column (300×7.8 mm, Bio-Rad, Watford, Herts, UK). Degassed 5 mM H_2SO_4 was used as eluent at a flow rate of 0.6 ml/min and an operating temperature of 50°C. Organic acids were detected by UV at a wavelength of 220 nm, and calibrated against standards of corresponding organic acids at concentrations of 12.5, 25, 50, 75 and 100 mM. Internal standard of 20 mM 2-ethylbutyric acid was included in the samples and external standards.

1H NMR Metabolomic profile of supernatants from fermentation

The fermentation supernatant from all time points was freeze-dried, dissolved in 600 µL of phosphate buffer 0.2 M (pH 7.4) in D_2O plus 0.001% TSP and 550 µL transferred into 5 mm NMR tubes for analysis. All NMR spectra were acquired on a Bruker Avance DRX 700 MHz NMR Spectrometer (Bruker Biopsin, Rheinstetten, Germany) operating at 700.19 MHz and equipped with a CryoProbe from the same manufacturer [28–29]. They were acquired using a standard 1-dimensional (1D) pulse sequence [recycle delay (RD)-90°-t1-90°-tm-90°-acquire free induction decay (FID)] with water suppression applied during RD of 2 s, a mixing time (tm) of 100 ms and a 90° pulse set at 7.70 µs. For each spectrum, a total of 128 scans were accumulated into 64 k data points with a spectral width of 14005 Hz. A range of 2D NMR spectra were performed on the same equipment for selective samples, including correlation spectroscopy (COSY), total correlation spectroscopy (TOCSY) and heteronuclear single quantum coherence (HSQC) NMR spectroscopy. The FIDs were multiplied by an exponential function corresponding to 0.3 Hz line broadening. All spectra were manually phased, baseline corrected and calibrated to the chemical shift of TSP (δ 0.00). Metabolites were assigned using our in house standard database, data from literature [37–38] and confirmed by 2D NMR experiments.

Gas production rate determinations

Sterile glass tubes (18×150 mm, Bellco, Vineland, New Jersey, USA) containing 13.5 mL pre-reduced basal medium [peptone water 2 g/L (Oxoid), yeast extract 2 g/L (Oxoid, Basingstoke, UK), NaCl 0.1 g/L, K_2HPO_4 0.04 g/L, $MgSO_4.7H_2O$ 0.01 g/L, $CaCl_2.6H_2O$ 0.01 g/L, $NaHCO3$ 2 g/L, Tween 80 2 mL (BDH, Poole, UK), haemin 0.05 g/L, vitamin K1 10 µL, cysteine-HCl 0.5 g/L, bile salts 0.5 g/L, pH 7.0)] were placed into the anaerobic cabinet and kept overnight. Pre-digested breads (1% w/v) were added to the fermentation tubes just prior to addition of the faecal inocula (1/10 w/v) [39]. The tubes were then sealed with a gas impermeable butyl rubber septum (Bellco, Vineland, New Jersey, USA) and aluminium crimp (Sigma Aldrich, Gillingham, Dorset, UK). Gas production was evaluated by recording the headspace pressure (pounds per square inch; psi) from each vial. Gas production experiments were performed in four replicates for each type of bread. Vials were incubated at 37°C and continuously shaken. Pressure readings were obtained

Table 1. Oligonucleotide probes used in this study for FISH analysis.

Probe	Target group	Reference
EUB338[§]	Most bacteria	[15]
EUB338II[§]	Most bacteria	[15]
EUB338III[§]	Most bacteria	[16]
Bac303	*Bacteroides* spp.	[17]
Bif164	*Bifidobacterium* spp.	[17]
Lab158	*Lactobacillus-Enterococcus* spp.	[18]
Erec482	Most of the *Clostridium coccoides-Eubacterium rectale* group (*Clostridium* cluster XIVa and XIVb)	[19]
Chis150	*Clostridium histolyticum* group	[19]
Prop853	*Clostridium* cluster IX	[20]
Delta496a-b-c	*Deltaproteobacteria-Gemmatimonadetes* group	[21]

[§]These probes are used together in equimolar concentrations.

every 3 h up to 36 h fermentation period by piercing the rubber caps with a U200/66 needle adaptor connected to a pressure transducer (type 2200BGF150WD3DA; Keller Ltd, Dorchester, Dorset, UK) with a T443A digital panel meter (Bailey and Makey Ltd, Birmingham, UK). Pressure readings (psi) were converted into gas volume (mL) using an established linear regression of pressure recorded in the same vials with known air volumes at the incubation temperature.

Statistical analysis

Differences between bacterial counts and SCFA profiles at 0, 4, 8 and 24 h fermentation for each substrate were tested for significance using paired t-tests assuming equal variances and considering a two-tailed distribution. To determine whether there were any significant differences in the effect of the substrates; differences at each time point were tested using 2-way ANOVA with Bonferroni post-test. All analyses were performed using GraphPad Prism 5.0 (GraphPad Software, La Jolla, CA, USA).

Metabolic profiles of fermentation supernatant were imported into Matlab version R2010b (Mathworks UK) and statistical algorithms were provided by Korrigan Sciences (Korrigan Sciences Ltd, UK). To minimise variability due to water pre-saturation, the water resonance region (δ 4.70–5.05) was removed. Data were then normalised to the probabilistic quotient as previously described [40]. All statistical models were performed using unit variance scaling. Principal component analyses (PCA) were performed on all spectra in order to detect any outliers and to identify patterns associated with volunteers, time, fermentation condition or donor group. In order to optimise statistical separation between samples derived from IBS and control donors at 24 h, a partial least square discriminant analysis was also performed using one predictive component. This later model was validated using 1000 random permutations, and a p value was calculated by rank determination of the model actual Q^2Y value (representing the goodness of prediction) among the Q^2Y values calculated for the permutated models. Finally, in order to focus on the ethanol production in control- and IBS-derived samples, the area under the ethanol triplet at 1.18 ppm was integrated and an ANOVA followed by a multiple comparison test (TukeyHSD) were performed in R (version 2.15).

Results

Compositions and properties of flour, doughs and breads

A sample of wheat cv Maris Widgeon was milled to 85% extraction rate to give a flour fraction enriched in fibre and other components derived from the bran and germ, compared to pure white flour which is derived solely from the starchy endosperm. This flour was similar to those used by many artisan bakers in the UK. Three types of bread were produced, with yeast but zero bulk fermentation (similar to the Chorleywood Breadmaking Process (CBP) (which is used widely for factory production of bread in the UK and many other countries) (type A), with yeast and 16 hours fermentation (type B) and a sourdough process using a "starter dough" and a total of 9 hours fermentation (type C). The composition of major soluble polar metabolites in the flour, doughs and bread samples were determined by ^{1}H NMR (Table 2). Typically, flour contained lower amounts of the abundant free sugars (maltose, glucose and fructose) that tended to be broken down in the dough samples by a longer fermentation process. Less abundant flour carbohydrates included sucrose and raffinose. Sucrose levels were markedly lower in the CBP dough and bread (type A) and in the dough and bread undergoing longer fermentation (type B) but remained stable in the sourdough samples. Sugars such as arabinose, xylose and galactose were not detected in the flour spectra but were present in all the bread and dough spectra. Increasing the fermentation time did not change the amounts of these carbohydrates (CBP vs dough/bread B). However, significantly higher levels of these sugars were released by the sourdough process. In addition, the dough and bread samples produced using the sourdough process contained higher levels of glycerol and mannitol, the latter not being present in bread B or that produced using the CBP. Organic acids showed striking differences between the samples. As expected, lactate levels were increased with longer fermentation and were very high in both the sourdough (405 µmoles/g) and the sourdough bread (111.4 405 µmoles/g). Other organic acid levels such as citrate and malate also discriminated the samples. While the malate levels fell to around 30% in the long fermentation samples compared with the CBP process, the citrate levels remained stable even with a longer fermentation process but were completely absent from the sourdough product spectra, Succinate generally showed an opposite profile, increasing during longer fermentation (B) but decreasing in the sourdough products. Interestingly, dough B was the only sample containing low levels of acetic acid. In general, the sourdough dough and resultant bread had significantly higher contents of many polar metabolites than the CBP and long fermentation doughs and breads. The majority of amino acid (alanine, valine, leucine, isoleucine, glutamate, glutamine, aspartate, phenylalanine, tryptophan, tyrosine and gamma-amino butyric acid (GABA, a non-protein amino acid) were present in higher concentrations in the sourdough samples and in some cases also the bread made using this process. Similarly, signals corresponding to methionine, whose levels were not detected in other flour, bread or dough samples, were clearly present in the spectra from the sourdough samples. Notable exceptions were asparagine, which showed significantly lower levels in dough and bread B but whose levels were unchanged in the sourdough samples and threonine whose levels were decreased with a longer fermentation process and which disappeared completely in the sourdough products. Choline and glycine-betaine, which are methyl donors, were elevated in both sourdough samples and those arising from process B compared to the CBP products. Ethanol, a product of the breadmaking process was present in all dough and bread samples and was typically higher in samples receiving a longer fermentation. All doughs had lower levels of raffinose and maltose than their corresponding bread products, which is consistent with their use as substrates during proofing.

Changes in faecal microbiota measured by FISH

Eight 16S rRNA-based fluorescence *in situ* hybridisation (FISH) probes were used to identify predominant groups, or species, of human faecal microbiota before and after incubation with digested bread samples (Table 1). Bacterial numbers of the samples from IBS donors were compared to the samples obtained from healthy subjects (Table 3). Numbers of bifidobacteria were higher in the control group compared to the IBS donors. A significant increase in bifidobacterial populations occurred (P< 0.005) after 8 hours of fermentation in bread produced with sourdough (type C) for healthy donors compared to breads produced with commercial yeast dough and no time fermentation (type A). No significant changes were also noted in *Bacteroides-Prevotella* group populations (detected by Bac303) at all time points in IBS donors. However, all type of breads stimulated the growth of bacteria detected by Bac303 at 8 h and over 24 hours fermentation in healthy donors, but there was no significant difference compared to the control substrate. No significant differences were detected for *Clostridium histolyticum* subgroup (detected by Chis150) and lactobacilli in IBS and healthy donors. Significant decreases in δ-Proteobacteria and most *Gemmatimonadetes* (enumerated by probe DELTA495 a-b-c), which are sulphate-reducing microorganisms, was observed after 24 h fermentation of type C bread in IBS and healthy donors. This may be due to the ability of the sourdough bread to enhance the growth of beneficial bacteria rather than undesirable microorganisms. Cluster IX representatives (detected by Prop853) were increased by bread type C at 8 h and 24 h in both donor types.

Short chain fatty acid analysis

Short chain fatty acids (SCFAs), which are the principal end products of gut bacterial metabolism, were measured after 0, 4, 8 and 24 h fermentation with the different test substrates using HPLC analysis. All substrates gave significant increases in total SCFA concentration after 8 h of fermentation in both donor types with fermentation of type C bread leading to significant increases in concentrations of butyrate after 8 h fermentation in both donor groups. Acetate was the dominant SCFA produced after 24 h

fermentation with all breads and in both IBS and healthy donors. Data are shown in Table 4.

Metabolic profiling

Metabolic profiles of fermentation supernatants (Type A, B and C breads) were acquired at 0, 4, 8 and 24 h post inoculation by High Resolution 700 MHz NMR spectroscopy. Principal component analysis (PCA) revealed a clear trajectory over time, mainly due to decreasing carbohydrate concentration and increased production of SCFAs (Figure 1). The cluster of samples isolated by PC2 displayed a higher polar lipid content (corresponding to medium chain fatty acids).

Supernatants from the different breads could not be statistically differentiated from one another but were all distinguished from the controls due to lower levels of polyethylene glycol (PEG), lipids and branched chain amino acids in the fermented bread samples (Figure 2). As expected, all supernatants incubated with bread samples displayed higher levels of SCFAs compared to controls (Figure 2).

While the fermentation supernatants (all types of bread) derived from the IBS and control patients could not be separated at 0 h post-fermentation, they were clearly separated at 24 h, as indicated by the PCA displayed in Figure 2A. An Orthogonal Partial Least Square (O-PLS) analysis also provided significant discrimination ($R^2Y = 0.82$, $R^2X = 0.10$, $Q^2Y = 0.40$; permutation test based on 1000 random permutations resulted in a p value of 0.003) (Figure 2B). This separation was due to a higher content of ethanol and taurine in the controls and of proline in the IBS samples. In order to determine more precisely the extent of ethanol production after 24 h of fermentation in these 2 groups, the area under the ethanol resonance of the methyl protons at 1.18 ppm was integrated at 0 h and 24 h (Figure 2C). This shows a 4 times increase in ethanol production in control-derived samples while almost no increase was observed in IBS-derived samples. The large standard error observed in IBS-derived samples at 24 h was due to the fact that the supernatants from only one donor contained ethanol.

Gas production kinetics

Gas production during the 36 h of non pH-controlled faecal batch culture is shown in Figure 3. The rates of gas production for type A and B breads were almost identical in IBS and healthy donors, peaking after 6 h, and continuing for up to 36 h (Figure 3 A, B). Type C bread resulted significantly in lower rates combined with lower total gas production (data not shown) compared to the control (P<0.05). This indicates that type C was fermented more slowly to produce a more gradual build-up of gas compared to other selected breads (Figure 3, B).

Discussion

Irritable bowel syndrome (IBS) is a common functional bowel disorder, with an estimated worldwide prevalence of 10%–20% among adults and adolescents. IBS is characterised by pain or discomfort, disturbed bowel habits and altered stool characteristics. The exact aetiology of IBS is likely to be multifactorial; moreover, patients diagnosed with the disorder may also be experiencing bowel symptoms due to different causes. Much attention has recently been focused on the impact of gastrointestinal microbiota on this disorder [20–26]. Indeed, in recent years, there has been much greater recognition that bloating results mainly from abnormal levels of gut fermentation. It is not known exactly which microbial agents contribute to excessive fermentation but there is evidence to support a role for both bacteria and

Table 2. Quantification (μmol/g dry wt) of selected metabolites in flour, dough A, B and C and breads A, B and C.

	Flour	Dough A (CBP)	Dough B	Dough C	Bread A (CBP)	Bread B	Bread C
Carbohydrates							
Glucose	2.693±0.577	34.072±1.185	22.133±1.702	129.00±11.68	30.873±9.117	13.339±1.297	45.036±1.062
Fructose	4.587±1.844	45.220±2.219	29.081±1.942	20.456±8.571	44.756±3.048	24.417±5.238	8.601±5.868
Maltose	7.660±1.007	41.521±3.981	27.471±1.014	31.160±2.643	101.420±0.171	53.900±5.684	78.210±9.105
Galactose	n.d.	1.168±0.345	1.196±0.544	9.506±1.017	0.814±0.234	0.799±0.660	1.911±0.723
Sucrose	2.199±0.302	0.414±0.073	0.281±0.052	1.912±0.256	0.421±0.087	0.441±0.087	2.413±0.072
Raffinose	2.194±0.245	2.176±1.548	1.129±0.824	2.754±1.312	2.487±0.363	1.607±0.396	5.407±0.386
Xylose	n.d.	0.992±0.596	1.324±0.640	11.280±3.900	1.009±0.432	0.879±0.408	2.443±0.569
Trehalose	n.d.	n.d.	n.d.	9.147±0.943	1.706±0.329	0.713±0.078	n.d.
Arabinose	n.d.	0.916±0.230	2.523±0.347	19.519±2.609	1.221±0.282	2.652±1.717	2.153±0.053
Sugar alcohols							
Glycerol	4.111±0.594	19.148±5.771	23.648±8.953	32.329±3.190	17.733±5.763	31.106±3.708	13.986±1.875
Mannitol	n.d.	5.944±5.421	7.488±1.305	86.124±7.017	n.d.	n.d.	35.170±0.751
Organic Acids							
3-Hydroxyisobutyrate	n.d.	0.167±0.012	n.d.	n.d.	0.207±0.003	n.d.	n.d.
Citrate	0.912±0.102	0.961±0.041	1.019±0.099	n.d.	1.472±0.187	0.876±0.363	n.d.
Fumarate	0.453±0.012	0.626±0.071	0.304±0.063	0.092±0.019	0.647±0.050	0.503±0.038	0.637±0.118
Malate	5.664±0.168	5.942±0.390	1.692±0.169	3.054±0.907	7.060±0.497	1.894±0.426	n.d.
Succinate	0.427±0.007	1.111±0.072	1.612±0.098	1.032±0.238	1.317±0.087	1.572±0.133	0.873±0.035
Formate	0.352±0.011	0.490±0.019	0.487±0.093	0.762±0.059	0.586±0.109	0.934±0.127	0.710±0.098
Lactate	0.857±0.057	1.058±0.199	11.638±0.624	404.89±46.022	1.048±0.166	11.233±1.206	111.417±3.443
Amino Acids							
Alanine	0.526±0.025	0.989±0.076	1.041±0.054	5.284±0.396	1.146±0.062	0.961±0.125	1.622±0.066
Asparagine	1.543±0.055	1.613±0.087	0.494±0.045	2.039±0.699	1.719±0.018	0.724±0.254	2.191±0.202
Aspartate	1.837±0.070	1.736±0.117	1.341±0.063	9.310±1.008	2.017±0.111	1.272±0.185	3.531±0.080
GABA	0.249±0.072	0.796±0.050	1.031±0.080	5.699±0.561	0.917±0.089	0.939±0.167	2.199±0.075
Glutamate	1.203±0.369	1.129±0.532	0.546±0.065	8.321±2.195	0.974±0.114	0.651±0.231	1.587±0.321
Glutamine	0.558±0.047	0.726±0.097	0.448±0.142	3.371±0.239	0.601±0.106	0.367±0.047	0.783±0.058
Leucine	0.139±0.045	0.250±0.075	0.221±0.057	14.050±1.599	0.342±0.091	0.149±0.021	2.246±0.733
Isoleucine	0.129±0.039	0.159±0.036	0.158±0.063	4.113±0.234	0.191±0.025	0.118±0.030	0.666±0.047
Methionine	n.d.	n.d.	n.d.	2.436±0.248	n.d.	n.d.	0.323±0.009
Phenylalanine	n.d.	0.178±0.022	0.124±0.069	4.912±0.446	0.186±0.050	n.d.	0.853±0.078
Threonine	0.132±0.047	0.487±0.166	0.378±0.098	n.d.	0.347±0.018	n.d.	n.d.
Tryptophan	0.800±0.056	0.782±0.014	0.509±0.132	2.570±0.327	0.921±0.071	0.473±0.094	1.281±0.042
Tyrosine	0.132±0.022	0.167±0.022	0.171±0.041	2.762±0.451	0.192±0.034	0.152±0.044	0.497±0.044

Table 2. Cont.

	Flour	Dough A (CBP)	Dough B	Dough C	Bread A (CBP)	Bread B	Bread C
Valine	0.173±0.032	0.284±0.034	0.270±0.036	6.993±0.219	0.384±0.017	0.230±0.033	1.344±0.035
Methyl Donors							
Betaine	8.464±0.658	8.919±1.826	9.903±0.458	22.930±1.860	6.262±0.289	8.070±4.125	15.688±0.401
Choline	1.103±0.066	1.798±0.092	2.139±0.076	3.266±0.239	1.570±0.098	1.917±0.185	2.508±0.060
Choline-O-Sulfate	0.301±0.044	0.530±0.038	0.613±0.054	2.941±0.297	0.424±0.049	0.519±0.040	0.908±0.005
Acetylcholine	0.048±0.005	0.042±0.002	0.042±0.008	1.999±0.092	0.049±0.005	0.043±0.010	0.358±0.008
Phosphocholine	0.062±0.007	0.053±0.003	0.060±0.006	2.593±0.120	0.204±0.250	0.056±0.015	0.463±0.009
Trigonelline	0.076±0.008	0.089±0.015	n.d.	n.d.	0.083±0.020	n.d.	n.d.
Miscellaneous							
Ethanol*	0.907±0.098	14.802±0.437	21.350±0.839	30.112±4.874	29.49±0.384	37.918±9.78	17.340±0.325
Putrescine	n.d.	n.d.	n.d.	n.d.	n.d.	n.d.	1.074±0.175
Adenine	n.d.	n.d.	n.d.	0.663±0.173	n.d.	n.d.	0.636±0.039
Adenosine	0.070±0.012	0.222±0.012	0.090±0.003	0.093±0.015	0.343±0.109	0.179±0.013	0.063±0.014
Uridine	0.109±0.027	0.263±0.057	0.156±0.010	n.d.	0.339±0.078	0.207±0.052	n.d.

n.d. denotes metabolite not detected at above 0.075 micromoles/g.

Errors are standard deviations of 3 replicates.

* denotes data obtained from an 80:20 D_2O:CD_3OD extraction rather than buffer at pH 6.5.

Table 3. Bacterial populations (Mean value Log$_{10}$ cells mL± SD) in pH controlled and stirred batch.

Probe	Time (h)	Negative control^h	Negative Control^{IBS}	Type A^h	Type A^{IBS}	Type B^h	Type B^{IBS}	Type C^h	Type C^{IBS}
Bif164	0	8.22±0.13	7.75±0.35	8.26±0.16	7.58±0.20	8.44±0.25	7.63±0.07*	8.48±0.11	7.94±0.06*
	4	8.34±0.13	7.81±0.33	8.63±0.24	7.78±0.47	8.46±0.23	7.98±0.36	8.60±0.08	8.22±0.09*
	8	8.44±0.17	7.95±0.31	8.69±0.32	8.33±0.29	8.71±0.38	8.28±0.26	8.86±0.15^A	8.29±0.28
	24	8.15±0.10	7.82±0.22	8.31±0.36^B	7.97±0.30	8.39±0.31	7.97±0.41	8.45±0.48	8.02±0.31
Lab158	0	7.92±0.15	8.04±0.51	8.12±0.26	8.00±0.40	8.14±0.17	7.83±0.40	8.09±0.01	7.89±0.40
	4	7.68±0.14	7.82±0.23	7.99±0.14	7.70±0.13	8.07±0.20	7.73±0.49	8.00±0.19	7.76±0.17*
	8	7.82±0.36	7.85±0.21	8.34±0.32	8.04±0.32	8.21±0.23	8.06±0.29	8.16±0.72	8.05±0.38
	24	7.99±0.30	7.82±0.18	8.22±0.24	7.76±0.20**	8.11±0.07	7.78±0.32	8.20±0.23	7.84±0.08
Eub338	0	9.06±0.20	8.82±0.12	8.91±0.11	8.75±0.28	8.81±0.22	9.06±0.29	9.09±0.25	8.95±0.35
	4	8.79±0.14	9.11±0.24	9.12±0.24	9.08±0.18	8.92±0.16	8.92±0.37	8.76±0.17	9.01±0.69
	8	9.11±0.39	9.04±0.23	9.48±0.08^A	9.41±0.26^A	9.46±0.28^A	9.24±0.21	9.22±0.62	9.25±0.21
	24	9.44±0.43	9.22±0.19^A	9.45±0.21	9.50±0.18^{AB}	9.76±0.19^{AC}	9.40±0.40^A	9.48±0.36	9.53±0.17^C
Erec482	0	8.26±0.18	8.60±0.55	8.18±0.60	8.37±0.21	8.52±0.56	8.22±0.21	8.05±0.13	8.08±0.17
	4	8.29±0.30	8.26±0.43	8.36±0.13	8.25±0.19	8.33±0.48	8.12±0.56	8.37±0.35	7.93±0.18
	8	8.14±0.53	8.26±0.36	9.07±0.04^B	8.17±0.21*	8.63±0.62	7.91±0.86	8.63±0.42	7.61±0.51
	24	8.43±0.36	8.60±0.10	8.96±0.32	8.29±0.14*	8.41±0.20	8.16±0.44	8.93±0.30	7.63±0.39
Prop853	0	7.19±0.30	8.06±0.38	7.37±0.44	8.15±0.35	7.53±0.44	8.19±0.39	7.50±0.39	7.74±0.20
	4	7.76±0.40	8.27±0.34	8.19±0.33	8.02±0.17	7.98±0.24	7.55±0.25	8.23±0.24	7.48±0.17^{A*}
	8	8.17±0.01^A	8.23±0.10	8.52±0.17	8.13±0.20	8.45±0.34^B	8.03±0.23	8.21±0.31	8.09±0.23^{AB}
	24	8.04±0.20^A	8.47±0.08^C	8.19±0.38	8.14±0.21	8.00±0.22	8.13±0.04	8.27±0.21^A	8.12±0.18^{AB}
Bac303	0	8.16±0.17	8.20±0.14	8.19±0.35	8.35±0.07	8.13±0.06	8.35±0.23	7.79±0.15	8.26±0.24
	4	7.98±0.47	8.46±0.22	8.50±0.14	8.30±0.29	8.14±0.40	8.45±0.11	8.16±0.51	8.43±0.33
	8	8.22±0.20	8.28±0.18	8.79±0.29	8.23±0.17	8.82±0.48^B	8.50±0.21	9.22±0.62^A	8.46±0.10
	24	8.43±0.34	8.29±0.02	9.20±0.15^{AB}	8.35±0.35*	8.61±0.59	8.29±0.20	8.84±0.26^A	8.48±0.24
Chis150	0	7.68±0.61	8.12±0.42	7.88±0.56	8.05±0.49	7.86±0.49	8.07±0.51	8.17±0.03	8.31±0.45
	4	8.18±0.27	8.31±0.39	8.10±0.06	8.02±0.51	7.88±0.28	7.98±0.45	8.04±0.39	8.22±0.09
	8	8.11±0.70	8.26±0.29	8.30±0.50	8.09±0.28	8.27±0.71	8.39±0.14	8.41±0.44	8.36±0.21
	24	7.93±0.68	8.14±0.08	8.08±0.67	8.34±0.36	8.14±0.62	8.07±0.40	8.26±0.64	7.82±0.68
Delta496a-b-c	0	7.43±0.50	6.95±0.13	7.31±0.51	7.12±0.13	7.36±0.32	7.11±0.27	7.30±0.58	6.89±0.49*
	4	7.67±0.15	7.09±0.09*	7.52±0.16	7.80±0.13^A	7.38±0.30	7.93±0.24*	7.49±0.24	7.48±0.28
	8	7.63±0.15	7.52±0.50	7.52±0.35	8.08±0.10^A	7.49±0.31	8.03±0.15	7.55±0.30	7.28±0.25
	24	7.50±0.58	7.18±0.34	7.38±0.28	7.83±0.28^A	7.20±0.07	8.05±0.33^C	7.16±0.23^C	7.45±0.24

cultures at 0, 4, 8 and 24 inoculated with healthy^h and IBS faecal microbiota^{IBS}.
^{A}Significantly different from 0 h for the same substrate.
^{B}significantly different from 4 h for the same substrate.
^{C}significantly different from 8 h for the same substrate (paired t-test, p<0.05).
* p<0.05, ** p<0.01, *** p<0.001 Significantly different from control (without any additional substrate) using two-way ANOVA with Bonferroni post-test.

Table 4. Short chain fatty acids production ± SD by bread fermentations in pH controlled and stirred batch cultures at 0, 4, 8 and 24 inoculated with healthy[h] and IBS faecal microbiota[IBS].

	Time (h)	Negative control[h]	Negative control[IBS]	Type A[h]	Type A[IBS]	Type B[h]	Type B[IBS]	Type C[h]	TypeC[IBS]
Total	0	0±0.00	0±0.00	0.22±0.38	0.15±0.26	0±0.00	0.26±0.40*	0±0.00	0.68±0.19*
production	4	8.20±10.5	12.26±17.0	6.30±6.10	7.01±6.20	1.30±1.75	0.00±0.36	0.00±0.08	1.90±1.85*
	8	8.44±0.17	7.95±0.31	8.69±0.32	8.33±0.29	8.71±0.38	8.28±0.26	8.86±0.15a	8.29±0.28
	24	6.50±1.63	7.01±1.63	50.83a±12.3***	43.0a±14.0***	49.15a±15.6***	51.98a±16.0***	52.00 a ±12.0***	59.0 a ±5.0 ***
Acetic acid	0	0.00±0	0.22±0.38	0.16±0.26	0.00±0.00	0.26±0.00	0.53±0.20	0.19±0.01	0.05±0.30
	4	1.40±1.22	1.33±1.22	7.20±7.14	7.70±0.13	0.07±0.20	7.73±0.49	4.10±3.19	9.67±9.16***
	8	1.30±0.36	1.30±0.30	5.34±8.32	25.01 a ±15.12**	18.21±7.23	8.06±0.29	12.65 a,b±17.72	27.38 a,b ±14.19
	24	3.99±0.30	5.82±0.18	40.22 a,c±8.24***	47.76 a,c±5.20***	36.41a,b ±7.07***	7.78±0.32	32.20 a,b,c ±12.23	34.85 a,b,c ±9.53***
Propionic acid	0	0.05±0.08	0.00±0.12	0.91±0.11	0.75±0.28	0.00±0.22	0.06±0.29	0.09±0.25	0.95±0.35
	4	0.50±0.61	1.11±3.12	3.26±3.88	7.26±7.88	0.92±0.16	1.26±1.88	1.59±2.76	1.26±1.88
	8	0.42±0.12	0.60±1.05	4.88±6.48	8.88±10.48	5.88±10.2	2.88±4.48	0.22±0.00	2.88±4.48
	24	1.70±0.59	5.22±3.19	9.45±0.21	14.54±13.56	7.21±7.05	8.54±7.56	14.48±12.7 a	18.54±17.56 a
Butyric acid	0	0.05±0.08	0.11±0.16	0.00±0.02	0.00±0.00	0.00±0.20	0.00±0.01	0.05±0.13	0.08±0.17
	4	0.50±0.61	0.00±0.00	8.36±0.13	0.25±0.19	0.33±0.48	2.12±0.56	0.37±0.35	0.93±0.18
	8	0.42±0.11	0.26±0.36	11.07±8.04b	18.17±0.21*	21.63±11.62	7.91±5.86	18.63±0.42	17.61±0.51
	24	1.61±0.59	0.34±0.31	18.96±11.32	28.29±0.14*	38.41±20.20	18.16±12.44	28.93±0.30 a,b,c	32.63±0.39***

Values are mmol/L concentrations in batch culture at 0, 4, 8 and 24 h fermentation as means of three experiments with different faecal donors.
[a]Significantly different from initial concentration (P<0.05).
[b]Significantly different from 4 h concentration (P<0.05).
[c]Significantly different from 24 h concentration (P<0.05) *P<0.05, **P<0.01, ***P<0.001.
Significantly different from control (cellulose) using 2-way ANOVA with Bonferroni post-test.

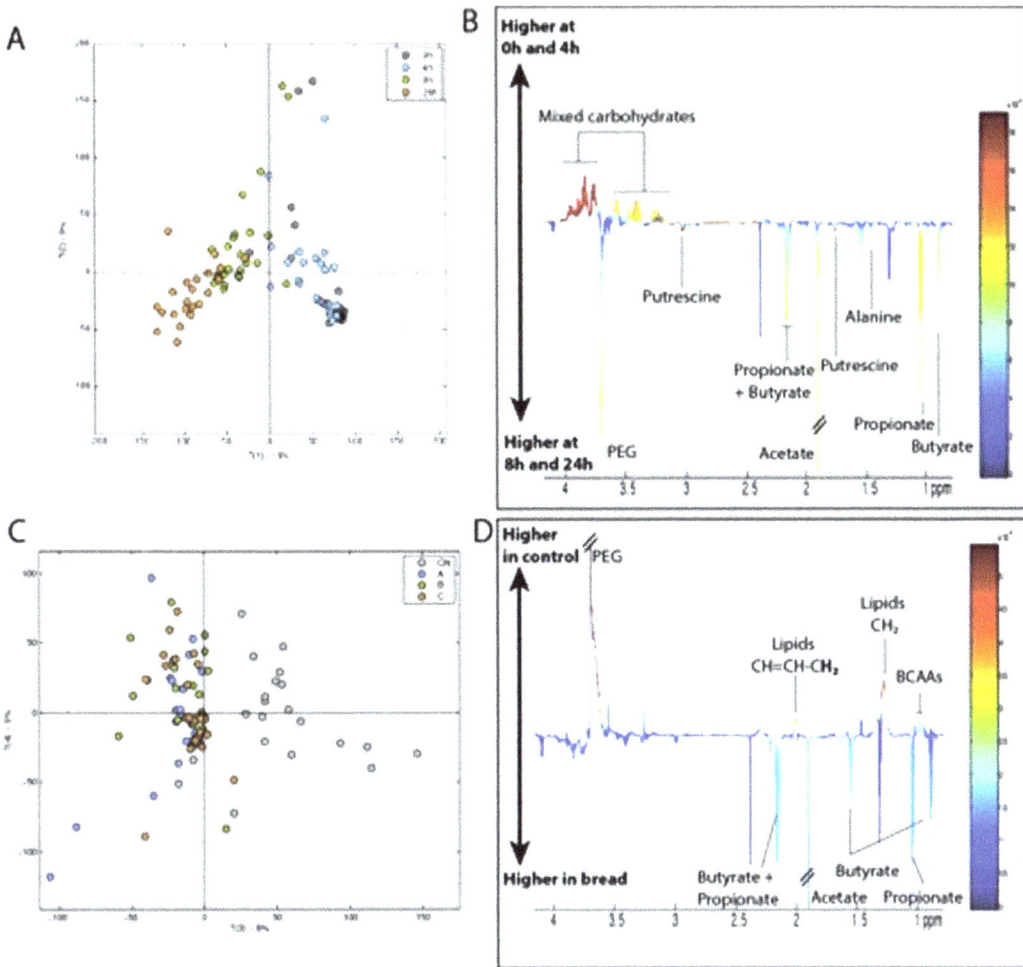

Figure 1. Metabolic trajectories of bread fermentated by gut bacteria obtained from both control and IBS patients (n = 3). PC1 versus PC2 scores plot (A) and PC1 loadings (B) derived from the 700 MHz ¹H NMR spectra of fermentation supernatants color coded for collection time-points. Key: Grey: 0 h, Blue: 4 h, Green: 8 h, Orange: 24 h. PC3 versus PC4 scores plot color coded for bread (C) and PC3 loadings (D). Key: Grey: control, Blue: bread A, Green: bread B, Orange: Bread C.

yeasts [16]. To date, wheat is frequently cited by patients as a trigger with exclusion of bread and other wheat products often leading to partial or complete resolution of symptoms [41–43]. Very few studies have investigated the effects of different bread-making processes on bloating or gastrointestinal symptoms. More specifically, a change in bread making processes from a traditional long fermentation process to a short, incomplete fermentation may have contributed to bread intolerance through its effects on fermentation in the colon. However, hitherto, there is no published evidence to support claims that bread made with the Chorleywood Bread Process (CBP) affects the gastrointestinal system in a different way compared with the more traditional Bulk Fermentation Process (BFP) or other commonly used bread-making processes [10–12].

Our hypothesis is that bread fermented by a traditional long fermentation technique is less likely to lead to IBS symptoms, especially gas and bloating, compared to bread made using the widely used short CBP. In this context, the overall aim of this study was to compare the fermentation properties of three breads prepared with different conditions using *in vitro* batch culture. Analysis of dough and breads showed clear effects of the production process on the concentrations of polar metabolites,

including carbohydrates which could affect the pattern of fermentation in the colon. For example, the sourdough process resulted in high levels of xylose, arabinose, galactose and mannitol, none of which are normally detected in flour samples. However, these differences may well be modulated by digestion and absorption in the upper part of the GI tract, and may not therefore represent the composition of the samples entering the colon.

In vitro studies of digested bread samples were therefore carried out to determine the impact of the processing system on the intestinal microbiota and to compare their ability to enhance faecal bifidobacteria. Bifidobacteria are of particular interest because this genus is used as a probiotic, does not produce gas, and has been tested for positive effects on IBS [5]. As expected, numbers of bifidobacteria were higher in healthy people compared to IBS donors. The increase in bifidobacteria population was also significantly higher (P<0.005) after 8 hours of fermentation of bread produced using a sourdough process (type C) for healthy people compared to breads produced with commercial yeasted dough and no time fermentation. In particular, the CBP (type A) bread showed significant increase in the bifidobacteria populations

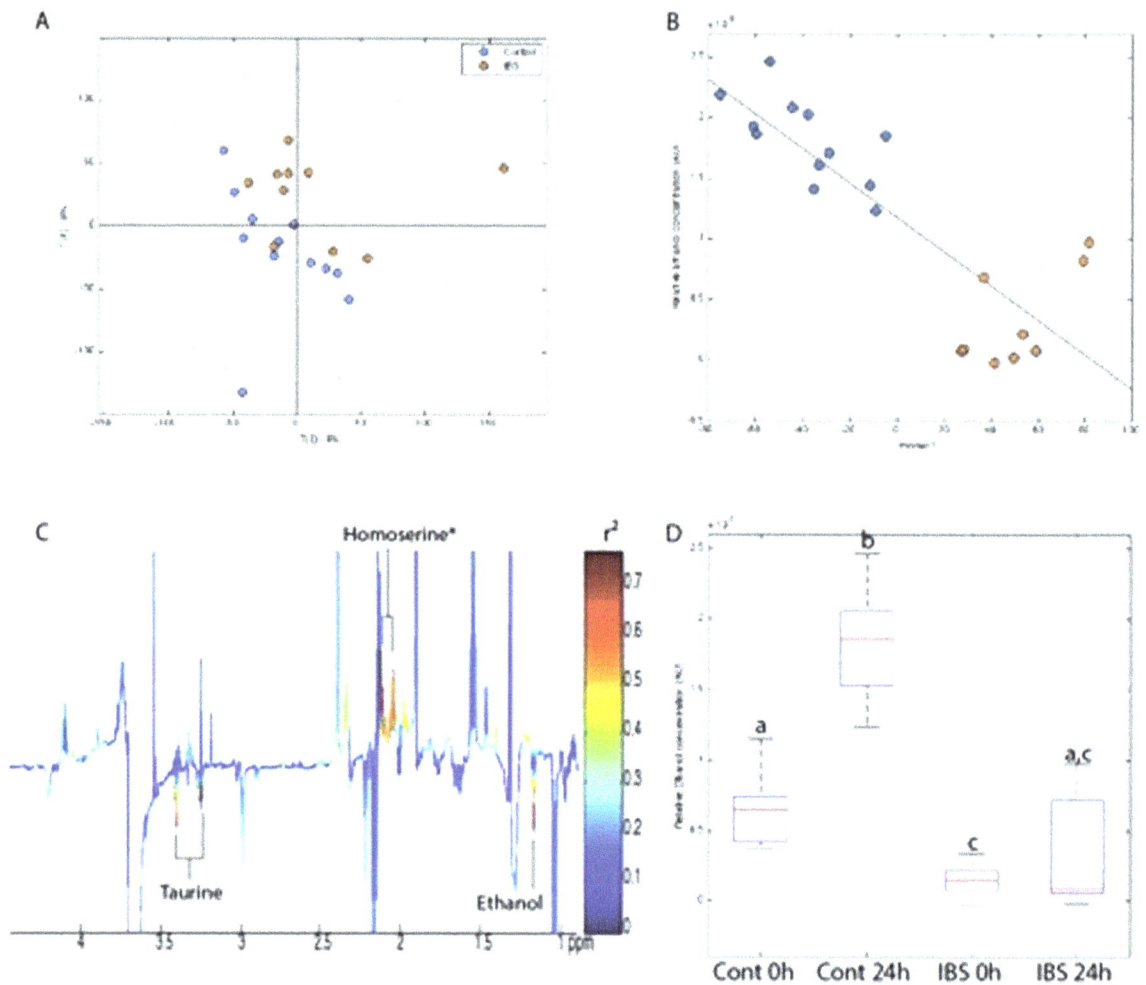

Figure 2. Divergent fermentation of bread samples by IBS and control microbiota. PCA scores plot (A) and O-PLS-DA scores (B) and associated loadings (C) derived from 700 MHz ^{1}H NMR spectra of fermentation supernatants. Relative ethanol production derived from the integrated area under the curve of original NMR spectra for the methyl protons at 1.18 ppm. ANOVA: $p < 2.10^{-16}$; Multiple comparison test: (a) $p < 0.0001$ different from b and c, (b) $p < 0.0001$ different from a and c. Key: Blue: control; Orange: IBS. *putative assignement.

(enumerated by probe Bif164) after 24 h. No significant change was recorded in bifidobacteria numbers in IBS patients.

Short chain fatty acids (SCFAs) were determined after 0, 4, 8 and 24 h fermentation with the different test substrates via HPLC and NMR techniques. All substrates gave significant increases in total SCFAs concentrations after 24 h fermentation in both donor types. Acetate was the dominant SCFA produced in all fermentations in both IBS and healthy donors. Fermentation of sourdough (type C) bread led to a significant increase in concentrations of butyrate. By focusing on faecal short-chain fatty acids (SCFAs) as the major end product of bacterial metabolism in the human large intestine, researchers have shown that SCFAs were increased in diarrhoea-predominant IBS patients and decreased in constipation-predominant IBS patients. However, another study reported the conflicting finding that SCFAs are decreased in diarrhoea-predominant IBS patients, suggesting that it is necessary to conduct a broader analysis of faecal microbiota, full profiles of organic acids and simultaneous GI symptoms in IBS patients [44].

Nevertheless, the suggested link between the SCFAs profile and GI symptoms can be discussed in the light of the contrasting biological activities of the SCFAs [45]. Acetate is a known

chemical irritant, and at high concentrations is used to induce mucosal lesions and abdominal cramps in experimental animals, while butyrate is considered as protective and able to dose-dependently reduce abdominal pain in humans *in vivo* [46]. Tana and co-workers [46] suggest that altered intestinal microbiota contributes to the symptoms of IBS through increased levels of organic acids. Furthermore, IBS patients with high acetic acid or propionic acid levels presented more severe symptoms, impaired quality of life and negative emotions. These results are in accordance with the concept that the gut microbiota influences the sensory, motor and immune system of the gut and interacts with higher brain centers [44–50]. Furthermore, metabolomic analysis of bread fermentation by gut bacteria did not distinguish between the different types of bread, although they all produced larger amounts of SCFAs compared to controls, as expected. Ethanol production was only consistently measured in control patients while only one IBS volunteer out of three was able to produce ethanol from bread fermentation. Ethanol is usually further metabolised to acetate but we could not detect any significant difference in acetate production between IBS and control patients and therefore conclude that the absence of ethanol in IBS patients could not be explained by an increased metabolism

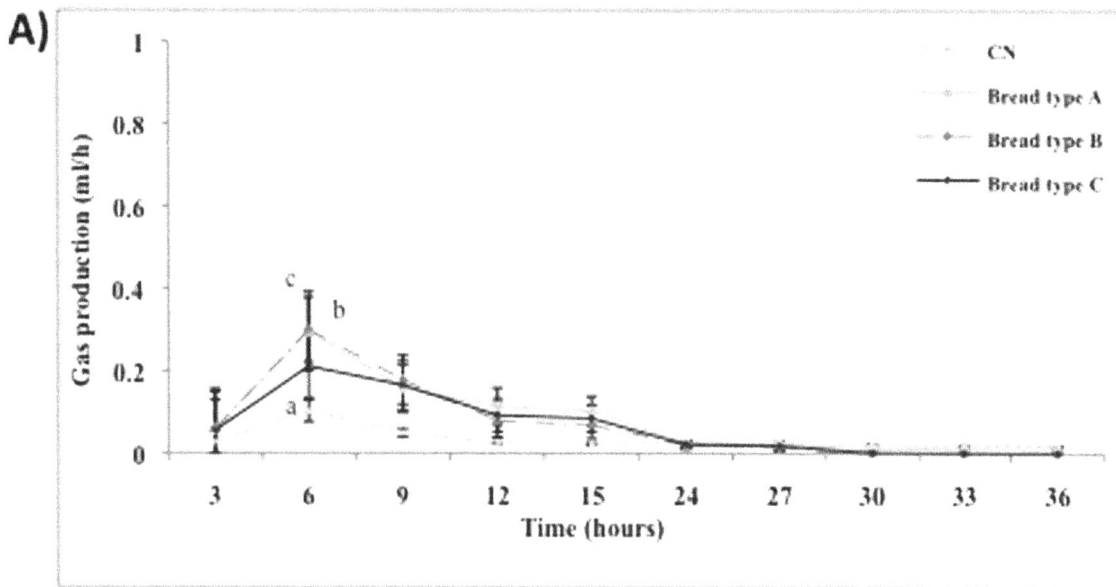

A)

Significant differences among treatment ($P < 0.05$) are indicated with different letters (a: slowest gas rate to c: fastest gas rate).

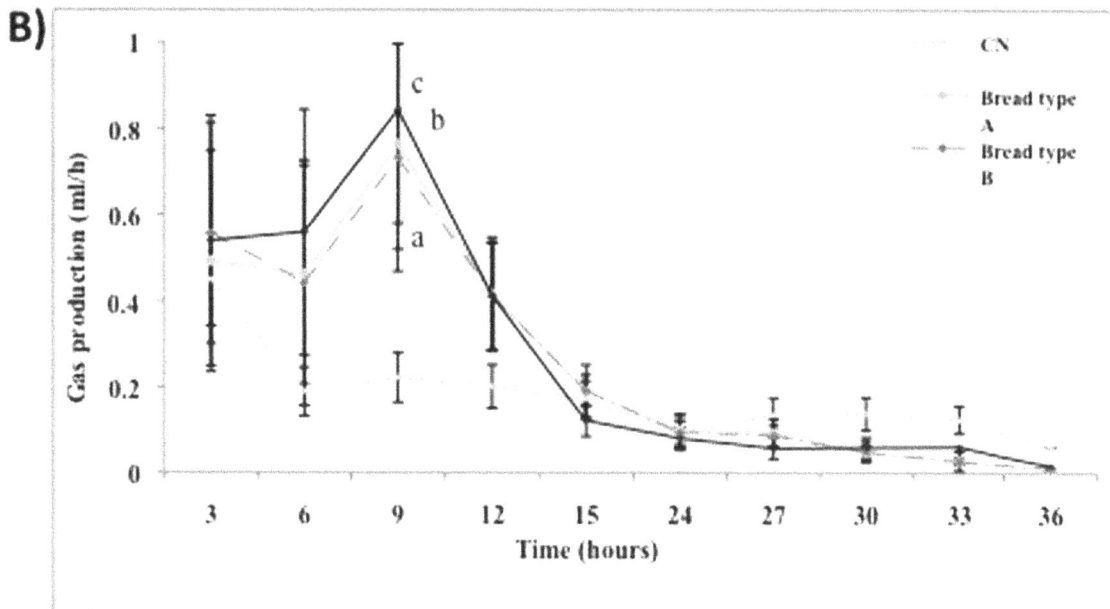

B)

Significant differences among treatment ($P < 0.05$) are indicated with different letters (a: slowest gas rate to c: fastest gas rate).

Figure 3. Gas production pattern expressed in mL per hour from non-pH controlled batch culture (average results ± standard deviation of 3 volunteers, $n=3$) inoculated with healthy faecal microbiota (A); Gas production pattern expressed in mL per hour from non-pH controlled batch culture (average results ± standard deviation of 3 volunteers, $n=3$) inoculated with IBS faecal microbiota (B).

to acetate. However, sulphate-reducing bacteria can convert ethanol to acetate, and this could therefore have occurred concomitant with the production of higher organic acids.

In vitro gas production was determined in non pH- controlled, 36 h faecal static batch culture tubes. In general, the IBS donors showed higher rates of gas production and total gas compared to healthy donors. Similarly, the rates of gas production for type A and B breads were almost identical in IBS and healthy donors.

Sourdough (type C) bread produced significantly lower cumulative gas after 15 h fermentation compared to the other types. This was also observed for the rate of gas production.

To conclude, significant changes were observed in the bacterial populations with sourdough (type C) bread, including lower numbers of sulphate-reducing bacteria, i.e., *Desulfovibrionales*, compared to types A and B.

All types of bread generated some gas after 9 h of fermentation and the patterns of gas production were similar for type A and B breads but were both significantly higher than for type C in IBS donors. In addition, an increase in the concentration of butyrate was the main impact of all breads on the overall SCFA production.

These findings suggest that sourdough products may be advantageous for patients suffering from IBS. This study provides findings supporting the utilization of breads fermented by the traditional long fermentation and sourdough with a positive effect on the composition and metabolic profile of the human intestinal microbiota.

References

1. Thompson WG, Longstreth GF, Drossman DA, Heaton KW, Irvine EJ, et al. (1999) Functional bowel disorders and functional abdominal pain. Gut 45, Suppl. 2, II43–II47.
2. Brenner DM, Moeller MJ, Chey WD, Schoenfeld PS (2009) The utility of probiotics in the treatment of irritable bowel syndrome: a systematic review. American Journal of Gastroenterology, 104(4), 1033–49, doi:10.1038/ajg.2009.25.
3. Camilleri M (2008) Probiotics and irritable bowel syndrome: rationale, mechanisms, and efficacy. Journal of Clinical Gastroenterology, 42 Suppl 3, S123–5. doi:10.1097/MCG.0b013e3181574393.
4. Hong Y-S, Hong KS, Park M-H, Ahn Y-T, Lee J-H, et al. (2011) Metabonomic understanding of probiotic effects in humans with irritable bowel syndrome. Journal of Clinical Gastroenterology, 45, 415–425.
5. Clarke G, Cryan JF, Dinan TG, Quigley EM (2012) Review article: probiotics for the treatment of irritable bowel syndrome–focus on lactic acid bacteria. Alimentary Pharmacology & Therapeutics, 35(4), 403–13. doi:10.1111/j.1365-2036.2011.04965.x.
6. Laparra JM, Sanz Y (2010) Interactions of gut microbiota with functional food components and nutraceuticals. Pharmacological Research 61(3), 219–225.
7. Hunter JO, Workman E, Jones AV (1985) The role of diet in the management of irritable bowel syndrome. In: Gibson PR, Jewell DP, eds. Topics in Gastroenterology. Oxford, Blackwell Scientific, pp. 305–313.
8. Stefanini GF, Saggioro A, Alvisi V, Angelini G, Capurso L, et al. (1995) Oral cromolyn sodium in comparison with elimination diet in the irritable bowel syndrome. diarrheic type. Multicenter study of 428 patients. Scandinavian Journal of Gastroenterology, 30, 535–441.
9. Bijkerk CJ, Muris JW, Knottnerus JA, Hoes AW, de Wit NJ (2004) Systematic review: the role of different types of fibre in the treatment of irritable bowel syndrome. Alimentary Pharmacology and Therapeutics 19, 245–251.
10. Gibson PR, Shepherd SJ (2010) Evidence-based dietary manage- ment of functional gastrointestinal symptoms: the FODMAP approach. Journal of Gastroenterology and Hepatology 25, 252–258.
11. Staudacher HM, Whelan K, Irving PM, Lomer MC (2011) Comparison of symptom response following advice for a diet low in fermentable carbohydrates (FODMAPs) versus standard dietary advice in patients with irritable bowel syndrome. Journal of Human Nutrition and Dietetics 24, 487–495.
12. Ford AC, Talley NJ, Spiegel BM, Foxx-Orenstein AE, Schiller L, et al. (2008) Effect of fibre, antispasmodics, and peppermint oil in the treatment of irritable bowel syndrome: systematic review and meta-analysis. British Medical Journal 337, a2313.
13. Spiller R, Aziz Q, Creed F, Emmanuel A, Houghton L, et al. (2007) Guidelines on the irritable bowel syndrome: mechanisms and practical management. Gut 56, 1770–1798.
14. NICE (National Institute for Health and Clinical Excellence) (2008) Irritable bowel syndrome in adults. Diagnosis and management of irritable bowel syndrome in primary care. NICE clinical guideline 61. National Collaborating Centre for Nursing and Supportive Care and National Institute for Health and Clinical Excellence, London.
15. Santelmann H, Howard JM (2005) Yeast metabolic products, yeast antigens and yeasts as possible triggers for irritable bowel syndrome. European Journal of Gastroenterology and Hepatology, Jan;17(1), 21–26.
16. Ghoshal UC, Park H, Gwee KA (2010) Bugs and irritable bowel syndrome: the good, the bad and the ugly. Journal of Gastroenterology and Hepatology, vol. 25, no. 2, 244–251.
17. Simre n M, Stotzer PO (2006) Use and abuse of hydrogen breath tests. Gut, 55, 297–303.
18. Scanu AM, Bull TJ, Cannas S, Sanderson JD, Sechi LA, et al (2007) Mycobacterium avium subspecies paratuberculosis infection in cases of irritable bowel syndrome and comparison with Crohn's disease and Johne's disease: common neural and immune pathogenicities. Journal of Clinical Microbiology, 45, 3883–3890.
19. Malinen E, Krogius-Kurikka L, Lyraetal A (2010). Association of symptoms with gastrointestinal microbiota in irritable bowel syndrome. World Journal of Gastroenterology, 16, 4532–4540.
20. Kassinen A, Krogius-Kurikka L, Mäkivuokko H, Rinttilä T, Paulin L, et al. (2007) The fecal microbiota of irritable Bowel syndrome patients differs significantly from that of healthy subjects. Gastroenterology, 133, 24–33.
21. Lyra A, Krogius-Kurikka L, Nikkilä J, Malinen E, Kajander K, et al. (2010) Effect of a multispecies probiotic supplement on quantity of irritable bowel syndrome-related intestinal microbial phylotypes. BMC Gastroenterology, 10, 110.
22. Parkes GC, Brostoff J, Whelan K, Sanderson JD (2008). Gastrointestinal microbiota in irritable bowel syndrome: their role in its pathogenesis and treatment. American Journal of Gastroenterology, 103, 1557–1567.
23. Lee BJ, Bak TJ (2011) Irritable bowel syndrome, gut microbiota and probiotics. Journal of Neurogastroenterology and Motility, 17, 252–266.
24. Ponnusamy K, Choi JN, Kim J, Lee SY, Lee LH (2011) Microbial community and metabolomic comparison of irritable bowel syndrome faeces. Journal of Medical Microbiology, 60, 817–827.
25. Ghoshal U, Shukla R, Ghoshal U, Gwee K, Ng S, et al.(2012) The gut microbiota and irritable bowel syndrome: friend or foe? International Journal of Inflammation, 151085.
26. Maccaferri S, Klinder A, Cacciatore S, Chitarrari R, Honda H, et al. (2012) In vitro fermentation of potential prebiotic flours from natural sources: impact on the human colonic microbiota and metabolome. Molecular Nutrition and Food Research 56, 1342–1352. doi:10.1002/mnfr.201200046. Epub 2012 Jul 2.
27. Baker JM, Hawkins ND, Ward JL, Lovegrove A, Napier JA, et al. (2006) A metabolomic study of substantial equivalence of field-grown GM wheat. Plant Biotechnology Journal 4, 381–392.
28. Howarth JR, Parmar S, Jones J, Shepherd CE, Corol DI, et al. (2008) Co-ordinated expression of amino acid metabolism in response to N and S deficiency during wheat grain filling. Journal of Experimental Botany 59, 3675–3689.
29. Martin-Pelaez S, Gibson GR, Martin-Orue SM, Klinder A (2008) In vitro fermentation of carbohydrates by porcine faecal inocula and their influence on Salmonella Typhimurium growth in batch culture systems. FEMS Microbiol Ecology 66(3), 608–619.
30. Amann RI, Binder BJ, Olson RJ, Chisholm SW, Devereux R, et al. (1990) Combination of 16S rRNA-targeted oligonucleotide probes with flow cytometry for analyzing mixed microbial populations. Applied Environmental Microbiology 56, 1919–1925.
31. Langendijk PS, Schut F, Jansen GJ, Raangs GC, Kamphuis GR, et al. (1995) Quantitative fluorescence in situ hybridization of Bifidobacterium spp. With genus-specific 16S rRNA-targeted probes and its application in fecal samples. Applied Environmental Microbiology 61, 3069–3075.
32. Manz W, Amann R, Ludwig W, Vancanneyt M, Schleifer KH (1996) Application of a suite of 16S rRNA-specific oligonucleotide probes designed to investigate bacteria of the phylum Cytophaga-Flavobacter- Bacteroides in the natural environment. Microbiology 142, 1097–1106.
33. Harmsen HJM, Elfferich P, Schut F, Welling GW (1999) A 16S rRNAtargeted probe for detection of lactobacilli and enterococci in fecal samples by fluorescent in situ hybridization. Microbiology Ecology in Health and Disease 11, 3–12.
34. Franks AH, Harmsen HJ, Raangs GC, Jansen GJ, Schut F, et al. (1998) Variations of bacterial populations in human feces measured by fluorescent in situ hybridization with group-specific 16S rRNA-targeted oligonucleotide probes. Applied and Environmental Microbiology 64, 3336–3345.
35. Walker AW, Duncan SH, McWilliam Leitch EC, Child MW, Flint HJ (2005) pH and peptide supply can radically alter bacterial populations and short-chain fatty acid ratios within microbial communities from the human colon. Applied and Environmental Microbiology 71, 3692–3700.
36. Lücker S, Steger D, Kjeldsen KU, MacGregor BJ, Wagner M, et al. (2007) Improved 16S rRNA-targeted probe set for analysis of sulfate-reducing bacteria by fluorescence in situ hybridization. Journal of Microbiology Methods 69, 523–528.
37. Fan T, Lane A (2008) Structure-based profiling of metabolites and isotopomers by NMR. Progress in Nuclear Magnetic Resonance Spectroscopy 52, 69–117.
38. Costabile A, Kolida S, Klinder A, Gietl E, Bäuerlein M, et al. (2010) A double-blind, placebo-controlled, cross-over study to establish the bifidogenic effect of a very-long-chain inulin extracted from globe artichoke (Cynara scolymus) in healthy human subjects. British Journal of Nutrition 104, 1007–1017.

Acknowledgments

We thank Miss Meline Bes and Dr Delia-Irina Corol for excellent technical assistance. Rothamsted Research receives strategic funding from the Biotechnology and Biological Sciences of the UK.

Author Contributions

Conceived and designed the experiments: JS BNH JB HEJ. Performed the experiments: SS AC SC JLW AL. Analyzed the data: AC SC BNH JLW AL PRS. Wrote the paper: AC SC PRS JS JB GRG. Contributed live sourdough cultures for baking experiments: AMW.

39. Dieterle F, Ross A, Schlotterbeck G, Senn H (2006) Probabilistic quotient normalization as robust method to account for dilution of complex biological mixtures. Application in 1H NMR metabonomics. Analytical Chemistry, 78, 4281–4290.
40. De Angelis M, Rizzello CG, Alfonsi G, Arnault P, Cappelle S, et al. (2007) Use of sourdough lactobacilli and oat fibre to decrease the glycemic index of white wheat bread. British Journal of Nutrition 98, 1196–1205.
41. Di Cagno R, Rizzello CG, De Angelis M, Cassone A, Giuliani G, et al. (2008) Use of selected sourdough for enhancing the nutritional and sensory properties of gluten-free bread. Journal of Food Protection 71, 113–117.
42. Katina K, Arendt E, Liukkonen KH, Autio K, Flander L, et al. (2005) Potential of sourdough for healthier cereal products. Trends in Food Science and Technology 16, 104–112.
43. Koide A, Yamaguchi T, Odaka T, Koyama H, Tsuyuguchi T, et al. (2000) Quantitative analysis of bowel gas using plain abdominal radiograph in patients with irritable bowel syndrome. American Journal of Gastroenteroly, 95, 1735–1741.

44. Mortensen PB, Andersen JR, Arffmann S, Krag E (1987) Short-chain fatty acids and the irritable bowel syndrome: the effect of wheat bran. Scandinavian Journal of Gastroenteroloy, 22, 185–192.
45. Tana C, Umesaki Y, Imaoka A, Handa T, Kanazawa M, et al. (2010) Altered profiles of intestinal microbiota and organic acids may be the origin of symptoms in irritable bowel syndrome Neurogastroenterology Motility 22, 512–519.
46. Valeur J, Norin E, Midtvedt T, Berstad A (2010) Assessment of microbial fermentation products in fecal samples. Neurogastroenteroly Motility, DOI:10.1111/j.1365-2982.2010. 01558.x.
47. Treem WR, Ahsan N, Kastoff G, Hyams JS (1996) Fecal short-chain fatty acids in patients with diarrhoea-predominant irritable bowel syndrome: in vitro studies of carbohydrate fermentation. Journal Pediatric Gastroenteroly and Nutrition 23, 280–236.
48. Kopecny J, Simunek J (2002) Cellulolytic bacteria in human gut and irritable bowel syndrome. Acta Veterinaria Brno 71, 421–427.
49. Lee KJ, Tack J (2010) Altered intestinal microbiota in irritable bowel syndrome. Neurogastroenterology Motility 22, 493–498.

14

The Influence of Whole Grain Products and Red Meat on Intestinal Microbiota Composition in Normal Weight Adults: A Randomized Crossover Intervention Trial

Jana Foerster[1]*, Gertraud Maskarinec[1,2], Nicole Reichardt[3], Adrian Tett[3], Arjan Narbad[3], Michael Blaut[4], Heiner Boeing[1]

1 Department of Epidemiology, German Institute of Human Nutrition Potsdam-Rehbruecke, Nuthetal, Germany, 2 University of Hawaii Cancer Center, Honolulu, Hawaii, United States of America, 3 Gut Health and Food Safety, Institute of Food Research, Norwich Research Park, Colney, United Kingdom, 4 Department of Gastrointestinal Microbiology, German Institute of Human Nutrition, Potsdam-Rehbruecke, Nuthetal, Germany

Abstract

Intestinal microbiota is related to obesity and serum lipid levels, both risk factors for chronic diseases constituting a challenge for public health. We investigated how a diet rich in whole grain (WG) products and red meat (RM) influences microbiota. During a 10-week crossover intervention study, 20 healthy adults consumed two isocaloric diets, one rich in WG products and one high in RM. Repeatedly data on microbiota were assessed by *16S rRNA* based denaturing gradient gel electrophoresis (DGGE). A blood sample and anthropometric data were collected. Mixed models and logistic regression were used to investigate effects. Microbiota showed interindividual variability. However, dietary interventions modified microbiota appearance: 8 bands changed in at least 4 participants during the interventions. One of the bands appearing after WG and one increasing after RM remained significant in regression models and were identified as *Collinsella aerofaciens* and *Clostridium sp.* The WG intervention lowered obesity parameters, while the RM diet increased serum levels of uric acid and creatinine. The study showed that diet is a component of major relevance regarding its influence on intestinal microbiota and that WG has an important role for health. The results could guide investigations of diet and microbiota in observational prospective cohort studies.

Trial registration: ClinicalTrials.gov NCT01449383

Editor: Vincent Wong, The Chinese University of Hong Kong, Hong Kong

Funding: The study was supported by the European Union funded project TORNADO (FP7 - KBBE 222720). Sources of financial support: 7th Framework Programme, EU. The funders had no role in study design, data collection and analysis, decision to publish, or preparation of the manuscript.

Competing Interests: The authors have declared that no competing interests exist.

* Email: Jana.Foerster@dife.de

Introduction

The composition of intestinal microbiota is gaining importance in human health as evidence is increasing that these bacteria play a role in disease aetiology [1]. Thus, it is important to increase knowledge in how lifestyle factors, e.g., diet, physical activity and smoking influence gut microbiota. This seems to be crucial since large observational studies, such as prospective cohort studies, aim at examination of the involvement of human microbiota in health and disease [2]. It is well established that the microbiome enables complex interactions between the 10^{14} cells of the intestinal microbiota and its host, including processes of fat storage, and maturation and maintenance of the immune system [3,4].

One important factor determining microbiota composition from birth besides the host's genome is habitual diet since different foods primarily serve as substrate for microbial growth in the gastrointestinal system [5]. A major function of gut microorganisms is to digest food compounds that are not degraded by human enzymes. In this manner, compounds like complex polysaccharides and starch promote proliferation of certain bacterial populations, which in turn provide degradation products, e.g.,

monosaccharides or short chain fatty acids (SCFA) for subsequent absorption. Results of a former study in mice indicate that in this way a changed microbiota composition contributes to a higher energy yield, weight gain, and possibly obesity [6].

Prospective studies have revealed two food groups with opposing effects on diet-related diseases. Whole grain (WG) products (and the associated intake of dietary fibre) appear to reduce risk for metabolic disorders, whereas red meat (RM) is more likely to constitute a risk factor for chronic diseases [7–10]. Therefore, the present study investigates the effects of diets high in WG products and RM, respectively, on gut microbiota composition by an isocaloric intervention approach in healthy adults. The results will help to concentrate the statistical efforts of risk evaluation in prospective cohort studies on those groups of bacteria that are known to be directly modified by dietary changes.

Methods

This crossover trial assessed the effects of WG and RM products on the composition of intestinal microbiota, anthropometry, several blood parameters and faecal biomarkers. The study protocol was approved by the ethics commission of the Brandenburg State Chamber of Physicians in October 2010 and the trial was conducted in accordance with the Declaration of Helsinki [11]. The study was conducted without deviation from the study protocol approved by the ethics commission. All participants signed an informed consent form. Because the trial was approved by an ethics commission already in October 2010 it was registered on ClinicalTrials.gov, NCT01449383, after starting the enrolment of participants in October 2011. The authors confirm that all ongoing and related trials for this intervention conducted by the department are registered. The protocol for this trial and the supporting CONSORT checklist are available as supporting information; see Checklist S1 and Protocol S1.

2.1 Study population and participant recruitment

Between November 2010 and August 2011 twenty healthy free-living participants (Figure 1) with equal portion men and women aged 20 to 60 years were recruited by advertisement and flyers, which were spread in the public in the surrounding of the institute. Additionally invitations were sent via email. Exclusion criteria were acute or chronic gastrointestinal diseases, disorders, or surgeries, prevalent chronic diseases such as diabetes mellitus and cancer, antibiotic treatment during the last 3 month, pregnancy, and breastfeeding. All participants were asked to report the consumption of pre- and probiotic foods, such as yogurts, and supplements during the last 2 weeks before the intervention.

2.2 Study plan and interventions

After recruitment and pre-study information sessions, the participants were instructed about their diet during the intervention. Successive order of invitation rendered an individual and detailed instruction possible. All participants obtained booklets containing information on the diet during the intervention periods. Every participant took part in both intervention periods lasting 3 weeks each seperated by a 3-week washout period (Figure 2). Sequence of diet was randomised via coincident allocation of the first diet and the two diets were isocaloric. The diet during the WG period required low intake of red meat products, i.e., not more than 30 g per day, and high amounts of WG products resulting in a daily intake of approximately 40 g dietary fibre (WG intervention). The other diet consisted of 200 g of RM per day and minimal amounts of dietary fibre (RM intervention) (Figure 2).

The participants were provided with the foods needed for the interventional diets free of charge. During WG, they received 3 types of bread rich in dietary fibre from the bakery 'Maerkisch Landbrot', ('Bergroggenbrot', 'Roggensonne' and 'Korn&Kraeuterbrot') with a fibre content between 12.2 and 15.2 g per 100 g. Additionally, a muesli from 'Seitenbacher' (Muesli 348 – Ballaststoffmischung, fibre content: 14.3 g/100 g) was supplied. For RM, portions of 200 g (fresh weight) of red meat, i.e., pork cutlet, beef steak and others were provided for every day of the intervention. In addition to receiving foods, participants were asked regularly about their consumption of the foods in order to assess adherence to the interventions. We decided against dietary assessment by questionnaire or recall due to the short intervention period and the highly conscientious participants. Instead, we relied on the detailed instructions, the food supplies and regular personal contact with the participants. Also we were only interested in the effects of red meat and whole grain consumption and not in the usual daily diet of the participants.

2.3 Data collection

Before and after each period participants have an appointment in the institute's study centre. At this body height, body weight, waist and hip circumference, and skin fold thickness on arm, back and hip were measured. Body mass index (BMI) [kg/m^2], waist to hip ratio (WHR) and body fat mass [%] [12] were calculated. At the same time, a blood sample was drawn, which was analysed immediately. Additionally, all participants provided a faecal sample, which was collected at the participants' home, frozen immediately after collection at $-20°C$ in a freezer, transported to the Institute without thawing within 3 hours, and stored at $-80°C$ until analyses. Participants were examined between November 2010 and September 2011 according to the above mentioned timeframes.

2.4 Laboratory analysis

Blood samples were analysed for clinical blood parameters (C-reactive protein [CRP], total cholesterol, high density lipoprotein [HDL], low density lipoprotein [LDL], triglycerides [TG], uric acid and creatinine) using standard clinical procedures. The cholesterol quotient was calculated as the proportion of LDL to HDL.

The faecal microbiota composition was analysed by PCR-DGGE. Briefly, total DNA from faecal samples was extracted using the FastDNA Spin Kit for Soil with a modified protocol [13] and stored at $-20°C$ until further analyses. The DNA was used as a template to amplify the variable regions V6-V8 of the bacterial *16S rRNA* genes with primers U968-GC-f (5'-<u>CGCCCGGGGGCGCGCCCCGGGCGGGGCGGGGGCACG</u>GGGGGAACGCGAAGAACCTTAC-3') and U1401-r (5'-CGGTGTGTACAAGACCC-3') [14]. The G+C clamp at the 5'-end of the primer U968-GC-f is underlined. PCR amplification was performed as follows: initial denaturation at 94°C for 5 min, denaturation at 94°C for 30 s, followed by 35 cycles with an annealing temperature of 50°C for 20 s, and primer extension at 72°C for 40 s, and a final extension for 7 min at 72°C.

For DGGE analysis 150 ng of PCR product was analysed by using the DCode Universal Mutation Detection System as described previously with minor modifications. Briefly, a 40–60% gradient of urea and formamide was used and the gels were run in 1 x TAE buffer at 200 V for 10 min and then at 55 V for 16 h. DGGE gels were stained for 45 min in a Sybr Green solution (5 μl/300 ml H$_2$O dest.), washed for 15 min in 300 ml H$_2$O dest. and scanned with a Pharos FX scanner. DGGE gel analysis and inter-gel-comparisons were performed using the Phoretix 1D and 1D Pro software package as described by Tourlomousis et al. [15]. The analysis of the scans provided dichotomous information on the presence or absence of each potentially detectable DGGE band in each sample. Additionally, bands that were found to be changed after intervention periods were identified by sequencing the corresponding V6-V8 regions using primers U968 and U1401 and PCR conditions given above.

Further, faecal concentrations of SCFA (acetic, propionic, butyric, iso-valeric and valeric acid) were measured with gas chromatography with iso-butyric acid as internal standard [16]. Total SCFA concentration was calculated as the sum of all acids. Faecal calprotectin was measured with a commercially available monoclonal antibody-based enzyme-linked immunosorbent assay following the manufacturer's instructions. This analytic method has coefficients for intra-assay variation between 3.2 and 5.6% and for inter-assay variation between 4.4 and 8.9% [17]. All analyses were performed with thawed and homogenized faecal samples.

2.5 Statistical analysis

Potential differences in baseline characteristics between the two intervention groups were investigated with Student's t test for continuous variables and x^2 tests for categorical variables. All

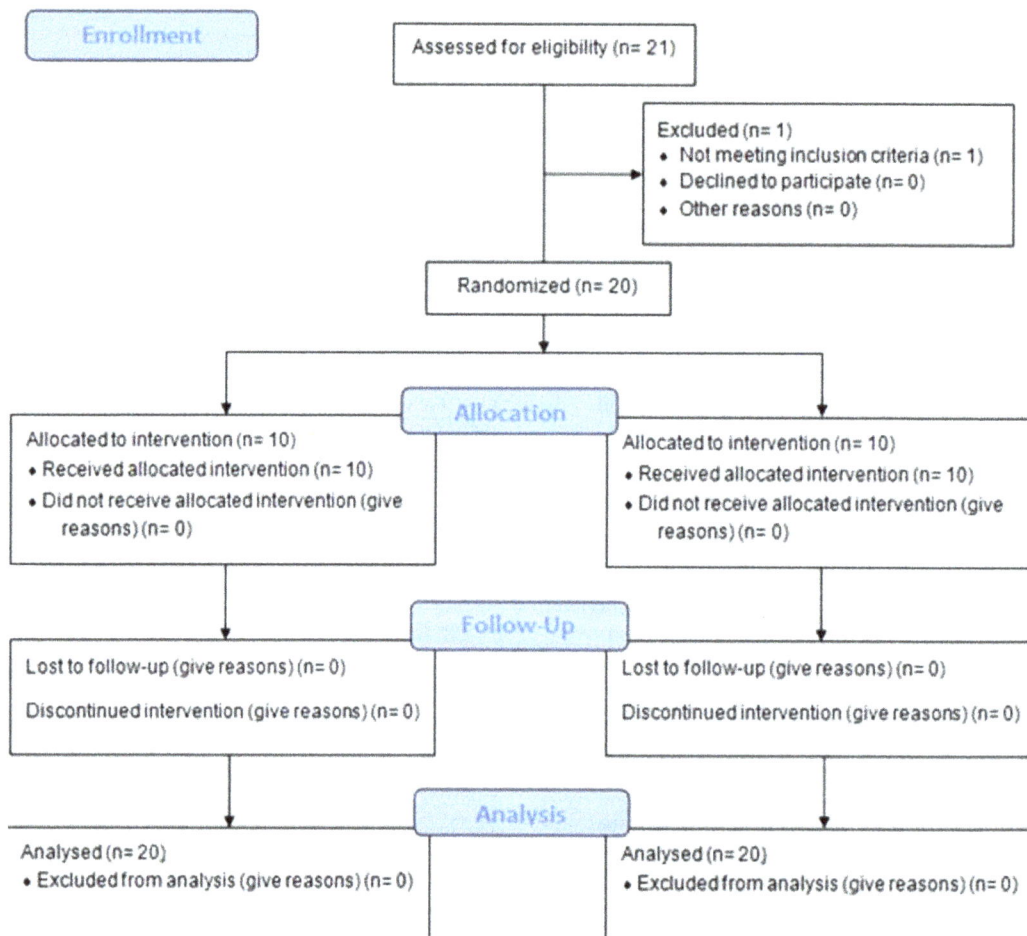

Figure 1. CONSORT flowchart of the study.

variables were tested for normal distribution with the Shapiro-Wilk test; non-normally distributed variables (WHR, uric acid, creatinine, calprotectin, CRP, triglycerides, total SCFA and butyric acid) were log-transformed for further analyses. To assess intervention effects, mixed linear models were applied for continuous variables taking into account the dependence of observations within-person due to the repeated measurement design and least-square means by diet status (no intervention, RM, WG) were computed. For 'no intervention' status characteristics from baseline and washout period were combined if there exist no differences.

The descriptive analysis of gut microbiota included the relative abundance of bands and the diversity within samples expressed as total number of bands per sample. Since gut microbiota data were binary, the appearance of bands was evaluated as bands present before an intervention and absent afterwards ($1 \rightarrow 0$) or vice versa ($0 \rightarrow 1$). To identify bands that were influenced by one of the dietary interventions, we searched for bands, for which absolute changes into one direction were observed in at least 4 (20%) participants. This cut-off value was coincidentally the value for the corresponding McNemar test to be significant. Then, we carried out multivariate logistic regressions for all bands that fulfilled the above mentioned criteria with diet as the dependent variable.

As an exploratory analysis to evaluate correlations among data of DGGE bands that changed after an intervention, we calculated tetrachoric correlation coefficients considering corresponding DGGE bands and performed a factor analysis based on the resulting correlation matrix [18]. We used algorithms for

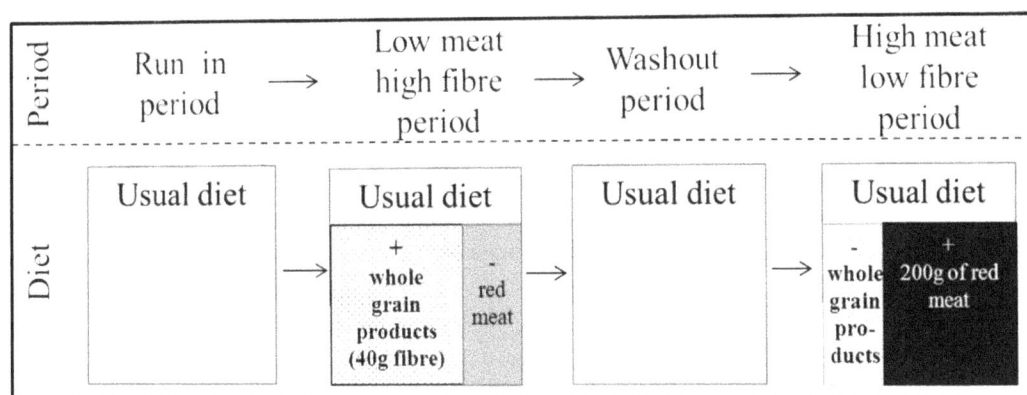

Figure 2. Diet plan for participants with diet succession: whole grain diet, washout period and red meat diet.

unweighted least squares and the varimax method as an orthogonal rotation procedure considering that the inter-factor correlation was not higher than 0.13. Factors were extracted according to the criterion of maximal explained variance and corresponding factors were calculated for each sample. To evaluate whether all 80 observations of the 20 participants can be used to generate the tetrachoric correlations and the factors despite their dependent character intraclass correlation coefficients (ICC) were calculated. The influence of the dietary treatment on the final factors was investigated in mixed models. Finally, the relations between the factors of interest and several measures of obesity were assessed. In additional analyses relations between factors of interest and anthropometric parameters were adjusted for sex.

All statistical analyses were done in SAS Enterprise guide 4.3 and the level of statistical significance was p<0.05.

Results

3.1 Baseline characteristics, anthropometric and blood parameters

Study participants had a mean age of 40 (11.6) years and a BMI of 24.4 (2.9) kg/m^2 (Table 1). The two groups that started either with WG or with RM did not differ significantly in their baseline characteristics. Therefore, they were analysed together. Further, no significant differences between the investigated characteristics at baseline and after washout could be observed (p-values range from 0.09 to 0.86), therefore baseline and washout values were combined to 'no intervention'.

After the WG intervention, the BMI, body fat mass and body weight were statistically significant lower by 0.14 kg/m^2, 0.77% and 0.45 kg, respectively (p=0.05, p=0.005 and p=0.05, respectively), compared to values at baseline and after washout (Tab. 2). All other measured parameters remained unchanged. In contrast, no alterations in anthropometric parameters were observed after the RM intervention.

The WG diet had no influence on blood levels of lipids and other blood parameters, but the RM intervention was associated with a significant increase in creatinine and uric acid of 6.8% and 10.0% (5.05 and 25.82 µmol/l; p=0.0003 and p=0.0002), respectively.

3.2 Intestinal microbiota and faecal parameters

Using DGGE, in total 128 different bands were detected. The microbial diversity ranged from 19 to 42 bands per participant at baseline with a mean of 31.1 (5.8) different bands (Table 1). A small proportion of 9 bands occurred frequently with a relative abundance of over 50%; most of the bands (96 bands) were less widespread and present in 10 to 50% of the participants (Figure 3).

When bands were compared before and after the intervention periods, 8 bands changed in at least 4 participants during the RM as well as during the WG diet. Considering the high inter-individual variation this change was a significant difference (pValue of McNemat test ≤0.05) in consequence of the intervention. After the RM period, 5 bands increased and 3 bands decreased in their appearance, and the WG diet caused the reduction of 3 bands and the rise of 5 bands (Table 3, Fig. 4). The group specific analysis incorporating the intervention sequence showed that for all but 2 bands changes occurred in both groups (b300 and b935 only changed in the WG → RM group). The multivariate logistic regression revealed for each diet intervention only one band, b812 for WG and b496 for RM, which can be regarded as the band containing key information in relation to dietary change (Table 3, Fig. 4). Sequencing identified b812 as *Collinsella aerofaciens* from the phylum of *Actinobacteria* and b496 as *Clostridium sp.* from the phylum of *Firmicutes*.

Separate analyses on the appearance of bands after the intervention periods revealed that after both intervention periods, more bands were present in a higher number of participants compared to baseline. After the WG and RM intervention 19 and 15 bands, respectively, were present in more than 50% of the participants. After the WG diet, diversity increased by 4 bands per sample, whereas it remained stable after the RM intervention (Table 2).

The interventions did not significantly change SCFA concentrations and calprotectin measured in faecal samples (Table 2).

3.3 Factor analysis

The factor analysis ascertained 5 factors with an eigenvalue greater 1. The proportions of explained variance were 31.1%, 30.9%, 19.8%, 12.4% and 12.2% for the factors 1 to 5 with respective eigenvalues of 4.3, 2.9, 1.8, 1.3 and 1.1. The ICCs for DGGE data were between 0 and 1 (mean (SD): 0.39 (0.25)); 52 out of 128 were significant indicating that the 80 observations cannot be treated as completely independent, a fact that has to be considered in the interpretation of the factors and their correlations.

The dietary interventions were significantly related to 3 of these factors; factor 1 to RM and factors 2 and 4 to WG (Table 4). After orthogonal rotation, factor 1 was negatively loaded by all bands

Table 1. Baseline characteristics of study participants, overall and by intervention sequence group after randomisation; # Mean (SD), *pValues for ttest comparing continous variables in the groups and χ^2test for categorical variables.

	Overall (n = 20)	WG → RM (n = 10)	RM → WG (n = 10)	p *
Age [years][#]	40.1 (11.6)	37.8 (9.5)	42.3 (13.5)	0.4
Sex (male) [n(%)]	10 (50)	6 (60)	4 (40)	0.37
Microbiota diversity [total nb of bands][#]	31.1 (5.8)	31.0 (4.1)	31.1 (7.4)	0.97
BMI [kg/m^2][#]	24.4 (2.9)	24.0 (2.5)	24.9 (3.4)	0.51
WHR[#]	0.86 (0.09)	0.85 (0.07)	0.87 (0.12)	0.66
Waist circumference category [n (%)]				
1 - ≤80/94 cm	13 (65)	6 (60)	7 (70)	
2 ->80/94 cm and ≤88/102 cm	5 (25)	3 (30)	2 (20)	
3 -> 88/102 cm	2 (10)	1 (10)	1 (10)	0.87
Body fat mass [%][#]	28.7 (5.9)	30.0 (7.0)	27.4 (4.7)	0.34
Cholesterol [mmol/l][#]	5.4 (0.8)	5.3 (0.8)	5.5 (0.9)	0.51
Quotient LDL/HDL[#]	2.4 (0.8)	2.3 (0.7)	2.6 (0.8)	0.33
Triglycerides [mmol/l][#]	1.5 (0.8)	1.5 (0.8)	1.4 (0.7)	0.76
Uric acid [µmol/l][#]	253.4 (81.0)	236.7 (69.6)	270.0 (91.6)	0.37
Creatinine [µmol/l][#]	75.0 (14.4)	75.1 (16.4)	74.9 (12.9)	0.98
CRP [mg/l][#]	2.1 (2.8)	2.3 (3.7)	2.0 (1.6)	0.76
Calprotectin [mg/kg][#]	25.4 (21.5)	21.6 (26.3)	29.3 (15.8)	0.44
Total SCFA [mmol/g][#]	37.0 (30.0)	47.1 (39.5)	26.9 (10.9)	0.14
Butyric acid [mmol/g][#]	6.4 (6.3)	7.7 (8.5)	5.0 (2.8)	0.37

that increased after RM intervention and it was positively loaded by bands that decreased after RM intervention. Factor 2 showed similar loading patterns since it was negatively loaded by bands rising after WG and positively loaded by bands declining after WG intervention. Despite the relation between factor 4 and the WG intervention no clear association between the factor and the 16 bands that changed after one of the diets could be detected.

Table 2. Intervention effects on selected parameters from mixed effect models.

	No intervention*	Intervention effect				
		Red meat			Whole grains	
	Mean (SD)	Mean (SD)	p[#]		Mean (SD)	p[#]
Diversity [total nb of bands]	31.6 (5.5)	31.7 (5.6)	0.96		35.7 (6.6)	0.01
BMI [kg/m^2]	24.5 (3.0)	24.4 (3.0)	0.64		24.3 (3.0)	0.05
Body fat mass [%]	28.6 (5.8)	28.6 (6.1)	0.91		27.9 (6.0)	0.005
Weight [kg]	75.4 (13.7)	75.3 (14.0)	0.66		74.9 (13.7)	0.05
WHR[§]	0.86 (0.09)	0.86 (0.09)	0.79		0.86 (0.09)	0.98
Uric acid [µmol/l][§]	260 (79)	285 (83)	0.0002		261 (76)	0.68
Creatinine [µmol/l][§]	74.4 (13.9)	79.4 (14.6)	0.0003		72.7 (12.2)	0.24
Calprotectin [mg/kg][§]	41.2 (63.9)	29.0 (32.8)	0.09		32.6 (44.3)	0.77
CRP [mg/l][§]	2.1 (2.4)	1.4 (1.2)	0.19		1.9 (3.8)	0.07
Cholesterol [mmol/l]	5.3 (0.8)	5.2 (0.8)	0.47		5.3 (1.0)	0.94
Quotient LDL/HDL	2.3 (0.7)	2.3 (0.6)	0.50		2.3 (0.8)	0.34
Triglycerides [mmol/l][§]	1.4 (0.7)	1.5 (0.8)	0.68		1.6 (1.0)	0.11
SCFA [mmol/g dryweight]						
Total SCFA[§]	41.8 (34.9)	33.4 (20.2)	0.34		40.7 (25.5)	0.69
Butyric acid[§]	7.0 (6.0)	5.5 (4.6)	0.12		7.6 (5.8)	0.91

* Data from examination at baseline and after washout period combined.
[#]p-values less than 0.05 were considered significant.
[§]p-values result from calculation of mixed models with log transformed variables.

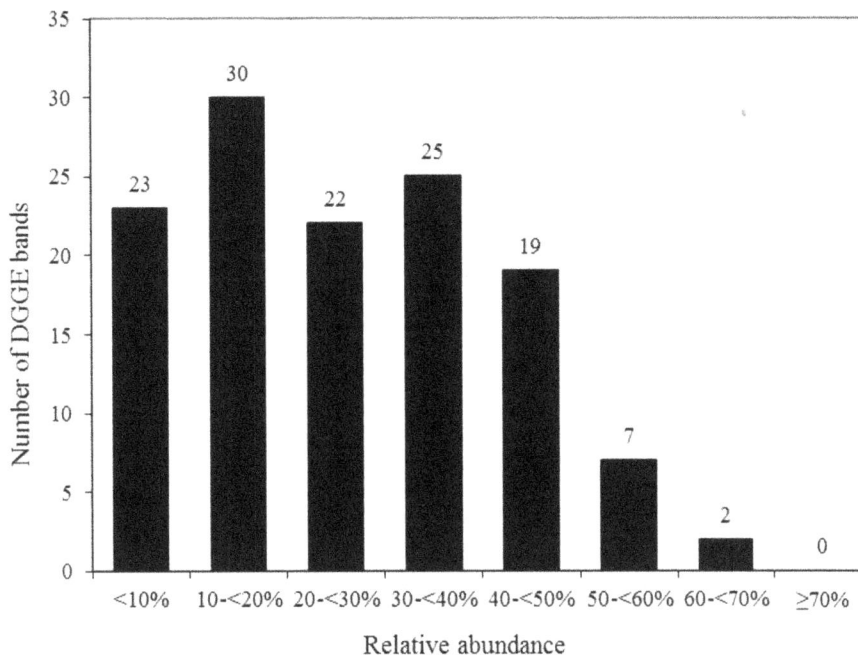

Figure 3. Relative abundance of DGGE-bands in human faecal samples from examinations before intervention periods.

Factor 1 was inversely associated with BMI, weight, and waist circumference and positively correlated with sex. Factor 2 was inversely associated with BMI only, and factor 4 was positively correlated with sex and negatively with BMI. None of these associations persisted after adjustment for sex (Table 4).

Discussion

In this crossover intervention study, we investigated the isocaloric exchange of RM with WG products for a direct influence on gut microbiota composition. For each diet, we

Table 3. Changes in microbiota due to intervention in all participants; shown are bands with a significant change (0→1: band is absent before and present after intervention; 1→0: vice versa); [#]pValue of McNemar test.

Band	Intervention	Change in appearance of band		$p^{\#}$
		0 → 1	**1 → 0**	
[1]b300°	Red meat	0	4	0.05
[5]b496*		0	4	0.05
b501		5	0	0.03
[6]b543		0	4	0.05
[8]b601		0	5	0.03
b664		4	0	0.05
[9]b710		0	4	0.05
b935°		5	0	0.03
	Whole grain			
[2]b323		7	1	0.03
[3]b337		6	0	0.01
[4]b378		4	0	0.05
b466		1	7	0.03
b557		1	7	0.03
[7]b594		7	1	0.03
[10]b812*		4	0	0.05
b968		0	4	0.05

* Bands remained significant in multivariate logistic regression model with intervention as dependent variable and all changed bands as independent variables.
°Change in microbiota only in group starting interventions with WG intervention.

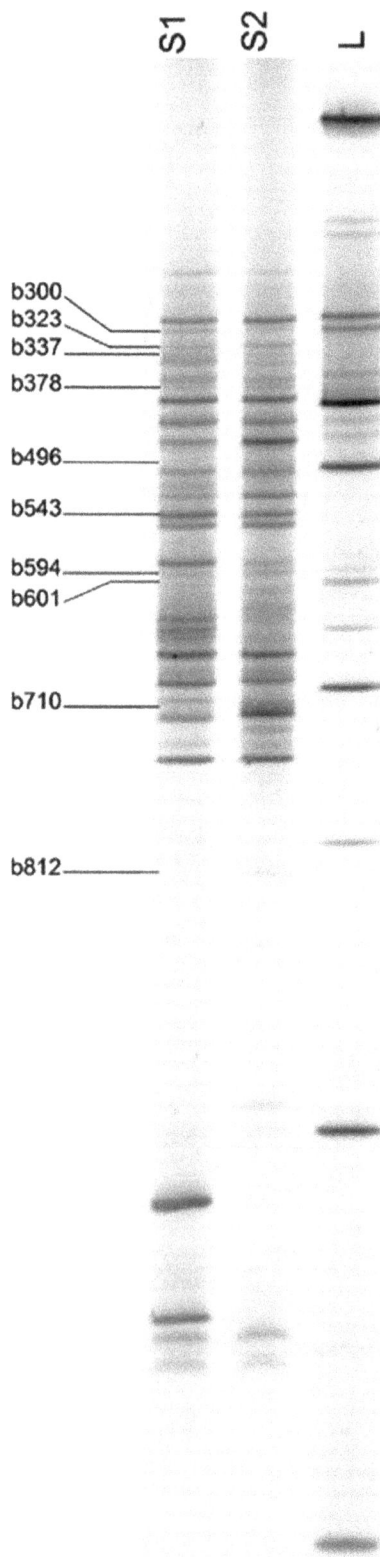

Figure 4. Position of DGGE bands with a significant change after intervention periods in a DGGE gel (L- Marker; S1/S2- Sample 1 and 2).

identified 8 DGGE bands with in both cases either one bacterium (*Collinsella aerofaciens* (*Actinobacteria*) for WG) or bacterial group (*Clostridium sp.* (*Firmicutes*) for RM) carrying the key information on the relation between the respective foods and intestinal microbiota. The study indicated that a diet rich in WG products increased microbial diversity. Further, we showed that the diets altered anthropometric and blood parameters.

The randomised crossover design enabled us to control inter-individual variability appropriately despite the small number of participants and to identify effects across individuals. High adherence of the study participants was secured by detailed instruction, provided food and regular contact and was also indicated by elevated serum levels of creatinine and uric acid after RM intervention, since both substances are components or metabolites of RM. Duration of the interventions was limited to 3 weeks, since study participants can only adhere to such regimens for a short period of time. However, the time was sufficient to observe changes in microbiota composition.

An interesting finding is the higher microbial diversity after the WG intervention which seems to be relevant in the context of other studies that reported a reduced diversity in obese and diseased individuals [19,20]. Also regarding colorectal cancer risk the microbial diversity seems to be relevant since other studies found the diversity decreased in cancer cases [21].

Agreeing with, but more far-reaching than studies showing effects of fibre supplements on intestinal microbiota [22], we demonstrated direct effects of common foods, i.e. WG bread and RM on gut microbiota composition. To our knowledge, previous food-based interventions in healthy, normal weight individuals have only been performed with fruits and vegetables [23,24]. We clearly associated the WG intervention with a higher intake of dietary fibre, a substrate for a large proportion of intestinal bacteria. It is known that non-starch polysaccharides, e.g., hemicellulose, promote the growth of commensal gut bacteria [25]. For RM intake, no direct effects on the proliferation of bacterial cells have been established so far, but meat-derived substances, such as nitrite, may also affect microbiota composition [26] and its effects on health [27].

The changed bands, particularly *Clostridium sp.*, belong to predominant bacterial groups in the intestinal microbiota [28,29]. It is not surprising, however that findings on the health effect of *Clostridium sp.* are inconsistent since the group is large and includes many species. The anaerobic bacterium is already present in the intestine of 30 days old newborns [30]. A study investigating the effects of prebiotics found *Clostridium sp.* decreased after a fibre rich diet, whereas this genus decreased after ingestion of red meat in our study [31]. However, from our data we cannot suggest the exact species from this group that was changed through a RM diet. Since many species from the *Clostridium sp.* are related to pathogenic conditions [32], the reduction of this genus of bacteria may also have beneficial effects. On the other hand studies found a lower abundance of *Clostridia* in colorectal cancer cases [21]. At least, we did not find any adverse results after a 3-week intervention with RM as assessed by our set of anthropometric and biochemical variables. In line with other reports, *C. aerofaciens* changed after the WG intervention. Walker and colleagues found *C. aerofaciens* decreased after a reduced carbohydrate, high protein diet [33]; in our study we could observe the bacterium more often after the WG period with low amounts of protein due to reduced RM intake. This bacterium is more prevalent in faeces of healthy subjects than in Crohn's disease patients [34] and it is inversely associated with irritable bowel syndrome symptoms supposing beneficial effect of this species on gut health [35].

Table 4. Intervention effects on factors retained after factor analysis (FA) with bands changed due to an intervention and correlations of corresponding factors with measures of obesity (BMI-body mass index, Waist circ.- waist circumference) and sex.

FA included changed bands (n = 16)	Intervention effect		Correlations of factors with			
			Measures of obesity			Sex
Factor (Eigenvalue/explained variance)	RM	WG	BMI	Weight	Waist circ.	
	p-Values		Correlation coefficients			
1 (4.3/31.3%)	0.001	0.29	−0.28*	−0.27*	−0.30*	0.23*
2 (2.9/30.9%)	0.22	<0.0001	−0.22*	−0.13	−0.12	0.06
3 (1.8/19.8%)	0.76	0.22	−0.06	−0.07	−0.07	0.05
4 (1.3/12.4%)	0.50	0.009	0.21*	0.18	0.19	−0.20*
5 (1.1/12.2%)	0.25	0.70	−0.05	−0.03	−0.02	−0.08

* Correlation significant on level $p < 0.05$.

A factor analysis of the 16 changed bands detected inter-correlations between microbiota bands and revealed 5 factors explaining most of the variance within microbiota data. Interestingly, the retained factors were related to measures of obesity and sex. Most of the associations between factors and measures of obesity could be explained by differences between men and women in regard to obesity. Calculated ICCs revealed that DGGE bands are not only correlated across individuals but also within participants. That may lead to an overestimation of the variance explained by the generated factors. It was not possible to clearly separate the respective proportions contributed by intra- and inter-individual covariance for each factor. To address this problem, future studies that apply advanced statistical methods with a larger number of participants are needed.

Although the interventions were designed to be isocaloric, we found a reduction in weight, BMI and body fat mass after the WG intervention indicating that WG products influence energy utilization. Similar effects of WG intake on anthropometry have been observed in cohort studies as well as in previous intervention studies [36,37]. Since the diets and changes in anthropometry were accompanied by a shift in microbiota, one can hypothesize that the altered microbiota composition is responsible for the energy utilization effect [6]. A reduction in total body fat was also observed in studies with mice harbouring different microbiota [3]. Additionally, it appears that individuals differing in microbiota composition possess different abilities of energy regulation and fat storage resulting in differential risks to gain weight [3,38]. Underlying mechanisms include distinct macronutrient exploitation from diet and differentially regulated gene expression [39].

On the other hand, our data on the simultaneous changes in microbiota composition and anthropometric parameters should induce a debate about the interpretation of the findings. We claim that diet effects obesity and microbiota and is the underlying factor inducing associations between microbiota composition and obesity. Thus, an association between gut bacteria and obesity may be due to WG intake. This hypothesis is in agreement with previous studies mostly conducted in mice that attributed the capability to play a role in the development of obesity to intestinal microbiota [39]. Further support is provided by cohort studies finding biomarkers of obesity reduced in participants with high WG intake [7]. In the present study, we did not detect any differences in biomarkers such as total cholesterol or serum triglycerides due to the WG intervention, which probably can be attributed to the brevity of 3-week intervention period.

A limitation of the study is the method of intestinal microbiota analysis. The DGGE bands characterise the microbial composition and diversity as well as shifts within the community, but they do not give information on the abundance and concentration of separate bacterial species [40]. We compensated this limitation by sequencing the two bands that carried the key information. A further limitation is the manner of collecting and analysing faecal samples representing microbiota composition from the gut lumen, but not from the entire gut [41].

In conclusion, this study showed that an elevated consumption of RM and WG products, respectively, modifies intestinal microbiota and is associated with a simultaneous change in measures of obesity. The increase of the appearance of *C. aerofaciens* through a fibre rich diet supports the assumption of health promoting effects of WG products. The finding of the study provides guidance to focus on particular microbiota bands (and strains) in observational studies to obtain further evidence on the relation of diet, gut microbiota and health outcomes within large cohorts.

Acknowledgments

Thanks to Bärbel Gruhl (Department of Gastrointestinal Microbiology, German Institute of Human Nutrition, Potsdam-Rehbruecke, Nuthetal, Germany) for assisting with the short chain fatty acid analysis and to Mattea Müller for analysing the samples regarding short chain fatty acids and calprotectin.

We also thank Dr. Ian Pagano (University of Hawaii Cancer Centre, Honolulu, Hawaii) for his help with the factor analysis.

Special thanks to all study participants enabling the present investigation.

Author Contributions

Conceived and designed the experiments: JF NR AN HB. Performed the experiments: JF NR HB. Analyzed the data: JF GM NR AT HB. Contributed reagents/materials/analysis tools: JF NR AT. Wrote the paper: JF GM NR AT AN MB HB.

References

1. Bischoff SC (2011) 'Gut health': a new objective in medicine? BMC Med 9: 1–14.
2. Cho I, Blaser MJ (2012) The human microbiome: at the interface of health and disease. Nat Rev Genet 13: 260–270.
3. Backhed F, Ding H, Wang T, Hooper LV, Koh GY, et al. (2004) The gut microbiota as an environmental factor that regulates fat storage. Proc Natl Acad Sci U S A 101: 15718–15723.
4. Cerf-Bensussan N, Gaboriau-Routhiau V (2010) The immune system and the gut microbiota: friends or foes? Nat Rev Immunol 10: 735–744.
5. Spor A, Koren O, Ley R (2011) Unravelling the effects of the environment and host genotype on the gut microbiome. Nat Rev Microbiol 9: 279–290.
6. Turnbaugh PJ, Ley RE, Mahowald MA, Magrini V, Mardis ER, et al. (2006) An obesity-associated gut microbiome with increased capacity for energy harvest. Nature 444: 1027–1031.
7. Montonen J, Boeing H, Fritsche A, Schleicher E, Joost HG, et al. (2012) Consumption of red meat and whole-grain bread in relation to biomarkers of obesity, inflammation, glucose metabolism and oxidative stress. Eur J Nutr 52: 337–345.
8. Pan A, Sun Q, Bernstein AM, Schulze MB, Manson JE, et al. (2011) Red meat consumption and risk of type 2 diabetes: 3 cohorts of US adults and an updated meta-analysis. Am J Clin Nutr 94: 1088–1096.
9. Ye EQ, Chacko SA, Chou EL, Kugizaki M, Liu S (2012) Greater whole-grain intake is associated with lower risk of type 2 diabetes, cardiovascular disease, and weight gain. J Nutr 142: 1304–1313.
10. Tuohy KM, Conterno L, Gasperotti M, Viola R (2012) Up-regulating the human intestinal microbiome using whole plant foods, polyphenols, and/or fiber. J Agric Food Chem 60: 8776–8782.
11. WMA (2000) World Medical Association Declaration of Helsinki - Ethical Principles for Medical Research Involving Human Subjects.
12. Durnin JV, Womersley J (1974) Body fat assessed from total body density and its estimation from skinfold thickness: measurements on 481 men and women aged from 16 to 72 years. Br J Nutr 32: 77–97.
13. Maukonen J, Matto J, Satokari R, Soderlund H, Mattila-Sandholm T, et al. (2006) PCR DGGE and RT-PCR DGGE show diversity and short-term temporal stability in the Clostridium coccoides-Eubacterium rectale group in the human intestinal microbiota. FEMS Microbiol Ecol 58: 517–528.
14. Nubel U, Engelen B, Felske A, Snaidr J, Wieshuber A, et al. (1996) Sequence heterogeneities of genes encoding 16S rRNAs in Paenibacillus polymyxa detected by temperature gradient gel electrophoresis. J Bacteriol 178: 5636–5643.
15. Tourlomousis P, Kemsley EK, Ridgway KP, Toscano MJ, Humphrey TJ, et al. (2010) PCR-denaturing gradient gel electrophoresis of complex microbial communities: a two-step approach to address the effect of gel-to-gel variation and allow valid comparisons across a large dataset. Microb Ecol 59: 776–786.
16. Schneider H, Schwiertz A, Collins MD, Blaut M (1999) Anaerobic transformation of quercetin-3-glucoside by bacteria from the human intestinal tract. Arch Microbiol 171: 81–91.
17. Immundiagnostik A (2012) PhiCal(R) Calprotectin ELISA Kit; For the in vitro determination of PhiCal(R) Calprotectin (MRP 8/14) in stool.
18. Christoffersson A (1975) Factor analysis of dichotomized variables. Psychometrika 40: 5–32.
19. Andoh A, Kuzuoka H, Tsujikawa T, Nakamura S, Hirai F, et al. (2012) Multicenter analysis of fecal microbiota profiles in Japanese patients with Crohn's disease. J Gastroenterol: 1297–1307.
20. Turnbaugh PJ, Hamady M, Yatsunenko T, Cantarel BL, Duncan A, et al. (2009) A core gut microbiome in obese and lean twins. Nature 457: 480–484.
21. Ahn J, Sinha R, Pei Z, Dominianni C, Wu J, et al. (2013) Human gut microbiome and risk for colorectal cancer. J Natl Cancer Inst 105: 1907–1911.
22. Hooda S, Boler BM, Serao MC, Brulc JM, Staeger MA, et al. (2012) 454 pyrosequencing reveals a shift in fecal microbiota of healthy adult men consuming polydextrose or soluble corn fiber. J Nutr 142: 1259–1265.

23. Paturi G, Mandimika T, Butts CA, Zhu S, Roy NC, et al. (2012) Influence of dietary blueberry and broccoli on cecal microbiota activity and colon morphology in mdr1a(-/-) mice, a model of inflammatory bowel diseases. Nutrition 28: 324–330.
24. Vendrame S, Guglielmetti S, Riso P, Arioli S, Klimis-Zacas D, et al. (2011) Six-week consumption of a wild blueberry powder drink increases bifidobacteria in the human gut. J Agric Food Chem 59: 12815–12820.
25. Kumar V, Sinha AK, Makkar HP, de Boeck G, Becker K (2012) Dietary roles of non-starch polysachharides in human nutrition: a review. Crit Rev Food Sci Nutr 52: 899–935.
26. Lundberg JO, Weitzberg E (2012) Biology of nitrogen oxides in the gastrointestinal tract. Gut: 1–14.
27. Mendelsohn AR, Larrick J (2013) Dietary modification of the microbiome affects risk for cardiovascular disease. Rejuvenation Res 16: 241–244.
28. Kageyama A, Benno Y (2000) Emendation of genus Collinsella and proposal of Collinsella stercoris sp. nov. and Collinsella intestinalis sp. nov. Int J Syst Evol Microbiol 50 Pt 5: 1767–1774.
29. Sghir A, Gramet G, Suau A, Rochet V, Pochart P, et al. (2000) Quantification of bacterial groups within human fecal flora by oligonucleotide probe hybridization. Appl Environ Microbiol 66: 2263–2266.
30. Brandt K, Taddei CR, Takagi EH, Oliveira FF, Duarte RT, et al. (2012) Establishment of the bacterial fecal community during the first month of life in Brazilian newborns. Clinics (Sao Paulo) 67: 113–123.
31. Linetzky Waitzberg D, Alves Pereira CC, Logullo L, Manzoni Jacintho T, Almeida D, et al. (2012) Microbiota benefits after inulin and partially hydrolized guar gum supplementation: a randomized clinical trial in constipated women. Nutr Hosp 27: 123–129.
32. Wells C, TD W (1996) Clostridia: Sporeforming Anaerobic Bacilli in. In: al BSe, editor. Baron's Medical Microbiology Chapter 18. 4th edition ed. Galveston (TX): Univ of Texas Medical Branch at Galveston.
33. Walker AW, Ince J, Duncan SH, Webster LM, Holtrop G, et al. (2011) Dominant and diet-responsive groups of bacteria within the human colonic microbiota. ISME J 5: 220–230.
34. Joossens M, Huys G, Cnockaert M, De Preter V, Verbeke K, et al. (2011) Dysbiosis of the faecal microbiota in patients with Crohn's disease and their unaffected relatives. Gut 60: 631–637.
35. Malinen E, Krogius-Kurikka L, Lyra A, Nikkila J, Jaaskelainen A, et al. (2010) Association of symptoms with gastrointestinal microbiota in irritable bowel syndrome. World J Gastroenterol 16: 4532–4540.
36. Hur IY, Reicks M (2011) Relationship between Whole-Grain Intake, Chronic Disease Risk Indicators, and Weight Status among Adolescents in the National Health and Nutrition Examination Survey, 1999–2004. J Acad Nutr Diet 112: 46–55.
37. Te Morenga LA, Levers MT, Williams SM, Brown RC, Mann J (2011) Comparison of high protein and high fiber weight-loss diets in women with risk factors for the metabolic syndrome: a randomized trial. Nutritional Journal 10: 1–9.
38. Backhed F, Ley RE, Sonnenburg JL, Peterson DA, Gordon JI (2005) Host-bacterial mutualism in the human intestine. Science 307: 1915–1920.
39. Tehrani AB, Nezami BG, Gewirtz A, Srinivasan S (2012) Obesity and its associated disease: a role for microbiota. Neurogastroenterol Motil 24: 305–311.
40. Gafan GP, Lucas VS, Roberts GJ, Petrie A, Wilson M, et al. (2005) Statistical analyses of complex denaturing gradient gel electrophoresis profiles. J Clin Microbiol 43: 3971–3978.
41. Zoetendal EG, von Wright A, Vilpponen-Salmela T, Ben-Amor K, Akkermans AD, et al. (2002) Mucosa-associated bacteria in the human gastrointestinal tract are uniformly distributed along the colon and differ from the community recovered from feces. Appl Environ Microbiol 68: 3401–3407.

15

Microbial Diversity of a Camembert-Type Cheese Using Freeze-Dried Tibetan Kefir Coculture as Starter Culture by Culture-Dependent and Culture-Independent Methods

Jun Mei, Qizhen Guo, Yan Wu, Yunfei Li*

Department of Food Science and Technology, School of Agriculture and Biology, Shanghai Jiao Tong University, Shanghai, P.R. China

Abstract

The biochemical changes occurring during cheese ripening are directly and indirectly dependent on the microbial associations of starter cultures. Freeze-dried Tibetan kefir coculture was used as a starter culture in the Camembert-type cheese production for the first time. Therefore, it's necessary to elucidate the stability, organization and identification of the dominant microbiota presented in the cheese. Bacteria and yeasts were subjected to culture-dependent on selective media and culture-independent polymerase chain reaction (PCR)-denaturing gradient gel electrophoresis (DGGE) analysis and sequencing of dominant bands to assess the microbial structure and dynamics through ripening. In further studies, kefir grains were observed using scanning electron microscopy (SEM) methods. A total of 147 bacteria and 129 yeasts were obtained from the cheese during ripening. *Lactobacillus paracasei* represents the most commonly identified lactic acid bacteria isolates, with 59 of a total of 147 isolates, followed by *Lactococcus lactis* (29 isolates). Meanwhile, *Kazachstania servazzii* (51 isolates) represented the mainly identified yeast isolate, followed by *Saccharomyces cerevisiae* (40 isolates). However, some lactic acid bacteria detected by sequence analysis of DGGE bands were not recovered by plating. The yeast *S. cerevisiae* and *K. servazzii* are described for the first time with kefir starter culture. SEM showed that the microbiota were dominated by a variety of lactobacilli (long and curved) cells growing in close association with a few yeasts in the inner portion of the grain and the short lactobacilli were observed along with yeast cells on the exterior portion. Results indicated that conventional culture method and PCR-DGGE should be combined to describe in maximal detail the microbiological composition in the cheese during ripening. The data could help in the selection of appropriate commercial starters for Camembert-type cheese.

Editor: Ali Al-Ahmad, University Hospital of the Albert-Ludwigs-University Freiburg, Germany

Funding: This work was funded by the Ministry of Science and Technology of the People's Republic of China with project reference no. 2013BAD18B02. The funders had no role in study design, data collection and analysis, decision to publish, or preparation of the manuscript.

Competing Interests: The authors have declared that no competing interests exist.

* Email: yfli@sjtu.edu.cn

Introduction

The main focus of research on cheese production in the past two decades has been on the improvement of quality characteristics and the production of healthier cheese [1]. An appropriate starter culture becomes increasingly important in cheese manufacturing and affects biochemical changes occurring during cheese ripening. The biochemical changes, including the metabolism of proteolysis, lipolysis and glycolysis, lead to the formation of key flavor and aroma components of cheese [2,3]. Consequently, an upsurge of interest in developing suitable starter cultures in cheese production has occurred. Many researchers have proposed a variety of cultures suitable for use as starters, including bifidobacteria, lactococcus, lactobacillus, leuconostoc and enterococcus species [4–9].

Recently, the kefir culture has gained researchers' attention with regarding to cheese manufacturing due to its potential effect on quality, health, and safety properties of the product. Resembling small cauliflower florets in kefir grains' appearance, they vary in size from approximately 3–30 mm, and contain a complex mixture of acetic acid bacteria, yeasts and lactic acid bacteria that are considered to have probiotic properties [10–13]. Lactic acid bacteria that exist in kefir grains have attracted a lot of attention because of their ability to inhibit the development of spoilage and the growth of pathogenic microorganisms, either by the production of lactic acid or by the expression of antimicrobial agents [14,15]. Kefir has been used as a starter in white pickled cheese [16], hard-type cheese [1], Feta-type cheese [17], and others. However, for the commercial production of cheese, direct use of kefir grains is impractical regarding transportation, storage, and cell dosage. Freeze-drying is a solution for long-term preservation of microorganisms and convenience for shipping [18].

Molecular culture-independent approaches have proven to be powerful tools in providing a more complete inventory of the microbial diversity in cheese [19]. As far as the development of molecular technology, PCR-DGGE has recently been shown to be a useful tool for studying community structure at the species level. 16S rDNA fragments from different microbial species have the same length but different DNA sequences therefore the species can be identified by the band positions on the DGGE gel. DGGE allows the simultaneous analysis of multiple samples and the comparison of microbial communities based on temporal and geographical differences [20]. Now PCR-DGGE has successfully been applied to analyze the microflora in various foods, such as chilled pork, wine, raw milk, dry fermented sausages, and so on [21–24].

The freeze-dried Tibetan kefir coculture has not been tested yet in Camembert-type cheese production where ripening periods are necessary for the product to acquire its microbial diversity. Therefore, the motivation of the present work was to elucidate the stability, organization and identification of the dominant microbiota present in the cheese using freeze-dried Tibetan kefir coculture as starter culture by culture-dependent and culture-independent methods.

Materials and Methods

Production of freeze-dried Tibetan kefir coculture

Tibetan kefir coculture isolated from a commercial Tibetan kefir beverage was used in the present study. It was grown on a synthetic medium [17] consisting of 4% lactose, 0.4% yeast extract, 0.1% $(NH_4)_2SO_4$, 0.1% KH_2PO_4, and 0.5% $MgSO_4 \cdot 7H_2O$ at 30°C. The synthetic medium was sterilized at 121°C for 20 min prior to use. Pressed wet-weight cells (about 0.5 to 1.0 g dry weight) were prepared and used directly in aerobic fermentations of whey for further production of kefir coculture. A kefir coculture was resuspended in the fermented whey and the whole suspension was freeze-dried overnight in a freeze-drying system (Free Zone Triad Cascade Benchtop Freeze Dry System, Labconco, Kansas City, U.S.A.).

Cheese Making

Pilot-scale cheese production (coagulation, cutting, draining, and molding of the curd) was carried out under aseptic conditions in a sterilized, 2 m^3 (2 m×1 m×1 m) cheese-making chamber in Technical Centre of Bright dairy & Food Co. The cheeses were made according to Leclercq-Perlat et al [25]. Then the cheeses were transferred to the ripening chamber and kept at 14±1°C with 85±2% RH, which was designated as the initial ripening time (day 0). After 24 h, the cheeses were changed to keep at 12±1°C and 95–97% RH for 14 days and at 4°C until day 35. Samples of mature cheese (0, 5, 10, 15, 25 and 35 days old; the date day 35 at which cheese is allowed to be sold according to PDO Council specifications) were taken following standard FIL-IDF procedures. The rind (1 mm thick all over the cheese surface) and body of each cheese (inner layer) were separated by the method of Le Graët and Brûlé [26].

DGGE analysis of the cheese

To sample the cheeses, 5 g cubes from the inside and 5 g strips from the outside were taken. These samples were then homogenized with 40 mL of a 2% (w/v) sterilized sodium citrate solution at 45°C for 1 min. DNA extraction was accomplished by using a commercial kit (SK8233 soil gDNA Miniprep kit; Songon, China) according to the manufacturer's instructions.

The bacterial community DNA was amplified with primers 338fgc (5′-CGC CCG CCG CGC GCG GCG GGC GGG GCG GGG GCA CGG GGG GAC TCC TAC GGG AGG CAG

CAG-3′) (the GC clamp is underlined) and 518r (5′-ATT ACC GCG GCT GCT GG-3′) spanning the V3 region of the 16S rDNA gene [27]. The D1 domain of the 26S rRNA gene of yeasts was amplified using the primers NL1-GC (5′-GCG GGC CGC GCG ACC GCC GGG ACG CGC GAG CCG GCG GCG GGC CAT ATC AAT AAG CGG AGG AAA AG-3′) (the GC clamp is underlined) and a reverse primer LS2 (5′-ATT CCC AAA CAA CTC GAC TC-3′), as reported by Cocolin et al. [23]. All GC primers contained a 39 bp GC-clamp sequence at their 5′ end to prevent the complete denaturation of amplicons. PCR was performed in 50 μL reaction volumes using a Taq-DNA polymerase master mix (Songon, China) with 100 ng of each DNA sample as a template and 0.2 mM of each primer. PCR conditions were as follows: 94°C for 4 min, 30 cycles of 94°C for 30 s, annealing at 56°C for 1 min and extension at 72°C for 30 s, followed by a melting curve.

The PCR products were analyzed by DGGE using a Bio-Rad DCode Universal Mutation Detection System (Bio-Rad, Richmond, CA, U.S.A.). Samples were applied to 8% (w/v) polyacrylamide gels in 1×TAE. Optimal separation was achieved with a 30–60% urea-formamide denaturing gradient (100% correspondent to 7 M urea and 40% (v/v) formamide). The gels were electrophoresised for 16 h at 60 V. Bands were visualized under UV light after staining with ethidium bromide (0.5 μg·mL⁻¹) and photographed.

DNA recovered from each DGGE band was reamplified with the primers 338fgc (5′-ACT CCT ACG GGA GGC AGC AG-3′) and 518r (5′-ATT ACC GCG GCT GCT GG-3′) for bacteria [27] and NL1-GC (5′-GCC ATA TCA ATA AGC GGA GGA AAA G -3′) and LS2 (5′-ATT CCC AAA CAA CTC GAC TC-3′) for yeasts [23]. DGGE bands were excised with a sterile scalpel and eluted in 30 mL sterile water, overnight at 4°C to allow diffusion of the DNA. Two microliters of the DNA of each DGGE band was reamplified as described above. The PCR amplicons were then sequenced (Applied Biosystems, Foster City, CA, USA). Sequences were used as a query sequence to search for similar sequences from GenBank by means of the blast program (http://www.ncbi.nlm.nih.gov/BLAST/) [28]. Sequences showing 97% similarity or higher were deemed to belong to the same species [29].

Culture-dependent approach

Bacteria and yeasts were enumerated by method of Magalhães et al [30] with some modifications. Representative 10g portions of duplicate samples taken from the cheese interior and outside were blended with 90 mL of sterilized Ringer's Solution (1/4 strength) and subjected to serial dilutions.

The following microbiological analyses were performed: (i) determination of total mesophilic bacteria on nutrient agar medium (Huankai, China) at 28°C for 48 h; (ii) enumeration of lactobacillus after incubation on acidified MRS agar (Huankai, China) at 37°C for 48 h anaerobically (Xinmiao YQX-II anaerobic incubator, China); (iii) enumeration of lactococcus after incubation on M-17 agar (Huankai, China) at 37°C for 48 h; (iv) enumeration of acetobacter after incubation on acetobacter medium (Huankai, China) at 30°C for 48 h; (v) enumeration of yeasts after incubation on PDA agar (Huankai, China) containing 100 mg chloramphenicol and 50 mg chlortetracycline (pH adjusted to 4.5 by sterile solution of 10% lactic acid) at 28°C for 5 days. All media for bacterial enumeration were supplemented with 0.4 mg/mL nystatin (Sigma-Aldrich, USA). Gram staining and catalase tests were performed for confirmation of lactic acid bacteria. Results are presented as the log of the mean number of

CFU on solid-medium culture plates containing between 30 and 300 colonies per g of cheese.

Observation of Tibetan kefir grains using SEM

Tibetan kefir grains were sliced to produce samples for microscopy according to Seydim et al. [31]. Samples were collected from the exterior and inner part. The grains were fixed (2.5% glutaraldehyde solution) at 4°C for 24 h. The grains were then transferred to 30% glycerol for 30 min and immersed in liquid nitrogen for subsequent fracture in the metal surface. The grains were post-fixed in 10 g/L osmium tetroxide in phosphate buffer for 1 h at 25°C and dehydrated in alcohol: 15%, 30%, 50%, 70%, 90% and 100%. Prior to observation, the grains were coated with a thin conducting layer of gold. Microstructure observations of the samples were carried out using a FEI Sirion 200 environmental scanning microscope (FEI Company, the Netherlands).

Results

Enumeration and identification of isolates by a culture-dependent method

In order to establish the different species of bacteria and yeasts present during ripening, a representative number of isolates from each culture medium were identified (Table 1). Lactococcus and lactobacillus were the most frequently found microorganism and showed an increase from 6.34 to 8.28 logCFU/g and 6.11 to 8.04 logCFU/g until day 35, respectively. The average number of yeasts, which was about 5.34 logCFU/g at day 0, increased to 7.17 logCFU/g at the end. Acetobacter and total mesophilic aerobic bacteria showed similar behavior for growth, ranging from about 5.00 to 7.00 logCFU/g. In general, lactic acid bacteria were much more numerous (10^8–10^9) than yeasts (10^5–10^7) and acetic acid bacteria (10^5–10^6) in the cheese during ripening.

A total of 276 isolates were obtained from the cheese (Table 2). Among the isolates, 147 isolates were bacteria and 129 isolates were yeasts. The bacteria contained lactic acid bacteria (104 isolates) and acetic acid bacteria (43 isolates). The culture-dependent approach indicated that *Lactobacillus paracasei* represents the largest and most commonly identified lactic acid bacteria isolates, with 49 of a total of 104 isolates, followed by *Lactococcus lactis* (31 isolates) and *Lactobacillus kefiri* (19 isolates). Isolates of *Lactobacillus kefiranofaciens* (5 isolates) was also sporadically identified. The only acetic acid specie, *Acetobacter lovaniensis*, was also identified (43 isolates). The lactose-fermenting yeasts (*Kluyveromyces lactis*) together with non-lactose-fermenting yeasts (*Trichosporon moniliiforme*, *Debaryomyces hansenii*, *Saccharomyces*

cerevisiae and *Kazachstania servazzii*) were found in the cheese during the ripening time. The yeast flora of the cheese was dominated by lactose-negative strains. Among them, *K. servazzii* predominated, with 51 of a total of 129 isolates, followed by *S. cerevisiae* (40 isolates). *K. lactis* (20 isolates) and *D. hansenii* (15 isolates) were the other identified members of yeasts. Isolates of *T. moniliiforme* (7 isolates) was also sporadically identified.

DGGE fingerprinting of bacterial and yeast communities

Traditionally, many culture methods are only partially selective and exclude members of the microbial community. Thus, to determinate the total composition of microbiota in the cheese, PCR-DGGE analysis was used. The V3 region of the 16S rDNA gene of the bacteria and D1 region of the 26S rRNA gene of yeasts were amplified, and representative DGGE fingerprints are shown in Figure 1a and 1b. No differences in community structure during fermentation were found for both bacteria and yeasts. To determine the composition of microbiota, individual bands observed in the DGGE profiles were excised from the acrylamide gel and re-amplified to provide a template for sequencing. After Blast analysis, sequence results showed 98–100% identity with sequences retrieved from GenBank accession numbers. DGGE band A was clearly identified as *Lb. kefiri*, band B as *Lb. kefiranofaciens*, band C as *A. lovaniensis*, band D as *Lc. lactis*, band E as *Lb. paracasei*, band F as *T. moniliiforme*, band H as *Geotrichum candidum*, band I as *D. hansenii*, band J as *K. lactis*, band K as *S. cerevisiae*, band L as *K. servazzii* and band M as *Penicillium crustosum*. Band G was excised from the gel, but it could not be recovered for sequencing. PCR-DGGE analysis showed that species of the genus lactobacillus were the dominant bacteria in the cheese, as already indicated by plating results. The representatives of lactobacillus could be differentiated according to the migration distances of their respective 16S rDNA fragments. *Lc. lactis* was another dominant microbe in the cheese. Band C in the DGGE analysis corresponded to *A. lovaniensis*. Interestingly, this was the only species of non-lactic acid bacteria found by culture-independent methods.

In fungal analysis, PCR-DGGE showed a good correlation with the culture-dependent methods. Band K represented the *Saccharomyces* sensu stricto group (Figure 1b). Among them, *S. cerevisiae* was the most probable strain identified because according to culture-based isolations; this species was the most commonly recovered yeast in the cheese (Table 2). The yeast *S. cerevisiae* and *K. servazzii* are described for the first time in the cheese with kefir starter.

Table 1. Mean counts (logCFU/g) of microorganisms in the Camembert-type cheese during ripening using freeze-dried Tibetan kefir coculture as starter.

Time (d)	TMAB	LB	LC	AC	YM[5]
0	5.09+0.07[a]	6.11+0.07	6.34+0.08	5.03+0.07	5.34+0.16
5	5.81+0.11	6.69+0.07	7.05+0.05	5.52+0.14	5.89+0.09
10	6.10+0.06	7.08+0.07	7.57+0.11	6.01+0.06	6.37+0.11
15	6.44+0.10	7.76+0.09	7.94+0.07	6.47+0.09	6.86+0.08
25	6.68+0.09	7.94+0.05	8.10+0.04	6.62+0.16	6.96+0.05
35	6.85+0.06	8.04+0.08	8.28+0.05	6.79+0.08	7.17+0.06

[a]Values reported are the means ± standard deviations.
[5]TMAB, total mesophilic aerobic bacteria; LB, Lactobacillus; LC, Lactococcus; AC, Acetobacter; YM, Yeasts.

Table 2. Distribution of bacteria and yeasts isolated during ripening of the cheese by the culture-dependent methods.

Closest relative	Identity (%)	Accession	Ripening time					
			Day 0	Day 5	Day 10	Day 15	Day 25	Day 35
Bacteria								
Lactobacillus kefiri	99%	AB362680.1	+(2)	+(*)	+(7)	+(2)	+(4)	+(4)
Lactobacillus kefiranofaciens	100%	NC015602.1	+(*)	+(3)	+(*)	+(*)	+(*)	+(2)
Lactococcus lactis	100%	NC002662.1	+(4)	+(6)	+(5)	+(6)	+(5)	+(5)
Lactobacillus paracasei	99%	AB368902.1	+(7)	+(12)	+(10)	+(9)	+(6)	+(5)
Acetobacter lovaniensis	98%	AB308060.1	+(5)	+(5)	+(7)	+(11)	+(7)	+(8)
Yeasts								
Trichosporon moniliiforme	99%	JN805525.1	−(*)	−(*)	−(*)	+(4)	+(3)	+(*)
Debaryomyces hansenii	99%	HF934036.1	−(*)	+(*)	+(4)	+(5)	+(3)	+(3)
Kluyveromyces lactis	99%	AJ229069.1	−(*)	+(4)	−(*)	+(10)	+(6)	+(*)
Saccharomyces cerevisiae	99%	EU649673.1	+(16)	+(10)	+(9)	+(5)	+(*)	+(*)
Kazachstania servazzii	99%	JQ808010.1	+(13)	+(16)	+(11)	+(11)	+(*)	+(*)

−, not detected by PCR-DGGE; +, detected by PCR-DGGE and sequencing of the DNA fragment upon excision from the gel;
* species not isolated by culturing methods; values in parentheses are numbers of colonies detected by culturing.

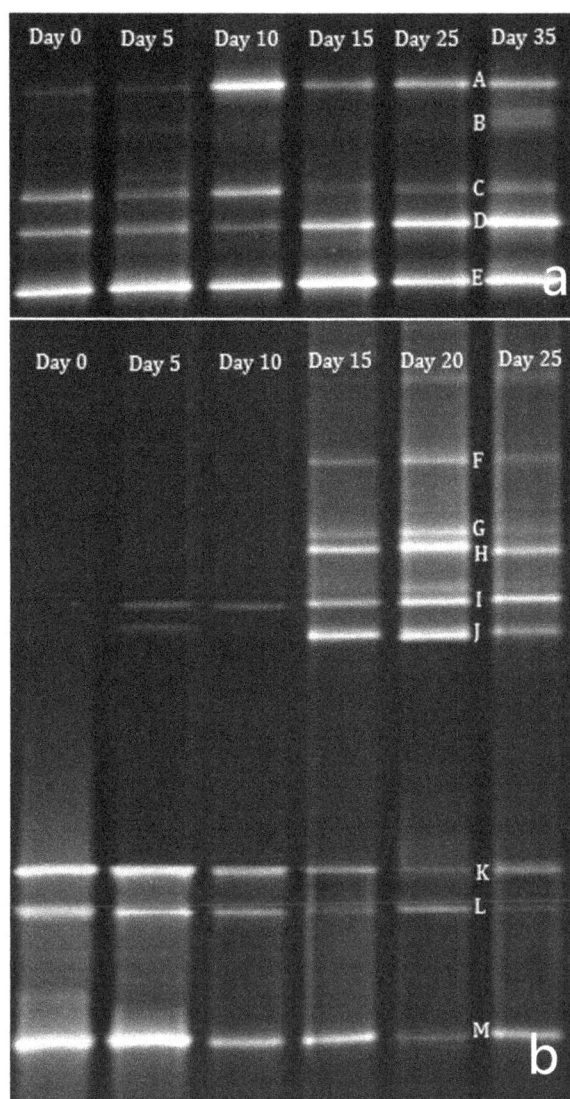

Figure 1. DGGE profiles of bacterial 16S rDNA gene V3 fragments (a) and fungal 26S rRNA gene D1 region (b) amplified from the cheese during ripening.

Distribution of the microflora in the kefir grains as observed using SEM

The exterior surfaces of the Tibetan kefir grains looked smooth and shiny with the naked eye (Figure 2a). However, the grain surfaces, under SEM at a magnification of ×5000, were revealed to be very rugged (Figure 2b). In the inner portion of the grain (Figures 2c, 2e, ×15 000; 2g, ×40 000), a variety of lactobacilli (long and curved), yeasts and fibrillar material were observed. Kefir grains had a spongy fibrillar structure that was branched and interconnected. Fibrillar material increased progressively towards the interior portions of the grain. On the inner portions of the grain, there were a variety of lactobacilli (long and curved) with only a few yeasts embedded in the fibrillar material. Lactobacilli, yeasts and fibrillar material were also observed at × 15 000 (Figures 2d, 2f) and × 40 000 (Figures 2h) on the exterior portion of the grain. The fibrillar material was most probably the polysaccharide kefiran. The short lactobacilli were observed embedded in the grain along with yeast cells.

Discussion

Descriptions of the different types of yeasts and bacteria present in cheese or beverage using kefir grains as a starter culture have been provided by different authors [17,32,33]. Using conventional culture techniques and culture-independent methods, we have monitored the development of bacterial and yeasts communities in the Camembert-type cheese using freeze-dried Tibetan kefir coculture as starter for the first time.

All the bacteria isolated in this study were Gram-positive and non-motile, except *Acetobacter* which is a Gram-negative bacterium. The groups of lactic acid bacteria cause rapid acidification of the milk through the production of organic acids, mainly lactic acid. Also, their production of acetic acid, ethanol, aroma compounds, bacteriocins, exopolysaccharides, and several enzymes are of importance. In this way, they enhance shelf life and microbial safety, improve texture, and contribute to the pleasant sensory profile of cheese [34]. *Lb. kefiri* is an important bacterium found in the cheese and easily observed from the kefir grains (Figure 2). It can produce NH_3 from arginine, which could explain the pH value increase in the cheese during ripening. *Lb. kefiranofaciens*, a heterofermentative bacterium, has also been reported as a kefiran (exopolysaccharide) producer [35] and is used as the starter for kefir beverage [36] and Caucasian cultured milk [37]. According to Cheirsilp et al. [37], *S. cerevisiae* could assimilate kefiran production rates of *Lb. kefiranofaciens* in a mixed culture of *Lb. kefiranofaciens* and *S. cerevisiae*. *S. cerevisiae* was also isolated from the cheese. The kefiran has been reported to have antibacterial and antitumor activities, modulate the gut immune system and protect epithelial cells against *Bacillus cereus* exocellular factors and infection [38]. So, the kefiran produced in the cheese could be regarded as one of the mechanisms contributing to health benefits for the consumer. According to Kesmen and Kacmaz, *Lb. kefiranofaciens* was the most dominant species in Turkey kefir grains, while *Lc. lactis* was found to be significantly prevalent in kefir beverages [39]. *Lc. lactis* was also the dominating species in the Camembert-type cheese analyzed in our study. This species was exceeded only by *Lb. paracasei* (Table 2). *Lc. lactis* is of great economic importance because of the world-wide use in cheese making. It is assumed that the lysis of *Lactococcus* during cheese ripening results in the release of intracellular proteolytic and esterolytic enzymes, which contribute to flavor development [40]. *Lb. paracasei* seems to find favorable conditions during the cheese-making process and its presence has already been reported in soft, and in many semi-hard and hard cheeses made from cow, ewe, and goat's milk, such as Caciocavallo [41], Ibores [42], Cheddar [43], and Salers [44]. The fact that *Lb. paracasei* is predominant in cheese is probably linked to its mesophilic properties and antimicrobial properties [45]. It is able to metabolize citrate; this could favor its development [46]. As citrate is gradually consumed in metabolism, *Lb. paracasei* is less detected at the end of ripening. The acetic acid species, *A. lovaniensis*, was also identified in our study. *A. lovaniensis* species belongs to the *A. pasteurianus* group. The species *A. pasteurianus* consists of five subspecies, and *A. pasteurianus* subsp. *lovaniensis* has been also described in fermented food from Indonesian sources [47]. Part of the ethanol produced by yeasts may be converted to acetic acid by the genus *Acetobacter*, which has alcohol dehydrogenase activity and converts ethanol to acetaldehyde [48]. Ethyl esters formed from the esterification of acetic acid and alcohols play quite an important role in cheese flavor.

The main contribution of yeasts to the cheese maturation process is the utilization of lactic acid which in turn increases the

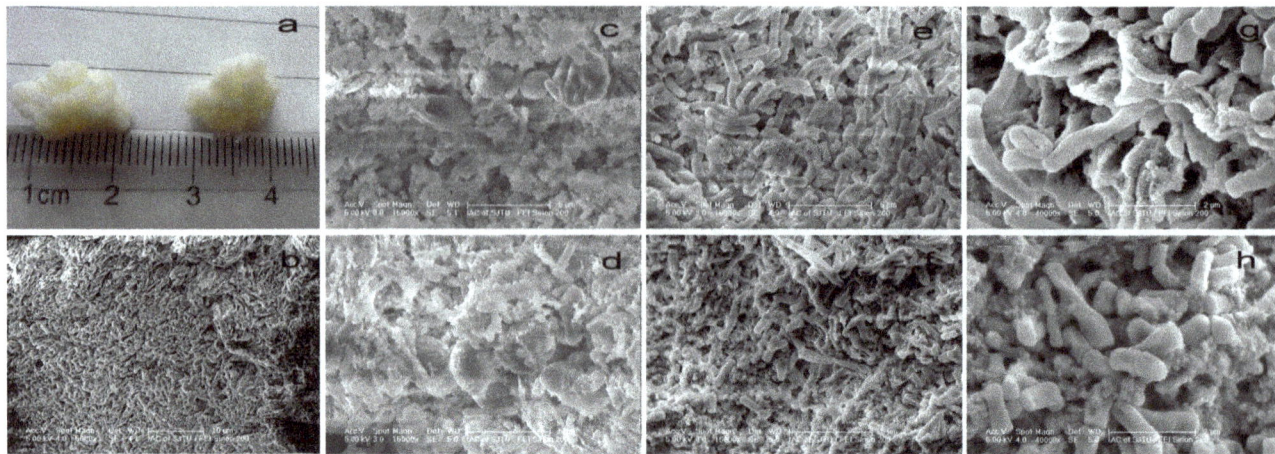

Figure 2. Electron micrographs of Tibetan kefir grains. (a) Tibetan kefir grains; (b) grain surfaces at ×5 000; (c, e) Inner portion of kefir grain at ×15 000; (g) Inner portion of kefir grain at ×40 000; (d, f) Exterior portion of kefir grain at ×15 000; (h) Exterior portion of kefir grain at ×40 000.

pH and therefore favors bacterial growth and initiates the second stage of cheese ripening [49]. Meanwhile, the yeasts partake in microbial interactions and contribute to the formation of aroma precursors such as amino acids and fatty acids [50]. *S. cerevisiae*, which exhibits strong fermentative metabolism and tolerance to ethanol, is known to be superior to non-*Saccharomyces* yeast in the process of alcohol fermentation, as regards Camembert and Blue-veined cheeses [51]. The presence of *S. cerevisiae* in the cheese contributes to enhancement of the organoleptic quality of the cheese, promoting a strong and typically yeasty aroma as well as its refreshing, pungent taste. Acetate esters such as ethyl acetate, isoamyl acetate and isobutyl acetate are mainly synthesized by acetyl-coenzyme A (acetyl-CoA) in *S. cerevisiae*. Since acetyl-CoA is an intermediate in lipid biosynthesis, ester production is closely linked to the metabolism of lipids. Lipids are accumulated at the beginning during cheese ripening and enhance significantly upon the increase of intracellular concentrations of acetyl-CoA [52]. As a result, the acetate esters production is low at the beginning even though *S. cerevisiae* was in a high level. This yeast also reduces the concentration of lactic acid, removes hydrogen peroxide by catalase activity and produces stimulators that stimulate the growth of other bacteria in the cheese [53]. It is also worth noting that the yeast species, *K. servazzii*, detected in this cheese for the first time (from day 0–15), could be associated with the presence of glucose and with the assimilation of some acids produced by lactic acid bacteria. In the present study, *D. hansenii* was frequently isolated from the cheese throughout the ripening period because of its tolerance of high concentrations of NaCl. This finding is in agreement with results obtained for surface-ripened Danish cheese [53]. *D. hansenii* increases pH, which enables the growth of less acid-tolerant coryneform bacteria. *K. lactis* growing on the cheese surface did not ferment lactose and did not therefore produce any lactic acid. At the beginning of ripening the growth of *K. lactis* was considerable. Then the occurrence of proteolysis was correlated with the growth of *K. lactis*, *G. candidum* and *Pen. Crustosum*, added to Camembert cheese production methods, known for their proteolytic and peptidolytic activities. In our study, the growths of *G. candidum* and *Pen. crustosum* was highly related to nitrogen formation. *K. lactis* in the cheese also produces esters through alcoholysis of acyl-CoA and esterification of an organic acid with an alcohol, which impart fruity flavors in cheese [54]. Ethyl acetate is by far the major ester produced by *K. lactis* and also

detected in more limited proportions in *G. candidum* [55]. The yeasts isolated were frequently detected during the first 15 days ripening and later decreased, even not detected. Two hypotheses can be postulated, i) competition with the ripening microorganisms, particularly *G. candidum*; ii) inhibition by the fatty acids liberated by *G. candidum* through lipolysis [25]. *T. moniliiforme* assimilates L-methionine as a sole nitrogen source and plays an important role in producing the volatile organic compounds [56].

Based on the results of the fingerprints of the bacterial community and the 16S rDNA sequencing, *L. lactis* was dominant microbe, while no lactococcus was found on SEM, which may be due to the bad attachment of lactococcus. This coincides with the results that Seydim et al. obtained on Turkey kefir using scanning electron microscope [31]. According to our observation, yeasts became less predominant further inside the grain. As reported, other researchers have observed that yeasts predominated at the centre part of granule while the exterior of the grain had mostly lactobacilli and only a few yeasts [57,58]. They reported a variety of lactobacilli (short, long, and curved) in all parts of the grain samples. However, in our study, long and curved lactobacilli were observed mainly in the interior portions, and short lactobacilli were found mainly in the exterior portions of the grains.

Conclusions

Based on culture-based detection methods, *Lactobacillus*, *Lactococcus*, *Acetobacter*, and yeasts represented the putative candidates in the cheese. In the latter stage of ripening, some yeasts were less frequently detected due to competition with *G. candidum* and *Pen. crustosum*. The culture based findings were confirmed using DGGE, the dominant microbiotas were composed by yeasts affiliated to *K. servazzii*, *S. cerevisiae*, and bacteria affiliated to *Lb. paracasei*, *Lc. lactis* and *A. lovaniensis*. As expected, the combination of culture dependant assays and PCR-DGGE analyses has allowed the microbial ecology of the Camembert-type cheese to be profiled. This application could provide an opportunity to better understand and control the transformation process during cheese ripening. Moreover the profiling of microbial populations occurring in cheese ripening can be useful to determine the technologically important strains to be employed as a suitable starter culture to obtain high quality and safety properties in the final product.

Author Contributions

The authors would like to express their profound gratitude to Instrumental Analysis Center of Shanghai Jiao Tong University for the technical assistance. Conceived and designed the experiments: JM QG YW YL. Performed the experiments: JM QG. Analyzed the data: JM QG YW. Contributed reagents/materials/analysis tools: JM YL. Wrote the paper: JM YL.

References

1. Katechaki E, Panas P, Rapti K, Kandilogiannakis L, Koutinas AA (2008) Production of hard-type cheese using free or immobilized freeze-dried kefir cells as a starter culture. J Agr Food Chem 56: 5316–5323.

2. Murtaza MA, Ur-Rehman S, Anjum FM, Huma N, Hafiz I (2013) Cheddar cheese ripening and flavor characterization: a review. Crit Rev Food Sci 54: 1309–1321. Crit Rev Food Sci

3. Lortal S, Chapot-Chartier M-P (2005) Role, mechanisms and control of lactic acid bacteria lysis in cheese. Int Dairy J 15: 857–871.

4. Boylston TD, Vinderola CG, Ghoddusi HB, Reinheimer JA (2004) Incorporation of bifidobacteria into cheeses: challenges and rewards. Int Dairy J 14: 375–387.

5. Broadbent J, Brighton C, McMahon D, Farkye N, Johnson M, et al. (2013) Microbiology of Cheddar cheese made with different fat contents using a Lactococcus lactis single-strain starter. J Dairy Sci 96: 4212–4222.

6. Cretenet M, Laroute V, Ulvé V, Jeanson S, Nouaille S, et al. (2011) Dynamic analysis of the Lactococcus lactis transcriptome in cheeses made from milk concentrated by ultrafiltration reveals multiple strategies of adaptation to stresses. Appl Environ Microb 77: 247–257.

7. Bove CG, De Dea Lindner J, Lazzi C, Gatti M, Neviani E (2011) Evaluation of genetic polymorphism among Lactobacillus rhamnosus non-starter Parmigiano Reggiano cheese strains. Int J Food Microbiol 144: 569–572.

8. Kleppen HP, Nes IF, Holo H (2012) Characterization of a Leuconostoc bacteriophage infecting flavor producers of cheese starter cultures. Appl Environ Microb 78: 6769–6772.

9. Jamet E, Akary E, Poisson M-A, Chamba J-F, Bertrand X, et al. (2012) Prevalence and characterization of antibiotic resistant Enterococcus faecalis in French cheeses. Food Microbiol 31: 191–198.

10. Zhou J, Liu X, Jiang H, Dong M (2009) Analysis of the microflora in Tibetan kefir grains using denaturing gradient gel electrophoresis. Food Microbiol 26: 770–775.

11. Leite A, Mayo B, Rachid C, Peixoto R, Silva J, et al. (2012) Assessment of the microbial diversity of Brazilian kefir grains by PCR-DGGE and pyrosequencing analysis. Food Microbiol 31: 215–221.

12. Marsh AJ, O'Sullivan O, Hill C, Ross RP, Cotter PD (2013) Sequencing-based analysis of the bacterial and fungal composition of kefir grains and milks from multiple sources. PloS one 8: e69371.

13. Zheng Y, Lu Y, Wang J, Yang L, Pan C, et al. (2013) Probiotic properties of lactobacillus strains isolated from Tibetan kefir grains. PloS one 8: e69868.

14. Settanni L, Gaglio R, Guarcello R, Francesca N, Carpino S, et al. (2013) Selected lactic acid bacteria as a hurdle to the microbial spoilage of cheese: application on a traditional raw ewes' milk cheese. Int Dairy J 32: 126–132.

15. Chen Y-P, Chen M-J (2013) Effects of Lactobacillus kefiranofaciens M1 Isolated from kefir grains on germ-free mice. PloS one 8: e78789.

16. Goncu A, Alpkent Z (2005) Sensory and chemical properties of white pickled cheese produced using kefir, yoghurt or a commercial cheese culture as a starter. Int Dairy J 15: 771–776.

17. Kourkoutas Y, Kandylis P, Panas P, Dooley J, Nigam P, et al. (2006) Evaluation of freeze-dried kefir coculture as starter in feta-type cheese production. Appl Environ Microb 72: 6124–6135.

18. Morgan C, Herman N, White P, Vesey G (2006) Preservation of microorganisms by drying; a review. J Microbiol Meth 66: 183–193.

19. Jany J-L, Barbier G (2008) Culture-independent methods for identifying microbial communities in cheese. Food Microbiol 25: 839–848.

20. Muyzer G, Smalla K (1998) Application of denaturing gradient gel electrophoresis (DGGE) and temperature gradient gel electrophoresis (TGGE) in microbial ecology. Antonie van Leeuwenhoek 73: 127–141.

21. Li M, Zhou G, Xu X, Li C, Zhu W (2006) Changes of bacterial diversity and main flora in chilled pork during storage using PCR-DGGE. Food Microbiol 23: 607–611.

22. Renouf V, Claisse O, Miot-Sertier C, Lonvaud-Funel A (2006) Lactic acid bacteria evolution during winemaking: use of rpoB gene as a target for PCR-DGGE analysis. Food Microbiol 23: 136–145.

23. Cocolin L, Aggio D, Manzano M, Cantoni C, Comi G (2002) An application of PCR-DGGE analysis to profile the yeast populations in raw milk. Int Dairy J 12: 407–411.

24. Fontana C, Vignolo G, Cocconcelli PS (2005) PCR-DGGE analysis for the identification of microbial populations from Argentinean dry fermented sausages. J Microbiol Meth 63: 254–263.

25. Leclercq-Perlat M-N, Buono F, Lambert D, Latrille E, Spinnler H-E, et al. (2004) Controlled production of Camembert-type cheeses. Part I: Microbiological and physicochemical evolutions. J Dairy Res 71: 346–354.

26. Le Graet Y, Brulé G (1988) Migration des macro et oligo-éléments dans un fromage à pâte molle de type Camembert. Le Lait 68: 219–234.

27. Muyzer G, De Waal EC, Uitterlinden AG (1993) Profiling of complex microbial populations by denaturing gradient gel electrophoresis analysis of polymerase chain reaction-amplified genes coding for 16S rRNA. Appl Environ Microb 59: 695–700.

28. Altschul SF, Gish W, Miller W, Myers EW, Lipman DJ (1990) Basic local alignment search tool. J Mol Biol 215: 403–410.

29. Chen L, Luo S, Chen J, Wan Y, Liu C, et al. (2012). Diversity of endophytic bacterial communities associated with Cd-hyperaccumulator plant Solanum nigrum L. grown in mine tailings. Appl Soil Ecol 62: 24–30.

30. Magalhães KT, Pereira GDM, Dias DR, Schwan RF (2010) Microbial communities and chemical changes during fermentation of sugary Brazilian kefir. World J Microb Biot 26: 1241–1250.

31. Guzel-Seydim Z, Wyffels JT, Seydim AC, Greene AK (2005) Turkish kefir and kefir grains: microbial enumeration and electron microscobic observation. Int J Dairy Technol 58: 25–29.

32. Dimitrellou D, Kourkoutas Y, Banat I, Marchant R, Koutinas A (2007) Whey-cheese production using freeze-dried kefir culture as a starter. J Appl Microbiol 103: 1170–1183.

33. Dimitrellou D, Kourkoutas Y, Koutinas A, Kanellaki M (2009) Thermally-dried immobilized kefir on casein as starter culture in dried whey cheese production. Food Microbiol 26: 809–820.

34. Leroy F, De Vuyst L (2004) Lactic acid bacteria as functional starter cultures for the food fermentation industry. Trends Food Sci Tech 15: 67–78.

35. Wang Y, Ahmed Z, Feng W, Li C, Song S (2008) Physicochemical properties of exopolysaccharide produced by Lactobacillus kefiranofaciens ZW3 isolated from Tibet kefir. Int J Biol Macromol 43: 283–288.

36. Hamet MF, Londero A, Medrano M, Vercammen E, Van Hoorde K, et al. (2013) Application of culture-dependent and culture-independent methods for the identification of Lactobacillus kefiranofaciens in microbial consortia present in kefir grains. Food Microbiol 36: 327–334.

37. Cheirsilp B, Shimizu H, Shioya S (2003) Enhanced kefiran production by mixed culture of Lactobacillus kefiranofaciens and Saccharomyces cerevisiae. J Biotechnol 100: 43–53.

38. Piermaria J, Bosch A, Pinotti A, Yantorno O, Garcia MA, et al. (2011) Kefiran films plasticized with sugars and polyols: water vapor barrier and mechanical properties in relation to their microstructure analyzed by ATR/FT-IR spectroscopy. Food Hydrocolloid 25: 1261–1269.

39. Kesmen Z, Kacmaz N (2011) Determination of lactic microflora of kefir grains and kefir beverage by using culture-dependent and culture-independent methods. J Food Sci 76: M276–M283.

40. De Ruyter P, Kuipers OP, Meijer WC, de Vos WM (1997) Food-grade controlled lysis of Lactococcus lactis for accelerated cheese ripening. Nat Biotechnol 15: 976–979.

41. Coppola R, Succi M, Sorrentino E, Iorizzo M, Grazia L (2003) Survey of lactic acid bacteria during the ripening of Caciocavallo cheese produced in Molise. Le Lait 83: 211–222.

42. Mas M, Tabla R, Moriche J, Roa I, Gonzalez J, et al. (2002) Ibores goat's milk cheese: Microbiological and physicochemical changes throughout ripening. Le Lait 82: 579–587.

43. Phillips M, Kailasapathy K, Tran L (2006) Viability of commercial probiotic cultures (L. acidophilus, Bifidobacterium sp., L. casei, L. paracasei and L. rhamnosus) in cheddar cheese. Int J Food Microbiol 108: 276–280.

44. Callon C, Millet L, Montel M-C (2004) Diversity of lactic acid bacteria isolated from AOC Salers cheese. J Dairy Res 71: 231–244.

45. Depouilly A, Dufrene F, Beuvier É, Berthier F (2004) Genotypic characterisation of the dynamics of the lactic acid bacterial population of Comté cheese. Le Lait 84: 155–167.

46. Weinrichter B, Luginbuhl W, Rohm H, Jimeno J (2001) Differentiation of facultatively heterofermentative lactobacilli from plants, milk, and hard type cheeses by SDS-PAGE, RAPD, FTIR, energy source utilization and autolysis type. LWT-Food Sci Technol 34: 556–566.

47. Lisdiyanti P, Kawasaki H, Seki T, Yamada Y, Uchimura T, et al. (2000) Systematic study of the genus Acetobacter with descriptions of Acetobacter indonesiensis sp. nov., Acetobacter tropicalis sp. nov., Acetobacter orleanensis (Henneberg 1906) comb. nov., Acetobacter lovaniensis (Frateur 1950) comb. nov., and Acetobacter estunensis (Carr 1958) comb. nov. J Gen Appl Microbiol 46: 147–165.

48. Beshkova D, Simova E, Frengova G, Simov Z, Dimitrov ZP (2003) Production of volatile aroma compounds by kefir starter cultures. Int Dairy J 13: 529–535.

49. Ferreira A, Viljoen B (2003) Yeasts as adjunct starters in matured Cheddar cheese. Int J Food Microbiol 86: 131–140.

50. Kaminarides S, Laskos N (1992) Yeasts in factory brine of Feta cheese. Aust J Dairy Technol 47: 68–71.

51. Roostita R, Fleet G (1996) The occurrence and growth of yeasts in Camembert and blue-veined cheeses. Int J Food Microbiol 28: 393–404.

52. Van Iersel M, Van Dieren B, Rombouts F, Abee T (1999) Flavor formation and cell physiology during the production of alcohol-free beer with immobilized *Saccharomyces cerevisiae*. Enzyme Microb Tech 24: 407–411.

53. Cheirsilp B, Shoji H, Shimizu H, Shioya S (2003) Interactions between *Lactobacillus kefiranofaciens* and *Saccharomyces cerevisiae* in mixed culture for kefiran production. J Biosci Bioeng 96: 279–284.

54. Petersen KM, Westall S, Jespersen L (2002) Microbial succession of *Debaryomyces hansenii* strains during the production of Danish surfaced-ripened cheeses. J Dairy Sci 85: 478–486.

55. Arfi K, Spinnler H, Tache R, Bonnarme P (2002) Production of volatile compounds by cheese-ripening yeasts: requirement for a methanethiol donor for S-methyl thioacetate synthesis by *Kluyveromyces lactis*. Appl Microbiol Biot 58: 503–510.

56. Jollivet N, Chataud J, Vayssier Y, Bensoussan M, Belin JM (1994) Production of volatile compounds in model milk and cheese media by eight strains of *Geotrichum candidum* Link. J Dairy Res 61: 241–248.

57. Buzzini P, Gasparetti C, Turchetti B, Cramarossa MR, Vaughan-Martini A, et al. (2005) Production of volatile organic compounds (VOCs) by yeasts isolated from the ascocarps of black (Tuber melanosporum Vitt.) and white (Tuber magnatum Pico) truffles. Arch Microbiol 184: 187–193.

58. Toba T, Arihara K, Adachi S (1990) Distribution of microorganisms with particular reference to encapsulated bacteria in kefir grains. Int J Food Microbiol 10: 219–224.

16

Towards the Determination of *Mytilus edulis* Food Preferences Using the Dynamic Energy Budget (DEB) Theory

Coralie Picoche[1]*, Romain Le Gendre[1¤], Jonathan Flye-Sainte-Marie[2], Sylvaine Françoise[1], Frank Maheux[1], Benjamin Simon[1], Aline Gangnery[1]

1 Laboratoire Environnement Ressources de Normandie, IFREMER, Port en Bessin, France, **2** Université de Bretagne Occidentale, Institut Universitaire Européen de la Mer, Laboratoire des sciences de l'Environnement Marin (LEMAR), UMR 6539 CNRS/UBO/IRD/IFREMER, Plouzané, France

Abstract

The blue mussel, *Mytilus edulis*, is a commercially important species, with production based on both fisheries and aquaculture. Dynamic Energy Budget (DEB) models have been extensively applied to study its energetics but such applications require a deep understanding of its nutrition, from filtration to assimilation. Being filter feeders, mussels show multiple responses to temporal fluctuations in their food and environment, raising questions that can be investigated by modeling. To provide a better insight into mussel–environment interactions, an experiment was conducted in one of the main French growing zones (Utah Beach, Normandy). Mussel growth was monitored monthly for 18 months, with a large number of environmental descriptors measured in parallel. Food proxies such as chlorophyll *a*, particulate organic carbon and phytoplankton were also sampled, in addition to non-nutritious particles. High-frequency physical data recording (*e.g.*, water temperature, immersion duration) completed the habitat description. Measures revealed an increase in dry flesh mass during the first year, followed by a high mass loss, which could not be completely explained by the DEB model using raw external signals. We propose two methods that reconstruct food from shell length and dry flesh mass variations. The former depends on the inversion of the growth equation while the latter is based on iterative simulations. Assemblages of food proxies are then related to reconstructed food input, with a special focus on plankton species. A characteristic contribution is attributed to these sources to estimate nutritional values for mussels. *M. edulis* shows no preference between most plankton life history traits. Selection is based on the size of the ingested particles, which is modified by the volume and social behavior of plankton species. This finding reveals the importance of diet diversity and both passive and active selections, and confirms the need to adjust DEB models to different populations and sites.

Editor: Adam J. Munn, The University of Wollongong, Australia

Funding: This work was achieved with the financial fundings of BlueDEB and OGIVE projects. BlueDEB was supported by the French Norwegian Foundation, Basse Normandie region and Hordaland County Council. OGIVE is supported by the European Fund for Fisheries, Seine-Normandie Water Agency, Basse Normandie region and the departments of Manche and Calvados. The funders had no role in study design, data collection and analysis, decision to publish, or preparation of the manuscript.

Competing Interests: The authors have declared that no competing interests exist.

* Email: cpicoche@gmail.com

¤ Current address: Unité Lagons, Ecosystèmes et Aquaculture Durable, IFREMER, Nouméa, New-Caledonia

Introduction

The blue mussel *Mytilus edulis* is common in Europe and North America and has been consumed by man for centuries. Aquaculture can be traced back to the 13th century and now exceeds fishing due to its stability and the possibility it offers to regulate harvests. Production has been increasing for the last 50 years in response to the rise in mussel consumption and trade. This economic significance has drawn attention to *M. edulis*. Understanding the behavior and physiological responses of this species may help maximize its productivity through optimization of rearing strategies. Several models have been developed to describe mussel growth in relation to the environment. The Dynamic Energy Budget [1] theory has been the most successful of these models to date [2].

DEB models allow the quantitative description of energy acquisition and use in living systems. These models quantify energy fluxes through organisms, from energy uptake to its allocation to growth, maintenance and reproduction. As for other bivalve species [3], one of the difficulties in applying DEB models to *M. edulis* growth is to link trophic resources available in the environment with energy uptake (assimilation). In order to solve this problem, various modeling strategies have been used: the first and simplest of these consists in re-estimating the food-ingestion parameters for each studied location [4], [5]; this has allowed the DEB model to be adapted for low [6], [7] or high [8] seston conditions, for example. Different food proxies have also been tested. Chlorophyll *a* (chl a) has often been used, either as a raw input [9] or a refined input taking into account Chl a/C ratio [10]. Phytoplankton also gives good results [11], as does total particulate matter (TPM) [12]. Nevertheless, no consensus has been reached when comparing all these quantifiers [12]. In order to develop a more generic approach, other authors formalized more detailed processes for food ingestion that could incorporate several food proxies [13].

Knowledge is also needed about mussel feeding processes. Physiological aspects of food uptake have been thoroughly studied [14], [15], [16], [17], but data on the differential effects of food quality on growth are scarce. Recent studies tend to indicate that mass growth is significantly affected by diet quantity and quality [18], and that mussels may even modify their feeding behavior according to local food composition [19]. Until now, investigation focused on food physical properties, such as size, that could affect feeding behavior [20]. New techniques are being developed that give more detailed insight into the diet of *M. edulis* [21], [22], [23], [24] in terms of *e.g.*, types of food, phytoplankton species and diversity. Results are presently contradictory, however, even for the most generic points such as the comparative roles of diatoms and dinoflagellates [21], or the effect of particulate matter and other food sources (*e.g.*, detritus) [25]. This is partly due to local variability and lack of knowledge about feeding mechanisms. In the present work, we suggest that the DEB model can provide this type of information, as well as benefit from such data.

Inversing the model makes food reconstruction possible from growth observations. This has been done for several bivalves using different approaches: shell length can be mathematically related to ambient food using DEB assumptions [26]; successive approximates can be made for food [27] and the fit of different simulations obtained with different food proxies can be used as an indicator of their accuracy [12]. All these methods have their advantages and drawbacks and, until now, have focused on global food quantifiers. Here, we compare these three approaches at the scale of the phytoplankton species. This article is thus structured in three main steps: (1) we test two existing parameter sets for mussels, each representing a different way of studying food; (2) we then develop and apply our own food reconstruction methods and (3) use results to deduce general properties for food quality assessment.

Materials and Methods

Study area

Mussels were raised on the French coast of the English Channel in the western part of the Bay of Seine, northwest of the Bay of Veys and south of Utah Beach (49°24.369′N/1°09.230′W) (Fig. 1). The Bay of Veys covers 37 km^2 and has a semi-diurnal macrotidal regime with maximum amplitudes ranging from 2.5 to 7 m during neap and spring tides respectively. Freshwater inputs enter this area from the Carentan and Isigny channels, supplied by four rivers, dominated by the Vire, which contributes up to 40% of total flow [28]. Jouenne *et al.* [29], [30] described the primary production dynamics at different timescales in this bay. This small catchment area is characterized by relatively high chl a content and low turbidity compared with other estuaries, making it an intermediately productive ecosystem. Nutrient deficiency is debated: while some authors consider that their input is sufficient in the Bay of Veys [30], others hypothesize that nitrogen might be limiting during spring [10]. Anthropogenic activities seem more important than climatic conditions in explaining temporal variability in ecosystem functioning. There is a significant level of shellfish farming in this area: cultured *Crassostrea gigas* amounted to 2262 tons on Utah Beach in 2006, accounting for 14% of oyster production in whole of the Bay of Veys; *M. edulis*, at about 1332 t, represented about 85% of total mussel production in this area.

Field measurements

Mussel seed used for the experiment was collected during spring 2009 in La Plaine-sur-Mer (French Atlantic coast), and first transferred to pregrowing structures in Agon (west coast of Normandy). Pregrown mussels (20.99±1.85 mm and 0.79±0.04 g) were finally installed on Utah Beach in mid-September 2009. This site is located at around 1 km offshore and could be accessed from land at low tide for mussel sampling and by sea for hydrological measurements at high tide. Tide variation in this area led to daily emersion of the mussels lasting around 39±5% of the day. All structures were located on private sites lent by professional oyster or mussel farmers. These sites are usually used for shellfish farming and the experiment was therefore not considered to alter the environment, flora or fauna. Mussels were put in 18 plastic baskets (35 cm in height and a triangular section of side 20 cm). The baskets were installed on the middle height of 3 adjacent poles at a rate of 6 baskets per pole. Poles were separated by about 1 meter. They are the common rearing structure used for mussel culture in Normandy. Sixty mussels were placed in each half-basket, giving an initial available volume of 50 cm^3 per individual. Ninety-six individuals were separated in 9 specific nets for individual shell length monitoring. These nets were distributed in 3 baskets at a rate of 3 nets per basket. Baskets were installed on a fourth adjacent pole. Monthly sampling began in January 2010 and continued for 18 months. Each sampling date comprised two sets of measurements: one from mussels that were alive throughout the study and one from mussels that were sacrificed. In the first set, the shell length (L in mm) of the 96 identified mussels was determined *in situ* using a manual caliper (FACOM, accuracy: 0.02 mm), allowing the acquisition of individual trajectories. For the second set, the content of one randomly selected basket was sampled at each date and brought back to the laboratory. Forty of the live mussels were randomly selected, measured using a digital caliper (MITUTOYO, accuracy: 0.02 mm) and sacrificed in order to separate flesh and shell. Dry flesh mass (DFM) was measured after a complete freeze-drying cycle (METTLER TOLEDO Balance, accuracy: 1 mg). Based on the assumption that average flesh mass is proportional to cubed mussel length (eq. 1), masses were corrected for size differences on the basis of the length obtained from the individual monitoring [31].

$$M_t = a_t L_t^3 \qquad (1)$$

where M_t is the reconstructed flesh mass, L_t is the measured length from the non-destructive sampling at date t and a_t is the coefficient relating length and mass, calculated with the destructive sampling. An example of such relation between mass and length is given in Fig. S1. This is different from the DEB formulation relating length and mass with a constant parameter set (eq. 13 in Text S1). Here, a_t is recalculated at each sampling date.

The two sets of mussels, corresponding to two different processes, were necessary because of the invertebrate nature of the mussels. Neither body length, nor flesh mass can be accessed without killing the animal. However, individuals should not be killed as their continuous monitoring alone can accurately assess growth. Indeed, the use of different individuals at each sampling date may introduce a bias: two different basket samples can show growth variations that should not be taken into account in the model. As a consequence, we used a first set of sacrificed mussels to calculate a proxy of the relation between shell length and flesh mass. We then used the shell length, the only measurement that we could obtain from the second set of non-sacrificed mussels, to

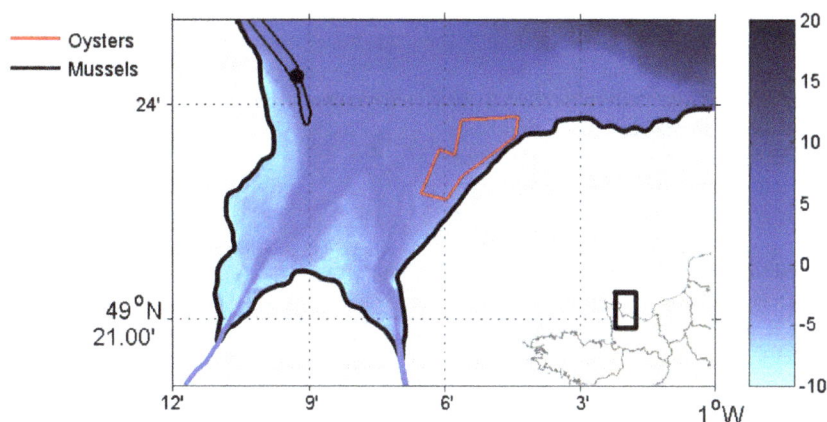

Figure 1. Location of the study area in the Bay of Veys.

reconstruct the dry flesh mass of the individuals. As a consequence, the dry flesh masses that are used as inputs in the model are a proxy of the masses of the same individuals at each sampling date.

Hydrological data (water temperature, water depth and salinity) were continuously recorded with an autonomous NKE data logger STPSO2-SI placed in a basket. Two data loggers were specifically dedicated to the experiment, used in rotation and changed every month. Observations were checked for outliers or drift, cleaned when necessary and filtered to account for immersed values only. As a consequence, temperature measurements were used only when they corresponded to immersion time. Metrological verifications were made before each deployment in the field. The frequency of data acquisition was set at 10 minutes.

Hydrobiological parameters were assessed fortnightly for a total of 33 sampling dates over 18 months. All water samplings were performed from the boat within the two hours around high tide. For nutrients (ammonia NH_4^+, silicates $Si(OH)_4$, phosphates PO_4^{3-} and nitrates-nitrites $NO_2+NO_3^-$), samples were taken 1 m below the surface using a Niskin bottle. Pre-filtrations were made on board with a 48 μm mesh for all nutrients plus a 0.45 μm mesh for silicates. Back in the laboratory, analyses were done with a Technicon Autoanalyzer III, according to the method described in [32]. For the other parameters, water samplings were performed by hand at around 50 cm below the surface. Particulate matter samples were filtered in duplicate through pre-combusted (450°C for 1 h) and pre-weighed Whatman GF/F filters, rinsed with distilled water to remove salts and stored at −20°C until analysis. Filters were dried at 70°C for at least 2 h and weighed for total particulate matter (TPM, mg dry mass l^{-1}). Inorganic matter (PIM, mg ash dry mass l^{-1}) was given by the mass of ash remaining after burning at 450°C for 5 h. Organic matter (POM, mg ash free dry mass l^{-1}) was given by losses at ignition. PIM and POM values are available from February to October 2010. In order to estimate trophic resources potentially available for mussels, phytoplankton biomass (chl a and pheopigment concentrations, μg l^{-1}) and composition (cell abundances of micro-, nano- and picoplankton, cell number l^{-1}) and particulate organic carbon (POC) and nitrogen (PON) were determined. For chl a and pheopigments, samples were filtered in duplicate through Whatman GF/F filters, which were frozen at −80°C for up to a month. Pigments were extracted in 90% acetone for 12 h and analyzed with the Lorenzen spectrophotometric method described by [33]. For POC and PON determination, water samples were filtered in duplicate through Whatman GF/F filters, rinsed with sodium sulfate and analyzed with a CHN analyzer according to [33].

Additional filtrations were also made for blank materials [33]. Water samples were fixed in a Lugol's solution for phytoplankton determination. The portion of phytoplankton between 20 and 200 μm in size was identified and counted in 10 ml tanks with the Ütermohl method [34] using a phase-contrast inverted microscope (Olympus IMT2 or IX71). The same analyst conducted the identification process for the entire experiment in order to maintain a consistent account of species evolution. Accuracy started at the family level and could reach species or groups of species depending on the plankton morphology. In agreement with the analyst, the level of detail was coarser in the post-processing than during the experiment: 53 out of the 77 initially identified groups of plankton were finally used. This approach discarded potential mistakes in the identification process. For pico- and nanoeukaryotes, as well as the bacteria *Synechococcus* and *Cryptophyceae*, water samples of 1 ml were fixed in 1.8 ml cryotubes with electron microscopy grade glutaraldehyde at a final concentration of 1% (vol/vol) and immediately stored in liquid nitrogen for a few months [35]. Samples were then counted by flow cytometry according to [36].

Finally, hourly rainfall measurements, wind direction and velocity and irradiance were provided by Météo France for the Sainte-Marie-Du-Mont station, located 6 km southwest of our sampling site.

Dynamic Energy Budget model

The model used in this study is based on the Dynamic Energy Budget (DEB) theory [1]. This model quantifies growth and energy allocated to reserves, structure and reproduction as a function of two forcing variables: water temperature and food availability. This type of model has been widely applied to the study of bivalve energetics [3], [4], [6], [10] and has been used to predict *M. edulis* growth and reproduction [5], [7], [9], [12], [13]. The main equations are given in Text S1. Implementation was taken from Rosland *et al.* [6], including a possible decrease in somatic mass during starvation (eq. 12 in Text S1), and adapted to take into account additions and adjustments concerning food assimilation and energy processing from Saraiva *et al.* [13] (eq. 1 to 4 in Text S1). Briefly, the model describes the energetics of an individual through the dynamics of three state variables: reserves (E), structure (V) and energy allocated to reproduction (E_R). Energy is taken from the environment and fuels the reserve compartment (eq. 1 to 6 in Text S1). A constant fraction κ of this energy is allocated to somatic maintenance and structural growth and the remaining 1-κ is allocated to maturity maintenance, development (in juveniles) and reproduction (in adults) (eq. 7 to 10 in Text S1). The energy content of the reproductive buffer is liberated at spawning. In the present application, spawning dates are forced and correspond to drops in dry flesh mass (DFM) [6]. Such a formulation is less flexible than an implementation based on gonado-somatic index (GSI) and water temperature thresholds [5] but was chosen for simplicity purposes. The reproductive buffer is assumed to be totally emptied at spawning, according to Sprung *et al.* [37].

The Arrhenius law was used to correct all rates for temperature. Main parameter values are given in Table 1. Recent food intake developments were included, requiring a calibration based on the carbon content of the particles [13]. In the absence of further knowledge, a conversion-to-carbon factor λ was added and calibrated for each food source (eq.1 in Text S1); some conversion factors are shown in Table 2. Additionally, due the same lack of knowledge, assimilation efficiency γ was assumed to be constant and equal to 0.75 [6].

Simulation and validation

Simulations were run with Matlab (Matlab R2012b), using the implementation of Rosland et al. [6]. Input daily temperature was the daily average of water temperature, measured during immersion time. Daily available food was linearly interpolated from the fortnightly observations. Chl a, abundance of micro-, nanophytoplankton and even smaller species like bacteria or *Synechococcus* (alone or as a group), POC and TPM were tested as model inputs one after the other. Composite variables were also tested by balancing chl a or phytoplankton abundance with the corresponding richness (number of species) or evenness (calculated as the ratio of the Shannon diversity index and the natural logarithm of the number of species). To account for inedible material, PIM was approximated as 80% of TPM. This is the maximum ratio of PIM to TPM that was measured during the experiment and was used to reduce assimilation [13]. Initial mass

was the average mussel mass at the beginning of the experiment. Shell length was calculated accordingly.

The effect of different food sources on growth was assessed by deviation d (eq. 2). This calculates the distance to mass measures.

$$d = \sum \frac{|M_{obs,t} - M_{sim,t}|}{M_{obs,t}} \qquad (2)$$

where d is the deviation of the simulation, and $M_{obs,t}$ and $M_{sim,t}$ are the respective observed and simulated DFM at observation time t.

A non-linear optimisation method (Nelder-Mead) was applied in the auto-calibration, which searched iteratively for the λ values that minimised the deviation d.

Food reconstruction

An inverse method was used to assess the quality of our food sources. Shell length and DFM were used to compute the corresponding functional response over the 18-month experiment. This response is taken as a function of both food availability and digestibility.

The first step was to reconstruct the evolution of the functional response over time from individual shell length time series, using a reversed DEB model as described in [38]. Briefly, temperature and shell length are taken as inputs of this reversed model to calculate the corresponding functional response. Different individual measurements of length are averaged and interpolated using a spline function so that the reversed model applies on the same daily time step as the standard one. The reconstructed functional response is based on the same equations and is therefore theoretically exact.

An iterative method was then used to compute a functional response corresponding to mass variation. The functional response was used with the modeling of Rosland *et al.* [6] (eq. 5 and 6 in Text S1) including the non-food related parameters that have been described by Saraiva *et al.* [13]. An initial functional response was built to vary randomly between 0 and 1. This was then used as an

Table 1. DEB parameters and values.

Symbol	Description	Value	Unit
TA	Arrhenius temperature	7022	K
$[\dot{p}_M]$	Volume-specific maintenance costs	11.6	J d^{-1} cm^{-3}
$[E_m]$	Maximum storage density	1438	J cm^{-3}
$[E_G]$	Specific cost for structure	5993	J cm^{-3}
κ	Fraction of reserves spent on somatic growth and maintenance	0.67	-
δ	Shape parameter	0.297	
$\{\dot{C}_{Rm}\}$	Maximum surface area specific clearance rate	96	L d^{-1} cm^{-2}
$\{\dot{J}_{aF}\}$	Algal max. s.a. specific filtration rate	0.00048	mol C d^{-1} m^{-2}
$\{\dot{J}_{iF}\}$	Inorganic material max. s.a. specific filtration rate	3.5	g d^{-1} cm^{-2}
ρ_a	Algal binding probability	0.99	-
ρ_i	Inorganic material binding probability	0.4	-
$\dot{j}_{aI}$	Algal max. ingestion rate	13000	mol C d^{-1}
$\dot{j}_{iI}$	Inorganic material maximum ingestion rate	0.11	g d^{-1}
C	Conversion factor	697000	mol J^{-1}
AE	Assimilation efficiency	0.75	-

Values were taken from Saraiva (2011a) and adapted to allow different food proxies.

Table 2. Food proxy calibration and use in the model.

Food source	Conversion parameter	Deviation (%)
Chl a	7.9×10^{-6} mol C.(µg Chl a)$^{-1}$	23.2
Chl a × richness	4.6×10^{-7} mol C.(µg Chl a)$^{-1}$	21.2
POC	8.3×10^{-8} mol C.(µg C)$^{-1}$	37.7
Phytoplankton	3.6×10^{-6} mol C.(Cell)$^{-1}$	35.8
Chaetoceros	1.8×10^{-9} mol C.(Cell)$^{-1}$	18.6

Food proxies had to be converted to mol C to adapt them to the rest of the model. Deviation corresponds to the relative difference between simulated and observed DFM, as described in eq. 2.

input for the simulation, skipping food ingestion and assimilation to affect growth directly. To do this, the classic DEB formulation was used ([6]), adjusting the maximum assimilation rate with the functional response term. Simulated growth was then compared with observed growth. For each time step between first and last sampling dates, the functional response was respectively increased or decreased as long as the simulated mass was below or above observed mass within a 1% range, and functional response was above 0. These stringent conditions ensured that variations were respected over time while maintaining a physiological sense. A functional response above 1 may indicate a bad parameter value for the maximum assimilation rate, which was expected from the results of the first simulations. On the contrary, a negative functional response can only indicate "negative" assimilation, which does not make sense.

Both reconstructed functional responses were then scaled by their maximum to vary between 0 and 1, and smoothed with a 4-day moving average to remove modeling bias.

Food quality assessment

The contribution of each food source was evaluated for each sampling date after being transformed with a Holling-type II function (eq. 3).

$$T_{i,t} = \frac{F_{i,t}}{F_{i,t} + c_i} \tag{3}$$

where $T_{i,t}$ is the transformed food source F_i at time t and c_i is a scaling coefficient, linked to saturation or satiation. This parameter regulates food absorption.

This transformation was meant to avoid signal distortion and to homogenize units between different food sources (phytoplankton abundance may vary between 0 and 10^6). It remains close to the usual DEB model formulation and helps to scale different food values. The same scaling constant was used for both functional responses (length and DFM) as mussels ingest food in the same way, with possibly different allocation. Values of both signals were similarly extracted. A linear combination of processed food measures was then used as a proxy for functional response signals. Each sampling date is thus characterized by a system of two equations (eq. 4).

$$\begin{cases} \sum_i a_i \dfrac{F_{i,t}}{F_{i,t} + c_i} = f_{l,t} \\ \sum_i b_i \dfrac{F_{i,t}}{F_{i,t} + c_i} = f_{m,t} \end{cases} \tag{4}$$

Where a_i and b_i represent the contribution of the food source F_i to growth in length or mass, respectively, and c_i is the scaling coefficient, which should be understood in this context as an index of the quantity of available food above which the contribution to total food input is maximal. It can be seen as the intensity with which mussels react to food presence: a lower c_i helps reaching a_i and b_i with lower food concentrations. f_l and f_m are the functional responses corresponding to length and mass, respectively.

Thirty-three sampling dates were available, which gave 66 equations. Each food source needed to be described by 3 parameters (a_i, b_i and c_i). An identifiability analysis ([39], used in [40], [41], [42]) showed that all three of these parameters could not be determined at the same time with a sufficient number of plankton species to study the ecological characteristics (see Text S2, Fig. S2 and S3 for more details about model assessment). c_i parameters were thus fixed, corresponding, for each species, to the median value of the abundance when the species was present in the field. This is close to the value that is obtained for X_k with the DEB model when using a single plankton species as the food input. The number of species for which a_i and b_i can be calculated is a trade-off between the condition index of the matrix model (indicating the quality of the formulation) and the final model error: in our study, 30 plankton species or groups of species was the maximum we could take into account while maintaining a reasonable condition index (3.3×10^3). These were chosen as the most abundant species that were also large enough to be efficiently retained [20].

Resolution was performed with Matlab's active-set sequential quadratic programming under constraints (Matlab R2012b). Plankton species used in hatchery were assumed to have a positive impact on growth (a_i and b_i are positive for *Chaetoceros*, *Skeletonema costatum, and Naviculaceae* for instance [43]). Initial a_i and b_i followed a uniform random distribution between -1 and 1, complying with the above constraints.

a_i and b_i were then transformed according to eq. 5 to represent relative contributions to total diet value.

$$w_i = \frac{x_i}{\sum_{j=1}^{n} |x_j|} * 100 \tag{5}$$

where w_i is the relative contribution of the food source i corresponding to the coefficient x_i (a_i or b_i) and n is the total number of food sources.

Statistical analysis

Relative contributions of each plankton species to total diet value were linked with plankton characteristics. Several databases

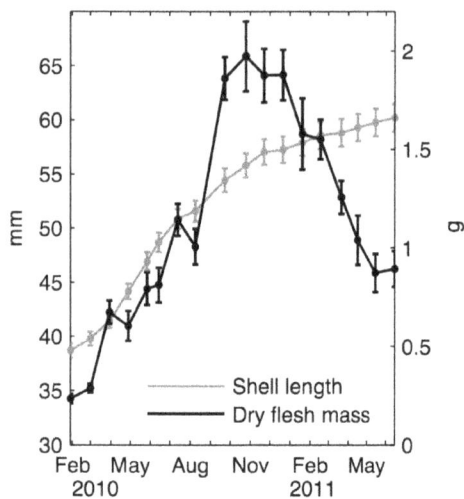

Figure 2. Observed mussel growth during the experiment. Shell length (grey) and dry flesh mass (DFM) (black) mean values and deviations correspond to monitoring with sacrifices, corrected for sampling bias.

([44], [45], [46] and M. Schapira's personal atlas) were collated and completed with data from the literature (see Table S1). Family was the first classification criterion, separating diatoms, dinoflagellates and others. Among diatoms, pennate and centric species were differentiated. Biovolume, surface area and their ratio were taken from [46], using the median value of the observations in our study area. The difference between smaller and larger species was qualitatively assessed by the plankton analyst to account for local variability. Biovolume and area were then log-10 transformed. Cell shape was also extracted from [46], using the conventions in [47]. Habitat values classified plankton according to their preference for coastal or pelagic areas ([28], [45]). Habitat was turned into an ordinal variable ranging from 1 to 3, 1 characterizing species that are specifically found in a coastal brackish environment, including those coming from freshwater inputs and 3 characterizing species that could mostly be found in the ocean. This was decided using classifications by [48], [49], [50] and [51] and evaluation of salinity and eutrophication tolerance. Plankton social behavior qualifies colony frequency. For some species, this has been quantified as the mean number of cells per colony. When this number is below 1, the species is considered as single. Above, it is considered colony forming.

Explanatory variables were chosen based on a stepwise approach using the AIC relative change as the selection criterion (LinearModel.stepwise routine in Matlab 2012b). For contribution based on DFM, social behavior and biovolume were the only relevant parameters while contributions based on length were influenced by social behavior only. To be consistent, an ANCOVA (analysis of covariance) was performed for both relative contributions, taking into account both explanatory variables.

Results

Mussel growth

Out of the 96 individually-monitored mussels, 22 individuals died during the experiment and shell-length measures of 18 individuals decreased at least once during the experiment. These trajectories were removed before data processing. Monitoring data obtained with and without sacrifice were consistent.

Mussel growth is shown on Fig. 2. During the experiment, shell length increased from 36 to 60 mm. Maximum relative length increase took place in April 2010 with a 7% gain. Shell length increased by 1.7 mm per month during 2010, while 2011 was characterized by a length gain of around 0.45 mm per month. Mass observations can be divided into two main periods: 2010 was characterized by a mass gain followed by a period of loss during year 2011. DFM increased by 1.7 g between February and October 2010. This high growth mainly took place during March, with a growth rate of $+13$ mg d^{-1} and between August and September ($+19$ mg d^{-1}). Mass was then stable until the end of the year. DFM was halved between January and May 2011.

Mussel length and mass growth were uncoupled: periods of maximum growth did not occur at the same time for shell length and flesh. In addition, the drop in flesh mass was obviously not reflected by shell length.

Environmental conditions

Variations of the main environmental descriptors over the studied period are shown in Fig. 3. During 2010, water temperature (Fig. 3A) varied between 3°C in February and 21°C at the beginning of July. During 2011, it varied between 4.4°C at the end of January and 17.6°C at the beginning of June. Average temperature from February to mid-June was 9.4°C in 2010 vs. 10.8°C in 2011. Water temperature varied 1.1 ± 0.6°C within a day, with a minimum variation of 0.2°C and a maximum of 3.3°C, which is low enough to consider temperature to be constant over a day in the DEB model. Rainfall, irradiance and nutrient concentration dynamics (Fig. 3B) were not significantly different between the two years (Kolmogorov-Smirnov test, $p > 0.05$) and patterns were consistent with previous records on this point.

Chl a dynamics can be divided into two periods (Fig. 3C). In 2010, two peaks appeared, in mid-March (16.5 µg L^{-1}) and mid-June (8.9 µg L^{-1}), followed by a longer period with a chl a level stabilized around 3 µg L^{-1} for a month at the end of summer. It reached its minimum value of 0.6 µg L^{-1} in December. Conversely, the first six months of 2011 were marked by a low chl a concentration for this period of the year. It only reached a maximum of 3.1 µg L^{-1} at the beginning of May and therefore showed a low spring plankton biomass. Between February and June, chl a was on average 66% higher in 2010 than in 2011. The maximum difference was in the intensity of the spring bloom.

Nanoeukaryote and phytoplankton dynamics were similar to one another (Fig. 3D). Nano- and microplankton abundance were low during the first four months, then tripled in mid-June 2010 to reach almost 5×10^6 C L^{-1} and nearly 2×10^6 C L^{-1} respectively. End of summer and autumn were also characterized by two smaller peaks of abundance. Winter had a low concentration of microorganisms. In 2011, phytoplankton growth resumed in April, resulting in a 13-fold increase in biomass. Nanoeukaryote abundance began increasing significantly from mid-May. Nanoeukaryotes were more abundant in 2010 than in 2011 throughout the common experiment period. Picoplankton abundance was about one order of magnitude higher than nanoeukaryote abundance but seemed to follow the same patterns. It reached a peak in July 2010 and remained high until November. After a sharp decrease, it bloomed again from April 2011, with values comparable to those observed in 2010. Overall, it may represent 86% of total chl a production. No difference was significant between the two years for the picoplankton populations.

Richness varied between 11 and 31 species in 2010, with a median value around 20, while it varied between 7 and 20 in 2011 with a median value of 14 and a standard deviation around 2.9 for both periods. It was positively correlated with abundance

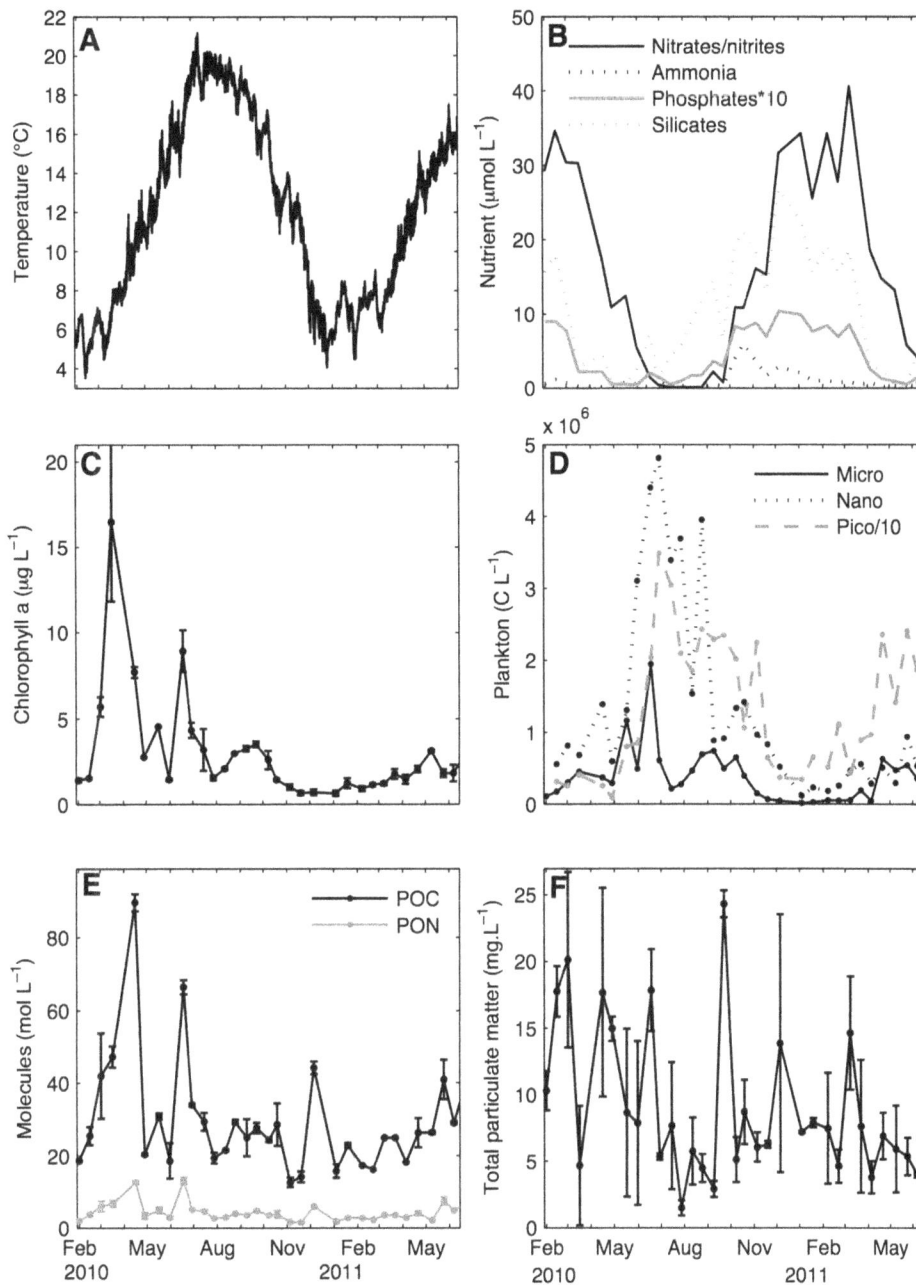

Figure 3. Observed environmental descriptors during the experiment. Average daily sea water temperature (A), nutrients (B), chlorophyll a (C), abundance of different sizes of plankton (D), particular organic carbon and nitrogen (E) and total particulate matter (F) were measured in 2010 and 2011.

$(R^2 = 0.31$, p = 0.08). Evenness varied between 0.27 and 0.86 with a deviation of 0.17 in 2010, and varied between 0.33 and 0.74 with a deviation of 0.13 in 2011; it was negatively correlated with total abundance $(R^2 = -0.64$, p<0.05). Chl a concentration could be related to these dynamics. The first chl a peak cannot be totally explained by phytoplankton or nanoeukaryote abundance but is confirmed by pheopigment concentration on the next sampling date. Phytoplankton counts explain the dynamics of last two blooms. In July, *Asterionellopsis glacialis* accounted for 76% of the total phytoplankton bloom while *Chaetoceros* amounted to 77% of the September phytoplankton biomass.

Chemical compounds did not vary as much between the two years as the variables mentioned above (Fig. 3E). POC ranged between 12.6 and 89.6 mol C L^{-1} during 2010, and between 16.1 and 40.9 mol C L^{-1} during 2011. PON ranged between 1.6 and 13.0 mol N L^{-1} during 2010, and between 1.7 and 7.4 mol N L^{-1} during 2011. Considering only the February–June period to compare the two years, there was a 33% decrease in POC and PON in 2011. This is only half the difference in chl a between 2010 and 2011. POC and PON are not only due to algal presence but also to detritus and river inputs.

Water quality was also impacted by TPM (Fig. 3F) as part of it is inorganic and may decrease food quality ([13]). TPM was highly

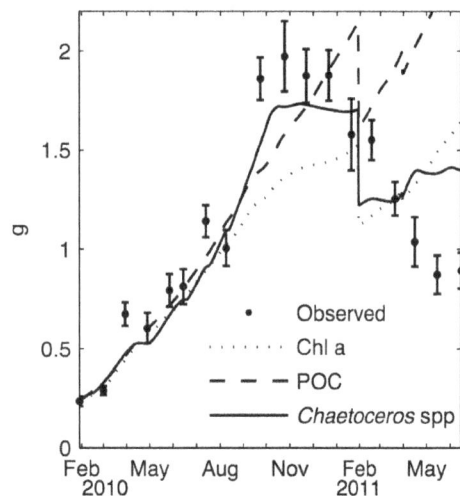

Figure 4. Comparisons of observed and modeled dry flesh mass (DFM) with different food inputs. Observations (dots) and DEB simulations (lines) are based on *in situ* measurements of food sources, with chl a (dashed line), POC (dotted line) and *Chaetoceros* spp. separately (solid line). The latter produced the best fit.

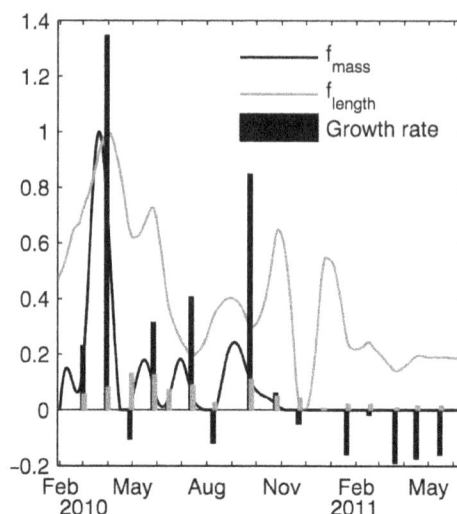

Figure 5. Comparisons of functional responses obtained with shell length (grey) and DFM (black) variation (see text for functional response computation). Bars represent the daily relative growth rates between two points. Both functional responses were standardized to vary between 0 and 1.

variable (CV = 63%), ranging from 1.49 mg L^{-1} in July 2010 to 24.3 mg L^{-1} two months later. 2011 showed less extremes with only one peak around 14.6 mg L^{-1} in March. The lower concentrations observed in 2011 should nevertheless be interpreted with caution as the strong northeasterly wind that predominated during this period may have suspended many more particles than the levels recorded at low frequency.

Relating food abundance and growth patterns is a first step towards simulations. Usual food proxies such as chl a and plankton abundance decreased in 2011, which can partly explain the observed mass loss. However, mass gain timing in 2010 cannot be totally explained by these food sources. Indeed, the first mass increase in March 2010 cannot be related to plankton abundance, but can be related to a chl a peak. Conversely, the extent of the mass gain in September 2010 cannot be related to a comparable chl a increase but it can be linked to a planktonic bloom. In contrast, the plankton bloom in June 2010 was not related to any great increase in mass. From the other point of view, POC and PON do not decrease enough in 2011 to totally explain the mass loss.

DEB model

The first simulations with the DEB model did not provide satisfactory results, especially when chl a and POC were used as food proxies, as shown on Fig. 4 and Table 2. The model failed to represent both mass gain and loss. The two steep slopes of the growth curve in March and August 2010 could not be reproduced and led to underestimates of DFM at the end of the growth season and during winter. On the contrary, food was always sufficient in 2011 to allow growth or, at least, to avoid the mass loss observed in the field. A different set of parameters (from [6] and [13]) may lead to different decrease due to spawning. The most satisfactory results were obtained using the *Chaetoceros* genus alone, which cannot represent the reality of mussel nutrition. However, when chl a was balanced by species richness, the second best fit was obtained with a total deviation of 21%; these contrasting results led us to consider the plankton species in more detail.

Figure 5 shows the functional responses computed with the reverse DEB model and compared to length and mass growth

rates. Length and DFM do not show the same patterns, as their increase and decrease do not match.

Figure 6 shows the difference between observed and simulated DFM, using the corresponding reconstructed functional response. This highlights problems in the parameter sets that could not be totally remediated by adjusting food input. Food availability is not the only explanation of these variations, as a functional response of more than one is necessary to reach the masses observed in 2010. This is especially true in March where the simulated functional response can reach 8. Mass loss was underestimated when no food was input in 2011.

Food quality

Table 3 shows the contributions of the 30 different plankton species to the reconstructed function responses based on length and soft tissue growth. Of the tested plankton groups 77% were diatoms.

Relative contribution coefficients associated with DFM/length variations ranged from 0.076/0.065 for *Ditylum* spp. to − 0.113/−0.087 for *Phaeocystis* spp; 57% of them were positive. Contributions obtained with shell lengths or DFM were not significantly different (Kolmogorov-Smirnov test, p>0.05). Consensus on the sign of the contributions was 73%. Species ranking was similar for length and DFM.

Cerataulina spp., *Euglenaceae, Guinardia striata* and *Thalassiosiraceae* play a positive role in growth in terms of mass but not in terms of length. On the contrary, *Bacillariaceae, Plagiogramma* spp, *Rhizosolenia setigera, R. pungens* and *Thalassionema nitzschioides* seem to have a positive effect on growth in terms of length, but not in terms of mass.

Both distributions were normal (Shapiro-Wilk test, p>0.05), enabling the use of the ANCOVA method. Social behavior has a significant effect on food quality, related to DFM (ANCOVA, $F_{1,27} = 2.9$, p<0.01), while this effect is unclear for length-related coefficients (ANCOVA, $F_{1,27} = 2.8$, p = 0.11). When plankton species tend to form colonies, mussel affinity decreases (Fig. 7A). Biovolume was not deemed significant in either group of contributions (ANCOVA, $F_{1,27} = 2.2$, p = 0.15 for DFM and

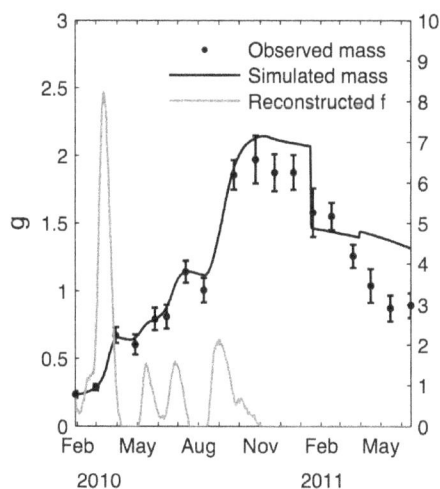

Figure 6. Comparisons of observed and modeled DFM with the reconstructed functional response. Functional response without standardization (grey) was obtained with DFM observations. Simulated DFM is shown by solid lines, observations are shown by dots.

$F_{1,27} = 0.4$, $p = 0.53$ for length). However, a slight effect can be seen, as mussels seem to grow better on larger cells (Fig. 7B). Shape effect was not significant in the analysis; however, it can be seen that large species with low contributions are all cylindrical. Among the positive contributions, 23% of plankton groups are cylinder-shaped, but these account for 46% of negative contributions. When this cylindrical shape is excluded from the analysis, size effect is significant at the 0.1 level.

Discussion

DEB performance

Mussel flesh and shell length variations reflect the ecosystem dynamics and seem to magnify them. The first mass increase (+135% in March and +85% at the end of summer) was higher than observed elsewhere for intertidal mussels, even on larger individuals [9], [52], [53]. Maximum mass at the end of the summer was surprisingly high for mussels of this size and age [4], [31], [54]. Relative gain remained coherent as it was still lower than values obtained with continuously immersed mussels [7], [55]. In all these cases, mass gain was observed much later in the year, with the more favorable climatic conditions. Total mass loss (58%) recorded in 2011 was among the highest values obtained during starvation studies [5], [6], [31].

The DEB model, as it is presently parameterized, is not able to reproduce these observations. Different parameter sets and food proxies have been tried but it has not been possible to obtain less than a 20% deviation in mass simulation, except when the input was made up of *Chaetoceros* spp only. However, diet diversity clearly improves the fit of the simulation to the observations when chl a was balanced by species richness (Table 2). These contrasted results are the first step towards a more specific analysis of the different plankton species.

More importantly, bias was always the same at the beginning of the simulation, underestimating mass during 2010. In 2011, only a total absence of food could lead to mass loss. Furthermore, while simulations were more satisfactory with the set of parameters from [13] for the first year, the one from [6] was more efficient at modeling the mass loss in 2011. When trying to avoid this problem by reconstructing the functional response, we found values over 7

for the beginning of 2010. Cardoso *et al.* [27] were faced with the same problems with *Macoma balthica* modeling, leading to functional responses higher than 1. We can conclude, as they did, that work is still needed on parameterization. δ and X_k are already known to depend on study site [5], [6] but other parameters may also be sensitive to phytoplankton ecotypes. We should therefore focus more on variations than on values.

Model quality is also problematic in recent works on *M. edulis*: while shell length is often correctly reproduced, this is not the case for DFM [7], [8] although mussel flesh is the most important aspect from an economic production viewpoint. The higher variability of flesh mass is mostly due to the losses that cannot be reproduced in shell length variation as this is an exoskeleton that is made of metabolically inert material. Gamete and reserve loss can be modeled with DEB theory, but structural loss cannot be modeled without altering the relationship between length and somatic energy [6]. This is due to the fact that we only have access to shell length and not body length. Studies have found no correlation between flesh and shell growth, even when both are increasing [12], [56]. This is partly due to differences in timing: shell length growth may precede soft tissue growth ([57], our data set) or succeed it [56]. Shell material is different from soft tissues and part of it comes from non-metabolic sources [58], [59]. This explains observed shell growth during starvation periods [60]. For the moment, the DEB model assumes that shell length is directly linked to structural flesh growth but, if this correlation does not always hold, other parameters will also need to be re-evaluated to obtain further knowledge about the species (*e.g.*, investment ratio κ). This also calls into question the use of shell length alone in functional response computation, as it can lead to the overestimation of ambient food conditions [26]. On the contrary, a comparison of both length and mass should be performed before conclusions are drawn. This is all the more difficult as shell length monitoring is preferable to avoid sampling bias that could emerge from the killing of animals to measure body length and dry flesh mass.

Food preferences

The first DEB modeling led us to choose several food quantifiers to test. We focused on phytoplankton abundance which was the closest available approximate of primary production. Chl a is commonly used, but its production inside each cell depends on varying environmental conditions [61]. POC and POM are composite elements that may overestimate available food [53]. Finally, Bracken *et al.* have found that mussels might depend more on phytoplankton than on other organic elements [62]. Plankton abundance was also the most flexible variable, allowing for several levels of detail. Pre-processing included the use of a transformation to homogenize values, which can be highly variable. We chose a Holling-type II transformation (eq. 3) mostly for its physiological grounding and closeness to the DEB formulation.

It is difficult to compare our plankton dataset with others in the literature because plankton assemblages and successions are highly variable. However, some species seem to bring about a consensus. For instance, our results agree with [24], showing that *Leptocylindrus* is not ingested by mussels, while *Pleurosigma* and *Gyrosigma* spp. are preferentially ingested and *Nitzschia longissima* and *Thalassiosiraceae* may have neutral roles.

Underlying patterns appeared among food preferences. Free-living cells seem to have a positive effect on mussel growth. Even if not significant in our dataset, a preference for larger species may be another component of food quality. Both characteristics may tend towards a passive selectivity, relying both on physical and chemical properties. Size is important, as put forward by [63], who

Table 3. Phytoplankton groups and contributions for flesh- and length-based growth.

Plankton group	Contribution to flesh growth (%)	Contribution to shell length growth (%)
Asterionellopsis glacialis	4.4	3.2
Bacillariaceae	−1.8	0.4
Biddulphia spp.	1.3	2.4
Cerataulina spp.	0.2	−1.3
Chaetoceros spp.	0.01	0.01
Ciliophora	3.5	4.1
Cryptophyceae	2.7	6.2
Dactyliosolen fragilissimus	−1.2	−5.2
Ditylum spp.	7.6	6.5
Euglenaceae	3.0	−2.1
Guinardia delicatula	−7.7	−5.1
Guinardia striata	2.5	−0.4
Gymnodiniaceae+Gymnodinium spp.	4.3	4.3
Leptocylindrus spp.	−4.5	−1.6
Melosiraceae	−5.3	−7.2
Navicula+Fallacia+Haslea+Lyrella+Petroneis spp.	1.4	1.3
Nitzschia longissima	−3.3	−2.2
Odontella spp.	−0.2	−2.5
Paralia sulcata	−4.9	−1.1
Plagiogramma spp.	−5.1	4.4
Pleurosigma+Gyrosigma spp.	4.1	5.7
Phaeocystis spp.	−11.3	−8.7
Prorocentrum spp.	4.2	3.2
Pseudo-nitzschia spp.	1.7	4.7
Rhizosolenia imbricata+styliformis	4.9	3.2
Rhizosolenia setigera+pungens	−5.5	1.7
Scrippsiella+Ensiculifera+Pentapharsodinium+Bysmatrum spp.	−3.5	−6.9
Skeletonema costatum	0.01	0.01
Thalassionema nitzschioides	−1.5	1.1
Thalassiosiracaea	2.0	−3.4

Groups were determined as described in the Material and Methods. Coefficients represent the contributions of each group to the total functional response of the DEB model, given by variation in length or DFM.

used the relative amount of small planktonic species as a depletion indicator because it is the only part of plankton that cannot be affected by bivalve consumption. Following [63] and [20], picoplankton was ignored and addition of nanoplankton to our model did not improve the fit. Even when targeting species in an accurate size range according to [20], smaller species were still less important for mussel growth. Cell size is modified by the ability to form a colony, increasing the actual volume that is filtered by mussels in the field. Our results could therefore be seen as contradictory.

The mechanism behind size preference needs to be clarified, as variability in retention efficiency cannot be attributed to larger species size only [20]. Colonies, which rely on other chemical components to bind themselves together, may overload the digestive system or the ciliary-gill pump, or may clog the gills [53]; this would trigger the ejection of pseudofaeces and/or feces consisting of undigested material [17]. Such a mechanism could explain the food assimilation decrease and mass loss, and seems all the more probable as *Phaeocystis* spp., forming colonies surround-

ed by an organic mucilage that can decrease clearance rate [64], are identified as the worst food source in our dataset. The potential role of shape also needs to be investigated.

Other plankton life history traits were not considered to make significant contributions to food quality, which may indicate the importance of a diverse diet. Regarding plankton ecological niche, Rouillon *et al.* [21] found more tychopelagic species in mussel stomachs than in ambient water, and Lefebvre *et al.* [65] showed that oyster growth in the Bay of Veys was dependent on microphytobenthos. Toupoint *et al.* [66] pointed out that pelagic cues overwhelmed biofilm ones, at least for mussel settlement. Our study cannot settle this argument, as few benthic species were found in our dataset and no biofilms could be observed at our sampling site. These films may not settle because of water mixing and sampling during flood tide leads to a bias towards pelagic species.

No preference was found for diatoms or dinoflagellates. The proportion of diatoms in food sources match that observed in the whole dataset. Previous studies are contradictory: some insist on

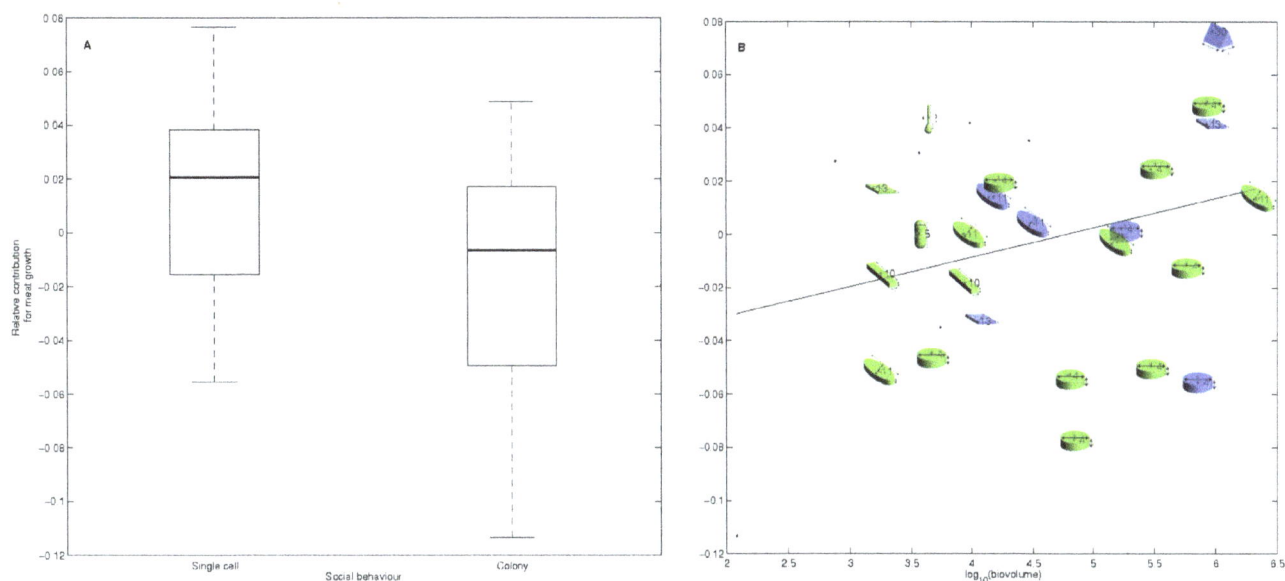

Figure 7. Effect of social behaviour (A) and biovolume (B) on mussel preference for plankton species. In the second panel, shapes correspond to the shape classification of each species, according to [47]. Blue color indicates that a species is free-living; green color indicates that it tends to form colonies. Black dots correspond to species with no defined shape.

the importance of diatoms in mussel diet [22], [67] while others emphasize the increase of dinoflagellates in the mussel diet compared with available species [68]; this has been discussed at length by Rouillon *et al.* [21]. Currently, our dataset can only show that both these groups are consumed and can play both positive and negative roles in food quality. Mussels may be sensitive to finer characteristics and/or may favor diversity [69].

Other characteristics that could have been investigated in our dataset include: how stickiness, electrostatic charge, mucopolysaccharides affect capture efficiency [17], and how morphology impacts palatability and digestibility [15]. Protein, carbohydrate and lipid contents are the key to assessing food effect on mussel growth by changing its composition [18] and even metabolism and reproductive cycle [70], [19]. Plankton composition is, however, too variable, and time- and site-dependent [14], [43], [62] to use values from the literature. Work on these aspects remains to be done and would certainly help in the search to find a structure behind our local species results. Once again, variability is high and the species that can be found in our area are certainly missing from the nearby sea [71] or will be in the coming years [30].

Finally, there are other food sources that have not been investigated and have already been found in the mussel diet: zooplankton [72], crustaceans, cnidarians, nematodes [22] and detritus [25] also contribute to the organic matter that can be ingested. Recent results tend to show that all of these sources are less influential than diatoms [23], but spatial and temporal variability would likely moderate any general conclusion [12].

Until now, we have tried to explain mass gain and loss with an emphasis on food availability and quality. Metabolism may have been altered by a significant switch between abundance and restriction. The former may have led to a decrease in growth efficiency that worsened the effect of restriction [73]. A high concentration of PIM may have altered assimilation efficiency while food was already low in quantity and quality for mussels.

However, two other, non-exclusive, explanations should be considered but cannot be proven with our dataset.

The reproductive cycle must have played a key role in mass regulation [54]. We cannot differentiate the loss due to starvation from that due to spawning. These must have coexisted as spawning alone cannot explain the mass loss in its entirety. Gonadosomatic ratio, although very variable, rarely reaches 55% [74], [75], [76], and recent studies tend to show that *M. edulis* might invest less in reproduction than was previously thought [13]. No mass recovery was recorded for 5 months, contrary to what is usually observed after spawning [56]. Conversely, this mass drop is too sharp to be due to metabolism alone.

Mussels have multiple and contradictory reproductive strategies depending on environmental conditions [19], [77], [78]. According to mussel farmers working in the study area, climatic conditions were very favorable to spawning in 2011. Metabolism might change during spawning time and requires more energy [79] or energy in a different form [24], which could have worsened mass loss that year. Loss of mass due to spawning in the DEB model depends on several parameters: allocation parameter κ, spawning efficiency and percentage of gametes left in the reproductive compartment after spawning. The latter has been discussed: mussels can spawn completely [37] or partially [76]. Furthermore, the use of GSI and temperature thresholds was not successful in reproducing growth and led to incoherent patterns (*e.g.* up to 5 or 6 spawning dates in 2010 at times where the actual reproduction cycle was not completed). This was not the main point of our study, which is why we decided, like [7] and [12], to empty the gonads at a fixed spawning date. Without further organ differentiation, it would have been presumptuous to model a reproductive cycle for our experiment, which is why this explanation has not been developed further; knowledge is still required in this area and needs to be improved.

The possibility of infection must also be presented. During 2010, the presence of *Mytilicola intestinalis* was recorded on Utah

Beach. Although infestation level was ranked low, there is still a possibility that the parasite developed and infected mussels in our experiments. Digestive disruptions may have resulted, decreasing assimilation efficiency; maintenance costs may also have been increased by the presence of a parasite. This could lead to differences in the physiological parameters describing mussel growth and explain the problems in DEB simulations that were observed even when food input was reconstructed.

To conclude, this article highlights the difficulty of representing different mussel growth patterns with a model smoothing tendency. Work needs to be done on DEB parameterization for *M. edulis* in this area. In the absence of further information, functional response reconstruction enabled us to get over the problem while still taking into account other environmental elements. This led to the selection of the preferred species that mussels have in the environments and to the identification of some patterns. A size gradient is noticeable in our dataset. Mussels tend to grow better on larger, single species. Plankton composition now needs to be studied further in order to relate it to mussel growth and investment.

Supporting Information

Figure S1 Comparisons of observed and modeled DFM as a function of cubed length.

Figure S2 Example of an eigenvector contribution to the model coefficients for the smallest eigenvalue, using a model optimizing the scaling coefficient c_i and simultaneous contributions of a_i and b_i.

Figure S3 Evolution of the parameter identifiability with the number of species taken into account. The condition number of the Hessian matrix of the model cost function

(in grey) is related to the convergence rate of the model. The fit is measured by the mean square model error (in black).

Table S1 Plankton ecological characteristics. Family affiliation and social behavior are from [44]. Position in the water column is mostly from [45] and [28]. Habitats are from [48], [49], [50] and [51]. Biovolumes are from [46].

Acknowledgments

The idea of this study originates from the BlueDEB project coordinated by Marianne Alunno-Bruscia and Øivind Strand. We are grateful to Mr Clouet for providing and storing young mussels in La-Plaine-sur-Mer and the Lefebvre company for loaning a part of their farming area for their rearing. We thank Aurore Lejolivet, Charlotte Mary and Sophie Parrad for their help on the field site and for taking care of our mussels, dead or alive; Xavier Philippon for CHN analyses; and Olivier Pierre-Duplessix and Emilie Rabiller for hydrological analyses. We would also like to thank Banyuls Oceanologic Observatory for the cytometry analyses. The data could not have been so fully understood without Mathilde Schapira's invaluable plankton atlas. English assistance was provided by Helen McCombie. Finally, we are especially grateful to both our reviewers for their insightful comments that have greatly improved this paper.

Author Contributions

Conceived and designed the experiments: AG FM BS. Performed the experiments: FM BS SF. Analyzed the data: CP JFSM AG RLG. Contributed reagents/materials/analysis tools: JFSM CP RLG AG. Wrote the paper: CP AG RLG.

References

1. Kooijman SALM (2010) Dynamic Energy Budget theory for metabolic organisation. Cambridge University Press. 532 pp. ISBN 9780521131919.
2. Beadman HA, Willows RI, Kaiser MJ (2002) Potential applications of mussel modeling. Helgol Mar Res 56: 76–85.
3. van der Veer HW, Cardoso JFMF, van der Meer J (2006) The estimation of DEB parameters for various Northeast Atlantic bivalve species. J Sea Res 56(2): 107–124.
4. Troost TA, Wijsman JWM, Saraiva S, Freitas V (2010) Modeling shellfish growth with Dynamic Energy Budget (DEB) models: an application for cockles and mussels in the Oosterschelde (SW Netherlands). Phil Trans R Soc B (365): 3567–3577.
5. Thomas Y, Mazurié J, Alunno-Bruscia M, Bacher C, Bouget JF, et al. (2011) Modelling spatio-temporal variability of *Mytilus edulis* (L.) growth by forcing a dynamic energy budget model with satellite-derived environmental data. J Sea Res 66(4): 308–317.
6. Rosland R, Strand Ø, Alunno-Bruscia M, Bacher C, Strohmeier T (2009) Applying Dynamic Energy Budget (DEB) theory to simulate growth and bio-energetics of blue mussels under low seston conditions. J Sea Res 62(2–3): 49–61.
7. Duarte P, Fernández-Reiriz MJ, Labarta U (2012) Modelling mussel growth in ecosystems with low suspended matter loads using a Dynamic Energy Budget approach. J Sea Res 67: 44–57.
8. Saraiva S, van der Meer J, Kooijman SALM, Witbaard R, Philippart CJM, et al. (2012) Validation of a Dynamic Energy Budget (DEB) model for the blue mussel *Mytilus edulis*. Mar Ecol Prog Ser 463: 141–158.
9. Dabrowski T, Lyons K, Curé M, Berry A, Nolan G (2013) Numerical modelling of spatio-temporal variability of growth of *Mytilus edulis* (L.) and influence of its cultivation on ecosystem functioning. J Sea Res 76: 5–21.
10. Grangeré K, Ménesguen A, Lefebvre S, Bacher C, Pouvreau S (2009) Modelling the influence of environmental factors on the physiological status of the Pacific oyster *Crassostrea gigas* in an estuarine embayment; The Baie des Veys (France). J Sea Res 62: 147–158.
11. Bourlès Y, Alunno-Bruscia M, Pouvreau S, Tollu G, Leguay D, et al. (2009) Modelling growth and reproduction of the Pacific oyster *Crassostrea gigas*: Advances in the oyster-DEB model through application to a coastal pond. J Sea Res 62 (2–3): 62–71.

12. Handå A, Alver M, Edvardsen CV, Halstensen S, Olsen AJ, et al. (2011) Growth of farmed blue mussels (*Mytilus edulis* L.) in a Norwegian coastal area; comparison of food proxies by DEB modeling. J Sea Res 66: 297–307.
13. Saraiva S, van der Meer J, Kooijman S, Sousa T (2011a) DEB parameters estimation for *Mytilus edulis*. J Sea Res 66: 289–296.
14. Widdows J, Fieth P, Worrall CM (1979) Relationships between Seston, Available Food and Feeding Activity in the Common Mussel *Mytilus edulis*. Mar Biol 50: 195–207.
15. Bougrier S, Hawkins AJS, Héral M (1997) Preingestive selection of different microalgal mixtures in *Crassostrea gigas* and *Mytilus edulis*, analysed by flow cytometry. Aquaculture 150: 123–134.
16. Saurel C, Gascoigne JC, Palmer MR, Kaiser MJ (2007) In situ Mussel Feeding Behavior in Relation to Multiple Environmental Factors: Regulation through Food Concentration and Tidal Conditions. Limnol Oceanogr 52(5): 1919–1929.
17. Riisgård HU, Egede PP, Saavedra IB (2011). Feeding Behaviour of the Mussel, *Mytilus edulis*: New Observations, with a Minireview of Current Knowledge. J Mar Biol DOI: 10.1155/2011/312459.
18. Pleissner D, Eriksen NT, Lundgreen K, Riisgård HU (2012) Biomass Composition of Blue Mussels, *Mytilus edulis*, is Affected by Living Site and Species of Ingested Microalgae. ISRN Zool doi:10.5402/2012/902152
19. Toupoint N, Gilmore-Solomon L, Bourque F, Myrand B, Pernet F, et al. (2012a) Match/mismatch between the *Mytilus edulis* larval supply and seston quality: effect on recruitment. Ecology 93(8): 1922–1934.
20. Strohmeier T, Strand Ø, Alunno-Bruscia M, Duinker A, Cranford PJ (2012) Variability in particle retention efficiency by the mussel *Mytilus edulis*. J Exp Mar Bio Ecol 412: 96–102.
21. Rouillon G, Guerra Rivas J, Ochoa N, Navarro E (2005) Phytoplankton composition of the stomach contents of the mussel *Mytilus edulis* L. from two populations: comparison with its food supply. J Shellfish Res 24(1): 5–14.
22. Maloy AP, Nelle P, Culloty SC, Slater JW, Harrod C (2013) Identifying trophic variation in a marine suspension feeder: DNA- and stable isotope-based dietary analysis in *Mytilus* spp. Mar Biol 160(2): 479–490.
23. Pernet F, Malet N, Pastoureaud A, Vaquer A, Quéré C., et al. (2012) Marine diatoms sustain growth of bivalves in a Mediterranean lagoon. J Sea Res 68: 20–32.

24. Lauringson V, Kotta J, Orav-Kotta H, Kaljurand K (2014) Diet of mussels *Mytilus trossulus* and *Dreissena polymorpha* in a brackish nontidal environment. Mar Ecol 35(1): 56–66.

25. Rodhouse PG, Roden CM, Burnell GM, Hensey MP, McMahon T (1984) Food resource, gametogenesis and growth of *Mytilus edulis* on the shore and in suspended culture: Killary Harbour, Ireland. J Mar Biol Assoc UK 64(3): 513–529.

26. Freitas V, Cardoso JFMF, Santos S, Campos J, Drent J, et al. (2009) Reconstruction of food conditions for Northeast Atlantic bivalve species based on Dynamic Energy Budgets. J Sea Res 62: 75–82.

27. Cardoso JFMF, Witte JIJ, van der Veer HW (2006) Intra- and interspecies comparison of energy flow in bivalve species in Dutch coastal waters by means of the Dynamic Energy Budget (DEB) theory. J Sea Res 56: 182–197.

28. Ubertini M, Lefebvre S, Gangnery A, Grangeré K, Le Gendre R, et al. (2012) Spatial Variability of Benthic-Pelagic Coupling in an Estuary Ecosystem: Consequences for Microphytobenthos Resuspension Phenomenon. PLoS ONE 7(8): e44155. doi:10.1371/journal.pone.0044155.

29. Jouenne F, Lefebvre S, Véron B, Lagadeuc Y (2005) Biological and physicochemical factors controlling short-term variability in phytoplankton primary production and photosynthetic parameters in a macrotidal ecosystem (eastern English Channel). Estuar Coast Shelf Sci 65: 421–439.

30. Jouenne F, Lefebvre S, Véron B, Lagadeuc Y (2007) Phytoplankton community structure and primary production in small intertidal estuarine-bay ecosystem (eastern English Channel, France). Mar Biol 151: 805–825.

31. Bayne BL, Worrall CM (1980) Growth and production of mussels *Mytilus edulis* from two populations. Mar Ecol Prog Ser 3: 317–328.

32. Aminot A, Kérouel R (2007) Dosage automatique des nutriments dans les eaux marines. Ed. Ifremer, 336 p.

33. Aminot A, Kérouel R (2004) Hydrologie des écosystèmes marins. Paramètres et analyses. Ed. Ifremer, 336p.

34. Utermöhl H (1958) Zur vervolkommung der quantativen phytoplankton methodik. Int Ver Theoret Angew Limnol 9: 1–38.

35. Vaulot D, Courties C, Partensky F (1989) A simple method to preserve oceanic phytoplankton for flow cytometric analyses. Cytometry, 10: 629–635.

36. Marie D, Partensky F, Jacquet S, Vaulot D (1997) Enumeration and cell cycle analysis of natural populations of marine picoplankton by flow cytometry using the nucleic acid stain SyBR Green I. Appl Environ Microbiol. 63: 186–193.

37. Sprung M (1983) Reproduction and fecundity of the mussel *Mytilus edulis* at Helgoland (North Sea). Helgol Mar Res 36: 243–255.

38. Flye Sainte Marie J, Alunno-Bruscia M, Gangnery A, Rannou E, Rosland R, et al. (2009). Individual mussel growth model using DEB (Dynamic Energy Budget) theory: revisiting the DEB parameter values for *Mytilus edulis*. Aquaculture Europe 2009, Trondheim, 14–17 August 2010.

39. Thacker WC (1989) The role of the Hessian matrix in fitting models to measurements. J Geophys Res 94: 6177–6196.

40. Fenner K, Losch M, Schröter J, Wenzel M (2001) Testing a marine ecosystem model: sensitivity analysis and parameter optimization. J Mar Syst 28: 45–63.

41. Faugeras B, Lévy M, Mémery L, Verron J, Blum J, et al. (2003) Can biogeochemical fluxes be recovered from nitrates and chlorophyll data? A case study assimilating data in the Northwestern Mediterranean Sea at the JGOFS-DYFAMED station. J Mar Syst 40: 99–125.

42. Dueri S, Faugeras B, Maury O (2012) Modelling the skipjack tuna dynamics in the Indian Ocean with APECOSM-E – Part 2: Parameter estimation and sensitivity analysis. Ecol Model 245: 55–64.

43. Leonardos N, Lucas IAN (2000) The nutritional value of algae grown under different culture conditions for Mytilus edulis L. larvae. Aquaculture 182: 301–315.

44. Horner RA (2002) A taxonomic guide to some common marine phytoplankton. Bristol: Biopress. 195p.

45. Guilloux L, Rigaut-Jalabert F, Jouenne F, Ristori S, Viprey M, et al. (2013) An annotated checklist of Marine Phytoplankton taxa at the SOMLIT-Astan time series off Roscoff (Western English Channel, France): data collected from 2000 to 2010. Cah Biol Mar 54: 247–256.

46. Leblanc K, Aristegui J, Armand L, Assmy P, Beker B, et al. (2012) A global diatom database – abundance, biovolume and biomass in the world ocean. Earth Syst Sci Data 4(1): 149–165.

47. Sun J, Liu D (2003) Geometric models for calculating cell biovolume and surface area for phytoplankton. J Plankton Res 25(11): 1331–1346.

48. Marshall HG, Burchardt L (2004) Phytoplankton composition within the tidal freshwater-oligohaline regions of the Rappahannock and Pamunkey Rivers in Virginia. Castanea 69(4): 272–283.

49. Marshall HG, Lacouture RV, Buchanan C, Johnson JM (2006) Phytoplankton assemblages associated with water quality and salinity regions in Chesapeake Bay, USA. Estuar Coast Shel Sci 69(1–2): 10–18.

50. Brand LE (1984) The salinity tolerance of forty-six marine phytoplankton isolates. Estuar Coast Shelf Sci 18(5): 543–556.

51. Jiang H (1996) Diatoms from the surface sediments of the Skagerrak and the Kattegat and their relationship to the spatial changes of environmental variables. J Biogeogr 23(2): 129–137.

52. de Zwaan A, Zandee DI (1972) Body distribution and seasonal changes in the glycogen content of the common sea mussel *Mytilus edulis*. Comp Biochem Physiol 43(A): 53–58.

53. Smaal AC, Vonck APMA (1997) Seasonal variation in C, N and P budgets and tissue composition of the mussel *Mytilus edulis*. Mar Ecol Prog Ser 153: 167–179.

54. Garen P, Robert S, Bougrier S (2004) Comparison of growth of mussel, *Mytilus edulis*, on longline, pole and bottom culture sites in the Pertuis Breton, France. Aquaculture 232: 511–524.

55. Dare PJ, Edwards DB (1975) Seasonal changes in flesh mass and biochemical composition of mussels (*Mytilus edulis* L.) in the Conwy estuary, North Wales. J exp Mar Biol Ecol 18: 89–97.

56. Hilbish TJ (1986) Growth trajectories of shell and soft tissue in bivalves: seasonal variation in *Mytilus edulis* L. J exp Mar Biol Ecol 96: 103–113.

57. Kautsky N (1982) Quantitative Studies on Gonad Cycle, Fecundity, Reproductive Output and Recruitment in a Baltic *Mytilus edulis* Population. Mar Biol 68: 143–160.

58. Tanaka N, Monaghan MC, Rye DM (1986) Contribution of metabolic carbon to mollusc and barnacle shell carbonate. Nature 320: 520–523.

59. Gillikin DP, Lorrain A, Bouillon S, Willenz P, Dehairs F (2006) Stable carbon isotopic composition of *Mytilus edulis* shells: relation to metabolism, salinity, $\delta^{13}C_{DIC}$ and phytoplankton. Org Geochem 37: 1371–1382.

60. Palmer AR (1981) Do carbonate skeletons limit the rate of body growth? Nature 292: 150–152.

61. Behrenfeld MJ, Halsey KH, Milligan AJ (2008b) Evolved physiological responses of phytoplankton to their integrated growth environment. Phil Trans R Soc B 363: 2687–2703.

62. Bracken MES, Menge BA, Foley MM, Sorte CJB, Lubchenco J, Schiel DR (2012) Mussel selectivity for high-quality food drives carbon inputs into open-coast intertidal ecosystems. Mar Ecol Prog Ser 459: 53–62.

63. Cranford PJ, Hargrave B, Li W (2009) No mussel is an island. ICES insight 46, 44–49.

64. Smaal AC, Twisk F (1997) Filtration and absorption of *Phaeocystis* cf. *globosa* by the mussel *Mytilus edulis* L. J Exp Mar Biol Ecol 209: 33–46.

65. Lefebvre S, Harma C, Blin JL (2009) Trophic typology of coastal ecosystems based on $\delta^{13}C$ and $\delta^{15}N$ ratios in an opportunistic suspension feeder. Mar Ecol Prog Ser 390: 27–37.

66. Toupoint N, Mohit V, Linossier I, Bourgougnon N, Myrand B, et al. (2012b) Effect of biofilm age on settlement of *Mytilus edulis*. Biofouling 28(9): 985–1001.

67. Pronker AE, Nevejan NM, Peene F, Geisjen P, Sorgeloos P (2007) Hatchery broodstock conditioning of the blue mussel *Mytilus edulis* (Linnaeus 1758). Part I. Impact of different micro-algae mixtures on broodstock performance. Aquacult Int 16(4): 297–307.

68. Trottet A, Roy S, Tamigneaux E, Lovejoy C, Tremblay R (2008) Impact of suspended mussels (*Mytilus edulis* L.) on plankton communities in a Magdalen Islands lagoon (Québec, Canada): A mesocosm approach. J exp Mar Biol Ecol 365: 103–115.

69. Strömgren T, Cary C (1984) Growth in length of *Mytilus edulis* L. fed on different algal diets. J Exp Mar Bio Ecol 76: 23:34.

70. Fearman JA, Bolch CJS, Moltschaniwskyj NA (2009) Energy Storage and Reproduction in Mussels, *Mytilus galloprovincialis*: The Influence of Diet Quality. J shellfish Res 28(2): 305–312.

71. Masquelier S, Foulon E, Jouenne F, Ferréol M, Brussaard CPD, et al. (2011) Distribution of eukaryotic plankton in the English Channel and the North Sea in summer. J Sea Res 66: 111–122.

72. Lehane C, Davenport J (2006) A 15-month study of zooplankton ingestion by farmed mussels (*Mytilus edulis*) in Bantry Bay, Southwest Ireland. Estuar Coast Shelf Sci 67: 645–652.

73. Bayne BL (1973) Physiological changes in *Mytilus edulis* L. induced by temperature and nutritive stress. J Mar Biol Ass UK 53: 39–58.

74. Toro JE, Thompson RJ, Innes DJ (2002) Reproductive isolation and reproductive output in two sympatric mussel species (*Mytilus edulis*, *M. trossulus*) and their hybrids from Newfoundland. Mar Biol 141: 897–909.

75. Doherty SD, Brophy D, Gosling E (2009) Synchronous reproduction may facilitate introgression in a hybrid mussel (*Mytilus*) population. J exp Mar Biol Ecol 378: 1–7.

76. Cardoso JFMF, Dekker R, Witte JIJ, van der Veer HW (2007) Is reproductive failure responsible for reduced recruitment of intertidal *Mytilus edulis* L. in the western Dutch Wadden Sea? Mar Biodivers 37(2): 83–92.

77. Newell RIE, Hilbish TJ, Koehn RK, Newell CJ (1982) Temporal variation in the reproductive cycle of *Mytilus edulis* L. (bivalvia, mytilidae) from localities on the East coast of the United States. Biol Bull 162: 299–310.

78. Thorarinsdóttir GG, Gunnarsson K (2003) Reproductive cycles of *Mytilus edulis* L. on the west and east coasts of Iceland. Polar Res 22(2): 217–223.

79. Hagger JA, Lowe D, Dissanayake A, Jones MB, Galloway TS (2010) The influence of seasonality on biomarker responses in *Mytilus edulis*. Ecotoxicology 19: 953–962.

17

The Value of Patch-Choice Copying in Fruit Flies

Shane Golden, Reuven Dukas*

Animal Behaviour Group, Department of Psychology, Neuroscience & Behaviour, McMaster University, Hamilton, Ontario, Canada

Abstract

Many animals copy the choices of others but the functional and mechanistic explanations for copying are still not fully resolved. We relied on novel behavioral protocols to quantify the value of patch-choice copying in fruit flies. In a titration experiment, we quantified how much nutritional value females were willing to trade for laying eggs on patches already occupied by larvae (social patches). Females were highly sensitive to nutritional quality, which was positively associated with their offspring success. Females, however, perceived social, low-nutrition patches (33% of the nutrients) as equally valuable as non-social, high-nutrition ones (100% of the nutrients). In follow-up experiments, we could not, however, either find informational benefits from copying others or detect what females' offspring may gain from developing with older larvae. Because patch-choice copying in fruit flies is a robust phenomenon in spite of potential costs due to competition, we suggest that it is beneficial in natural settings, where fruit flies encounter complex dynamics of microbial communities, which include, in addition to the preferred yeast species they feed on, numerous harmful fungi and bacteria. We suggest that microbial ecology underlies many cases of copying in nature.

Editor: Johan J. Bolhuis, Utrecht University, Netherlands

Funding: This research has been funded by the Natural Sciences and Engineering Research Council of Canada, Canada Foundation for Innovation, and Ontario Innovation Trust. The funders had no role in study design, data collection and analysis, decision to publish, or preparation of the manuscript.

Competing Interests: The authors have declared that no competing interests exist.

* Email: dukas@mcmaster.ca

Introduction

In many animal species, individuals copy the choices of others. Examples include choices of feeding sites [1–3], territories [4,5], egg laying substrates [6–8] and mates [9–11]. Depending on the system, copying can have substantial effects on organismal ecology and evolution. For example, conspecific aggregation at feeding and egg laying sites can promote species coexistence [12,13] and mate choice copying can influence the intensity and direction of sexual selection [14–16].

While it is widely agreed that copying can influence animal ecology and evolution, it is often unclear how the possible fitness benefits from copying outweigh the likely costs. For example, patch-choice copying typically involves a focal individual choosing a feeding or egg laying site that is either occupied by other individuals (models) or contains products left by these individuals. There are probably only two non-mutually exclusive explanations for such copying. The first explanation involves pure information: a focal can either find a satisfactory patch faster, or locate a better patch among the available alternatives by copying others than by exploring on its own [4,17–19]. That is, the first explanation focuses on two related difficulties that animals have in locating optimal resource patches. Either the patches are hidden, so it takes time to find them, or it is difficult and time consuming to assess the multitude of features that determine patch quality. Given individuals' limited time horizon, focals that copy others can shorten the time devoted to exploration and hence increase the time spent exploiting without compromising on the quality of the patch utilized. This proposition, of course, is based on the tenuous assumption that the models indeed have chosen the optimal patch.

The other explanation for patch-choice copying involves material benefits that focals can gain from joining others, which include reduced per capita risk of attack by predators and parasitoids, and enhanced foraging efficiency and thermoregulation [20–25]. It is worth noting that, when patch-choice copying involves joining others, focals and models might face asymmetric payoffs: while a focal can gain more from joining than from settling alone, the models might lose from having another individual joining [26]. The obvious costs from joining others are competition for resources and reduced patch quality caused by accumulating waste products [7,19,20]. Competition can cause another possible asymmetric payoff that is size dependent. For example, newly hatched larvae may lose more from competition than the older resident larvae.

While there are numerous reports of copying in a wide variety of species and contexts, the value of copying has been rarely quantified. We have recently developed protocols for quantifying patch-choice copying in fruit flies (*Drosophila melanogaster*). Larvae and adults from both established laboratory strains and recently caught wild populations copy the choices of others: adult females prefer the egg laying substrates chosen by other females [27,28], both male and female adults are attracted to volatiles emanating from conspecific larvae, females show a strong preference for laying eggs in patches with larvae over unoccupied alternatives, and larvae also show significant attraction to patches already occupied by larvae [29–31]. The establishment of fruit flies as a model system for research on patch-choice copying offers new opportunities. First, the fruit fly system allows one to conduct highly controlled experiments assessing the factors that influence patch choice copying. Second, findings from the behavioral analyses of patch-choice copying can be extended to research on the genetics and neurobiology of such behavior in a highly amenable model system. Indeed there has recently been increased interest in establishing simple model systems for research on the mechanisms that control social behavior as well as behavioral decisions in general [32–35].

To elucidate the value of patch-choice copying in fruit flies, we conducted a series of experiments. We began with a titration experiment designed to quantify the perceived value that females assign to food occupied by larvae. This involved testing female preferences between reference patches and test patches of varying food qualities, which were either occupied or unoccupied by larvae. In follow-up experiments, we compared larval success on occupied and unoccupied patches of relevant food qualities. This allowed us to translate patch-choice copying by females into the consequent success of their offspring. Because females showed strong patch-choice copying even when nutritionally superior patches were readily available and in spite of the expected costs owing to larval competition, we wished to assess whether females would moderate their strong tendency to copy when the occupied patches either contain numerous larvae or have already experienced heavy consumption by larvae. Finally, to assess possible informational benefits to females, we tested whether larvae were better than adult females at assessing food quality.

Materials and Methods

Nutritional Titration

We maintained two population cages of several hundred *Drosophila melanogaster Canton-S* following standard protocol [27]. To quantify the value that females assign to patches already occupied by larvae, we placed each of 192 recently mated female inside a 60 mm Petri dish. The bottom of the dish contained agar, which provided moisture. On top of the agar, we placed two discs cut from a thin layer of fly medium. Both discs were 1.1 cm in diameter and each contained 0.5 ml food (Fig. 1a). The reference disc always had standard food in which 1 litre contained 60 g dextrose, 30 g sucrose, 32 g yeast, 75 g cornmeal, 20 g agar, 2 g methyl-paraben and water. The test disc was either fresh (non-social) or contained five early second instar larvae that had fed on that disc for 24 h (social). The test disc had standard food or one of two lower food concentrations containing either 33% or 11% of the nutrients (dextrose, sucrose, yeast, and cornmeal) available in the standard food and a larger proportion of water. The reference and test discs were 3 cm apart with the central 2 cm being a trough filled with fine sand (Fig. 1a) to prevent larvae located on the social discs from crossing to the reference discs. We housed all dishes in a chamber kept at 25°C and 90% RH and allowed the females in the Petri dishes to lay eggs overnight for 14 h. Then we discarded the females and counted the number of eggs laid on each disc. We used a generalized linear model with a Tweedie distribution and identity link function and conducted pairwise comparisons with Bonferroni corrections and 95% Wald confidence intervals [36]. See Data S1 for the raw data for all experiments.

Larval Success on Social vs. Non-Social Food

Our nutritional titration experiment indicated that females perceive social food with about one third the nutrients as equally valuable as the non-social reference food (Fig. 1b). We thus wished to quantify the success of females' eggs on social vs non-social food discs of distinct nutritional concentrations. To assess the value of laying eggs on currently versus previously occupied patches, we also included a previously social treatment. We had a total of 6 treatments involving 2 food concentrations, 100% and 33%, and 3 social treatments, non-social, social and previously social. We omitted the 11% food concentration because females in the titration experiment mostly avoided it even when it was social (Fig. 1b). The food discs were identical in constitution and volume to the 100% and 33% food discs in the titration experiment.

The non-social discs contained unmodified food. To generate the social and previously social discs, we placed on each disc 5 24-hour old first instar larvae and allowed these larvae to feed for 24 hours. In the social disc treatment, we kept the now second instar, 48 h old larvae on each disc. In the formerly social disc, we removed the larvae. That is, both the social and previously social discs were equally modified by the five larvae prior to the placement of focal eggs. Then the focal larvae emerging on the formerly social disc could reap potential benefits from such previous food modification without experiencing competition with the older larvae. Thus the formerly social disc gave us a greater power for quantifying possible benefits of prior food modification by larvae.

We placed each food disc inside a 35 mm Petri dish lined with agar, added to each disc five focal eggs and housed all the dishes in a chamber kept at 25°C and 90% RH. When the five older larvae in the social dishes pupated, we removed these pupae. We then monitored the number of focal larvae reaching pupation and calculated the larval developmental rate as the cumulative proportion of larvae reaching pupation while taking the final pupal number as 1. We counted all eclosing adults and calculated the proportion of eggs that produced adults. Because females are heavier than males, we sexed the adults, dried them in an oven at 70°C for 3 days and weighed groups of five flies of the same sex on a microbalance.

Because no larvae survived in the social 33% treatment, we conducted two separate analyses. First, we omitted the social treatment and compared larval performance in the four treatments of non-social and formerly social on 33% and 100% food. Second, we compared larval performance in all three treatments of non-social, formerly social and social on the 100% food.

We analyzed larval development rate and the proportion of eggs surviving to adulthood using a generalized estimating equation with a gamma distribution and log link function [29]. We had sufficient sample sizes for analyzing adult dry mass only for the 100% food (Fig. 2E, F). These data met ANOVA assumptions and we thus used a two-way ANOVA with a Tukey HSD. We conducted all post-hoc pairwise comparisons using the sequential Bonferroni method adjusting for multiple comparisons.

Larval Success on Abundant Food

In our previous larval success experiment, larvae were reared on 0.5 ml of food. Because the results indicated strong effects of competition, we tested larval success on social and non-social discs each containing 2.5 ml of 100% food. As a reference, fruit fly laboratories typically rear a few dozen flies per vial containing 5 ml of similar food [37,38]. By providing abundant food, we wished to maximize our ability to detect possible benefits that larvae may gain from developing on social food. All other protocol details were similar to those detailed above. That is, The social food contained 5 larvae and the non-social food had no larvae.

Females' Patch Choice When the Social Patches Have Had High Larval Densities

In our titration experiment (Fig. 1B), females showed a strong preference for laying eggs near larvae even though this reduced their offspring success in our laboratory settings (Figs 2, 3). Because larval crowding and the consequent lower larval success

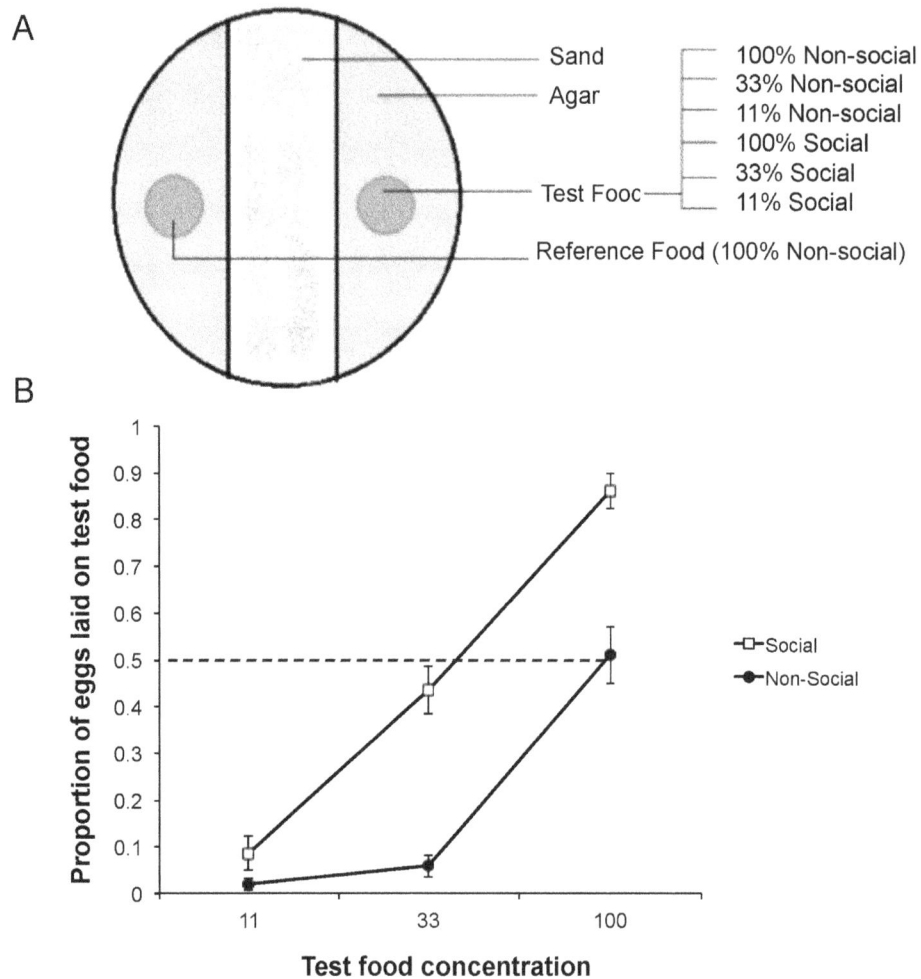

Figure 1. Nutritional titration. (A) Each dish always contained a reference disc and one of six types of test discs varying in nutritional concentration and larval presence. Sand at the centre of the dish prevented larval crawling to the reference food. (B) The average proportion of eggs (±1 SE) laid on the test disc as a function of its nutrient concentration and presence or absence of larvae (social or non-social). The horizontal dashed line indicates random choice. N = 30 replicates per treatment. Females laid more eggs on the test food in the presence than absence of larvae.

are prevalent in nature as well [39,40], we expected females to make egg laying decisions that balance their perceived benefit from laying next to larvae versus the expected cost due to larval overcrowding. We thus allowed females to choose between either a non-social patch and a social patch occupied by 5 larvae, or a non-social patch and a social patch occupied by 20 larvae. We predicted that females would lay a lower proportion of eggs on the social food when it was more crowded.

We used a protocol modified from Durisko et al [30]. We placed each recently mated female inside a plastic cage (15 cm wide, 30 cm long, and 15 cm high), which contained two 35 mm Petri dishes placed at the opposite far corners of each cage. One dish was non-social and the other was social. Both dishes contained 0.5 ml food discs composed of 100% standard lab diet. The dishes were lined with a layer of agar to prevent desiccation. Non-social food discs were unoccupied. The social food discs had either 5 or 20 middle second instar larvae, which we had added 6 h before the addition of females.

We allowed the females to lay eggs overnight. In the following morning, we removed the females from the cages, counted the number of eggs on each food disc and analyzed the proportion of eggs laid in the social dish out of the total number of eggs that a

female laid. Based on preliminary data indicating effects of larval density on egg location, we also counted the number of eggs laid on the agar layer within 1 cm of the food disc and calculated the proportion of eggs laid on agar versus food in the social dish. We analyzed the data using a generalized linear model with a Tweedie distribution and log link function.

The experiment above tested females' sensitivity to larval density. It is possible however, that females are more sensitive to the condition of food as indicated by the microbial community and waste products rather than to the number of larvae already on the food. To test this possibility, we allowed females to choose between either a non-social patch and a social patch that had been previously occupied by 5 larvae, or a non-social patch and a social food patch that had been previously occupied by 20 larvae. Again, we predicted that females would lay a lower proportion of eggs on the social food that had been more crowded.

Forty-eight hours before the experiment, we transferred groups of either 5 or 20 middle second instar larvae to social food discs and kept them in 35 mm Petri dishes lined with agar. We also kept unoccupied food discs in Petri dishes lined with agar. All food discs contained 0.5 ml of 100% standard lab diet. By the day of the experiment, all larvae on the social discs had pupated. We then

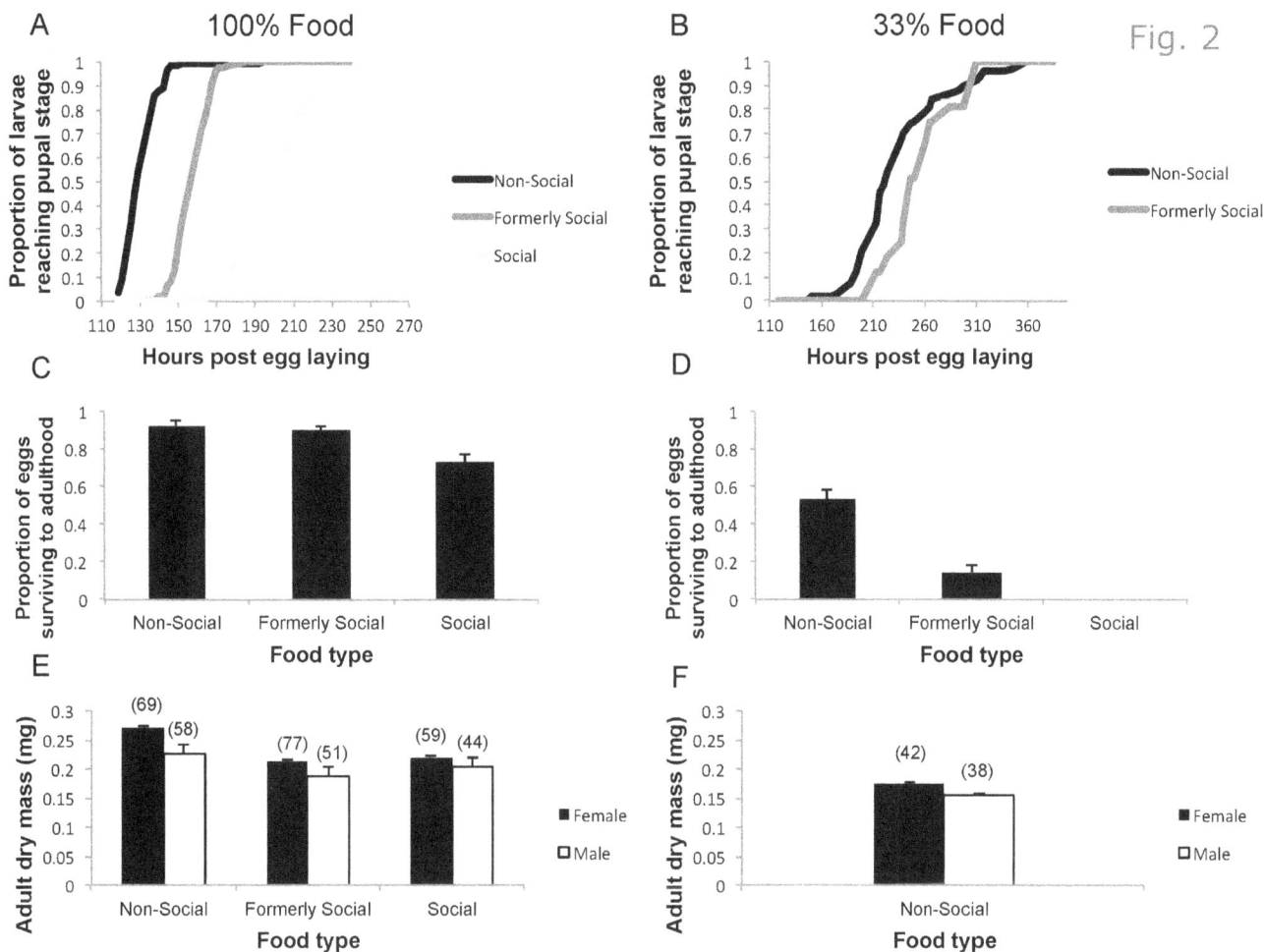

Figure 2. Larval performance as function of a disc's nutritional and social status. The left panels (A, C, and E) refer to the 100% nutrients while the right panels (B, D, and F) refer to the 33% nutrients. (A) and (B) show the time it takes for the larvae to develop from eggs into pupae. (C) and (D) show the proportions of eggs that survived to adulthood (mean+SE). In (B) and (D), survival in the social treatment was 0. (E) and (F) show the adult dry mass (mean+SE). N = 30 replicates for each treatment. The number of eclosing adults is shown above the bars in panels E and F.

placed one non-social food disc and one social disc in 60 mm Petri dishes lined with agar. The social disc had been previously consumed by either 5 or 20 larvae but was free of larvae and pupae by the time of the test. Discs were placed 2 cm apart. We then added a recently mated female to each 60 mm dish through a hole in the lid, which was then plugged with foam. We allowed the females to lay eggs overnight. In the morning, we removed the females from the dishes and counted the number of eggs on the social and non-social food discs. We analyzed the proportion of eggs on each type of social food using a generalized linear model with a Tweedie distribution and log link function.

Adult vs. Larval Abilities to Detect Differences in Yeast Concentration of Food

Because we documented a lower larval success of eggs laid at social patches, we wished to test whether the benefit of patch choice copying is related to information rather than to joining. To this end, we tested whether larvae could detect pertinent patch characteristics that adult females could not. We had two treatments testing larval and adult females' abilities to detect differences in yeast content between adjacent patches. We focused on yeast rather than sugar because larval and adult perception of

sweetness is well documented [41–44]. One test involved a reference 100% standard fly medium vs standard medium with only 33% of the yeast content, and the other test involved a reference 100% standard medium vs standard medium with only 50% of the yeast content. All other medium ingredients were identical.

We added either one recently mated adult female or five mid-second instar larvae to Petri dishes containing one reference food disc and one food disc with lower yeast concentration (either 33% or 50%) placed 2 cm apart. We added the adults and focal larvae in the evening at an identical location 1 cm between the food discs. We gave them 14 hours to decide where to lay eggs or feed. In the following morning, we counted the number of eggs laid on each food disc in the adult female treatments and counted the number of larvae on each food disc in the larval treatments. We then calculated the proportion of eggs laid and the proportion of larvae on the reference 100% disc and analyzed the data with a generalized linear model with a Tweedie distribution and identity link function.

A

B

C

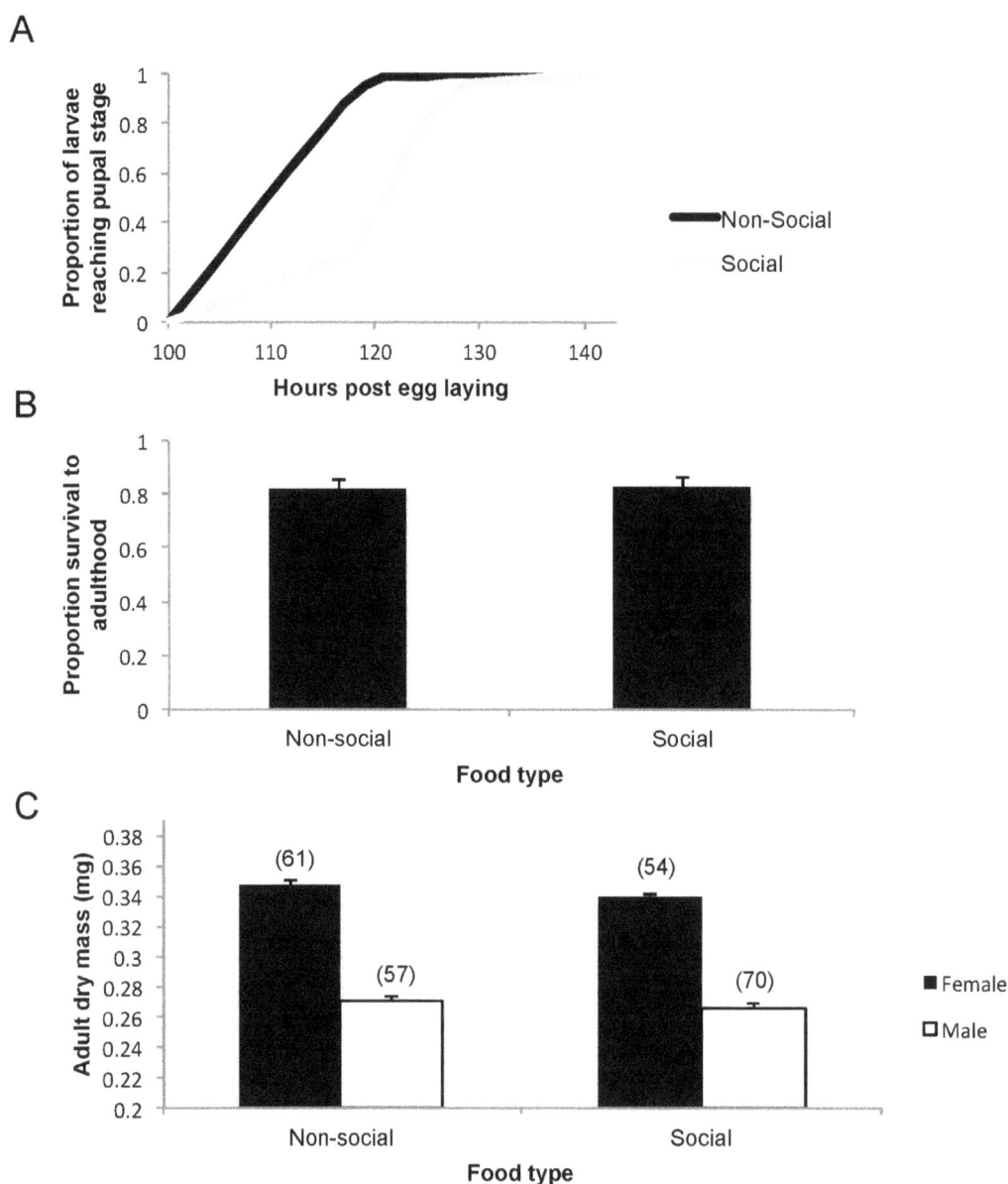

Figure 3. Performance measures of focal larvae on abundant food. Discs were either social or non-social (n = 30 replicates per treatment). (A) Time from egg laying to pupal formation (B) The proportion of eggs surviving to adulthood (mean+SE). (C) The adult dry mass of females and males in both conditions (mean+SE). Numbers in brackets above the bars indicate the number of adults in each group.

Results

Nutritional Titration

Females laid significantly higher proportions of eggs on the test food when it was social than non-social at all three food concentrations (Wald $\chi^2_1 = 49$, P<0.001 for the main effect and P<0.01 for the three pairwise comparisons with Bonferroni corrections, Fig. 1).

Larval Success on Social vs. Non-Social Food

Larval performance across food qualities. Owing to 100% mortality in the social 33% food treatment, we could compare larval performance across food qualities only for the non-social and previously social treatments. Larvae developed much faster (Wald $\chi^2_1 = 474.74$, P<0.001; Fig. 2A,B) and had higher survival rates on the 100% than 33% food (Wald $\chi^2_1 = 75.6$, P<

0.001; Fig. 2C,D). Similarly, larvae developed much faster (Wald $\chi^2_1 = 33.361$, P<0.001; Fig. 2A,B) and had higher survival rates in the non social than formerly social treatments (Wald $\chi^2_1 = 75.769$, P<0.001; Fig. 2C,D).

Because survival rates in the 33% food treatment were low, we could only compare adult body mass across food qualities in the non-social treatments. Adults in the 100% food quality were much heavier than those in the 33% food quality (Wald $\chi^2_1 = 512.96$, P<0.001; Fig. 2E,F).

Larval performance across social treatments. This analysis could include only the 100% food owing to 100% mortality in the social 33% food treatment. Larvae developed significantly faster in the non-social treatment, intermediate in the formerly social treatment, and slowest in the social treatment (Wald $\chi^2_2 = 1700$, P<0.001; Fig. 2A). Post-hoc pairwise comparisons showed that each treatment was significantly different from the other two (P<0.001).

Survival to adulthood was significantly affected by the social treatment (Wald $\chi^2_2 = 13.9$, P = 0.001; Fig. 2C). Survival was similar in the non-social and formerly social treatment (post-hoc pairwise comparison, P = 0.709) but higher in each of these treatments than in the social treatment (post-hoc pairwise comparisons, P = 0.002 and 0.005 for the non-social and formerly social treatment respectively).

Adult mass was significantly affected by the social treatment ($F_{2,61} = 85.2$, P<0.001; Fig. 2E). In both males and females, adults of the non social treatment were heavier than those of the social and formerly social treatments (Tukey HSD, P<0.001). While males of the formerly social treatment were lighter than those in the social treatment (P = 0.007), females of the formerly social and social treatments had similar masses (P = 0.438).

Larval Success on Abundant Food

Larvae developed faster in the non-social condition than in the social condition (Wald $\chi^2_1 = 34.683$, P<0.001; Fig. 3A). However, the same proportion of focal eggs survived to adulthood (Wald $\chi^2_1 = 0.014$, P = 0.905; Fig. 3B). Adult flies in the non-social condition were heavier than adults in the social condition (Wald $\chi^2_1 = 4.515$, P = 0.034; Fig. 3C).

Females' Patch Choice When the Social Patches Have Had High Larval Densities

Females laid similar proportions of eggs in the social dishes occupied by 5 and 20 larvae (Wald $\chi^2_1 = 0.204$, P = 0.651; Fig. 4A). However, females placed a greater proportion of their eggs on the agar in the social dishes with 20 than 5 larvae (Wald $\chi^2_1 = 4.649$, P = 0.031; Fig. 4B). When females had a choice between non-social and previously occupied social discs, they laid a similar proportion of their eggs on the social disc regardless of the number of larvae that had previously occupied it (Wald $\chi^2_1 = 0.472$, P = 0.492; Fig. 4C).

Adult vs. Larval Abilities to Detect Differences in Yeast Concentration Of Food

The proportion of eggs that females laid on the 100% food and the proportion of larvae choosing the 100% food were similar when the alternative had only 33% of yeast concentration (Wald $\chi^2_1 = 0.227$, P = 0.634; Fig. 5). When the alternative was 50% yeast concentration, females showed a greater preference than larvae for the higher quality food (Wald $\chi^2_1 = 3.835$, P = 0.05; Fig. 5).

Discussion

Our titration experiment (Fig. 1) indicated that, while females were highly sensitive to the nutritional values of alternative patches, they perceived low-nutrition patches occupied by larvae (social patches with 33% of the nutrients) as suitable as the reference, unoccupied patches (non-social patches with 100% of the nutrients). The larval success experiment (Fig. 2) indicated that the females' sensitivity to nutrient concentration was highly justified: their larvae developed significantly faster, had higher survival rates and produced larger adults on the non-social 100% than non-social 33% patches. Because females were willing to trade the nutritional quality of patches for the opportunity to lay eggs at patches already occupied by larvae, we expected that such choice would translate into some larval benefit. However, we did not find such an advantage. First, in all cases, larval success on social patches was lower than that on non-social patches (Fig. 2).

Second, in the previously social treatment, we removed the larvae that had occupied the patches before placing focal eggs. This allowed us to test for possible benefits that females could gain from laying eggs at patches that have been occupied by larvae while eliminating the negative effects of competition from such larvae. Even in this case, however, we found a cost rather than benefit from laying on previously occupied patches (Fig. 2). Finally, one could argue that our larval to food-volume ratio was too high so that larval competition obscured a gain occurring when food is abundant. To address this possibility, we repeated the larval success experiment with a much lower larval to food-volume ratio. Even in this case, however, larvae performed better under the non-social then social treatment (Fig. 3). The mechanism underlying this negative social effect is unknown and will require close examination in the future.

To further assess the egg laying decisions by females, we wished to quantify females' responses to clear signs of competition in social patches due to either the previous or current presence of many larvae. Although we expected females to reduce their preferences for the social patches when they were either crowded or heavily exploited, we found no such moderation (Fig. 4). Finally, although the sense of taste provides important information about the nutritional quality of food, it is insufficient for assessing whether all nutrients required for optimal larval development are available [41,43–45]. We thus proposed that the presence of feeding larvae is the best cue indicating to females that a substrate is nutritionally sufficient. First, the substrate is adequate for sustaining the larvae as indicated by the fact that they are alive. Second, the larvae are highly mobile and are adept at exploring and settling at the best locally available food [29,46]. Contrary to our expectation, however, we found in two experiments that adult females were as sensitive as larvae to realistic variations in nutritional qualities (Fig. 5).

To summarize our key results, we have strong evidence that females assign high values to patches already occupied by larvae as we quantified by titrating the nutritional quality of the patches (Fig. 1) and we could translate these values into the relevant currency of larval success (Figs 2, 3). Our data, however, indicated neither informational gain (Fig. 5) nor direct benefits from patch choice copying (Figs 2, 3). How can this puzzle be resolved? We propose four non-mutually exclusive explanations related to fruit flies' ecology under natural settings. The first three explanations deal with microbial ecology while the last one focuses on fruit fly parasitoids, which, alongside microbes, are the prominent natural enemies of fruit fly larvae. While the third explanation (microbial information) pertains to the informational benefits of patch choice copying, all other three explanations relate to the direct benefits to larvae from joining other larvae.

Competition with Microbes

While fruit flies feed on yeast species growing on fallen fruit [47], such fruit are also consumed by numerous other fungi as well as bacteria. This means that the other microbes can adversely impact yeast through exploitation competition. Furthermore, microbial interference competition involves a rich arsenal of compounds toxic to other microbes as well as to animals. That is, such compounds can either hamper yeast growth, thus reducing the amount of food available to larvae, or have direct negative effects on larval survival and growth [48–53]. Although highly pertinent for our understanding of the behavior of larval and adult fruit flies, the microbial ecology relevant to fruit flies remains mostly unexplored. A notable exception is work by Rohlfs and colleagues [54,55], which quantified negative effects of three mold species on fruit fly larvae and indicated that groups of five and 10

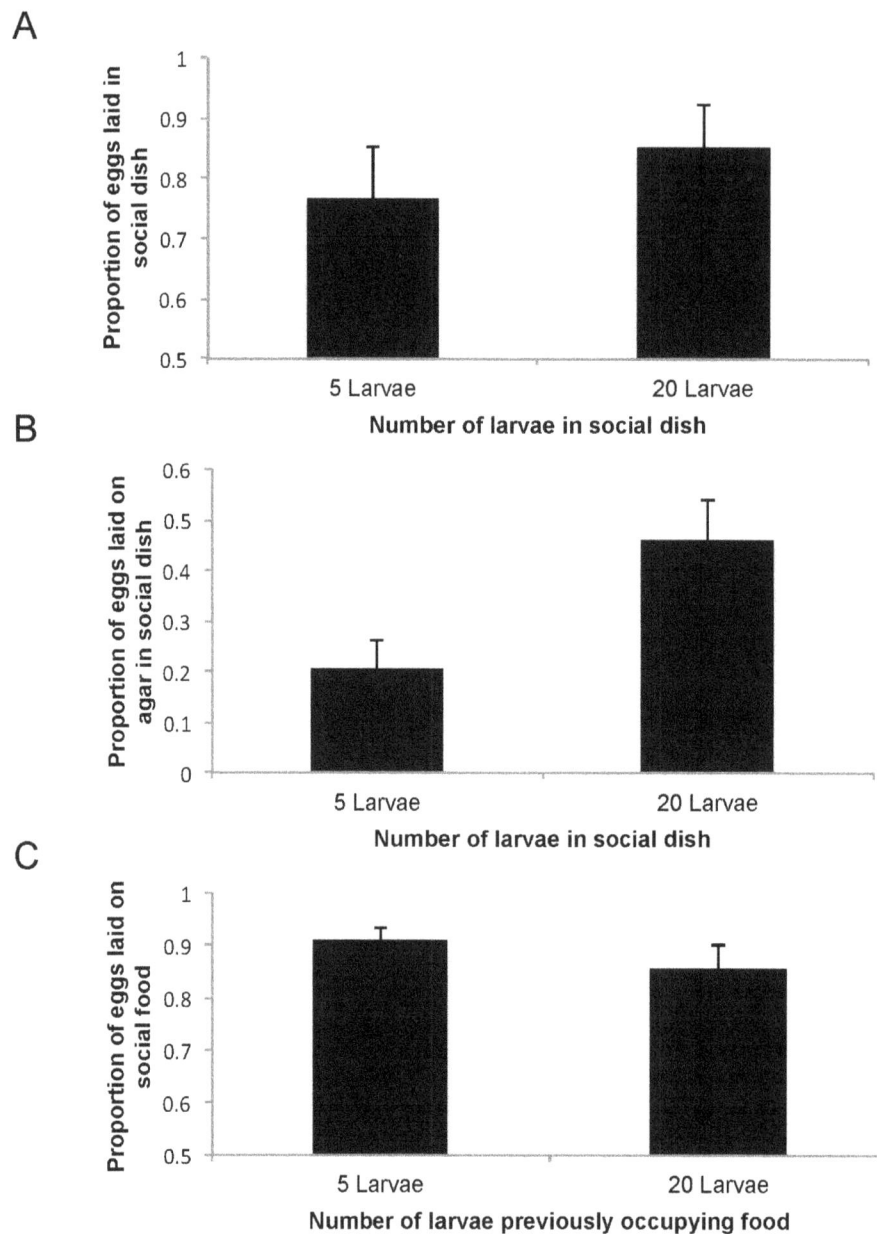

Figure 4. Social patch choice under high larval densities. The proportion (mean+SE) of eggs laid at the social disc, which currently (A, B) or previously (C) contained either five or 20 larvae. In each case, females could choose between laying at a social or non-social disc. (A) The proportion of eggs laid in the social dish out of all eggs laid. (B) The proportion of eggs laid on agar rather than on the food disc out of the eggs laid in the social dish. N = 24 replicates per treatment. (C) The proportion of eggs laid on the social disc, which had been previously consumed by either 5 or 20 larvae, out of all eggs laid. No larvae were present on the food at the time of egg laying. N = 28 replicates per treatment.

larvae were more effective at suppressing mold growth than single larvae. Another relevant observation is that fruit flies possess a dedicated olfactory circuit tuned to geosmin. Fruit flies rely on this circuit to avoid feeding and egg laying on substrates containing geosmin-producing microbes, which are harmful to fruit flies [56]. This indicates that fruit flies are sensitive to the constitution of microbial communities at prospective egg laying sites. It is thus likely that, by preferring to lay eggs at patches already occupied by larvae over unoccupied patches, females in natural settings ensure that their newly hatched larvae will be better protected from microbes harmful either to their larvae or to their larval yeast-food.

Group Enhancement of Favourable Yeasts

There appear to be mutualistic interactions between some yeast species and fruit flies. Adults and larvae inoculate fruit with yeast and larval activity promotes the growth of certain yeast species [57–59]. While some of the positive effects of larvae on yeast can be modulated through churning of the substrate, the larval gut bacteria also produce antifungals, which could selectively suppress mold and thus enhance the growth of the preferred yeast food [31,60–62]. Intriguingly, adult and larval fruit fly attraction to food inhabited by larvae is mediated by volatiles emitted from gut bacteria [31]. Hence it is likely that females in nature lay eggs in

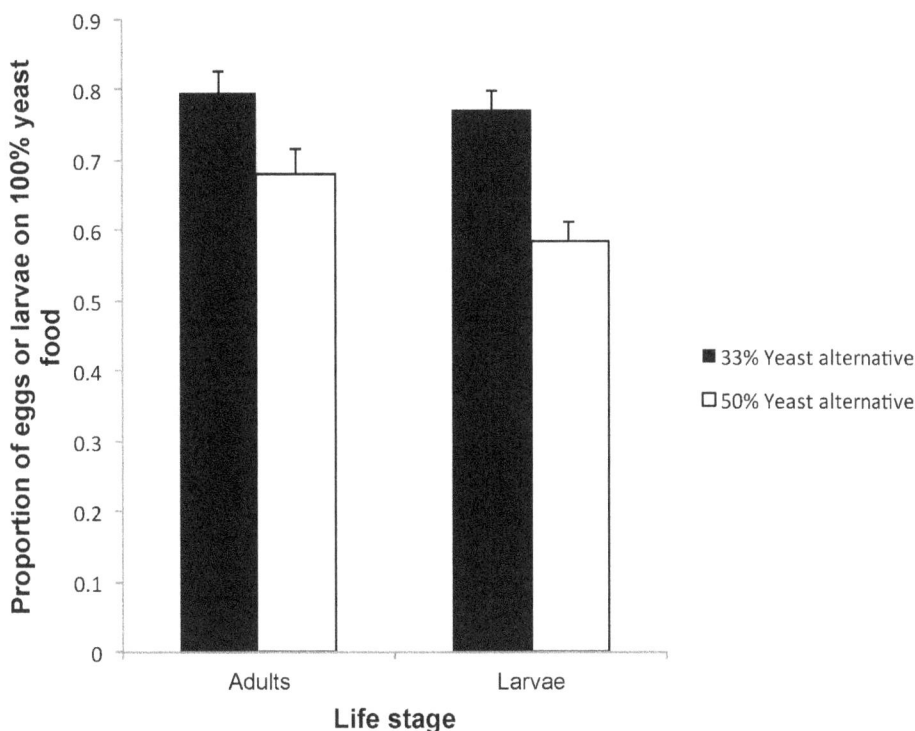

Figure 5. Patch choice by adult females versus larvae. In one experiment (black bars), adult females or larvae had a choice between a disc containing the regular yeast concentration (100%) or a disc containing 33% of the regular yeast concentration. In the other experiment (white bars), adult females or larvae had a choice between a disc containing the regular yeast concentration (100%) or a disc containing 50% of the regular yeast concentration. N = 80 replicates per nutrition treatment for larvae, and N = 60 replicates per nutrition treatment for the adult females.

occupied patches because such patches are more favourable for further growth of yeast food than are unoccupied patches.

Microbial Information

While we found no evidence that larval presence provides superior nutritional information about patch quality that females cannot readily assess, the discussion above suggests that larval presence is the best indicator that the microbial ecology is favourable to larval growth. That is, it is likely that different fruit patches allow for the optimal growth of different microbial species with only some of them being hospitable to fruit flies. For example, substrates may vary in their ability to sustain the growth of harmful mold and bacteria versus the yeast species favoured by fruit flies. Assuming that females cannot assess all the relevant ecological settings that would influence fungal growth, the presence of thriving larvae may be the best cue indicating that a patch is providing the appropriate microbial environment.

Parasitoid Avoidance

Larval parasitoids are a major source of fruit fly mortality in natural settings and fruit flies possess a suite of behavioral and physiological adaptations for reducing parasitoid success [63–66]. One way by which larvae can avoid parasitism is through hiding in micro-sites inaccessible to parasitoids. Although newly hatched larvae are not proficient at burrowing, older larvae, especially ones in the third instar stage, have stronger and larger mandibular hooks containing several teeth [67] and they spend much of their time tunnelling deep inside the substrate [34]. It is thus possible that, by laying eggs close to larvae, females ensure that their hatching offspring can hide in burrows dug by the older larvae.

Limited evidence indeed indicates that larvae hidden deep in natural fruit experience lower rates of parasitoid attacks [24].

Patch Choice Copying in Other Species

Our work on the value of patch choice copying in fruit flies can inform and be informed by research on copying in other species. Perhaps the best studied and most relevant system involves the economically important bark beetles (Scolytidae), which aggregate at host trees. While there are many species of bark beetles, we focus here on obligate parasites, which attack and kill trees [68]. Long-distance attraction to host trees in bark beetles is mediated by pheromones. Early colonizers benefit from attracting others because a critical mass of beetles and perhaps associated fungi are necessary for overcoming the massive defence mounted by the host tree [68–70]. Because prospective females gain from joining patch occupiers, the adaptive function of patch choice copying is clear.

There are at least two major differences between the fruit fly and bark beetle systems. First, in the bark beetles, there is active recruitment by early colonizers, which is crucial for their success [69,71]. In fruit flies, cis-vaccenyl acetate (cVA), has been referred to as an aggregation pheromone [57,72]. However, cVA is produced only by males, who transfer it during copulation to females [73], in which it signals to prospective males that the females are recently mated and unreceptive. Indeed females emitting cVA are much less attractive to males than females with no cVA [74–76]. It is thus likely that cVA has a relatively negligible role in long-distance attraction compared to the dominant role of microbial volatiles [31,77,78]. That is, there is no critical evidence indicating active recruitment of conspecifics in fruit flies.

The second and somewhat related difference between the bark beetle and fruit fly systems is the change in patch attractiveness with density. In the bark beetle system, there is a clear decline in tree attractiveness once a threshold beetle density has been reached. Such decline can readily be explained. Functionally, the occupiers no longer require further individuals once the tree is dying. Mechanistically, the occupiers can readily modulate patch attractiveness by ceasing to emit the aggregation pheromone [69,71]. In the fruit fly system, we failed to identify the predicted lower patch attractiveness under higher density. It is likely, however, that, in natural settings, cues from microbes associated with high density could decrease patch attractiveness or even repel females, as does geosmin discussed above.

Most other systems in which patch choice copying occurs are not as well studied as bark beetles. We suggest, however, that fruit flies can serve as an excellent general model system for further research on the topic owing to their amenability to research in the ecological, evolutionary and mechanistic domains. Our work so far suggests that direct benefits from joining others are likely in many systems even when such benefits are not observed under controlled settings. The most likely reason for such discrepancies is an involvement of harmful microbes in natural settings, which a group is more likely to overcome than an individual. Similarly, because the microbial ecology and dynamics is complex, prospective individuals probably gain the best available information from relying on others, because the others' presence indicates a suitable microbial setting. Our proposition about the central importance of microbes will require extensive experimental work in collaboration with microbial ecologists.

Acknowledgments

We thank J. Holmes, V. Mavandadi, and T. Padilla for assistance and K. Abbott L. Dukas and two anonymous referees for comments on the manuscript.

Author Contributions

Conceived and designed the experiments: SG RD. Performed the experiments: SG. Analyzed the data: SG. Wrote the paper: SG RD.

References

1. Thorpe W (1963) Learning and Instinct in Animals. London: Methuen and Co. 56–81 p.
2. Waite RK (1981) Local enhancement for food finding by Rooks (*Corvus frugilegus*) foraging on grassland. Zeitschrift fur Tierpsychologie 57: 15–36.
3. Krebs JR (1973) Social learning and the significance of mixed-species flocks of chickadees (*Parus* spp.). Canadian Journal of Zoology 51: 1275–1288.
4. Stamps JA (1987) Conspecifics as cues to territory quality: a preference of juvenile lizards (*Anolis aeneus*) for previously used territories. The American Naturalist 129: 629–642.
5. Betts MG, Hadley AS, Rodenhouse N, Nocera JJ (2008) Social information trumps vegetation structure in breeding-site selection by a migrant songbird. Proceedings of the Royal Society B: Biological Sciences 275: 2257–2263.
6. Fletcher RJ, Miller CW (2008) The type and timing of social information alters offspring production. Biology Letters 4: 482–485.
7. Prokopy RJ, Roitberg BD (2001) Joining and avoidance behavior in nonsocial insects. Annual Review of Entomology 46: 631–665.
8. Raitanen J, Forsman JT, Kivelä SM, Mäenpää MI, Välimäki P (2014) Attraction to conspecific eggs may guide oviposition site selection in a solitary insect. Behavioral Ecology 25: 110–116.
9. Dugatkin LA (1992) Sexual selection and imitation: females copy the mate choice of others. American Naturalist 139: 1384–1389.
10. Galef BG, White DJ (1998) Mate-choice copying in Japanese quail, *Coturnix coturnix japonica*. Animal Behaviour 55: 545–552.
11. Alonzo SH (2008) Female mate choice copying affects sexual selection in wild populations of the ocellated wrasse. Animal Behaviour 75: 1715–1723.
12. Krijger CL, Sevenster JG (2001) Higher species diversity explained by stronger spatial aggregation across six neotropical *Drosophila* communities. Ecology Letters 4: 106–115.
13. Shorrocks B, Sevenster JG (1995) Explaining Local Species Diversity. Proceedings of the Royal Society of London Series B: Biological Sciences 260: 305–309.
14. Wade MJ, Pruett Jones SG (1990) Female Copying Increases the Variance in Male Mating Success. Proceedings of the National Academy of Sciences of the United States of America 87: 5749–5753.
15. Kirkpatrick M, Dugatkin LA (1994) Sexual selection and the evolutionary effects of mate copying. Behavioral Ecology and Sociobiology 34: 443–449.
16. Agrawal AF (2001) The evolutionary consequences of mate copying on male traits. Behavioral Ecology and Sociobiology 51: 33–40.
17. Valone TJ, Templeton JJ (2002) Public information for the assessment of quality: a widespread social phenomenon. Philosophical Transactions of the Royal Society of London Series B-Biological Sciences 357: 1549–1557.
18. Danchin E, Giraldeau L-A, Valone TJ, Wagner RH (2004) Public information: from nosy neighbors to cultural evolution. Science 305: 487–491.
19. Danchin E, Wagner RH (1997) The evolution of coloniality: the emergence of new perspectives. Trends in Ecology & Evolution 12: 342–347.
20. Allee WC (1931) Animal Aggregations. A Study in General Sociology. Chicago: University of Chicago Press.
21. Arnold W (1988) Social thermoregulation during hibernation in alpine marmots (*Marmota marmota*). Journal of Comparative Physiology B 158: 151–156.
22. Willis CR, Brigham RM (2007) Social thermoregulation exerts more influence than microclimate on forest roost preferences by a cavity-dwelling bat. Behavioral Ecology and Sociobiology 62: 97–108.
23. Beauchamp G (2014) Social Predation: How Group Living Benefits Predators and Prey. London: Elsevier Academic Press.
24. Rohlfs M, Hoffmeister TS (2004) Spatial aggregation across ephemeral resource patches in insect communities: an adaptive response to natural enemies? Oecologia 140: 654–661.
25. Wertheim B, van Baalen E-JA, Dicke M, Vet LEM (2005) Pheromone-mediated aggregation in nonsocial arthropods: an evolutionary ecological perspective. Annual Review of Entomology 50: 321–346.
26. Pulliam HR, Caraco T (1984) Living in groups: is there an optimal group size? In: Krebs JR, Davies NB, editors. Behavioural Ecology. 2nd ed. Oxford: Blackwell. pp. 122–147.
27. Sarin S, Dukas R (2009) Social learning about egg laying substrates in fruit flies. Proceedings of the Royal Society of London B-Biological Sciences 276: 4323–4328.
28. Battesti M, Moreno C, Joly D, Mery F (2012) Spread of social information and dynamics of social transmission within *drosophila* groups. Current Biology 22: 309–313.
29. Durisko Z, Dukas R (2013) Attraction to and learning from social cues in fruit fly larvae. Proceedings of the Royal Society of London B-Biological Sciences 280: 20131398.
30. Durisko Z, Anderson B, Dukas R (2014) Adult fruit fly attraction to larvae biases experience and mediates social learning. Journal of Experimental Biology 217: 1193–1197.
31. Venu I, Durisko Z, Xu JP, Dukas R (2014) Social attraction mediated by fruit flies' microbiome. Journal of Experimental Biology 217: 1346–1352.
32. Sokolowski MB (2010) Social interactions in "simple" model systems. Neuron 65: 780–794.
33. Robinson GE, Fernald RD, Clayton DF (2008) Genes and social behavior. Science 322: 896–900.
34. Durisko Z, Kemp B, Mubasher A, Dukas R (2014) Dynamics of social interactions in fruit fly larvae. PLoS ONE 9: e95495.
35. Yang C-H, Belawat P, Hafen E, Jan LY, Jan Y-N (2008) *Drosophila* egg-laying site selection as a system to study simple decision-making processes. Science 319: 1679–1683.
36. IBM-Corp. (2011) IBM SPSS Statistics for Windows, Version 21.0. Armonk, NY: IBM Corp.
37. Ashburner M (1989) *Drosophila* a Laboratory Handbook. Cold Spring Harbor: Cold Spring Harbor Laboratory Press.
38. Roberts DB, editor (1998) *Drosophila*: a Practical Approach. 2nd ed. Oxford ; New York: IRL Press at Oxford University Press. xxiv, 389 p.
39. Atkinson WD (1979) A field investigation of larval competition in domestic *Drosophila*. Journal of Animal Ecology 48: 91–102.
40. Grimaldi D, Jaenike J (1984) Competition in natural populations of mycophagous *Drosophila*. Ecology 65: 1113–1120.
41. Masek P, Scott K (2010) Limited taste discrimination in *Drosophila*. Proceedings of the National Academy of Sciences 107: 14833–14838.

42. Burke CJ, Waddell S (2011) Remembering nutrient quality of sugar in *Drosophila*. Current Biology 21: 746–750.
43. Vosshall LB, Stocker RF (2007) Molecular architecture of smell and taste in *Drosophila*. Annual Review of Neuroscience 30: 505–533.
44. Yarmolinsky DA, Zuker CS, Ryba NJP (2009) Common sense about taste: from mammals to insects. Cell 139: 234–244.
45. Stafford JW, Lynd KM, Jung AY, Gordon MD (2012) Integration of taste and calorie sensing in *Drosophila*. The Journal of Neuroscience 32: 14767–14774.
46. Schwarz S, Durisko Z, Dukas R (2014) Food selection in larval fruit flies: dynamics and effects on larval development. Naturwissenschaften 101: 61–68.
47. Begon M (1982) Yeasts and *Drosophila*. In: Ashburner M, Carson HL, Thompson JN, editors. The Genetics and Biology of *Drosophila*. London: Academic Press. pp. 345–384.
48. Janzen DH (1977) Why fruits rot, seeds mold, and meat spoils. The American Naturalist 111: 691–713.
49. Demain A, Fang A (2000) The natural functions of secondary metabolites. In: Fiechter A, editor. History of Modern Biotechnology I: Springer Berlin Heidelberg. pp. 1–39.
50. Janisiewicz WJ, Korsten L (2002) Biological control of postharvest diseases of fruits. Annual Review of Phytopathology 40: 411–441.
51. Sharma R, Singh D, Singh R (2009) Biological control of postharvest diseases of fruits and vegetables by microbial antagonists: A review. Biological Control 50: 205–221.
52. Lacey LA, Shapiro-Ilan DI (2008) Microbial control of insect pests in temperate orchard systems: potential for incorporation into IPM. Annual Review of Entomology 53: 121–144.
53. Arndt C, Cruz MC, Cardenas ME, Heitman J (1999) Secretion of FK506/FK520 and rapamycin by *Streptomyces* inhibits the growth of competing *Saccharomyces cerevisiae* and *Cryptococcus neoformans*. Microbiology 145: 1989–2000.
54. Rohlfs M (2005) Density-dependent insect-mold interactions: effects on fungal growth and spore production. Mycologia 97: 996–1001.
55. Rohlfs M, Obmann B, Petersen R (2005) Competition with filamentous fungi and its implication for a gregarious lifestyle in insects living on ephemeral resources. Ecological Entomology 30: 556–563.
56. Stensmyr Marcus C, Dweck Hany KM, Farhan A, Ibba I, Strutz A, et al. (2012) A conserved dedicated olfactory circuit for detecting harmful microbes in *Drosophila*. Cell 151: 1345–1357.
57. Wertheim B, Dicke M, Vet LEM (2002) Behavioural plasticity in support of a benefit for aggregation pheromone use in *Drosophila melanogaster*. Entomologia Experimentalis Et Applicata 103: 61–71.
58. Stamps JA, Yang LH, Morales VM, Boundy-Mills KL (2012) *Drosophila* regulate yeast density and increase yeast community similarity in a natural substrate. PLoS ONE 7: e42238.
59. Wertheim B, Marchais J, Vet LEM, Dicke M (2002) Allee effect in larval resource exploitation in *Drosophila*: an interaction among density of adults, larvae, and micro-organisms. Ecological Entomology 27: 608–617.
60. Crowley S, Mahony J, van Sinderen D (2012) Comparative analysis of two antifungal *Lactobacillus plantarum* isolates and their application as bioprotectants in refrigerated foods. Journal of Applied Microbiology 113: 1417–1427.
61. Mauch A, Dal Bello F, Coffey A, Arendt EK (2010) The use of *Lactobacillus brevis* PS1 to in vitro inhibit the outgrowth of *Fusarium culmorum* and other common *Fusarium* species found on barley. International Journal of Food Microbiology 141: 116–121.
62. Schnürer J, Magnusson J (2005) Antifungal lactic acid bacteria as biopreservatives. Trends in Food Science & Technology 16: 70–78.
63. Carton Y, Bouletreau M, Alphen JJMv, Lenteren JCv (1986) The *Drosophila* parasitic wasps. In: Ashburner M, Carson HL, Thompson JN, editors. The Genetics and Biology of *Drosophila*. London: Academic Press. pp. 347–934.
64. Fleury F, Ris N, Allemand R, Fouillet P, Carton Y, et al. (2004) Ecological and genetic interactions in Drosophila–parasitoids communities: a case study with D. *Melanogaster, D. Simulans* and their common *Leptopilina* parasitoids in south eastern France. Genetica 120: 181–194.
65. Kacsoh BZ, Lynch ZR, Mortimer NT, Schlenke TA (2013) Fruit flies medicate offspring after seeing parasites. Science 339: 947–950.
66. Hwang RY, Zhong L, Xu Y, Johnson T, Zhang F, et al. (2007) Nociceptive neurons protect *Drosophila* larvae from parasitoid wasps. Current Biology 17: 2105–2116.
67. Bodenstein D (1950) The postembryonic development of *Drosophila*. In: Demerec M, editor. Biology of *Drosophila*. Cold Spring Harbor: Cold Spring Harbor Laboratory Press. pp. 275–367.
68. Paine TD, Raffa KF, Harrington TC (1997) Interactions among Scolytid bark beetles, their associated fungi, and live host conifers. Annual Review of Entomology 42: 179–206.
69. Wood DL (1982) The role of pheromones, kairomones, and allomones in the host selection and colonization behavior of bark beetles. Annual Review of Entomology 27: 411–446.
70. Raffa KF, Berryman AA (1983) The role of host plant resistance in the colonization behavior and ecology of bark beetles (Coleoptera: Scolytidae). Ecological Monographs 53: 27–49.
71. Raffa KF, Aukema BH, Bentz BJ, Carroll AL, Hicke JA, et al. (2008) Cross-scale drivers of natural disturbances prone to anthropogenic amplification: the dynamics of bark beetle eruptions. BioScience 58: 501–517.
72. Bartelt RJ, Schaner AM, Jackson LL (1985) cis-vaccenyl acetate as an aggregation pheromone in *Drosophila melanogaster*. Journal of Chemical Ecology 11: 1747–1756.
73. Brieger G, Butterworth FM (1970) *Drosophila melanogaster*: identity of male lipid in reproductive system. Science 167: 1262.
74. Ejima A, Smith BPC, Lucas C, van der Goes van Naters W, Miller CJ, et al. (2007) Generalization of courtship learning in *Drosophila* is mediated by cis-vaccenyl acetate. Current Biology 17: 599–605.
75. Dukas R, Dukas L (2012) Learning about prospective mates in male fruit flies: effects of acceptance and rejection. Animal Behaviour 84: 1427–1434.
76. Keleman K, Vrontou E, Kruttner S, Yu JY, Kurtovic-Kozaric A, et al. (2012) Dopamine neurons modulate pheromone responses in *Drosophila* courtship learning. Nature 489: 145–149.
77. Becher PG, Flick G, Rozpędowska E, Schmidt A, Hagman A, et al. (2012) Yeast, not fruit volatiles mediate *Drosophila melanogaster* attraction, oviposition and development. Functional Ecology 26: 822–828.
78. Stökl J, Strutz A, Dafni A, Svatos A, Doubsky J, et al. (2010) A deceptive pollination system targeting Drosophilids through olfactory mimicry of yeast. Current Biology 20: 1846–1852.

18

Did School Food and Nutrient-Based Standards in England Impact on 11–12Y Olds Nutrient Intake at Lunchtime and in Total Diet?

Suzanne Spence[1,2], **Jennifer Delve**[1,2], **Elaine Stamp**[1,2], **John N. S. Matthews**[3], **Martin White**[1,2,4], **Ashley J. Adamson**[1,2,4]*

1 Institute of Health and Society, Newcastle University, Newcastle upon Tyne, England, 2 Human Nutrition Research Centre, Newcastle University, Newcastle upon Tyne, England, 3 School of Mathematics and Statistics, Newcastle University, Newcastle upon Tyne, England, 4 Fuse, UKCRC Centre for Translational Research in Public Health, Newcastle upon Tyne, England

Abstract

Introduction: In September 2009, middle and secondary schools in England were required to comply with food and nutrient-based standards for school food. We examined the impact of this policy change on children's lunchtime and total dietary intake.

Methods: We undertook repeat cross-sectional surveys in six Northumberland middle schools in 1999–2000 and 2009–10. Dietary data were collected from 11–12 y olds ($n = 298$ in 1999–2000; $n = 215$ in 2009–10). Children completed two consecutive 3-day food diaries, each followed by an interview. Linear mixed effect models examined the effect of year, lunch type and level of socio-economic deprivation on children's mean total dietary intake.

Results: We found both before and after the introduction of the food and nutrient-based standards children consuming a school lunch, had a lower per cent energy from saturated fat (-0.5%; $p = 0.02$), and a lower intake of sodium (-143 mg; $p = 0.02$), and calcium (-81 mg; $p = 0.001$) in their total diet, compared with children consuming a home-packed lunch. We found no evidence that lunch type was associated with mean energy, or absolute amounts of NSP, vitamin C and iron intake. There was marginal evidence of an association between lunch type and per cent energy NMES ($p = 0.06$). In 1999–2000, children consuming a school lunch had a higher per cent energy from fat in their total diet compared with children consuming a home-packed lunch (2.8%), whereas by 2009–10, they had slightly less (-0.2%) (year by lunch type interaction $p < 0.001$; change in mean differences -3%).

Conclusions: We found limited evidence of an impact of the school food and nutrient-based standards on total diet among 11–12 year olds. Such policies may need to be supported by additional measures, including guidance on individual food choice, and the development of wider supportive environments in school and beyond the school gates.

Editor: Martin Young, University of Alabama at Birmingham, United States of America

Funding: This work was undertaken as part of the research programme of the Public Health Research Consortium (PHRC). The Public Health Research Consortium is funded by the Department of Health (DH) Policy Research Programme. The views expressed in this publication are those of the authors and not necessarily those of DH. Information about the wider programme of the PHRC is available from http://phrc.lshtm.ac.uk/. The funders had no role in the study design, data collection or analysis, interpretation of findings, writing of, or the decision to submit for publication. All authors had access to data, and take responsibility for the integrity of the data and the accuracy of the data analysis. MW is director and AJA is associate director of Fuse, the Centre for Translational Research in Public Health, a UK Clinical Research Collaboration (UKCRC) Public Health Research Centre of Excellence. Funding for Fuse from the British Heart Foundation, Cancer Research UK, Economic and Social Research Council, Medical Research Council, and the National Institute for Health Research, under the auspices of the UKCRC, is gratefully acknowledged. AJA is funded by the National Institute of Health Research as an NIHR Research Professor. Opinions expressed are not necessarily those of the funders.

Competing Interests: The authors have declared that no competing interests exist.

* Email: ashley.adamson@ncl.ac.uk

Introduction

Reducing childhood overweight and obesity are public health priorities [1]; improving diet is central to achieving a healthier lifestyle and losing weight [2,3]. Although there is some evidence of a levelling off in childhood obesity [4,5], in 2011–12, the National Child Measurement Programme in England identified a third of 10–11 y olds as overweight or obese [6], and socio-economic disparities persist [4,7].

Obesity has been found to track from adolescence to adulthood [8,9]; one potentially contributing factor is poor dietary patterns [9]. The English National Diet and Nutrition Survey found per cent energy from saturated fat and non-milk extrinsic sugar (NMES) exceeded the Dietary Reference Value of 11%; per cent energy from NMES was highest in 11–18 y olds (15.3%) [10]. Only 11% of boys and 8% of girls met the recommended '5-a-day' for fruit and vegetables [10]. Certain micronutrients, for example iron, were below the Reference Nutrient Intake.

Improving dietary intake in this age group is complex. During adolescence there is increasing independence in food choice [11] with social factors playing a crucial role[12–14]. For adolescents, food and drink consumption is related to 'identity' and 'status' [12,13]. One effort to tackle adolescent's diets has been a change in government policy requiring middle and secondary schools in England to comply with food and nutrient-based standards for school food from September 2009 [15]. These specify the provision of certain foods and the average nutrient content school lunches must provide over a three week menu cycle [16]. The majority of studies exploring the impact of the food and nutrient-based standards have focused on change in lunchtime intake in primary schools[17–24]; few have reported on middle and secondary schools [25,26]. Following the implementation of nutritional standards, Fletcher et al. reported the increased selling of junk food by students and suggested these standards ignore the wider contextual issues associated with food choice [14]. Studies have also highlighted negative aspects of school lunches, for example pricing [14] and a preference to socialise with friends at lunchtime [12]. Findings also reveal negative aspects of the dining environment, for example overcrowding, queuing [12,14,27] and noise [14].

With limited findings from quantitative studies, it is important to examine whether the food and nutrient-based standards could potentially affect nutrient intake among adolescents. In this paper we report research which examined the impact of the introduction of food and nutrient-based standards for school lunch on the lunchtime and total diet of a representative sample of children aged 11–12 years, between 1999–2000 (before) and 2009–10 (after) introduction of the policy in England.

Methods

Ethics statement

Ethical approval was granted by Newcastle University ethics committee (reference 000011/2007). In 2009–10, Newcastle University ethics committee granted approval for opt-out to be used as the method of consent (reference 00011/2009). Parents were provided with a written information letter about the study and a consent form, however, they were only required to return the consent form if they did not wish their child to participate. Newcastle University ethics committee approved our study design, methods and the consent procedure used for this study. All the data in this study were anonymised.

Study design, setting and participants

Cross-sectional studies were undertaken in middle schools in Morpeth, Ashington and Newbiggin-by-the-Sea in Northumberland, North East England over two academic years: 1999–2000 (before) and 2009–10 (after implementation of the standards). These areas were previously selected to be representative of schools with catchment populations across the socio-economic spectrum [28,29]. The 1999–2000 data were collected as part of a series of studies conducted in Northumberland[11,30–32] to track changes in dietary patterns and used as the baseline in this study. The same schools were invited by letter in 2009 to participate in this study. This was followed up with a school visit to answer questions and ascertain interest. During discussions with heads of schools they suggested consent should be changed from 'opt-in' (as used in the previous studies in these schools) to 'opt-out'. The rationale was that by using opt-in we excluded children whose parents failed to return forms sent by schools, rather than just those children whose parents actively did not want their child to participate. After obtaining documented support from heads and school governors, an amendment to the Newcastle University Ethics approval was granted for the use of opt-out in 2009–10 (reference 00011/2009). One head preferred that his school continued to use opt-in (this was the smallest school) and the decision was taken to retain this school despite a different method used in the consent process. Children could still exclude themselves by not completing food diaries and were free to leave the study at any time.

All children in year 7 were eligible to participate. A presentation was given at individual schools and each child received a parental information letter and a consent form to return if they did not wish to participate. Participating children received a unique identification number to anonymise data. All data were stored securely according to Newcastle University policies and regulations.

Data

Dietary consumption. We used dietary assessment methods identical to those used in the previous Northumberland studies [11,30]. This method has been described in detail [11,27,30,33] and validated [29,34]; a brief overview is provided here. Verbal

Table 1. Number of children consenting and reasons for exclusion in 1999–2000 and 2009–10.

	1999–2000	2009–10
Number consenting	**n = 424**	**n = 295**
Reasons for exclusion:		
From non-comparable school*	19	–
Mixed lunch[†]	96	73
No postcode	6	7
Completed less than 6 food diary days	5	0
Number included in analysis	298	215

*Non-comparable school: one school had closed from 1999–2000 to 2009–10.
[†]Mixed lunch means a child having both a school and home-packed lunch.

Table 2. Number (percentage) of children consuming a school lunch by year and level of deprivation.

Level of deprivation	1999–2000			2009–10			[1999–00]-[2009–10]
	No. having school lunch	Total *	(%)	No. having school lunch	Total	(%)	Decrease in %
Quintile 1 (least)	44	54	(81)	12	55	(20)	61
Quintile 2	43	55	(78)	11	41	(27)	51
Quintile 3	40	50	(80)	15	34	(44)	36
Quintile 4	38	49	(78)	10	26	(38)	40
Quintile 5 (most)	75	90	(83)	30	59	(51)	32
All children	240	298	(81)	78	215	(36)	45

*Total = no. having school and home-packed lunch.

instructions on how to complete the diary were given to each participating child; the diary also included an example page with instructions. Children recorded the day, date and time when food or drink was consumed, descriptions of items and amounts of foods/drinks for two consecutive three-day periods (for example Thursday, Friday, Saturday and Sunday, Monday, Tuesday). On the fourth day the child was interviewed by a trained researcher to clarify information recorded and estimate portion size using food models and a photographic food atlas for 11–14 y olds [35]. Foods were coded using McCance and Widdowson's Integrated Composition of Food dataset [36]. If available, school recipes were used to code school lunch, and if not, foods were coded as above. Foods were categorised into 'school lunch', 'home-packed lunch' and 'food consumed outside of school hours'. In common with the large majority of secondary schools in England [37]none of the schools permitted pupils to leave school premises at lunchtime. The macro- and micronutrients examined in this paper relevant to the nutrient-based standards are: energy (kcals), per cent energy from fat, saturated fat, and non-milk-extrinsic sugars (NMES); and absolute amounts of non-starch polysaccharides (NSP) (g), sodium (mg), vitamin C (mg), calcium (mg) and iron (mg).

Socio-economic status. Socio-economic status was estimated using the English Index of Multiple Deprivation (IMD) 2007 [38], allocated using individual children's postcodes. IMD is calculated at lower layer super output areas in England and provides a single deprivation score based on seven domains: income, employment, health and disability, education, skills and training, barriers to housing and services, crime and living environment [38]. The IMD scores were categorised into quintiles for the analyses: quintile 1 included children living in the 20% least deprived areas, quintile 5 included children living in the 20% most deprived areas.

Main outcome measures

Main outcome measures were mean daily intakes of macro- and micronutrients in 'school lunch', 'home-packed lunch' and total diet, measured as indicated below.

Statistical analysis

We undertook three sets of analyses. The first considered the change in school lunch take-up. A linear model was fitted directly to the proportions taking school lunch using maximum likelihood (fitted in R using optim), which allowed for differences between IMD quintiles, between years and their interaction. The second examined the change at lunchtime in children's mean macro- and micronutrient intake from a school or home-packed lunch on school days only between 1999–2000 and 2009–10. The third analysis considered the intake of macro- and micronutrients in children's total diet: this explored the effect of year (before and after the food and nutrient-based standards), lunch type (school or home-packed lunch) and level of deprivation. We used linear mixed effect models to examine the effect of these variables; interactions between variables were considered (year by lunch type, year by level of deprivation and lunch type by level of deprivation). Where there was no evidence for a particular interaction for a given nutrient, the interaction was excluded from the final model. All analyses adjusted for the effect of gender and day type (week or weekend day). Within each model random effects were included for school and child. Data were analysed using Stata version 11 and models were fitted using *xtmixed*. Vitamin C was log transformed for analysis, and for this variable geometric means and ratios are reported in tables.

Table 3. Lunchtime: Change in children's mean daily nutrient intake from school lunch between 1999–2000 and 2009–10, and nutrient-based standards [16].

Nutrient	Standard	Consumption from school lunch				
		1999–2000	2009–10	[2009–10]–[1999–2000]		
		n = 240	n = 78			
		mean*		mean difference	95% CI for difference	p-value[†]
Energy (kcals)	610	729	497	−232	−276; −189	<0.001
% energy fat	–	40.6	30.7	−9.9	−11.4; −8.6	<0.001
% energy saturated fat	–	12.5	10.6	−1.9	−2.7; −1.3	<0.001
% energy NMES	–	11.9	13.0	1.1	−0.4; 2.7	0.2
NSP (g)	min 4.9	3.9	3.2	−0.7	−1.0; −0.4	<0.001
Sodium (mg)	max 714	908	518	−390	−453; −328	<0.001
Vitamin C (mg) [‡]	min 12.3	28.8	28.2	1.0	0.9; 1.1	0.7
Calcium (mg)	min 350	206.5	184.2	−22.3	−44.4; −0.3	0.05
Iron (mg)	min 5.2	2.8	2.1	−0.7	−0.9; −0.5	<0.001

*Mean adjusted for gender.
[†]P-value derived from a linear mixed effects model.
[‡]Vitamin C log transformed; geometric means and ratios reported.

Results

Study sample characteristics

Table 1 shows the number of children who consented to take part by year and reasons for exclusion. There was a similar percentage of males and females participating in 1999–2000 (m = 47%; f = 53%) and 2009–10 (m = 50%; f = 50%), and there was no evidence of a statistically significant difference in children's mean IMD score (p = 0.3).

From Table 2 it can be seen that school lunch take-up was similar across all IMD quintiles in 1999–2000: between 1999–2000 and 2009–10 there was a decrease in the percentage of children consuming a school lunch, with evidence that the decrease differed across the IMD quintiles. The fall in school lunch take-up decreased linearly across the IMD quintiles (linear by year interaction p = 0.01, likelihood ratio test), with a fall of 61 percentage points in the least deprived group compared with a mean reduction of 32 percentage points in the most deprived group.

Lunchtime diet

Tables 3 and 4 show the change in children's mean daily nutrient intake in school and home-packed lunches respectively between 1999–2000 and 2009–10, compared with the nutrient-based standards [16]. In school lunches, between 1999–2000 and

Table 4. Lunchtime: Change in children's mean daily nutrient intake in home-packed lunch between 1999–2000 and 2009–10, and nutrient-based standards [16].

Nutrient	Standard	Consumption from home-packed lunch				
		1999–2000	2009–10	[2009–10]–[1999–2000]		
		n = 58	n = 137			
		mean*		mean difference	95% CI for difference	p-value[†]
Energy (kcals)	610	605	578	−27	−77; 23	0.3
% energy fat	–	34.0	32.3	−1.7	−4.0; 0.7	0.2
% energy saturated fat	–	14.1	14.2	0.1	−1.3; 1.5	0.8
% energy NMES	–	17.8	17.1	−0.7	−3.0; 1.7	0.6
NSP (g)	min 4.9	2.9	3.4	0.5	0.04; 1.0	0.03
Sodium (mg)	max 714	954	889	−65	−165; 34	0.2
Vitamin C (mg) [‡§]	min 12.3	26.9	34.7	1.3	1.1; 1.6	0.006
Calcium (mg)	min 350	223.2	292.1	68.9	21.1; 116.7	0.005
Iron (mg)	min 5.2	2.6	2.4	−0.2	−0.5; 0.1	0.3

*Mean adjusted for gender.
[†]P-value derived from a linear mixed effects model.
[‡]Vitamin C log transformed; geometric means and ratios reported.

2009–10, there was strong evidence of a decrease in mean energy intake (mean difference −232 kcals; p<0.001), per cent energy from fat (−9.9%; p<0.001) and saturated fat (−1.9%; p<0.001), and in absolute amounts of sodium (−390 mg; p<0.001), but also a decrease in mean NSP (−0.7 g; p<0.001) and iron intake (−0.7 mg; p<0.001). We found no evidence of a change in per cent energy from NMES (1.1%; p = 0.2), mean vitamin C (ratio 1.0; p = 0.7) and marginal evidence of a change in calcium intake (−22.3 mg; p = 0.05) (Table 3). In 1999–2000, children's mean energy and sodium intake from school lunch were above the target for the current school nutrient-based standards. By 2009–10, mean intakes were below these targets [16]. In 1999–2000, mean intakes of NSP, calcium, iron and vitamin C intake were below the nutrient-based standards [16]; these deficits persisted in 2009–10 (Table 3).

In packed lunches, between 1999–2000 and 2009–10, there was a statistically significant increase in absolute amounts of mean NSP (mean difference 0.5 g; p = 0.03), calcium (68.9 mg; p = 0.005) and vitamin C intake (1.3; p = 0.006) (Table 4). We found no evidence of a change in mean energy (−27 kcals; p = 0.3), per cent energy from fat (−1.7%; p = 0.2), saturated fat (0.1%; p = 0.8), NMES (−0.7%; p = 0.6), or absolute amounts of sodium (−65 mg; p = 0.2) or iron intake (−0.2 mg; p = 0.3) (Table 4).

Total diet

The results from the total diet analysis are shown in Tables 5, 6, 7 and Figure 1. Table 5 shows the effect of year (before and after the food and nutrient-based standards), Table 6 the effect of lunch type (school or home-packed lunch) and Table 7 the effect of level of deprivation. There was evidence of a year by lunch type interaction only for per cent energy from fat (Figure 1).

In total diet, between 1999–2000 and 2009–10, there was strong evidence of a decrease in mean energy intake (mean difference −259 kcals; p<0.001), and absolute amounts of sodium (−475 mg; p<0.001), but also a decrease in NSP (−0.9 g; p = 0.002), and iron intake (−1.0 mg; p<0.0001). Mean calcium and vitamin C intake increased (104 mg; p<0.001 and ratio 1.2; p<0.001 respectively) (Table 5). We found no evidence of a change in per cent energy from saturated fat (−0.2%; p = 0.4) or NMES (−0.5%; p = 0.3) (Table 5). In 2009–10, children's per cent energy from saturated

fat and NMES remained above the recommendation of ≤11% [39]. Mean vitamin C intake was the only micronutrient to meet the Reference Nutrient Intake [39].

Table 6 shows the effect of lunch type (school or home-packed lunch) on children's mean total dietary intake, with data from before and after the introduction of the legislation combined. There was clear evidence that children who consumed a school lunch both before and after the implementation of the food and nutrient-based standards had a lower per cent energy from saturated fat (mean difference −0.5%; p = 0.02), and absolute amounts of sodium (−143 mg; p = 0.02), and calcium intake (−81 mg; p = 0.001) compared with children who consumed a packed lunch (Table 6). We found no evidence of a statistically significant effect of lunch type on mean energy, or absolute amounts of NSP, vitamin C and iron intake in total diet. We found marginal evidence of an effect on per cent energy from NMES (−0.9%; p = 0.06) (Table 6).

In both 1999–2000 and 2009–10, we found strong evidence of a level of deprivation effect on mean vitamin C intake. Mean intakes were lowest for children in the most deprived quintile (test for the effect of level of deprivation: p<0.001, Table 7). We found no evidence of an effect on mean energy, per cent energy from fat, saturated fat, NMES, or absolute amounts of NSP and sodium intake. We found marginal evidence of an effect on mean calcium and iron intake. Mean intakes were lowest for those in the most deprived quintile (test for the effect of level of deprivation: p = 0.04 and p = 0.08 respectively) (Table 7).

For one nutrient, per cent energy from fat, we found a statistically significant year by lunch type interaction on children's total dietary intake (p<0.001; Figure 1). This was because there was a markedly higher per cent energy from fat in school lunches compared with packed lunches in 1999–2000 (35.9% and 33.1% respectively; mean difference 2.8%), whereas the corresponding difference in 2009–2010 was very small (31.9% and 32.1% respectively; −0.2%). The change in these differences: (2009/10–1999/00) is (−0.2) −2.8 = −3% (95% CI −4.4 to −1.4; see Figure 1). We found no evidence of any statistically significant year by level of deprivation or lunch type by level of deprivation interactions.

Table 5. Total diet: The effect of year on children's mean daily nutrient intake and Dietary Reference Values/Reference Nutrient Intakes (DRV/RNI) [39].

Nutrient	DRV/RNI	1999–2000*	2009–10	[2009–10]–[1999–2000]		
		Mean[†]		Mean difference	95% CI for difference	p-value[‡]
Energy (kcals)	M[§] = 2220; F[§] = 1845	1924	1665	−259	−332; −185	<0.001
% energy saturated fat	≤11	12.9	12.7	−0.2	−0.6; 0.2	0.4
% energy NMES	≤11	16.5	16.0	−0.5	−1.3; 0.4	0.3
NSP (g)	–	10.8	9.9	−0.9	−1.5; −0.3	0.002
Sodium (mg)	1600	2593	2118	−475	−590; −361	<0.001
Vitamin C (mg)[ǁ]	35	67.6	79.4	1.2	1.1; 1.3	<0.001
Calcium (mg)	M = 1000; F = 800	698	802	104	60; 149	<0.001
Iron (mg)	M = 11.3; F = 14.8	9.6	8.6	−1.0	−1.6; −0.5	<0.001

*Number of children participating in 1999–2000 (n = 298) and 2009–10 (n = 215).
[†]Mean adjusted for gender, day-type, lunch type and level of deprivation.
[‡]95% CI and p-value derived from a linear mixed effects model.
[§]M (male) F (female).
[ǁ]Vitamin C log transformed; geometric means and ratios reported.

Table 6. Total diet: The effect of lunch type (school or home-packed lunch) on children's mean daily nutrient intake and Dietary Reference Values/Reference Nutrient Intakes (DRV/RNI) [39].

Nutrient	DRV/RNI	Packed (PL)*	School (SL)	[SL-PL]		
		Mean†		Mean difference	95% CI for difference	p-value‡
Energy (kcals)	M§ = 2220; F§ = 1845	1792	1788	−4	−78; 71	0.9
% energy saturated fat	≤11	13.2	12.7	−0.5	−0.9; −0.1	0.02
% energy NMES	≤11	16.9	16.0	−0.9	−1.8; 0.0	0.06
NSP (g)	–	10.1	10.2	0.1	−0.5; 0.7	0.8
Sodium (mg)	1600	2490	2347	−143	−261; −26	0.02
Vitamin C (mg)$^∥$	35	70.8	72.4	1.0	0.9; 1.1	0.5
Calcium (mg)	M = 1000; F = 800	778	697	−81	−127; −35	0.001
Iron (mg)	M = 11.3; F = 14.8	9.2	8.8	−0.4	−0.9; 0.2	0.2

*Number of children participating in 1999–2000 (n = 298) and 2009–10 (n = 215).
†Mean adjusted for year, gender, day-type and level of deprivation.
‡95% CI and p-value derived from a linear mixed effects model.
§M (male) F (female).
$^∥$Vitamin C log transformed; geometric means and ratios reported.

Discussion

Summary of key findings

Between 1999–2000 and 2009–10, the number of children consuming a school lunch decreased with the greatest decline in children from more affluent families. At lunchtime, in 2009–10, we found that children eating school lunches consumed a healthier diet with regard to per cent energy from fat, saturated fat, NMES and sodium, but had a lower mean micronutrient intake than children consuming packed lunches. In total diet, between 1999–2000 and 2009–10, there was a statistically significant decrease in mean intakes of energy and sodium, but also a decrease in NSP and iron, while vitamin C and calcium intake increased. We found no evidence of a change in per cent energy from NMES or saturated fat. There was limited evidence that a child's lunch type was associated with a change in children's mean total dietary intake. The only association found between year (before and after the introduction of the food and nutrient-based standards) and a child's lunch type (school or home-packed lunch) was in relation to per cent energy from fat consumed. By 2009–10, children who consumed a school lunch had a slightly lower intake of per cent energy from fat in their total diet compared with those who consumed a home-packed lunch. We found little evidence that mean nutrient intakes were associated with level of deprivation.

Relationship to other studies

In 2009–10, school lunch take-up in the six Northumberland middle schools participating in this study was 36%. A study in English academies and city technology colleges found school lunch take-up was 37.6% in 2010–11 [40].

There is limited research examining the impact (before and after implementation) of the food and nutrient-based standards in England on dietary intake at lunchtime and the impact of this policy change on total diet in 11–12 y olds. A number of studies have examined nutritional intake in this age group at school or in their total diet. What this study adds is a consideration of school and home-packed lunch both separately and in the context of total diet, prior to and following a major change in school food policy.

At lunchtime, we found mean energy, NSP, calcium and iron intakes were below the nutrient-based standards in both school and home-packed lunches; however, vitamin C was above. These findings are similar to those from a national survey of 80 secondary schools in England [26]. In school lunch, per cent energy from fat, saturated fat and NMES were comparable with the national survey. In home-packed lunch, we found a lower per cent energy from fat, but a higher per cent energy from saturated fat and NMES compared with the national survey. In contrast to other studies, [26,41,42] we found that a school lunch provided a lower mean energy, NSP, and micronutrient intake than a home-packed lunch. Our findings concur with those by Hur et al [43] and Taylor et al [44] who found children who consumed a school lunch had a lower mean energy intake than children consuming a home-packed lunch. Similarly Taylor et al [44] also found lower intakes of some micronutrients, such as iron and vitamin C. The lower mean intakes of micronutrients for children consuming a school lunch in our study may be due to the lower mean energy intake which highlights the need for increased nutrient quality with lower energy intakes. These findings show some inconsistencies in energy and some micronutrient intakes in studies that have investigated what children eat in a school or home-packed lunch. These differences may be due to a number of factors, for example: age of children studied and variation in food provision and wider support to which children are exposed, however, differences due to dietary data collection methods cannot be excluded. A study by Pearce et al [45] showed that some portion sizes of foods on offer had decreased since the implementation of the policy; variation in portion sizes served across schools may also explain inconsistencies in findings.

A study by Fung et al [46] that examined change in children's total diet pre to post-school lunch policy in Canada (Grade 5 children) reported similar findings to our study. For example, they found a decrease in per cent energy from fat and absolute amounts of sodium; and also a decrease in mean fibre intake. In contrast to our study they found mean iron intake increased. [46] In total diet, we found children's mean energy, calcium and iron intake were below recommended intakes [39]; per cent energy from saturated fat and NMES, and absolute amounts of sodium were above. This is similar to findings from 11–18 y olds in the National Dietary and Nutrition Survey (NDNS) [47]. Between 1999–2000 and 2009–10, we found a decrease in energy, per cent energy from fat and saturated fat, and little change in per cent energy from NMES. Mean vitamin C and calcium intake increased, but iron

Table 7. Total diet: The effect of level of deprivation on children's mean daily nutrient intake and Dietary Reference Values/Reference Nutrient Intakes (DRV/RNI) [39].

Nutrient	DRV/RNI	Level of deprivation					
		1 (least deprived) *	2	3	4	5 (most deprived)	
		Mean[†]					p-value[‡]
Energy (kcals)	M[§] = 2220; F[§] = 1845	1773	1830	1813	1821	1747	0.4
% energy fat	≤35	33.5	33.6	33.9	34.1	34.4	0.2
% energy saturated fat	≤11	12.6	12.9	12.9	13.1	12.8	0.7
% energy NMES	≤11	16.3	16.7	16.2	15.8	16.4	0.8
NSP (g)	–	10.1	10.4	10.4	10.5	9.8	0.5
Sodium (mg)	1600	2421	2452	2407	2484	2310	0.2
Vitamin C (mg)[‖]	35	81.3	79.4	77.6	67.6	61.7	<0.001
Calcium (mg)	M = 1000; F = 800	744	763	753	723	680	0.04
Iron (mg)	M = 11.3; F = 14.8	8.9	9.4	9.4	8.8	8.5	0.08

*Number of children participating in 1999–2000 ($n = 298$) and 2009–10 ($n = 215$).
[†]Mean adjusted for year, lunch type, gender and day-type.
[‡]P-value derived from a linear mixed effects model.
[§]M (male) F (female).
[‖]Vitamin C log transformed; geometric means and ratios reported.

% energy fat

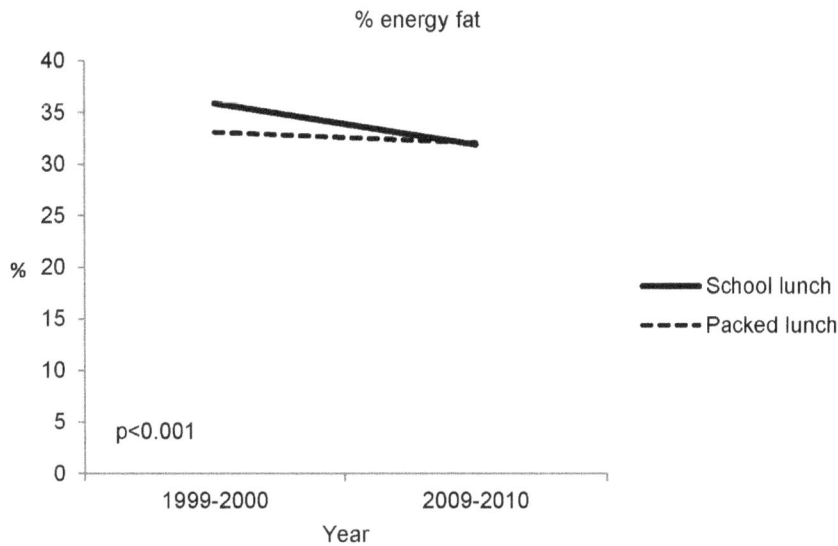

Figure 1. Total diet: The effect of year and lunch type interaction on children's per cent energy from fat (adjusted for gender, level of deprivation and day type).

decreased; these findings are also similar to the trends observed in the NDNS [47,48]. This decrease in mean energy and per cent energy from fat was also observed in a previous study in Northumberland examining the macronutrient intake in 11–12 y olds between 1980 and 2000 [11]. In contrast, in this later study we found no evidence of a change in per cent energy from NMES which remained above recommended intakes [38] (16% compared with 11%). This suggests products with a high sugar content, such as breakfast cereals, confectionery and fruit juices, remain a constant element of children's dietary intake.

Strengths and Limitations

This is the first study in a middle school setting to use a natural experimental, repeat cross-sectional design before and after the implementation of the standards to evaluate the impact both at lunchtime and in total diet [49]. A limitation of this approach is attributing causality [24]. National implementation of the food and nutrient-based standards in primary, middle and secondary schools prevented the use of a stronger study design with a control group and prospective follow-up of individual children [24]. This study was limited to the North East of England, so, findings may not be generalisable [24]. Socio-economic status was estimated using IMD, which does not measure individual levels of deprivation, and is therefore subject to potential misclassification bias [50]. We used identical prospective dietary data collection methods at both time points to ensure consistency. The data collection method relied on self-report and was potentially subject to misreporting [51]. We collected two three-day periods of dietary data to limit this bias.

Conclusions and implications

The school environment offers an opportunity to influence dietary intake. Yet, our findings have shown limited evidence of the food and nutrient-based standards affecting total diet in this age group, which is in contrast to the results among younger children [24]. Reasons for this may be a reduction in the proportion of children consuming a school lunch, less than full compliance with the food and nutrient-based standards, or individual food choice. School lunches have potential to improve

children's dietary intake but only if they are consumed. This study found a decrease in school lunch take-up which suggests the importance of addressing the wider social aspects of overcrowding, noise and queues in school dining rooms [12,14,27] to provide an attractive environment conducive to healthy eating. Other factors may also be associated with a decrease in school lunch take-up. The standards limit the frequency of serving of certain foods and also restrict what food and drink can be served. A process evaluation undertaken parallel to this study highlighted that parents of younger children (4–7 y olds) supported the restriction of food choice. However, there was more *'ambivalence* in the parents of middle school children (11–12 y olds) for who personal preference was an important issue. In the 11–12 y olds some parents were more concerned about value for money and that children had enough to eat, therefore, some parents preferred to give their children a home-packed lunch as this was considered cheaper and *'less risky'*. [27] This may be reflected in the lower decline of take-up in children from more deprived families who would be more likely to be in receipt of free school meals.

We noted variation in provision between schools and not all of the middle schools that participated in this study were fully compliant with the standards. For policy changes to be implemented effectively in schools and achieve the potential impact, support needs to be available for all stakeholders, including catering suppliers, head teachers and school catering staff. Policies affecting the provision of school food should also take account of the views of students using these facilities, [12,14] both at policy development and implementation stages. Strategies to support and guide food choice by pupils remains important; on a positive note children consuming school lunches were shown to eat a lower per cent energy from fat, saturated fat, NMES and sodium than those consuming home-packed lunches, but fewer micronutrients, which is a cause for concern. This study shows improvements are needed in the nutritional content of both school lunches and home-packed lunches. Our findings highlight a persistent need to improve dietary intake in this age group both at school and throughout the day. Across the socio-economic spectrum, children's consumption of saturated fat and NMES remain above the recommended limits, while micronutrients remain below. In 1984, Hackett et al. noted the need for a focus on nutrient density in children's diets due to falling energy intakes [33]. This remains relevant today. These findings reiterate the importance of considering the influence of the wider environment in this age group, and also, the need for both policy and societal approaches.

Acknowledgments

We thank the schools, parents/guardians and children who provided us with extensive data. Thanks to all members of the research advisory group including representatives from the Department of Health, Northumbria County Council, Newcastle primary care trust, the School Food Trust (now Children's Food Trust), and Professor Andrew Rugg-Gunn (Professor Emeritus Newcastle University) for his invaluable expertise and guidance from the inception of the Northumberland dietary surveys.

Author Contributions

Conceived and designed the experiments: AJA MW JNSM SS. Performed the experiments: SS JD. Analyzed the data: SS JNSM ES. Contributed to the writing of the manuscript: SS JD ES JNSM MW AJA.

References

1. HM Government (2007) PSA Delivery Agreement 12: Improve the health and wellbeing of children and young people. London:The Stationery Office.
2. Department of Health (2004) Choosing health: making healthy choices easier. London: Department of Health.
3. Department of Health (2005) Choosing a Better Diet: a food and health action plan. London: Department of Health.
4. Stamatakis E, Wardle J, Cole TJ (2010) Childhood obesity and overweight prevalence trends in England: evidence for growing socioeconomic disparities. Int J Obes 34: 41–47.
5. Rokholm B, Baker J, Sorensen T (2010) The levelling off of the obesity epidemic since the year 1999-A review of evidence and perspectives. Obes Rev 11: 835–846.
6. The Health and Social Care Information Centre (2012) National Child Measurement Programme: England, 2011/2012 school year. London:Depart-:Department of Health.
7. Stamatakis E, Zaninotto P, Falashetti E, Mindell J, Head J (2010) Time trends in childhood and adolescent obesity in England from 1995 to 2007 and projections of prevalence to 2015. J Epidemiol Community Health 64: 167–174.
8. Craigie A, Matthews JNS, Rugg-Gunn AJ, Lake A, Mathers JC, et al. (2009) Raised adolescent body mass index predicts the development of adiposity and a central distribution of body fat in adulthood:A longitudinal study. Obes Facts 2: 150–156.
9. Craigie A, Lake A, Kelly S, Adamson AJ, Mathers JC (2011) Tracking of obesity-related behaviours from childhood to adulthood:A systematic review. Maturitas.doi 10.1016/j.maturitas.2011.08.005.
10. Bates B, Lennox A, Prentice A, Bates C, Swan G (2008/9-2010/11) National Diet and Nutrition Survey:Headline results from Years 1, 2 and 3 (combined) of the Rolling Programme. Available: http://transparency.dh.gov.uk/2012/07/25/ndns-3-years-report/ Accessed 22nd Jan 2014.
11. Fletcher ES, Rugg-Gunn AJ, Matthews JNS, Hackett A, Moynihan PJ, et al. (2004) Changes over 20 years in macronutrient intake and body mass index in 11- to 12-year old adolescents living in Northumberland. Br J Nutr 92: 321–333.
12. Wills W, Milburn-Backett K, Gregory S, Lawton J (2005) The influence of the secondary school setting on the food practices of young teenagers from disadvantaged backgrounds in Scotland. Health Educ Res 20: 458–465.
13. Stead M, McDermott L, Mackintosh A, Adamson AJ (2011) Why healthy eating is bad for young people's health:identity, belonging and food. Soc Sci Med 72: 1131–1139.
14. Fletcher A, Jamal F, Fitzgerald-Yau N, Bonell C (2013) We've got some underground business selling junk food:Qualitative evidence of the unitended effects of English school food policies. Sociology.doi: 10.1177/0038038513500102.
15. The Education (Nutritional Standards and Requirements for School Food) (England) (Amendment) Regulations 2008 Available: http://www.opsi.gov.uk/si/si2008/pdf/uksi_20081800_en.pdf Accessed 28th Mar 2012.
16. School Food Trust: A Guide to introducing the Government's food-based and nutrient-based standards for school lunches. Available: http://www.childrensfoodtrust.org.uk/assets/sft_nutrition_guide.pdf Accessed September 2014.
17. Rogers IS, Ness AR, Hebditch K, Jones LR, Emmett PM (2007) Quality of food eaten in English primary schools:school dinners vs packed lunches. Eur J Clin Nutr 61: 856–864.
18. Rees GA, Richards CJ, Gregory J (2008) Food and nutrient intakes of primary school children: a comparison of school meals and packed lunches. J Hum Nutr Diet 21: 420–427.
19. Evans CEL, Cleghorn CL, Greenwood DC, Cade JE (2010) A comparison of British school meals and packed lunches from 1990 to 2007:meta-analysis by lunch type. Br J Nutr 104: 474–487.
20. Haroun D, Harper C, Wood L, Nelson M (2010) The impact of the food-based and nutrient-based standards on lunchtime food and drink provision and consumption in primary schools in England. Public Health Nutr 14: 209–218.
21. Golley R, Pearce J, Nelson M (2010) Children's lunchtime food choices following the introduction of food-based standards for school meals:observations from six primary schools in Sheffield. Public Health Nutr 14: 271–278.
22. Harrison F, Jennings A, Jones A, Welch A, van Sluijs E, et al. (2011) Food and drink consumption at schooltime:the impact of lunch type and contribution to overall intake in British 9-10-year-old children. Public Health Nutr doi.10.1017/s1368980011002321.
23. Pearce J, Harper C, Haroun D, Wood L, Nelson M (2011) Key differences between school lunches and packed lunches in primary schools in England in 2009. Public Health Nutr 14: 1507–1510.
24. Spence S, Delve J, Stamp E, Matthews JNS, White M, et al. (2013) The Impact of Food and Nutrient-Based Standards on Primary School Children's Lunch and Total Dietary Intake:A Natural Experimental Evaluation of Government Policy in England. PLoS One 8 (10): e78298.doi.10.1371/journal.pone.0078298.
25. Prynne C, Handford C, Dunn V, Bamber D, Goodyer I, et al. (2011) The quality of midday meals eaten at school by adolescents:school lunches compared with packed lunches and their contribution to total energy and nutrient intakes. Public Health Nutr doi.10.1017/s1368980011002205.
26. Stevens L, Nicholas J, Wood L, Nelson M (2011) Secondary school food survey 2. School lunches versus packed lunches. Available: http://www.childrensfoodtrust.org.uk/research/schoolfoodstandardsresearch/secondaryschoolfoodsurvey/secondary-school-meals-versus-packed-lunches-2011 Accessed 4th Feb 2014.
27. Adamson AJ, White M, Stead M, Delve J, Stamp E, et al. (2011) The process and impact of change in the school food policy on food and nutrient intake of children aged 4–7 and 11–12 years both in and out of school:a mixed methods approach. Available http://phrc.lshtm.ac.uk/papers/PHRC_B5-07_Final_Report.pdf.
28. Hackett AF, Rugg-Gunn AJ, Appleton DR, Parkin JM, Eastoe JE (1984) A 2-year longitudinal study of dietary intake in relation to the growth of 405 English children initially aged 11–12 years. Ann Hum Biol 11: 545–553.
29. Hackett AF, Rugg-Gunn AJ, Appleton DR (1983) Use of a dietary diary and interview to estimate the food intake of children. Human Nutr:Appl Nutr 37: 293–300.
30. Adamson AJ, Rugg-Gunn AJ, Butler TJ, Appleton DR, Hackett AF (1992) Nutritional intake, height and weight of 11–12-year-old Northumbrian children in 1990 compared with information obtained in 1980. Br J Nutr 68: 543–563.
31. Adamson AJ, Rugg-Gunn AJ, Butler TJ, Appleton DR (1996) The contribution of foods from outside the home to the nutrient intake of young adolescents. J Hum Nutr Diet 9: 55–68.
32. Rugg-Gunn AJ, Fletcher ES, Matthews JNS, Hackett AF, Moynihan PJ, et al. (2007) Changes in consumption of sugars by English adolescents over 20 years. Public Health Nutr 10: 354–363.
33. Hackett AF, Rugg-Gunn AJ, Appleton DR, Eastoe JE, Jenkins GN (1984) A 2-year longitudinal nutritional survey of 405 Northumberland children initially aged 11.5 years. Br J Nutr 51: 67–75.
34. Hackett AF, Appleton DR, Rugg-Gunn AJ, Eastoe J (1985) Some influences on the measurement of food intake during a dietary survey of adolescents. Hum Nutr:Applied Nutr 39A: 167–177.
35. Foster E, Hawkins A, Adamson AJ (2010) Young Person's Food Atlas. London:Food Standards Agency.
36. Food Standards Agency (2002) McCance and Widdowson's the Composition of Foods, Sixth Summary Edition. Cambridge:The Royal Society of Chemistry.
37. Kitchen S, Poole E, Reilly N (2013) School food:Head Teachers' and Senior Managers' Perceptions Survey. London: Department of Education.
38. Communities and Local Government Indices of Deprivation 2007. Available: http://webarchive.nationalarchives.gov.uk/+/http://www.communities.gov.uk/communities/neighbourhoodrenewal/deprivation/deprivation07/ Accessed 30th June 2011.
39. Department of Health (1991) Dietary Reference Values for Food Energy and Nutrients for the United Kingdom. London:HMSO.
40. Nelson M, Nicholas J, Wood L, Riley K, Russell S (2011) Statistical release:Take up of school lunches in England 2010–2011. Available: http://www.childrensfoodtrust.org.uk/research/annual-surveys Accessed 4th Feb 2014.
41. Pearce J, Wood L, Nelson M (2012) Lunchtime food and nutrient intakes of secondary-school pupils:a comparison of school lunches and packed lunches following the introduction of mandatory food-based standards for school lunch. Public Health Nutr 16: 1126–1131.
42. Stevens L, Nicholas J, Wood L, Nelson M (2012) School lunches v. packed lunches:a comparsion of secondary schools in England following the introduction of compulsory food standards. Public Health Nutr 16: 1037–1042.

43. Hur I, Burgess-Champoux T, Reicks M (2011) Higher Quality Intake from School Lunch Meals Compared with Bagged Lunches. Infant, Child, & Adolesc Nutr 3: 70–75.

44. Taylor J, Hernandez K, Caiger J, Giberson D, MacLellan D, et al. (2012) Nutritional quality of children's school lunches:differences according to food source. Public Health Nutr 15: 2259–2264.

45. Pearce J, Wood L, Stevens L (2013) Portion weights of food served in English schools:have they changed following the introduction of nutrient-based standards? J Hum Nutr Diet 26: 553–562.

46. Fung C, McIsaac JLD, Kuhle S, Kirk SFL, Veugelers PJ (2013) The impact of a population-level school food and nutrition policy on dietary intake and body weights of Canadian children. Prev Med 57: 934–940.

47. Bates B, Lennox A, Bates C, Swan G (2011) National Diet and Nutrition Survey:Headline results from years 1 and 2 (combined) of the Rolling Programme 2008/9-2009/10. London: Department of Health.

48. Gregory J, Lowe S (2000) National Diet and Nutrition Survey:young people aged 4 to 18 years. London:HMSO.

49. Craig P, Cooper C, Gunnell D, Haw S, Lawson K, et al. (2012) Using natural experiments to evaluate population health interventions: new Medical Research Council guidance. J Epidemiol Community Health 66: 1182–1186.

50. Pockett R, Adlard N, Carroll S, Rajoriya F (2011) Paediatric hospital admissions for rotavirus gastroenteritis and infectious gastroenteritis of all causes in England:an analysis of correlation with deprivation. Cur Med Res Opin 27: 777–784.

51. Livingstone M, Robson P, Wallace J (2004) Issues in dietary intake assessment of children and adolescents. Br J Nutr 92: S213–S222.

19

Bloom-Forming Cyanobacteria Support Copepod Reproduction and Development in the Baltic Sea

Hedvig Hogfors[1¤], Nisha H. Motwani[1], Susanna Hajdu[1], Rehab El-Shehawy[2], Towe Holmborn[3], Anu Vehmaa[4,5], Jonna Engström-Öst[4], Andreas Brutemark[4,5], Elena Gorokhova[1,6]*

1 Department of Ecology, Environment and Plant Sciences, Stockholm University, Stockholm, Sweden, 2 IMDEA Agua, Alcalá de Henares, Madrid, Spain, 3 Calluna AB, Stockholm, Sweden, 4 ARONIA Coastal Zone Research Team, Novia University of Applied Sciences & Åbo Akademi University, Ekenäs, Finland, 5 Tvärminne Zoological Station, University of Helsinki, Hangö, Finland, 6 Department of Applied Environmental Science, Stockholm University, Stockholm, Sweden

Abstract

It is commonly accepted that summer cyanobacterial blooms cannot be efficiently utilized by grazers due to low nutritional quality and production of toxins; however the evidence for such effects *in situ* is often contradictory. Using field and experimental observations on Baltic copepods and bloom-forming diazotrophic filamentous cyanobacteria, we show that cyanobacteria may in fact support zooplankton production during summer. To highlight this side of zooplankton-cyanobacteria interactions, we conducted: (1) a field survey investigating linkages between cyanobacteria, reproduction and growth indices in the copepod *Acartia tonsa*; (2) an experiment testing relationships between ingestion of the cyanobacterium *Nodularia spumigena* (measured by molecular diet analysis) and organismal responses (oxidative balance, reproduction and development) in the copepod *A. bifilosa*; and (3) an analysis of long term (1999–2009) data testing relationships between cyanobacteria and growth indices in nauplii of the copepods, *Acartia* spp. and *Eurytemora affinis*, in a coastal area of the northern Baltic proper. In the field survey, *N. spumigena* had positive effects on copepod egg production and egg viability, effectively increasing their viable egg production. By contrast, *Aphanizomenon* sp. showed a negative relationship with egg viability yet no significant effect on the viable egg production. In the experiment, ingestion of *N. spumigena* mixed with green algae *Brachiomonas submarina* had significant positive effects on copepod oxidative balance, egg viability and development of early nauplial stages, whereas egg production was negatively affected. Finally, the long term data analysis identified cyanobacteria as a significant positive predictor for the nauplial growth in *Acartia* spp. and *E. affinis*. Taken together, these results suggest that bloom forming diazotrophic cyanobacteria contribute to feeding and reproduction of zooplankton during summer and create a favorable growth environment for the copepod nauplii.

Editor: Adrianna Ianora, Stazione Zoologica Anton Dohrn, Naples, Italy

Funding: This work was financially supported by Stockholm University's strategic marine environmental research program Baltic Ecosystem Adaptive Management, the Swedish Environmental Protection Agency, SYVAB (The Southwestern Stockholm Region Sewage Company), Ivar Bendixsons Stipendiefond and the Swedish Research Council for the Environment, Agricultural Sciences and Spatial Planning (FORMAS), the Academy of Finland (projects 125251 and 255566), the Maj and Tor Nessling Foundation, Walter and Andrée de Nottbeck Foundation, and the Research and Development Institute ARONIA. The funders had no role in study design, data collection and analysis, decision to publish, or preparation of the manuscript.

* Email: elena.gorokhova@itm.su.se

¤ Current address: AquaBiota Water Research, SE-11550, Stockholm, Sweden

Introduction

Toxic blooms of filamentous cyanobacteria are proliferating worldwide due to the climate change and eutrophication [1]. These cyanobacteria are commonly considered to impair survival, growth and reproduction of grazers [2]. The negative effects of cyanobacteria on zooplankton are usually related to a combination of (*i*) low nutritional value due to inadequate dietary fatty acid composition [3,4], (*ii*) production of toxins and feeding deterrents [5,6], and (*iii*) poor manageability of the colonies [7]. However, there are also studies showing that cyanobacteria have neutral or positive effects on zooplankton egg production and growth [8–12]. Application of stable isotopes, fatty acids and DNA-based methods

[13–16] for zooplankton diet analysis suggest relatively high *in situ* grazing on colony-building cyanobacteria in freshwaters and estuaries, where summer cyanobacterial blooms are a regular feature. One can speculate that when preferred prey is scarce, zooplankton may increase feeding on the abundant cyanobacteria, despite their inadequate biochemical composition and toxicity. We know, for example, from terrestrial ecology that rainforest-dwelling parrots consume toxic foods during dry seasons when other, non-toxic, food is limited [17]. Supporting these contradictory reports, meta-analysis of cyanobacteria effects on various grazers [18,19] suggest that cyanobacterial effects on biota are multifactorial and species- and system-specific [20].

In the Baltic Sea, summer blooms of diazotrophic filamentous cyanobacteria are dominated by *Aphanizomenon* sp., *Nodularia spumigena* and *Dolichospermum* spp. [21]. Biomass of Baltic mcsozooplankton peaks during the same period as that of cyanobacteria, with calanoid copepods being the most important group [22,23]. The evidence is accumulating that cyanobacterial blooms may have a more important role in the Baltic food webs than previously assumed, and it has been suggested that utilization of cyanobacteria as a food source by zooplankton in this system is underestimated [24]. For example, a strong nitrogen isotopic signal in of various size following summer bloom in the northern Baltic Sea [25], indicates that nitrogen (N) fixed by diazotrophic cyanobacteria is directly or indirectly utilized in the food web. Indeed, using both isotopic signals [26] and cyanobacterial pigments as tracers, copepods in the Baltic proper have been found to assimilate cyanobacteria in measurable quantities [24]. Also, lipid signal in the copepod *Pseudocalanus acuspes* in the southern Baltic Sea indicates that cyanobacteria contribute substantially to its diet during summer [27]. Finally, studies employing DNA-based analysis of stomach content provide unequivocal evidence that Baltic copepods and mysids ingest filamentous cyanobacteria *in situ* even when alternative food is present [14].

Similarly, effects of filamentous cyanobacteria on survival and development are often contradictory, with both negative [12,28] and positive [9,10] effects being observed. As a monospecific food, cyanobacteria do not seem to support neither egg production [11,28,29] nor egg hatching [28] in copepods, with nodularins and microcystins being the most commonly implicated in these effects. These hepatotoxins have been reported to cause oxidative stress in various invertebrates and fish [30,31] by increasing formation of reactive oxygen species, decreasing the antioxidant capacity (e.g., inhibiting detoxification enzymes) and oxidation of macromolecules (proteins, lipids and DNA). However, when offered in mixtures with alternative prey, filamentous cyanobacteria may have positive effects on copepod egg production, hatching and juvenile development [9,11,32]. Due to this conflict of information, more studies on the occurrence and mechanisms of these effects in ecologically relevant settings are needed.

Here, we explored linkages between Baltic Sea diazotrophic filamentous cyanobacteria and fitness-related responses in copepods: reproductive output, juvenile development and growth, and oxidative status. To evaluate these connections, we conducted a set of interrelated studies in the northern Baltic proper. First, in a field survey, we used a correlative approach to relate reproduction (egg production and their viability) of the copepod *Acartia tonsa* to phytoplankton community structure, with specific focus on the effects of the bloom-forming cyanobacteria. Then, we conducted a follow-up laboratory experiment, where ingestion of the cyanobacterium was measured in *A. bifilosa* using molecular diet analysis and related to the copepod oxidative status, and recruitment. Finally, we analyzed long-term data on growth indices in nauplii of *Acartia* spp. and *Eurytemora affinis* in relation to cyanobacteria bloom intensity.

Methods

Ethics Statement

The sampling was conducted within Swedish National Marine Monitoring in the Baltic Sea and no specific permissions were required for the sampling locations of this study. Also, we did not require ethical approval to conduct this research as no animals considered in any animal welfare regulations and no endangered or protected species were involved in either field or experimental studies.

Summer field survey

Study sites. Sampling was conducted biweekly during cyanobacterial bloom (July 4 to September 26, 2007) at three stations in the north-west Baltic proper (58°49'N, 17°39'E; Figure 1). Two of these stations are located in the Himmerfjärden Bay (stns H2 and H4; SYVAB's marine monitoring program) and stn B1 (Swedish National Marine Monitoring Program, SNMMP) is outside the bay (Figure 1).

Sampling. Sea surface temperature (SST) was measured on each sampling occasion. Integrated phytoplankton samples were collected by a hose (inner diameter 19 mm) at stns H2 and H4 (0–14 m) and stn B1 (0–20 m), and preserved with acidic Lugol's solution. Using an inverted microscope (Wild M40), cells counts (>2 μm) and biovolume analysis were conducted according to HELCOM monitoring guidelines [33,34]; see also *http://www.ices.dk/marine-data/Documents/ENV/PEG_BVOL.zip*. Based on size measurements, phytoplankton were divided into size fractions of 2–5 μm, 5–15 μm, 15–30 μm and >30 μm (filamentous cyanobacteria excluded); each species of filamentous cyanobacteria was treated as a separate category regardless of size.

Copepods were sampled by vertical hauls (0–10 m) using a WP2 net (diameter 58 cm, mesh size 200 μm) and brought to the laboratory in 20 L insulated containers. Using a wide-mouth pipette, adult *A. tonsa* females were sorted upon arrival to the laboratory; the choice of species was based on relative abundance of copepod species in the samples and availability of adult females. They were incubated individually for 24 h in darkness at ambient (±2°C) temperature, in microplate wells (12 wells; Corning Costar, Corning NY, USA) filled with 5 mL of 6-μm filtered seawater. The mortality of the females during the incubation was <5%.

Reproductive variables. The egg production rate (EPR; eggs female^{-1} day^{-1}) was recorded by counting number of eggs in each well under a stereomicroscope (Wild Heerbrugg, 6–50×). No crumbled or empty egg shells were found, indicating absence of egg cannibalism. Live females were individually placed in Eppendorf tubes, frozen at −80°C and stored for a few weeks before RNA analysis. For egg viability (EV%; percentage of viable eggs) analysis, all eggs from each well were transferred to a depression slide and stained with TO-PRO-1 iodide (Molecular Probes) [35]. The viable egg production rate (VEPR; viable eggs female^{-1} day^{-1}) was calculated by multiplying EV% with EPR.

RNA quantification. As a proxy for growth rate, individual RNA content of the females was used [36] measured by the high-range RiboGreen (Molecular Probes, Inc., Eugene, OR, USA) assay after extraction with N-laurylsarcosine followed by RNase digestion [37]; fluorometer FLUOstar Optima (BMG Labtechnologies; 485 nm for excitation and 520 nm for emission) and black flat bottom microplates (Greiner Bio-One GmbH) were used. Before the analysis, the females (PL, prosome length mm) were measured using an inverted microscope (80×; Wild 40, Heerbrugg) with an ocular micrometer.

Laboratory experiment

Study animals and algal cultures. In August 2010, copepods were collected, with a 150 μm plankton net in the Storfjärden Bay, Western Finland (59°51'20"N, 23°15'42"E; Figure 1). Adult females and males of *Acartia bifilosa* which dominated the copepod community during the study period and area, were gently sorted under a stereo microscope and placed in 1.2 L bottles. Cultures of the green alga *Brachiomonas submarina*

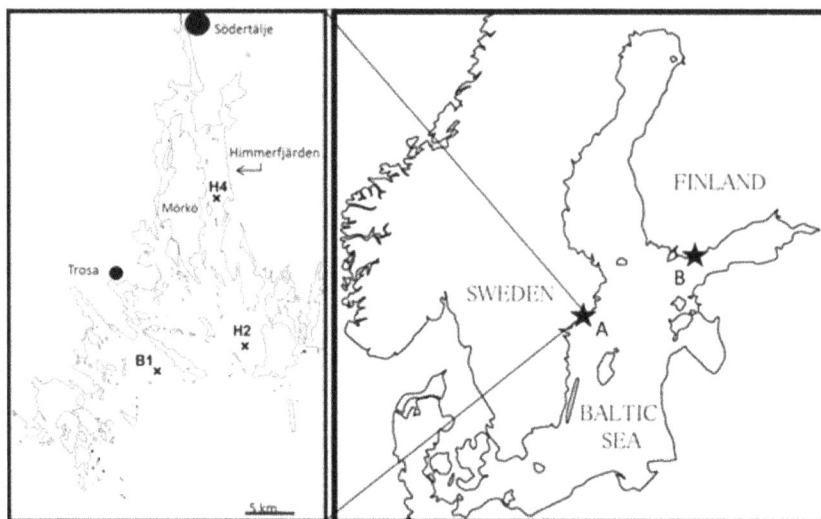

Figure 1. Schematic map over the Baltic Sea and the sampling sites. (A) Himmerfjärden Bay (sampling stations H2 and H4) and a coastal area near Askö laboratory (station B1) in the western part of the northern Baltic proper, where field data for the summer field survey and the long term data (stn H4 and B1) were collected, and (B) Storfjärden Bay, at the entrance to the Gulf of Finland, where study animals for the laboratory experiment were collected.

(strain TV15, collection of Tvärminne Zoological Station, University of Helsinki) and the cyanobacterium *Nodularia spumigena* (strain AV1 obtained from Prof. K. Sivonen, University of Helsinki) were grown under irradiance of 13.7 µmol photons m^{-2} s^{-1} for 16 h a day, and at 18°C, in f/2 medium (without silica) and Z8 (without nitrogen), respectively [38].

Experimental set-up. The copepods (17 females and 3 males per bottle, 12 replicates per treatment) were incubated in two alternative feeding regimes, with or without *N. spumigena* in the media, at 20°C. In the treatments without cyanobacteria, a monoculture of *B. submarina* (1061±87 µg C L^{-1}; average ± SD) was used as a sole food, and in the treatments with cyanobacteria, *B. submarina* (971±208 µg C L^{-1}) were mixed with *N. spumigena* (102±18 µg C L^{-1}); this mixture approximates average *N. spumigena* contribution to the summer phytoplankton community in the northern Baltic proper (Figure 2). The bottles were incubated on a plankton wheel (1 rpm), with 16:8 h light:dark cycle. Prior to the experimental incubation, the copepods were acclimatized under the same conditions for 36 h. At the end of the experiment, the copepods were filtered through a 120 µm mesh and examined under a stereo microscope. Live copepods were transferred to Eppendorf tubes and stored at −80°C for stomach content analysis using qPCR (quantitative real-time polymerase chain reaction) and oxidative status analysis. To get sufficient amount of material for these analyses, all material collected from consecutive replicates were pooled two and two within a treatment, giving 6 replicates for each treatment. Eggs and nauplii from each bottle were collected with a 48 µm mesh, stored overnight at 3°C in dark, and used to estimate EPR, EV% and development index (DI).

Sample preparation for molecular and biochemical assays. To determine amount of *N. spumigena* in the copepod gut content and to characterize oxidative status of the copepods, adult animals recovered from the incubation bottles were placed into microcentrifuge tubes with 0.7 mL phosphate buffer saline (PBS) and 100 µm glass beads and homogenized for 4 minutes using FastPrep with cooling. The tubes were thereafter centrifuged at 4°C for 5 min with 10 000×*g*.

Molecular diet analysis. For DNA extraction, 50 µL of the homogenate were used by mixing with 50 µL 20% Chelex [39]. These samples were heated at 105°C for 30 min, centrifuged and the supernatant was used to quantify *N. spumigena* using qPCR. The *Nodularia*-specific primers were used to amplify ~200 bp fragment of *N. spumigena* 16S rDNA in the guts of copepods [15]. All qPCR reactions were performed in a final volume of 25 µL reaction mixture with a StepOne real-time cycler (Applied Biosystems) using the KiCqStart SYBR Green qPCR Ready Mix (Sigma). To prepare a standard curve of five step 10-fold dilutions (8.2 to 8.2×10^{-4} ng), DNA extracted from a culture of *N. spumigena* was used [40]; the linearity of the standard curve was high (R^2>0.99), with amplification efficiency of 95–100%. As a reference sample, the feeding media containing *N. spumigena* but no copepods following the same procedure as for the test samples was used. A standard curve and no template controls (water) in triplicates were included in all runs. To check for non-specific products, DNA melt-curve analysis was performed after each qPCR experiment.

Oxidative stress biomarkers. To measure the intracellular soluble antioxidant capacity, the homogenized copepod samples were analyzed using oxygen radical absorption capacity (ORAC; µM trolox equivalents) assay [41]; see Protocol S1. To measure oxidative damage, the lipid peroxidation assay using Quanti-Chrom thiobarbituric acid reactive substances (TBARS; mol MDA Assay Kit; DTBA-100; BioAssay Systems, USA) was used following the manufacturer's directions. The ratio between ORAC and TBARS (ORAC:TBARS ratio) was used as a proxy for oxidative balance [32].

Reproductive state variables. To estimate copepod reproductive output, the female EPR, the EV% and early nauplia development (development index; DI) were used. To determine EPR, all eggs were counted and related to the number of live females in the corresponding bottle. For EV% measurements, ~50 eggs per bottle were analyzed with TO-PRO-1 iodide staining [35]. The remaining eggs and nauplii were preserved with acidic Lugol's solution for calculating DI, which incorporates survival and metamorphosis success in copepods:

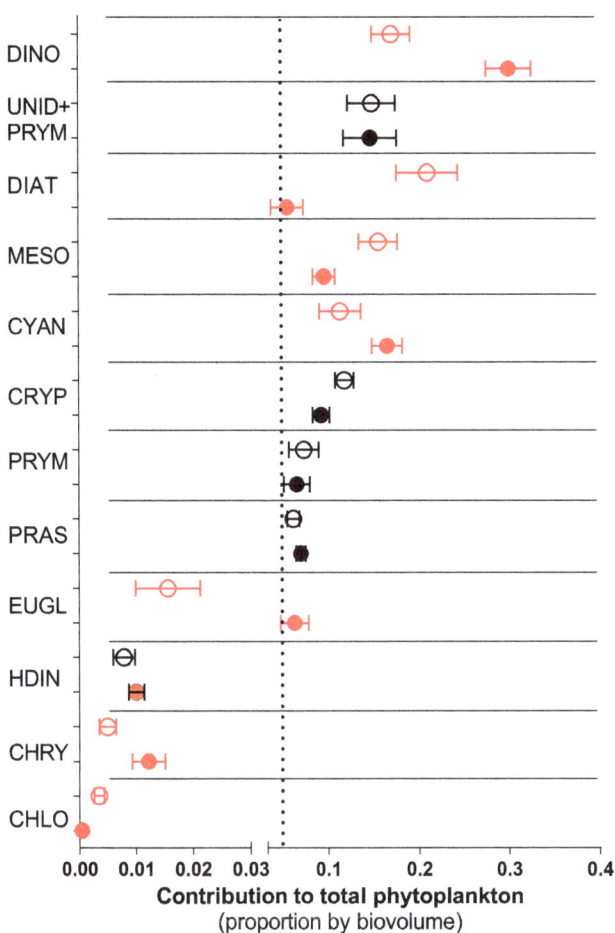

Figure 2. Contribution of the main taxonomic groups to the phytoplankton communities (by biovolume) at stations B1 (closed symbols) and H4 (open symbols) in June-August (mean ± SD; years 1986–2009). Significant differences between the stations (paired t-test, p<0.05) are indicated in red. Dotted line indicates 5% threshold for including a phytoplankton group in the GLM analysis of the field survey study.

$$DI = \frac{\sum k_i \times n_i}{NS},$$

where k_i is assigned stage value (in our study: 0 for egg, 1 for nauplii NI, and 2 for nauplii NII; no nauplii had developed beyond NII), n_i number of copepods at that stage, and NS – total number of individuals [42]. All nauplii from the incubations (i.e., Lugol-preserved samples for DI-analysis, and nauplii hatched in the TO-PRO-stained samples) were included in the EV calculation. VEPR was estimated by multiplying EV and EPR.

Long term data analysis

Zooplankton collected in the northern Baltic proper within SNMMP and SYVAB's marine monitoring program in Himmerfjärden Bay (Himmerfjärden Eutrophication Study; www.2.ecology.su.se) were used to study effects of various phytoplankton groups and physical factors on copepod growth and recruitment, with particular focus on the effects of cyanobacteria. As a proxy for copepod growth during early life stages, we used RNA:DNA ratio in the nauplii (feeding stages, NIII–NVI) of Acartia spp. (A.

bifilosa, A. tonsa and A. longiremis) and Eurytemora affinis. In the study areas, A. bifilosa dominate in summer Acartia communities, contributing 62 to 94% on the long-term basis (E. Gorokhova, pers. obs.). Specifically, we considered that biotic (biovolumes of total phytoplankton, phytoplankton excluding filamentous cyanobacteria, and specific phytoplankton groups) and abiotic (North Atlantic Oscillation [NAO] indices, SST and salinity) variables integrate local environmental variability, and could define an adequate environmental framework for copepod growth.

Sampling. We used 34 and 36 samples for stations H4 and B1, respectively, collected every other week in July and August, i.e., when cyanobacteria are abundant in the study area, during 1999–2009, resulting in 3–4 samples per station and year. On each sampling occasion, zooplankton and phytoplankton were collected in concert; salinity and SST were measured by CTD (Meeresstechnik Elektronik GmbH). Phytoplankton samples were collected as integrated hose samples, preserved and analyzed in the same way as in the summer field survey [33,34]. Zooplankton samples were taken by vertical tows from near bottom to surface using a WP2 net (diameter 57 cm, 90 µm mesh size). From each tow, randomly selected zooplankton were preserved in bulk using RNAlater and stored for 12 to 24 months at −20°C until the nucleic acid analysis [43].

RNA:DNA ratio. RNA and DNA contents were measured in nauplii (NIII–NVI) of Acartia spp. and E. affinis. The younger nauplial stages (NI–NII) were considered non-feeding [44] and thus not included in the analysis. Individual specimens were picked from the RNAlater preserved samples, rinsed and transferred to Eppendorf tubes (5–7 ind. sample^{-1}); the two copepod genera were treated separately. Nucleic acids were quantified with fluorometric high-range RiboGreen (Molecular Probes, Inc., Eugene, OR assay) using the same instrumentation as for RNA analysis in the summer field survey [37]. Mean standard curve slope ratio (mDNA/mRNA) was 1.87.

Data analyses

The environmental data and phytoplankton biovolume data used in all analyses are available from www.smhi.se (SHARK database) and www.2.ecology.su.se/dbhfj/index.htm, and data on copepod growth- and reproduction-related variables are provided as Supporting Information (Summer field survey: Table S1; Laboratory experiment: Table S2; and Long-term data analysis: Table S3).

To evaluate effects of feeding environment with particular focus on cyanobacteria on copepod reproduction, growth and oxidative status, Generalized Linear Models (GLMs) with normal error structure and either identity or log-link function were applied using STATISTICA 10 (StatSoft, Inc., 2010). For all GLMs, the response variables and biovolume data were Box-Cox transformed to approach normal distribution. Akaike Information Criterion (AIC) was used to optimize the number and combination of predictive variables included. To validate the models, the Wald statistic was used to check the significance of the regression coefficient for each parameter, a likelihood ratio test was used to evaluate the statistical significance of including or not including each parameter and model goodness of fit was checked using deviance and Pearson χ^2 statistics. Residual plots for each model were assessed visually to exclude remaining unattributed structure indicative of a poor model fit.

Summer field survey. GLMs were used to examine relationships between the dependent variables (reproductive/growth variables: VEPR, EPR, EV% and RNA) and the explanatory variables (biovolumes of filamentous cyanobacteria species [Aphanizomenon sp., N. spumigena and Dolichospermum

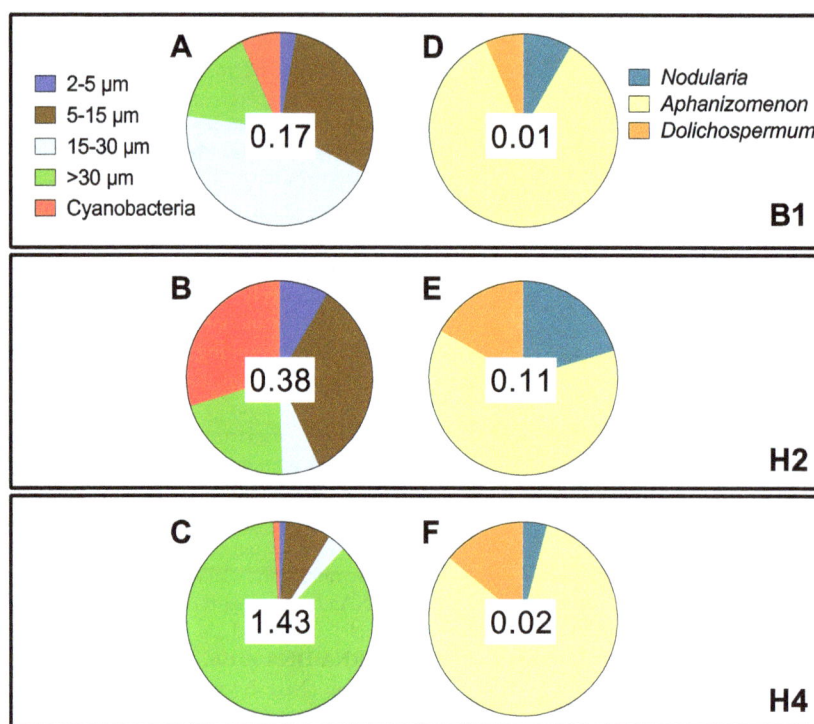

Figure 3. Composition of phytoplankton (A–C) and cyanobacterial (D–F) communities during the study period at: (A, D) stn B1, (B, E) stn H2, and (C, F) stn H4. Number in the middle of each pie chart indicates total biovolume (mm^3 L^{-1}) of the contributing categories.

spp.], autotrophs in each size fraction [2–5 μm; 5–15 μm; 15–30 μm and>30 μm], total phytoplankton biovolume, salinity and SST]. Copepod PL was also included as independent variable for RNA and EPR [45].

Laboratory experiment. Based on the positive effects of *N. spumigena* observed in the summer field survey and recent reports on high levels of antioxidants in cyanobacteria [46] as well as stimulation of antioxidant defenses in grazers feeding on hepatotoxic cyanobacteria [47], GLMs were used to test whether copepod reproductive output (EPR, EV%, VEPR and DI), antioxidant levels (ORAC), lipid peroxidation (TBARS), and oxidative balance (ORAC:TBARS ratio) were positively affected by grazing on *N. spumigena*. The amount of *Nodularia* in copepod guts determined by qPCR was used as a measure of grazing.

Long term data analysis. GLMs were used to examine relationships between the nauplial RNA:DNA ratio and biovolume of each phytoplankton group and abiotic variables that have been reported to affect zooplankton in the Baltic Sea (NAO, SST and salinity). Wilcoxon signed rank test was used to determine the significance of differences in nauplial RNA:DNA ratio between the stations (B1 and H4) and between the months (July and August) within the stations, whereas paired t-test was used to evaluate these differences in phytoplankton. The models were constructed for each taxa (*Acartia* spp. and *E. affinis*) and station (B1 and H4), because of the significant differences in RNA:DNA ratio between the species and differences in phytoplankton community structure between the stations. Only those phytoplankton groups that contributed>5% to the total phytoplankton biovolume were considered (Figure 2); these were: (1) dinoflagellates, (2) pooled group of small (<10 μm) unidentified flagellates and Prymnesiophyceae species, (3) diatoms, (4) *Mesodinium rubrum* (autotrophic ciliate), (5) nitrogen fixing filamentous Cyanobacteria, (6) Crypto-

phyceae, (7) large (>10 μm) Prymnesiophyceae, and (8) Prasinophyceae. We also included biovolumes of total phytoplankton and phytoplankton without cyanobacteria. The monthly NAO index values were taken from the Climate Prediction Center, Washington, DC (ftp://ftp.cpc.ncep.noaa.gov/wd52dg/data/indices/nao_index.tim). A high, positive winter NAO index (wNAO; December – March) indicates mild and rainy winters, while low, negative values occur during cold winters over Europe. A warmer winter is generally followed by a warmer spring (sNAO; March – May) with early onset of the growth season.

Results

Summer field survey

Phytoplankton communities differed among the stations, with the highest total phytoplankton (1.43 mm^3 L^{-1}) and highest contribution of filamentous cyanobacteria (29%) observed at stns H4 and H2, respectively (Figure 3). All cyanobacterial communities dominated by *Aphanizomenon* sp. (62–85% of total cyanobacteria; Figure 3). There were significant, albeit weak, negative correlations between the cyanobacteria and the two largest size classes of phytoplankton, 15–30 μm and>30 μm (Pearson r: −0.29 and −0.36 respectively; $p<0.05$), and a significantly positive correlation between the cyanobacteria and the smallest phytoplankton size class (2–5 μm; Pearson r: 0.38; $p<0.05$), whereas no correlation was found for the intermediate size class (5–15 μm; Pearson r: 0.13; $p>0.05$). Among the cyanobacteria, *Aphanizomenon* sp. had the highest significant negative correlation with phytoplankton>15 μm (Pearson $r = −0.38$; $p<0.05$) and *N. spumigena* had a moderate positive correlation with phytoplankton 2–5 μm (Pearson $r = 0.53$; $p<0.05$).

The maximum values for EPR and individual RNA content were observed in the first half of July, whereas their minima

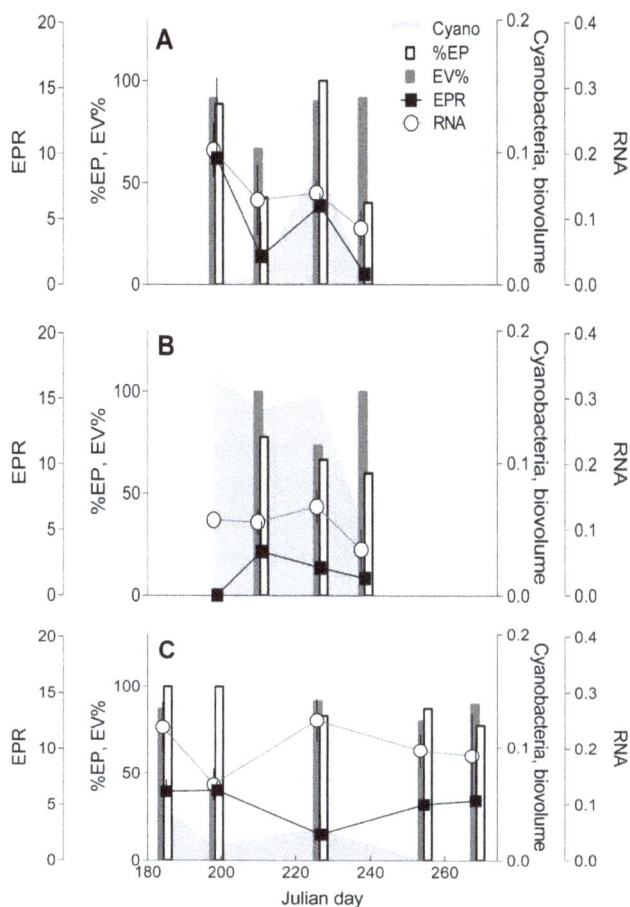

Figure 4. Seasonal dynamics of reproductive and growth variables in the copepod _Acartia tonsa_ in relation to cyanobacteria bloom progression (biovolume, mm^3 L^{-1}) at stations B1 (A), H2 (B) and H4 (C). Egg production rate (EPR; eggs female^{-1} day^{-1}), egg viability (EV%; % viable eggs), percentage of females producing eggs during 24-h incubation (%EP) and individual RNA content (μg ind^{-1}).

occurred in the end of August. The opposite trend was observed for EV%, with generally low values in July and high in the end of August (Table S4, Figure 4). The GLMs indicated that _N. spumigena_ was a significant positive predictors for all reproductive variables (EPR, EV% and VEPR) but not the RNA content (Table 1). _Aphanizomenon_ sp., on the other hand, significantly affected only EPR (stns B1 and H4) and the effect was negative. No statistically significant model for EPR was found for stn H2, which had significantly lower EPR compared to stns B1 and H4 (Table 1). Significant positive relationships were observed between phytoplankton 15–30 μm and all response variables. Also, phytoplankton>30 μm was significantly positively related to VEPR and RNA, albeit the relationships were relatively weak (Table 1).

Laboratory experiment

In copepods, incubated in the presence of _N. spumigena_ in the feeding media, amount of the cyanobacterium per stomach estimated by qPCR varied ~3-fold (0.47 to 1.32 ng _Nodularia_ DW ind.$^{-1}$); none of the controls were positive (Table S4). The amount of _N. spumigena_ per copepod had significant positive effects on all response variables tested, except EPR where it was

negative (Table 2). The concurrent opposite effects on EPR and EV% resulted in no significant influence on VEPR (Table 2). Also, the significant positive effect of _N. spumigena_ on ORAC in combination with negative but not significant effect on TBARS resulted in the significant positive effect on the ORAC:TBARS ratio (Table 2).

Long term data

Significant differences were found between the stations in terms of the phytoplankton community structure over the years (Figures 2 and 5). Cyanobacteria contribution to the total phytoplankton biovolume during the bloom period (July-August) varied from less than 1%, with the lowest values observed in 2009 at both stations to 40% at B1 and 45% at H4, in 2007 and 2003, respectively. Other important phytoplankton groups were diatoms (up to 39% of the total biovolume), dinoflagellates (up to 47%) and prymnesiophyceans (up to 40%). Total phytoplankton biovolume also varied both over time and between the stations, reaching its peak in 2008–2009 and 2007–2008 at B1 and H4, respectively (Figure 5), with significantly higher values at stn H4 compared to stn B1 (Wilcoxon signed rank; $p<0.0017$, $n = 33$). Moreover, the nauplial RNA:DNA ratio was also significantly higher at stn H4 than at stn B1 (Wilcoxon signed rank; _Acartia_ spp.: $p<0.004$; _E. affinis_: $p<0.002$). Therefore, to describe responses of nauplial RNA:DNA ratio (Figure 6) to variations in phytoplankton groups over time (Figure 5), the GLMs were fit for each station separately.

For _Acartia_ spp., only models based on filamentous cyanobacteria, diatoms, cryptophyceans, and abiotic parameters (SST, salinity, sNAO and wNAO) were significantly predictive. For _E. affinis_, only models incorporating total phytoplankton, filamentous cyanobacteria, diatoms and SST were significant (Table 3). Thus, in both species, the RNA:DNA ratio was significantly positively associated with the amount of cyanobacteria (Figure 6); moreover, parameter estimates indicated that these effects were the strongest in all models that included cyanobacteria (Table 3). Effects of other phytoplankton groups (cryptophyceans, total phytoplankton and diatoms) were all negative (Table 3). At stn B1, these groups correlated significantly negatively with cyanobacteria (Pearson r; -0.56, -0.44 and -0.62 for cryptophyceans, total phytoplankton and diatoms, respectively; Table S5). At stn H4, the effects of diatoms were also negative, but no significant correlations between the cyanobacteria and other phytoplankton groups were observed (Table S5).

Among the abiotic parameters, wNAO and sNAO indices were repeatedly indicated as significant, with positive effects of wNAO and negative of sNAO (Table 3). In addition, salinity had positive effects on RNA:DNA ratio in _Acartia_ spp. and negative in _E. affinis_, whereas SST effect was significant only for _Acartia_ spp. at B1 (Table 3).

Discussion

Both positive and negative effects of Baltic filamentous cyanobacteria on copepod biochemical and physiological responses were observed in the field and laboratory settings (Table S6). Moreover, these effects were species-specific and differed, in some cases, among the studies. However, contrary to the widely reported harmful effects of these cyanobacteria on the copepod reproduction [2], no negative effects on the net reproductive output, i.e. viable egg production, were found. In fact, viable egg production in _A. tonsa_ was positively related to _N. spumigena_ abundance in the field (Table 1). Although no effect of _N. spumigena_ on viable egg production was observed in the laboratory experiment with _A. bifilosa_, the development of nauplii

Table 1. GLM results of the summer field study.

Response variable	Stn	Explanatory variables	Beta	p	R^2	Adj. R^2	SE estimate
EPR	B1, H4*	*Nodularia*	124.5	<0.0001	0.39	0.35	3.29
		Aphanizomenon	−18.6	<0.0001			
		2–5 µm	42.1	0.001			
		15–30 µm	7.6	0.0001			
EV%	All	*Nodularia*	7.0	0.006	0.28	0.19	0.21
				0.008			
VEPR	All	*Nodularia*	7.6	<0.0001	0.39	0.35	3.29
		15–30 µm	6.4	0.034			
		>30 µm	1.3	<0.0001			
				0.004			
RNA	All	15–30 µm	2.1	<0.0001	0.64	0.62	0.04
		>30 µm	0.6	<0.0001			
		PL	0.1	<0.0001			
				0.001			

Response variables: egg production rate (EPR; eggs female^{-1} day^{-1}), egg viability (EV%; % viable eggs), viable egg production rate (VEPR; viable eggs female^{-1} day^{-1}), individual RNA content (µg ind^{-1}). Explanatory variables: biovolume of different size fractions of phytoplankton and cyanobacteria species, copepod prosome length (PL) and station (B1: $n = 28$; H2: $n = 29$; and H4: $n = 33$). When a significant station effect was found, separate models were fitted for stations without significant differences. In EV% analysis, only females that produced eggs during incubation were included ($n = 58$).
*No significant model was found for stn H2.

Table 2. GLM results of the experimental data.

Response variable	Estimate	Wald stat	p
EPR	−0.686	13.73	**0.0002**
EV%	1.560	10.643	**0.0011**
VEPR	−0.293	1.783	0.1817
DI	1.687	4.279	**0.0385**
ORAC	0.004	7	**0.0101**
TBARS	−0.814	2.391	0.1220
ORAC:TBARS	0.00003	4.299	**0.0381**

All models tested effects of grazing on *Nodularia spumigena* measured as amount of the cyanobacterium in the copepod guts (ng *Nodularia* DW ind^{-1}) on reproductive output (EPR, EV%, and VEPR), nauplial development (DI), antioxidant (ORAC) and lipid peroxidation (TBARS) levels as well as the resulting oxidative balance (ORAC:TBARS ratio). Abbreviations and units for all response variables as in Tables 1 and 2. Significant models are in bold face.

(non-feeding stages) was significantly advanced if their mothers were feeding on the cyanobacterium, thus effectively increasing recruitment (Table 2). Also, the long term data analysis also identified diazotrophic filamentous cyanobacteria as significant positive growth predictors for the copepod nauplii (Table 3).

Specific mechanisms behind the observed stimulating and suppressing effects of diazotrophic filamentous cyanobacteria on copepod reproduction and early development are largely unknown. We suggest three, not mutually exclusive, pathways involved in stimulating effects of cyanobacteria on copepod reproduction: (1) direct increase of macro- and micronutrient intake by copepods that use cyanobacteria as an additional food source; (2) supplementing copepod's diet with phytochemicals, such as polyphenols, vitamins and various antioxidants, that may enhance physiological responses [46], including hormesis effect, which is a generally favorable biological response to low exposures of toxins and other stressors [47]; and (3) indirect increase of nutrient intake by stimulation of the microbial loop through cyanobacterial exudates, resulting in increase of bacterio- and nanoplankton prey for smaller copepods [48]. The qPCR-based gut content analysis in the laboratory experiment confirmed that the copepods were actively feeding on *N. spumigena* (Table S4), and thus, the observed variations in the reproductive variables are – at least partly – the result of the direct grazing. Some filamentous

cyanobacteria could be nutritionally valuable for grazers, particularly in combination with other food sources [49]. Moreover, as pointed out by Jiang and colleagues [50], "*a putatively harmful alga is not always deleterious to grazers, and its ecological effects may be distinctly different during bloom and non-bloom periods*". They found that toxicity and nutritional value of the dinoflagellate *Cochlodinium polykrikoides* went from deleterious to beneficial for *A. tonsa* when the algal density decreased. Cyanobacteria may contain complimentary nutrients and microelements, e.g. amino acids, antioxidants, vitamins, proteins, phosphorus and nitrogen [3,49,51,52] that are of particular value for copepods. Nitrogen, for instance, is frequently depleted in the pelagia during summer in the Baltic Sea [53] and can thus be limiting for copepod reproduction [54]. Therefore, feeding on nitrogen fixing cyanobacteria may relax this limitation.

In addition to the positive effects on egg viability and nauplial development, the presence of *N. spumigena* increased the antioxidant defenses and improved oxidative status as indicated by the ORAC:TBARS ratio (Table 2). This is in contrast to other studies implicating hepatotoxic cyanobacteria as pro-oxidants for grazers [31]. However, the nature of cyanobacteria-induced oxidative stress is poorly understood; in fact, cyanobacteria produce dietary antioxidants that may directly improve antioxidative capacity of consumers [52]. Moreover, microcystins have

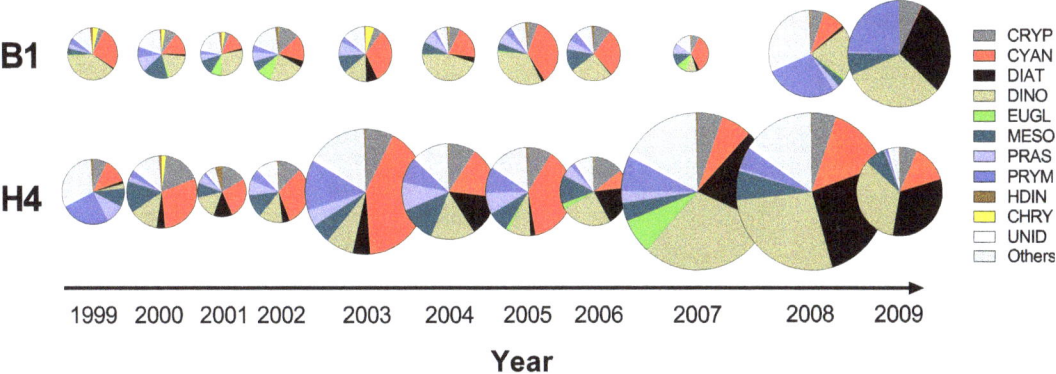

Figure 5. Long-term (1999–2009) dynamics of phytoplankton community structure at stations B1 (A) and H4 (B). Data are averages for July and August, weeks 27 to 35 (e.g., period of the summer cyanobacterial bloom, $n = 3$ or 4). Circle diameter is proportional to the average total phytoplankton biovolume (mm^3 L^{-1}) observed during the same period (0.43 ± 0.15, $n = 70$). Abbreviations: CHLO – Chlorophyceae, CHRY – Chrysophyceae, CRYP – Cryptophyceae, CYAN – Cyanophyceae, DIAT – Diatoms, DINO – Dinophyceae, EUGL – Euglenophyceae, MESO – *Mesodinium rubrum* (Myrionecta rubrum), PRAS – Prasinophyceae, PRYM – Prymnesiophyceae, UNID – unidentified flagellates, and HDIN – heterotrophic dinoflagellates.

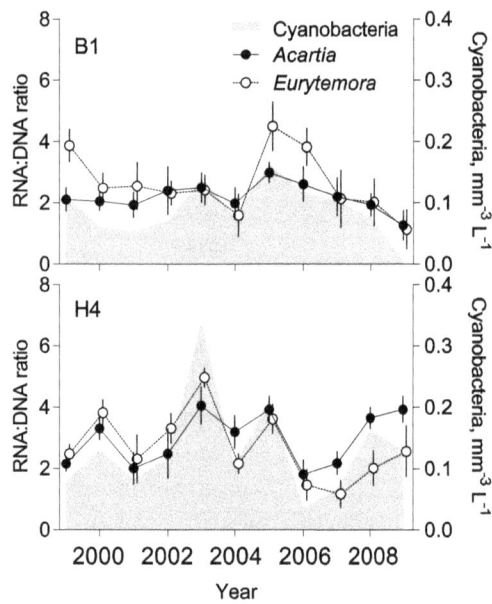

Figure 6. Long-term (1999–2009) dynamics of RNA:DNA ratio (mean ± SD, n=3 or 4) in the nauplii (NIII-NVI) of copepods *Acartia* spp. and *Eurytemora affinis* in relation to cyanobacteria biovolume at stations B1 (A) and H4 (B). Data are averages for July and August, weeks 27 to 35.

been shown to increase activity of antioxidant enzymes in estuarine crabs [55]. The observed increase in antioxidant capacity may be indicative of the hormetic response to the cyanobacterial toxins and bioactive compounds, known as *The Xenohormesis Hypothesis* [56]. The latter has been suggested to be an evolutionary adaptation to sustain fitness in a changing environment, where the presence of a toxin in low concentrations acts as a signal for the organism to mobilize metabolic reserves to prepare itself for higher environmental stress [47,56,57]. Some of phytochemicals have evolved as toxins to intimidate grazers, triggering adaptive stress responses by, for instance, stimulating the production and/or activity of antioxidant enzymes in the consumer [56,58]. Nodularins and microcystins produced by *N. spumigena* and *Dolichospermum* spp., respectively, could have contributed to the observed positive linkages between these cyanobacteria and copepod egg viability. One can speculate that an increase in the antioxidant capacity of a female triggered by a hepatotoxin and/or another secondary metabolite could enhance allocation of maternal antioxidants to the eggs, which improved embryonic and post-embryonic development.

The positive linkages between nauplial growth inferred from the RNA:DNA ratio and total diazotrophic filamentous cyanobacteria in the long term data set (Table 3) are not likely to be due to the grazing on cyanobacteria by the nauplii. Although some grazing on *N. spumigena* colonies by copepod nauplii has been observed [59], the pathway involving fueling of microbial communities is much more likely. During summer, when inorganic nitrogen is depleted, bloom-forming cyanobacteria are responsible for a significant proportion of N_2-fixation in the Baltic Sea [60–63]. As much as one third of the fixed N_2 leaks out as ammonium NH_4^+ from these cyanobacteria [62,63] and is further utilized by other organisms, such as heterotrophic bacteria and picoautotrophs [64]. The microbial loop is also stimulated by dissolved organic matter (DOM) and detritus derived from cyanobacteria and utilized by bacteria [64,65]. Nauplii and, to a lesser extent,

copepodites, graze on pico- and bacterioplankton [48,65,66], which also support dietary nano- and microzooplankton, and these communities flourish in association with *N. spumigena* [67]. Therefore, N-leakage from diazotrophic cyanobacteria, which stimulate microbial prey, could explain the positive association between nauplial growth and filamentous cyanobacteria during summer blooms (Table 3). Remarkably, cyanobacteria were the only phytoplankton group positively related to the nauplial RNA:DNA ratio (Table 3, Figure 6). Although cryptophyceans, total phytoplankton and diatoms were all negatively associated with nauplial growth at stn B1 (Table 3), these negative effects may, at least partially, be explained by the negative cross-correlations between these groups and the cyanobacteria (Table S5); see also [68]. However, the negative effect of the diatoms on the nauplial growth at stn H4 cannot be explained by such correlation (Table S5). In this case, diatoms may have affected nauplii either directly, e.g., via deleterious effects [69], or by interspecific interaction with some other prey that was beneficial for growth. The observed effects of salinity (Table 3) are rather expected as *A. bifilosa* is a species of marine origin, whereas *E. affinis* is a brackish water copepod thriving in a broad salinity range [22]. Moreover, high genetic diversity of Baltic *E. affinis* [70] may have contributed to the varying responses to winter and spring NAO (Table 3).

While stimulation of the microbial food web is possible in the field, it cannot explain the positive effects of *N. spumigena* on copepod egg viability and non-feeding nauplial development in the experiment (Table 2), because the copepods were incubated in 0.2 µm filtered sea water, where bacteria were heavily reduced and flagellates and ciliates were largely eliminated. Hence, both the direct and indirect pathways are likely to be responsible for the positive relationships between *N. spumigena* and copepod reproduction observed in the summer field survey (Table 1). The microbial communities thriving in cyanobacterial bloom may nourish the copepods either via the microbial loop [65,71] or as epibions on the ingested cyanobacterial colonies. Moreover, the mechanisms may differ between the cyanobacteria species as well as among zooplankton grazers.

Although the net effects of cyanobacteria on copepod reproduction (i.e., VEPR) appear either neutral or positive, there is some variation among the species and studies. For example, in the summer field survey, effects of *N. spumigena* on egg production in *A. tonsa* were positive (Table 1), whereas in the experimental study, this effect in *A. bifilosa* was negative, albeit these opposite responses resulted in no significant net effect on viable egg production (Table 2). This discrepancy between the laboratory experiment and the summer field survey suggest that variations in the feeding environment may be crucial for cyanobacteria-copepod interactions. In the field, *N. spumigena* had a significant positive correlation with biovolume of edible phytoplankton (15–30 µm), which was also a positive predictor for egg production; whereas these were not available for the copepods in the experiment. There might also be differences between the test species, *A. tonsa* (field survey) and *A. bifilosa* (experiment), in their sensitivity to the cyanobacterium effects and dependency on other food sources and environmental factors. In particular, *A. bifilosa* might have been avoiding to ingest larger filaments in the experiment due to food selection against this cyanobacterium [72]. This would effectively decrease food availability in the incubations as food (*Brachiomonas* with or without *Nodularia*) was provided at the same amount. As a result, lower food intake when feeding in mixtures containing cyanobacteria would translate into lower egg production, which was observed in this copepod species.

Table 3. Best-fit GLMs relating RNA:DNA ratio in the copepod nauplii (*Acartia* spp. and *E. affinis*) to biovolumes of total phytoplankton (TotPhyto) and specific phytoplankton groups as well as climatic factors (SST, salinity and NAO indices) in the long-term dataset for stations B1 and H4.

Station (*n*)	Phytoplankton group tested	Explanatory variables	Estimate	Wald statistics	*p*
Acartia spp.					
B1	Cyanobacteria	Cyanobacteria	3.788	170.892	<0.0001
(36)		SST	0.045	21.118	<0.0001
	Cryptophyceans	Cryptophyceans	−7.002	18.845	<0.0001
		SST	0.149	36.445	<0.0001
H4	Cyanobacteria	Cyanobacteria	1.583	9.736	0.0018
(34)		wNAO	0.566	17.104	<0.0001
		sNAO	−0.304	15.284	<0.0001
		Salinity	1.240	13.593	0.0003
	Diatoms	Diatoms	−0.820	5.838	0.0157
		wNAO	0.603	18.495	<0.0001
		sNAO	−0.376	22.139	<0.0001
		Salinity	1.189	11.680	0.0007
Eurytemora affinis					
B1	TotPhyto	TotPhyto	−1.326	9.077	0.003
(36)		sNAO	−0.346	26.641	<0.0001
	Cyanobacteria	Cyanobacteria	5.164	22.059	<0.0001
		sNAO	−0.142	5.718	0.0282
	Diatoms	Diatoms	−3.248	4.796	0.0295
		sNAO	−0.152	13.031	0.0012
H4	Cyanobacteria	Cyanobacteria	2.611	47.899	<0.0001
(34)		wNAO	0.327	10.008	0.0016
		Salinity	−0.778	6.870	0.0087
	Diatoms	Diatoms	−1.646	18.088	<0.0001
		sNAO	−0.400	24.733	<0.0001
		wNAO	0.745	32.529	<0.0001
		Salinity	−1.384	16.06	<0.0001

sNAO and wNAO are NAO indices averaged for spring (March – May) and winter (December – March), respectively; *n* – number of observations.

The observed differences among the cyanobacteria species could also be related to their relative abundance, morphology, nutritional value and/or biochemistry of toxins and metabolites. *N. spumigena* colonies do not form bundles, are not as rigid [73], and could, therefore, be easier to handle for mesozooplankton grazers than *Aphanizomenon* sp., which was found to negatively affect EPR (Table 1), and which has been reported to impair copepod reproduction to a higher extent than *N. spumigena* [29]. These cyanobacteria species also differ in their abundance (Figure 3) and nutritional value [49,51], including the value of their microbial epibionts [67,74]. *Aphanizomenon* sp. has lower colonization by heterotrophic bacteria compared to *N. spumigena* [59,63], which hosts a lucrative microenvironment for microorganisms [62,67,75]. This was also supported by significant positive correlation between *N. spumigena* and phytoplankton 2–5 µm observed in the summer field survey. This size fraction might be more N-sufficient than larger phytoplankton and thereby provide the more balanced food for the copepods. In line with this, significant positive effects of the phytoplankton 2–5 µm on the copepod reproductive variables were observed (Table 1). The highest egg production has been reported to occur when the diet consisted of alternative food with small contribution of cyanobac-

teria, mainly *N. spumigena* [29]; this resembles the *in situ* feeding conditions in our field study. The negative correlation between *Aphanizomenon* sp. and EPR (Table 2) could also be a result of poor feeding conditions for the copepods, due to scarcity of edible phytoplankton during the cyanobacteria bloom. Indeed in the field survey, *Aphanizomenon* sp., among the filamentous cyanobacteria, had the highest significant negative correlation with phytoplankton>15 µm. For *A. tonsa*, the optimal food size is 2–5% of the prosome length [76], which implies that optimal food size for the females sampled here would be 15–40 µm, explaining the observed positive relationships between phytoplankton 15–30 µm and growth related variables (RNA content and EPR; Table 1). Moreover, station had a significant influence on egg production, indicating presence of other factors, not accounted for in our analysis (Table 1).

In conclusion, our results indicate that summer blooms of diazotrophic filamentous cyanobacteria in the Baltic Sea could be important for copepod growth and reproduction by providing complementary food, supporting high antioxidant levels and fueling growth of microbial prey. In particular, *N. spumigena*, which are the most conspicuous bloom forming toxic filamentous cyanobacteria that build surface accumulations [20], is grazed upon and have positive effects on copepod recruitment and growth (Table S6). This was however not the case for the most abundant cyanobacterium, *Aphanizomenon* sp., which appears to decrease total egg production, with, however, neutral effects on the production of viable eggs. Given that dominant Baltic zooplankters are selectively feeding on non-toxic cyanobacteria or can avoid cyanobacterial filaments altogether [72], it is possible that by supporting growth and recruitment in grazer populations, these cyanobacteria may gain a competitive advantage over other phytoplankton [77]. In this case, the effects of cyanobacteria on grazers will be highly system-specific, depending on evolutionary trajectories of the species and populations in question. Our results, together with findings reporting high incorporation of diazotrophic nitrogen in pelagic and benthic food webs in this and other systems experiencing blooms of these cyanobacteria [78,79], have important implications for understanding impacts of these blooms on secondary, and, ultimately, fish production in the Baltic Sea. Further investigations on the associated ecological and evolutionary trade-offs behind these interactions are needed, if we are to understand and manage eutrophication and fishery in this system.

Supporting Information

Table S1 *Acartia tonsa*: **Data on reproductive and growth variables obtained in the summer field survey conducted in July - September 2007, at three coastal stations (B1, H2 and H4) in the northern Baltic proper.** Variables measured: EPR (egg production rate), VEPR (viable egg production rate), RNA (individual ribonucleic acid content), and PL (prosome length).

Table S2 *Acartia bifilosa*: **Data on feeding, reproduction and nauplii development obtained in the laboratory experiment conducted in August 2010, in the Storfjärden Bay, Western Finland.** Variables measured: EPR (egg production rate), VEPR (viable egg production rate), RNA (individual ribonucleic acid content), and PL (prosome length).

Table S3 *Acartia* sp. and *Eurytemora affinis*: **Long-term data (mean ± SD; period 1999–2009) on RNA:DNA ratio in nauplii at stns B1 and H4 in the northern Baltic proper.** The mean values are based on 3–4 replicate samples analyzed per year (July-August).

Table S4 Growth and reproduction indices observed in the (A) field survey, (B) experimental study, and (C) long-term dataset. Data are presented as means with their standard deviations; abbreviations as in Tables S1 and S2.

Table S5 Pearson *r* for correlations among phytoplankton groups at stn B1 and stn H4. Significant correlations are in bold face (*p*<0.05). Abbreviations: CHLO – Chlorophyceae, CHRY – Chrysophyceae, CRYP – Cryptophyceae, CYAN – Cyanophyceae, DIAT – Diatomea, DINO – Dinophyceae, EUGL – Euglenophyceae, MESO – *Mesodinium rubrum* (*Myrionecta rubrum*), PRAS – Prasinophyceae, PRYM – Prymnesiophyceae, UNID – unidentified flagellates, and HDIN – heterotrophic dinoflagellates.

Table S6 Summary of the observed effects linking filamentous cyanobacteria to reproductive output, growth indices and oxidative status in copepods. Abbreviations as in Tables S1 and S2.

Acknowledgments

We thank Askö Laboratory (Sweden) and Tvärminne Zoological Station (Finland) for permission to use facilities and for technical assistance in the field and laboratory studies. We also thank Dr. Ulf Larsson (Stockholm University, Sweden) for providing scientific advice on data interpretation, Leif Lundgren, Bengt Abrahamsson, Stefan Svensson, Lisa Mattsson, Carl Mattsson and Marika Tirén (Stockholm University, Sweden) for collecting samples within Swedish National Monitoring Programme and assistance in the laboratory, and Prof. K. Sivonen (University of Helsinki, Finland) for providing *Nodularia spumigena* (AV1) strain for our experiments.

Author Contributions

Conceived and designed the experiments: EG HH TH JEÖ AV AB. Performed the experiments: HH NM EG TH RE AV AB JEÖ SH. Analyzed the data: EG HH NM SH. Contributed reagents/materials/analysis tools: EG JEÖ. Wrote the paper: HH EG NM.

References

1. Paerl HW, Fulton RS (2006) Ecology of harmful cyanobacteria, In: Granéli E, Turner J (Eds.) Ecology of harmful marine algae. Springer-Verlag, Berlin, pp. 95–107.
2. Wiegand C, Pflugmacher S (2005) Ecotoxicological effects of selected cyanobacterial secondary metabolites a short review. Toxicol Appl Pharm 203: 201–218.
3. Müller-Navarra DC (2008) Food web paradigms: The biochemical view on trophic interactions. Int Rev Hydrobiol 93: 489–505.
4. Müller-Navarra DC, Brett MT, Liston AM, Goldman CR (2000) A highly unsaturated fatty acid predicts carbon transfer between primary producers and consumers. Nature 403: 74–77.
5. DeMott WR, Moxter F (1991) Foraging on cyanobacteria by copepods – responses to chemical defenses and resource abundance. Ecology 72: 1820–1834.
6. Carmichael WW (1992) Cyanobacteria secondary metabolites – the cyanotoxins. J Appl Bacteriol 72: 445–459.
7. Webster EK, Peters RH (1978) Some size-dependent inhibitions of larger cladoceran filterers in filamentous suspensions. Limnol Oceanogr 23: 1238–1245.
8. DeMott WR (1998) Utilization of a cyanobacterium and a phosphorus-deficient green alga as complementary resources by daphnids. Ecology 79: 2463–2481.
9. Koski M, Schmidt K, Engström-Öst J, Viitasalo M, Jónasdóttir S, et al. (2002) Calanoid copepods feed and produce eggs in the presence of toxic cyanobacteria *Nodularia spumigena*. Limnol Oceanogr 47: 878–885.
10. Schmidt K, Koski M, Engström-Öst J, Atkinson A (2002) Development of Baltic Sea zooplankton in the presence of a toxic cyanobacterium: a mesocosm approach. J Plankton Res 24: 979–992.
11. Kozlowsky-Suzuki B, Karjalainen M, Lehtiniemi M, Engström-Öst J, Koski M et al. (2003) Feeding, reproduction and toxin accumulation by the copepods *Acartia bifilosa* and *Eurytemora affinis* in the presence of the toxic cyanobacterium *Nodularia spumigena*. Mar Ecol Prog Ser 249: 237–249.
12. Schmidt K, Jónasdóttir SH (1997) Nutritional quality of two cyanobacteria: How rich is 'poor' food? Mar Ecol Prog Ser 151: 1–10.
13. Oberholster PJ, Botha AM, Cloete TE (2006) Use of molecular markers as indicators for winter zooplankton grazing on toxic bentic cyanobacteria colonies in an urban Colorado lake. Harmful Algae 5: 705–716.
14. Gorokhova E (2009) Toxic cyanobacteria *Nodularia spumigena* in the diet o Baltic mysids: Evidence from molecular diet analysis. Harmful Algae 8: 264–272.
15. Gorokhova E, Engström-Öst J (2009) Toxin concentration in *Nodularia spumigena* is modulated by mesozooplankton grazers. J Plankton Res 31 1235–1247.

16. Davis TW, Gobler CJ (2011) Grazing by mesozooplankton and microzooplankton on toxic and non-toxic strains of *Microcystis* in the Transquaking River, a tributary of Chesapeake Bay. J Plankton Res 33: 415–430.

17. Gilardi JD, Toft CA (2012) Parrots eat nutritious foods despite toxins. PLoS One 7: e38293.

18. Wilson AE, Sarnelle O, Tillmanns AR (2006) Effects of cyanobacterial toxicity and morphology on the population growth of freshwater zooplankton: Meta-analyses of laboratory experiments. Limnol Oceanogr 51: 1915–1924.

19. Ibelings BW, Havens KE (2008) Cyanobacterial toxins: a qualitative meta-analysis of concentrations, dosage and biota, Chapter 32, In: Hudnell HK (Ed.) Cyanobacterial harmful algal blooms: State of the science and research needs, pp. 675–732.

20. El-Shehawy R, Gorokhova E (2013) The bloom-forming cyanobacterium *Nodularia spumigena*: a peculiar nitrogen-fixer in the Baltic Sea food webs, In: Cyanobacteria: Ecology, Toxicology and Management. Nova publisher, USA.

21. Wasmund N (1997) Occurrence of cyanobacterial blooms in the Baltic sea in relation to environmental conditions. Int Rev Ges Hydrobiol 82: 169–184.

22. Viitasalo M, Katajisto T, Vuorinen I (1994) Seasonal dynamics of *Acartia bifilosa* and *Eurytemora affinis* (Copepoda: Calanoida) in relation to abiotic factors in the northern Baltic Sea. Hydrobiologia 292/293: 415–422.

23. Johansson M, Gorokhova E, Larsson U (2004) Annual variability in ciliate community structure, potential prey and predators in the open northern Baltic Sea proper. J Plankton Res 26: 67–80.

24. Meyer-Harms B, Reckermann M, Voss M, Siegmund H, von Bodungen B (1999) Food selection by calanoid copepods in the euphotic layer of the Gotland Sea (Baltic Proper) during mass occurrence of N_2-fixing cyanobacteria. Mar Ecol Prog Ser 191: 243–250.

25. Rolff C (2000) Seasonal variation in $\delta^{13}C$ and $\delta^{15}N$ of size-fractionated plankton at a coastal station in the northern Baltic proper. Mar Ecol Prog Ser 203: 47–65.

26. Loick-Wilde N, Dutz J, Miltner A, Gehre M, Montoya JP, et al (2012) Incorporation of nitrogen from N_2 fixation into amino acids of zooplankton. Limnol Oceanog 57: 199–210.

27. Peters J, Renz J, van Beusekom J, Boersma M, Hagen W (2006) Trophodynamics and seasonal cycle of the copepod *Pseudocalanus acuspes* in the Central Baltic Sea (Bornholm Basin): evidence from lipid composition. Mar Biol 149: 1417–1429.

28. Koski M, Engström J, Viitasalo M (1999) Reproduction and survival of the calanoid copepod *Eurytemora affinis* fed with toxic and non-toxic cyanobacteria. Mar Ecol Prog Ser 186: 187–197.

29. Sellner KG, Olson MM, Olli K (1996) Copepod interactions with toxic and non-toxic cyanobacteria from the Gulf of Finland. Phycologia 35: 177–182.

30. Smith JL, Boyer GL, Zimba PV (2008) A review of cyanobacterial odorous and bioactive metabolites: Impacts and management alternatives in aquaculture. Aquaculture 280: 5–20.

31. Amado LL, Monserrat JM (2010) Oxidative stress generation by microcystins in aquatic animals: Why and how. Environ Int 36: 226–235.

32. Vehmaa A, Hogfors H, Gorokhova E, Brutemark A, Holmborn T, et al.(2013) Projected marine climate change: effects on copepod oxidative status and reproduction. Ecol Evol 3: 4548–4557.

33. HELCOM (2008) Manual for marine monitoring in the COMBINE Programme of HELCOM, Annex 6: Guidelines concerning phytoplankton species composition, abundance and biomass..

34. Olenina I, Hajdu S, Edler L, Andersson A, Wasmund N, et al. (2006) Biovolumes and size-classes of phytoplankton in the Baltic Sea. Balt Sea Environ Proc 106, 144 pp..

35. Gorokhova E (2010) A single-step staining method to evaluate egg viability in zooplankton. Limnol Oceanogr-Meth 8: 414–423.

36. Gorokhova E (2003) Relationships between nucleic acid levels and egg production rates in *Acartia bifilosa*: implications for growth assessment of copepods *in situ*. Mar Ecol Prog Ser 262: 163–172.

37. Gorokhova E, Kyle M (2002) Analysis of nucleic acids in *Daphnia*: development of methods and ontogenetic variations in RNA-DNA content. J Plankton Res 24: 511–522.

38. Guillard RRL (1975) Culture of phytoplankton for feeding marine invertebrates, In: Smith WL, Chanley MH (Eds.) Culture of marine invertebrate animals. Plenum publishing corp, New York, pp. 29–60.

39. Giraffa G, Rossetti L, Neviani E (2000) An evaluation of Chelex-based DNA purification protocols for the typing of lactic acid bacteria. J Microbiol Meth 42: 175–184.

40. Engström-Öst J, Hogfors H, El-Shehawy R, De Stasio B, Vehmaa A, et al. (2011) Toxin producing cyanobacterium *Nodularia spumigena*, potential competitors and grazers: testing mechanisms of reciprocal interactions in mixed plankton communities. Aquat Microb Ecol 62: 39–48.

41. Prior RL, Hoang H, Gu LW, Wu XL, Bacchiocca M, et al. (2003) Assays for hydrophilic and lipophilic antioxidant capacity (oxygen radical absorbance capacity (ORAC(FL))) of plasma and other biological and food samples. J Agr Food Chem 51: 3273–3279.

42. Knuckey RM, Semmens GL, Mayer RJ, Rimmer MA (2005) Development of an optimal microalgal diet for the culture of the calanoid copepod *Acartia sinjiensis*: Effect of algal species and feed concentration on copepod development. Aquaculture 249: 339–351.

43. Gorokhova E (2005) Effects of preservation and storage of microcrustaceans in RNAlater on RNA and DNA degradation. Limnol Oceanogr-Meth 3: 143–148.

44. Baud A, Barthelemy RM, Nival S, Brunet M (2002) Formation of the gut in the first two naupliar stages of *Acartia clausi* and *Hemidiaptomus roubaui* (Copepoda, Calanoida): comparative structural and ultrastructural aspects. Can J Zool 80: 232–244.

45. Holmborn T, Lindell K, Holcton C, Hogfors H, Gorokhova E (2009) Biochemical proxies for growth and metabolism in *Acartia bifilosa* (Copepoda, Calanoida). Limnol Oceanogr Methods 7: 785–794.

46. Guedes AC, Gião MS, Seabra R, Ferreira ACS, Tamagnini P, et al. (2013) Evaluation of the antioxidant activity of cell extracts from microalgae. Mar Drugs 11: 1256–1270.

47. Parsons PA (2001) The hormetic zone: An ecological and evolutionary perspective based upon habitat characteristics and fitness selection. Q Rev Biol 76: 459–467.

48. Motwani NH, Gorokhova E (2013) Metazooplankton grazing on picocyanobacteria as inferred from molecular diet analysis. PLoS ONE 8(11): e79230.

49. Singh S, Kate BN, Banerjee UC (2005) Bioactive compounds from cyanobacteria and microalgae: An overview. Crit Rev Biotechnol 25: 73–95.

50. Jiang XD, Lonsdale DJ, Gobler CJ (2010) Density-dependent nutritional value of the dinoflagellate *Cochlodinium polykrikoides* to the copepod *Acartia tonsa*. Limnol Oceanogr 55: 1643–1652.

51. Vargas MA, Rodriguez H, Moreno J, Olivares H, Del Campo JA, et al. (1998) Biochemical composition and fatty acid content of filamentous nitrogen-fixing cyanobacteria. J Phycol 34: 812–817.

52. Pandey U, Pandey J (2009) Enhanced production of delta-aminolevulinic acid, bilipigments, and antioxidants from tropical algae of India. Biotechnol Bioproc E 14: 316–321.

53. Stal LJ, Staal M, Villbrandt M (1999) Nutrient control of cyanobacterial blooms in the Baltic Sea. Aquat Microb Ecol 18: 165–173.

54. Kiørboe T (1989) Phytoplankton growth rate and nitrogen content: implications for feeding and fecundity in a herbivorous copepod. Mar Ecol Prog Ser 55: 229–234.

55. Pinho GLL, da Rosa CM, Maciel FE, Bianchini A, Yunes JS, et al. (2005) Antioxidant responses and oxidative stress after microcystin exposure in the hepatopancreas of an estuarine crab species. Ecotox Environ Safe 61: 353–360.

56. Howitz KT, Sinclair DA (2008) Xenohormesis: Sensing the Chemical Cues of Other Species. Cell 133: 387–391.

57. Forbes VE (2000) Is hormesis an evolutionary expectation? Funct Ecol 14, 12–24.

58. Mattson MP, Cheng AW (2006) Neurohormetic phytochemicals: low-dose toxins that induce adaptive neuronal stress responses. Trends Neurosci 29: 632–639.

59. Sellner KG (1997) Physiology, ecology, and toxic properties of marine cyanobacteria blooms. Limnol Oceanogr 42: 1089–1104.

60. Degerholm J, Gundersen K, Bergman B, Söderbäck E (2008) Seasonal significance of N_2 fixation in coastal and offshore waters of the northwestern Baltic Sea. Mar Ecol Prog Ser 360: 73–84.

61. Larsson U, Hajdu S, Walve J, Elmgren R (2001) Baltic Sea nitrogen fixation estimated from the summer increase in upper mixed layer total nitrogen. Limnol Oceanogr 46: 811–820.

62. Ploug H, Adam B, Musat N, Kalvelage T, Lavik G, et al. (2011) Carbon, nitrogen and O_2 fluxes associated with the cyanobacterium *Nodularia spumigena* in the Baltic Sea. ISME J 5: 1549–1558.

63. Ploug H, Musat N, Adam B, Moraru CL, Lavik G, et al. (2010) Carbon and nitrogen fluxes associated with the cyanobacterium *Aphanizomenon* sp. in the Baltic Sea. ISME J 4: 1215–1223.

64. Ohlendieck U, Stuhr A, Siegmund H (2000) Nitrogen fixation by diazotrophic cyanobacteria in the Baltic Sea and transfer of the newly fixed nitrogen to picoplankton organisms. J Mar Syst 25: 213–219.

65. de Kluijver A, Yu J, Houtekamer M, Middelburg JJ, Liu Z (2012) Cyanobacteria as a carbon source for zooplankton in eutrophic Lake Taihu, China, measured by ^{13}C labeling and fatty acid biomarkers. Limnol Oceanogr 57: 1245–1254.

66. Wilson SE, Steinberg DK (2010) Autotrophic picoplankton in mesozooplankton guts: evidence of aggregate feeding in the mesopelagic zone and export of small phytoplankton. Mar Ecol Prog Ser 412: 11–27.

67. Hoppe HG (1981) Blue-green algae agglomeration in surface water: a microbiotope of highly bacterial activity. Kieler Meeresforsch Sonderh 5: 291–303.

68. Suikkanen S, Laamanen M, Huttunen M (2007) Long-term changes in summer phytoplankton communities of the open northern Baltic Sea. Estuar Coast Shelf Sci 71: 580–592.

69. Miralto A, Barone G, Romano G, Poulet SA, Ianora A, et al. (1999) The insidious effect of diatoms on copepod reproduction. Nature 402: 173–176.

70. Gorokhova E, Lehtiniemi M, Motwani NH (2013) Trade-offs between predation risk and growth benefits in copepods *Eurytemora affinis* with contrasting pigmentation. PLoS ONE 8(8): e71385. doi:10.1371/journal.pone.0071385.

71. Gifford DJ (1991) The protozoan-metazoan trophic link in pelagic ecosystems. J Protozool 38: 81–86.

72. Engström J, Koski M, Viitasalo M, Reinikainen M, Repka S, et al. (2000) Feeding interactions of the copepods *Eurytemora affinis* and *Acartia bifilosa* with the cyanobacteria *Nodularia* sp. J Plankton Res 22: 1403–1409.

73. Komárek J, Anagnostidis K (1989) Modern approach to the classification system of Cyanophytes 4 — Nostocales. Algol Stud Arch Hydrobiol, Suppl 56: 247–345.

74. Breteler WCM, Schogt N, Baas M, Schouten S, Kraay GW (1999) Trophic upgrading of food quality by protozoans enhancing copepod growth: role of essential lipids. Mar Biol 135: 191–198.

75. Tuomainen J, Hietanen S, Kuparinen J, Martikainen PJ, Servomaa K (2006) Community structure of the bacteria associated with *Nodularia* sp. (cyanobacteria) aggregates in the Baltic Sea. Microb Ecol 52: 513–522.

76. Berggreen U, Hansen B, Kiørboe T (1988) Food size spectra, ingestion and growth of the copepod *Acartia tonsa* during development — implications for determination of copepod production. Mar Biol 99: 341–352.

77. Holland DP, van ErpI, Beardall J, Cook PLM (2012) Environmental controls on the nitrogen-fixing cyanobacterium *Nodularia spumigena* in a temperate lagoon system in SE Australia. Mar Ecol Prog Ser 461: 47–57.

78. Wannicke N, Korth F, Liskow I, Voss M (2013) Incorporation of diazotrophic fixed N_2 by mesozooplankton — Case studies in the southern Baltic Sea. J Mar Syst 117–118: 1–13.

79. Woodland RJ, Holland DP, Beardall J, Smith J, Scicluna T, et al. (2013) Assimilation of diazotrophic nitrogen into pelagic food webs. PLoS ONE 8(6): e67588.

20

Observations of the "Egg White Injury" in Ants

Laure-Anne Poissonnier[1], Stephen J. Simpson[2], Audrey Dussutour[1]*

1 Research Center on Animal Cognition, The National Center for Scientific Research and Toulouse University, Toulouse, France, **2** Charles Perkins Centre, The University of Sydney, Sydney, New South of Wales, Australia

Abstract

A key determinant of the relationship between diet and longevity is the balance of protein to carbohydrate in the diet. Eating excess protein relative to carbohydrate shortens lifespan in solitary and social insects. Here we explored how lifespan and behavior in ants was affected by the quality of protein ingested and the presence of associated antinutrients (i.e. compounds that interfere with the absorption of nutrients). We tested diets prepared with either egg white protein only or a protein mixture. Egg white contains an anti-nutrient called avidin. Avidin binds to the B vitamin biotin, preventing its absorption. First, we demonstrate that an egg-white diet was twice as deleterious as a protein-mixture diet. Second, we show that ingestion of egg-white diet drastically affected social behavior, triggering elevated levels of aggression within the colony. Lastly, we reveal that by adding biotin to the egg white diet we were able to lessen its detrimental effects. This latest result suggests that ants suffered biotin deficiency when fed the egg white diet. In conclusion, anti-nutrients were known to affect health and performance of animals, but this is the first study showing that anti-nutrients also lead to severe changes in behavior.

Editor: James A.R. Marshall, University of Sheffield, United Kingdom

Funding: AD was supported by a grant from the "Agence Nationale de la Recherche 2011, Programme Jeunes Chercheurs, Reference number: JSV7-0009-01". The funders had no role in study design, data collection and analysis, decision to publish, or preparation of the manuscript.

Competing Interests: The authors have declared that no competing interests exist.

* Email: dussutou@gmail.com

Introduction

In insects and mammals it has been shown that a determinant of the relationship between diet and longevity is the balance of protein to carbohydrate [1]. The quantity of dietary protein ingested has been recognized as an important factor for the reduction of lifespan [1], but the nutritional quality of protein is also of critical importance [2–5]. Protein quality depends on two main factors: the profile and concentration of amino acids [1,6–7] and the presence of associated compounds that may act as 'antinutrients' [8]. Antinutrients have been defined as "substances, which by themselves, or through their metabolic products arising in living systems, interfere with food utilization and affect the health and performance of animals" [9].

In many experiments conducted on omnivorous insect such as ants, eggs white are used as a main source of proteins [10–12]. Eggs are considered high quality in their amino acid composition, but the inclusion of large amounts of egg white in special experimental diets causes a definite nutritional disease in animals. This disorder is commonly called egg white injury. Egg whites contain avidin – an antinutrient that reduces longevity of numerous insects [13] mammalian species and fish species [14]. Most insects fed on avidin-containing diets show retarded growth and high mortality rate [13] and cannibalism [15]. Avidin is a glycoprotein, which irreversibly binds biotin and renders it unavailable [16]. Biotin, as a cofactor of major carboxylases, is involved in key process such as gluconeogenesis, lipogenesis, fatty acid and amino acid catabolism, making it essential for insect growth [13].

In this paper, we explored how lifespan and behavior in ants are affected by the quantity of egg white ingested. Ants offer some unique opportunities to tackle this question. First, ants share some fundamental features with other non-social insects. For instance, ants regulate their intake of both protein and carbohydrate when suitable foods are available, but live less long when confined to diets that have an elevated proportion of protein [17–21]. Second, ants with a pronounced division of labour between similarly sized sterile worker castes provide a unique opportunity to study the direct effect of egg white ingestion on the evolution of lifespan, independently of reproductive effort [22]. Third, in ants, interactions among individuals coordinate the activities of the entire colony so that it acts as a nutrient acquiring, distributing and digesting 'superorganism' [17] and offer the intriguing possibility to consider the relationship between egg white ingestion and social behaviour.

Methods

1- Species studied and rearing conditions

Eight colonies of the black garden ant *Lasius niger*, each comprising ca. 10,000 workers, were collected in Marquefave (South-West France, 43°19′23.7″N 1°15′11.2″E) in 2009 and 2010. No specific permissions were required for these locations/ activities. The field studies did not involve endangered or protected species. Experimental colonies of 200 workers without brood were constructed from these mother colonies. In *Lasius niger* nurses are younger than foragers [23]. In order to harmonize the experimental colonies in term of age and task, all workers were collected from the foraging arena rather than within the nest of the mother colonies and were probably all foragers.

Each experimental colony was housed in a plastic box of 100 mm diameter, the bottom of which was covered by a layer of cotton moistened by a cotton plug soaking in a water reservoir underneath. The box was connected to an arena (diameter 150 mm) with walls coated with Fluon to prevent ants from escaping. The nests were regularly moistened and the colonies were kept at room temperature (24–26°C) under a 12:12 L/D photoperiod. Before starting the experiments, colonies were fed *ad libitum* with honey solution (15%) and prey (mealworms *Tenebrio molitor*) for a week.

2- Synthetic foods

In the field, black garden ants scavenge for dead insects, as well as tend and collect honeydew from sap-feeding Homoptera. Accordingly, these ants are confronted with foods varying widely in their ratio of protein to carbohydrate, from near pure sources to mixtures. For the experiment described below, we used synthetic foods varying in the ratio and concentration of protein (P) and digestible carbohydrate (C). The protein content of all the foods consisted of a mixture of casein, whey protein, egg powder, and glucose was used as a digestible carbohydrate source. The quantity of whole egg was kept constant in each food to maintain the same quantity of associated lipid and minerals. Each food contained 0.5% of vitamins (Vanderzant vitamin mixture for insects). The foods were presented to the ants in a 1% agar solution at a 4.5:1 ratio of agar solution to dry mass of ingredients. Further preparation details are given in [11].

3- Experiments

Experiment 1: Survival effect of protein type. We investigated if protein source could affect longevity by confining experimental colonies to one of four diets differing in their protein source and in their protein to carbohydrates ratio, with a fixed total P+C of 200 g L^{-1}. The two P:C ratios used were 5:1 and 1:5. On one hand, we used protein biased diet (5:1) to amplify any effects of protein quality on variables quantified and on the other hand we used carbohydrate biased diet (1:5) typically used in most

laboratories to rear ants [11]. The diets were either prepared with a mixture of whey protein 18.3%, casein 73.2%, egg yolk 3.1% and egg white 5.4% (1:5 MIX or 5:1 MIX) or only with egg white powder 96.9% and egg yolk 3.1% (1:5 EGG or 5:1 EGG) (Figure S3 in File S1). The diet 1:5 EGG was very similar to the most commonly used artificial diet for ant breeding: the chemically undefined Bhatkar and Whitcomb diet [10]. The egg white was pasteurized at a temperature of 56°C for 3–5 min and cooled at 3°C. Since temperatures greater than 70°C are required to denature avidin, a constituent of egg white [24], this was not denatured in our diet. For each synthetic diet, we tested 8 experimental colonies originating from 8 different mother colonies i.e. for each mother colony, one daughter colony was confined to 1:5 MIX, one to 5:1 MIX, one to 1:5 EGG and the last one to 5:1 EGG. All colonies had ad libitum access to food that was replenished daily. Colonies never collected all the food offered before it was renewed. To assess mortality, the number of dead ants within each colony was counted every day and removed from the colony until all ants had died.

Experiment 2: Foraging effects of protein type. We conducted a second experiment to monitor ant foraging behavior. In the first experiment, we showed that an egg-white diet was twice as deleterious as a protein-mixture diet only for the high protein biased diet so we kept only three treatments 1:5 MIX, 5:1 MIX and 5:1 EGG for the following experiment, 1:5 MIX becoming our "control" for which we expect standard foraging behavior. For each dietary regime, we tested 4 experimental colonies originating from 4 different mother colonies i.e. for each mother colony, one daughter colony was confined to 1:5 MIX, one to 5:1 MIX and the last one to 5:1 EGG. All colonies were starved for 2 days before the beginning of the experiment to encourage food consumption from the first day of the experiment. Afterward, from day 1, all colonies had ad libitum access to food that was replenished daily. The colonies were filmed for 3 hours per day after replenishing the food during 5 days. The film analysis was conducted without knowledge of treatments.

First, we measured mortality rates to ensure that they were comparable to experiment 1. The number of dead ants within each colony was counted every day for 40 days and removed from the colony until all ants died.

Second, we measured colony intake for the first five days of the experiment. The food was placed in the foraging arena in two small containers (diameter, 15 mm; height, 5 mm). The ants had access only to one container; the second was used as a control for measuring and correcting for evaporation. We also provisioned the nest with moistened cotton wool to minimize the water loss. In order to evaluate the colony's intake, the small containers with the food were weighed within 0.01 mg every day before they were placed in the foraging arena and again after they were removed. We adjusted colony intake to the number of ants in each colony, to take into account differences in mortality between colonies.

Third, we analyzed foraging behavior to quantify how ants responded to each dietary regime. The total number of ants in the foraging arena and the total number of ants feeding were counted every minute for 3 h for five days, providing measures of 'exploratory behaviour' and 'foraging behaviour' respectively. We also counted the total number of meals eaten over the 3 h recording period for five days and their respective duration. A meal consisted of a contact with the food that lasted more than 10 seconds. Brief contacts with the food (less than 5 seconds) are usually referred as food probing [25].

Lastly, we observed ant inactivity. We extracted from the videos the coordinates of each ant in the foraging arena every second using ImageJ. We assumed that an ant remained inactive when the

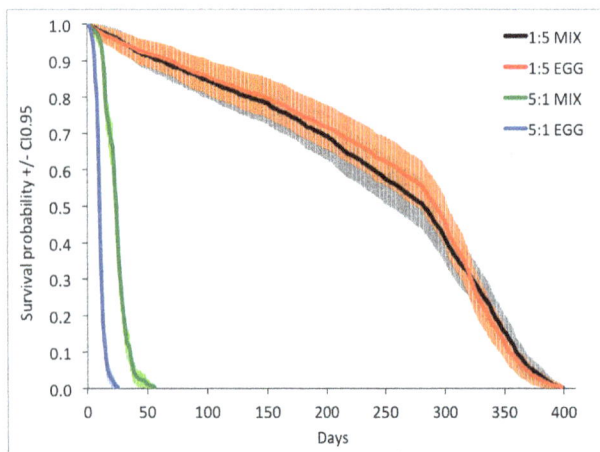

Figure 1. Ant survival according to the protein type and protein to carbohydrate ratio. N = 8 experimental colonies of 200 individuals per treatment. Mortality dynamics were consistent between colonies of the same treatment. The dotted lines represent the confidence intervals.

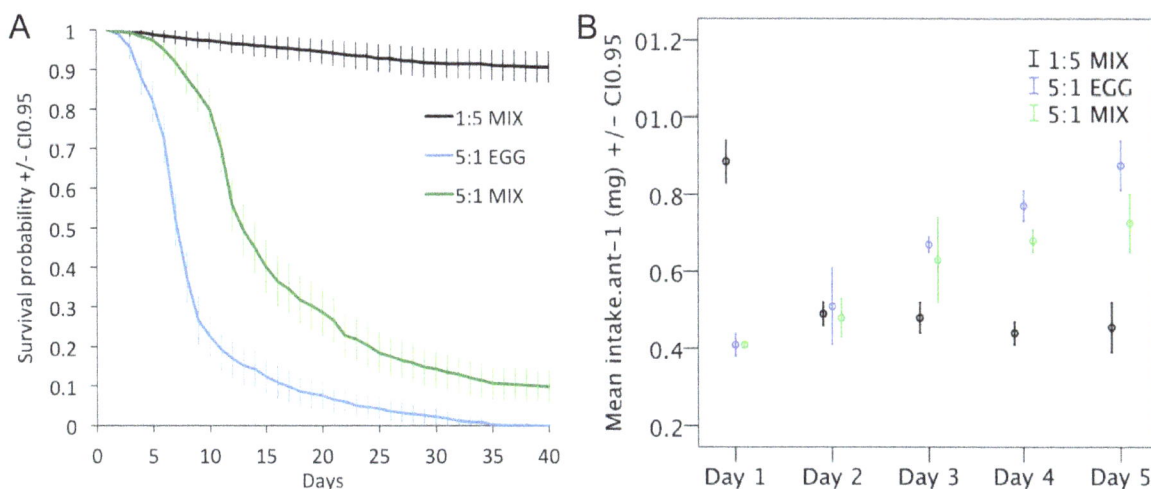

Figure 2. A. Ant survival according to protein type and protein to carbohydrate ratio. Mortality dynamics were consistent between colonies of the same treatment. The dotted lines represent the confidence intervals. **B.** Diet intake per ant (colony intake divided by colony size) according to the diet. N = 4 experimental colonies of 200 individuals per treatment. Colony sizes were adjusted to account for ant mortality.

two following conditions were fulfilled: (1) its displacement between two successive seconds was less than 0.1 mm and (2) the stop duration was at least 10 seconds (corresponding to ten successive observations). The total number of inactive ants in the foraging arena was counted every minute for 3 h for five days. We also recorded the time spent inactive in the foraging arena for each inactive ant.

Experiment 3: Behavior effects of protein type. During experiment 1 and 2, we observed that ants were biting each other on the EGG_5:1 diet in all colonies tested. We conducted a third experiment to specifically monitor ant interactions. We focused our attention on the two protein biased diets (5:1 MIX and 5:1 EGG). For each dietary regime we tested 4 experimental colonies originating from 4 different mother colonies i.e. for each mother colony, one daughter colony was confined to 5:1 MIX and one to 5:1 EGG. All colonies had ad libitum access to food that was replenished daily. First, the total number of biting events was counted every 3 min for 3 h (after replenishing the food) for five days. Second, at day 6, a pair of ants was placed in a Petri dish (Ø 35 mm, H 10 mm) to observe changes in the behavior of ants with regard to nestmate recognition. We observed the ants for 5 min after their first encounter and classified worker behavior into one of three categories: (i) Indifference: no visible response to one another, (ii) hostility: antagonistic interactions such as mandible gaping, avoidance or intense antennation, (iii) aggression: a physical attack by one or both workers, including lunging, biting, holding or pulling legs or antennae. Each behavior was given an arbitrary score related to increasing aggression: 0, Indifference; 1, hostility and 2, aggression. Overall aggression index for each dyadic encounter was computed as the sum of the aggression scores divided by the total number of scores. To avoid resampling, workers were not returned to their colonies. The ants used represented four treatment combinations, being (i) from the same colony and fed the same diet (N = 12 pairs for both diets 5:1 MIX and 5:1 EGG), (ii) from different colonies and fed the same diet (N = 12 pairs for both diets 5:1 MIX and 5:1 EGG), (iii) from the same colony and fed different diets (N = 12 pairs for both colonies), or (iv) from different colonies and fed different diets (N = 24 pairs).

Experiment 4: Survival and behavior effects of biotin supplementation. Egg white contains avidin – an antinutrient that has been shown to have detrimental effects that are mediated through its biotin-binding activity [13]. Based on the avidin and biotin content of egg white published in [24,26], we estimated the avidin content of the 5:1 EGG diet to be 0.4 g.Kg-1 (5.7 µM/L^{-1}), the biotin content to be 69 µg.Kg-1 (0.29 µM/L^{-1}) and the avidin/biotin ratio on molar basis to be 20:1. Avidin contains four identical subunits. Each subunit binds one molecule of biotin; thus a total of four biotin molecules can bind a single avidin molecule, i.e the avidin:biotin binding ratio is 1:4. We investigated if biotin supplementation could affect longevity by confining experimental colonies to the two high protein diets used in the third experiment (5:1 MIX and 5:1 EGG) in which we added 0.025 g.Kg-1 of biotin (102.4 µM/L^{-1}). As a result, the avidin/biotin ratio on molar basis in both diets became about 1:20 preventing any biotin deficiency. For each diet we tested eight experimental colonies originating from eight different mother colonies. To assess mortality, the number of dead ants within each colony was counted every day and removed from the colony until all ants died. On day 5, four colonies per treatment were chosen at random and observed for 3 hours and the total number of biting events was counted every 3 min for 3 h (after replenishing the food).

4- Statistical Analysis

All statistical tests were conducted with R version 3.1.1. For all experiments, normality was assessed for each variable using a Kolmogorov-Smirnov test. Data were transformed prior to analysis in order to normalize it if needed. For experiment 1, 2 and 4, longevity data across treatment were compared using Cox regression analysis, with protein source, ratio and colony as categorical variables. Protein and ratio were included in the analysis as factors, whereas colony was included as a clustered term (nested factor). For experiment 2, we used linear models with repeated measures to test for the effect of diet and time on intake, proportion of ants in the foraging arena, proportion of foragers feeding, number of meals and proportion of inactive ants. Intakes and proportions were computed using the daily survivors. We used

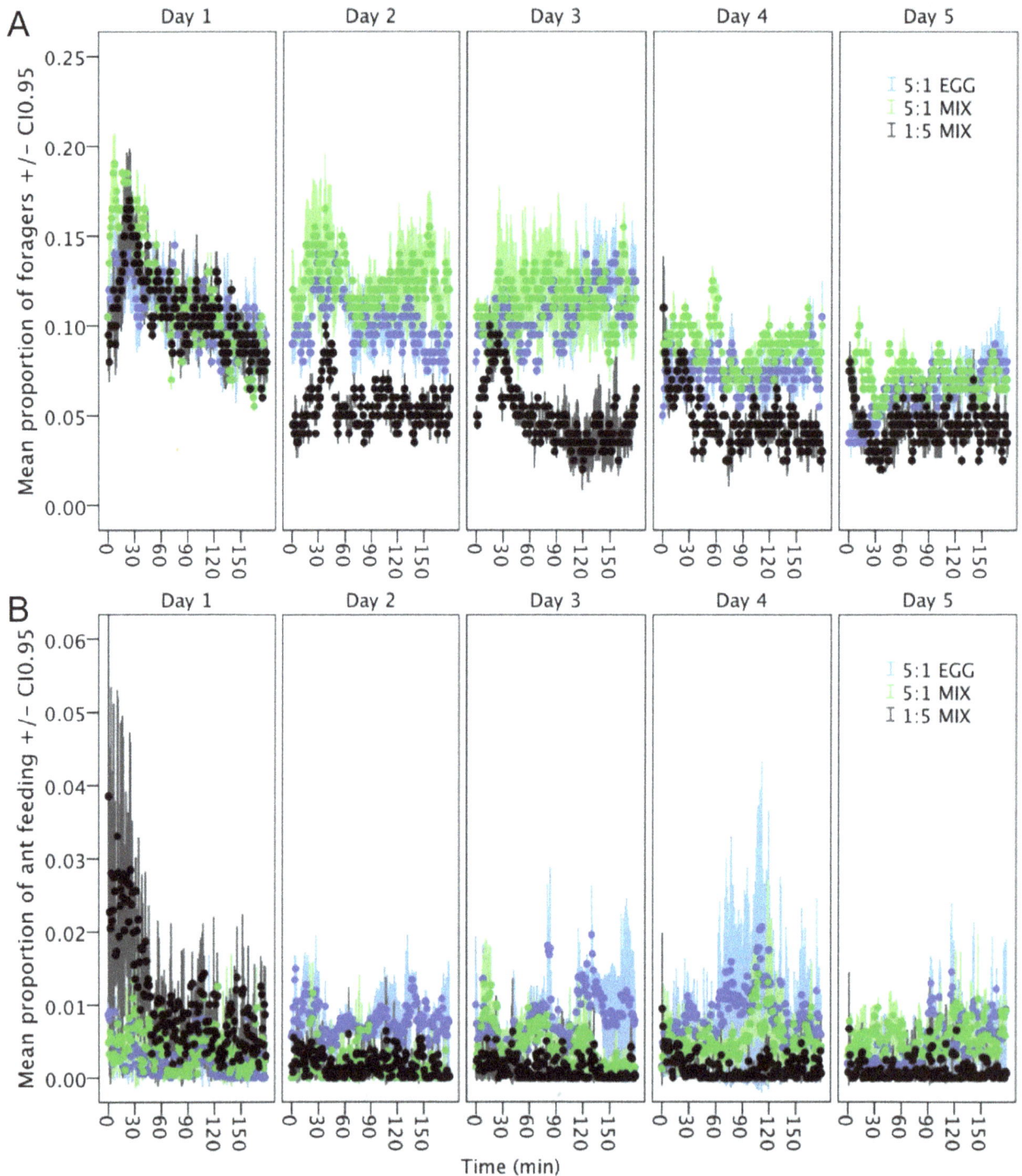

Figure 3. A Proportion of ants in the foraging arena according to the diet (number of ants observed in the foraging arena divided by colony size). **B** Proportion of ants in the foraging arena that are feeding according to the diet (number of ants observed feeding divided by colony size). N = 4 experimental colonies of 200 individuals per treatment. Colony sizes were adjusted to account for ant mortality.

colony of origin as random factor to correct for possible intrinsic behavioral differences among colonies. The time spent feeding (meal duration) and the time spent inactive across treatment were compared using Cox regression analysis, with diet and day as categorical variables. For experiment 3, we conducted a linear model testing how ants' behavior (aggression index) was affected by the colony of the ant encountered (same or different), the diet of the ant encountered (same or different), and the ant's own food regime (protein mix or egg protein).

Table 1. Results of a three-way ANOVA to test for the effect of the treatment, the time (1 min to 180 min) and the day (N = 5) at which the measures were done on the proportion of ants in the foraging arena per colony.

Source of variation	Mean squares	df	F	p
Between colonies				
Treatment	0.028	2	16.09	<0.001
Within colonies				
Time	0.002	179	7.71	<0.001
Time x Treatment	0.001	358	3.05	<0.001
Day	0.908	4	14.12	<0.001
Day x Treatment	0.146	8	2.28	0.044
Time x Day	0.001	716	7.03	<0.001
Time x Day x Treatment	0.001	1432	3.50	<0.001

Results

Experiment 1: Survival effect of protein type

Ants lived longest on a diet comprising a 1:5 ratio of protein to carbohydrate and died earlier on a 5:1 diet (ratio effect Wald = 1701.09, P<0.001). Colonies fed the EGG diet live half as long as colonies fed the MIX diet but only on highly protein biased diets (protein effect Wald = 33.45, P<0.001, protein x ratio effect Wald = 188.58, mean lifespan $\pm CI_{0.95}$:243.21$\pm$5.28, 248$\pm$6.23, 23.34$\pm$0.44, 10.06$\pm$0.19 days for 1:5 MIX, 1:5 EGG, 5:1 MIX and 5:1 EGG respectively, Figure 1).

Experiment 2: Foraging effects of protein type

The pattern of longevity observed for experiment 2 was similar to the one observed for experiment 1 (Figure 2A, diet effect Wald = 986.24 P<0.001).

On the high carbohydrate diet (1:5 MIX) food collection decreased from day 1 to day 2 and remained constant until day 5, whereas on both high protein diets (5:1 MIX and 5:1 EGG) food collection increased from day 1 to day 5 (Figure 2B) (diet effect $F_{2,9}$ = 7.08 P = 0.014, day effect $F_{4,36}$ = 16.33 P<0.001, Interaction diet x day $F_{8,36}$ = 39.51 P<0.001). In our previous study conducted in the same exact conditions, we found that the regulation point of protein and carbohydrate intake, known as the

intake target, was P:C 1:4 [18]. Thus, the increase in intake across days observed for both protein diets indicates presumably an effort to reach a certain carbohydrate intake [17–18].

The proportion of ants in the foraging arena decreased across days from day 1 for the high carbohydrates diet (1:5 MIX) and only from day 3 for both high protein diets (5:1 MIX and 5:1 EGG) (Table 1, Figure 3A). For the high carbohydrate diet (1:5 MIX) the proportion of ants in the foraging arena decreased throughout each day while it remained almost constant throughout the day for the high protein diets (5:1 MIX and 5:1 EGG).

The proportion of ants feeding differed according to the diet received (Table 2, Figure 3B). For the high carbohydrate diet (1:5 MIX), the proportion of ants feeding decreased strongly from day 1 to day 2 and then remained constant. When the high carbohydrate diet (1:5 MIX) was introduced to the colony on day 1, the number of ants feeding increased exponentially over the first hour, indicating a strong recruitment process due to the food deprivation experienced before, and then decreased. This pattern was not observed during the following days because the colonies were fed ad libitum from day 1. By contrast, the proportion of ants feeding remained constant throughout each day for both high protein diets (5:1 MIX and 5:1 EGG) but increased from day to day, presumably in an effort to maintain a constant carbohydrate intake.

Table 2. Results of a three-way ANOVA to test for the effect of the treatment, the time (1 min to 180 min) and the day (N = 5) at which the measures were done on the proportion of ants feeding per colony.

Source of variation	Mean squares	df	F	p
Between colonies				
Treatment	0.008	2	9.43	<0.001
Within colonies				
Time	0.0005	179	2.78	<0.001
Time x Treatment	0.0006	358	3.63	<0.001
Day	0.002	4	1.39	0.257
Day x Treatment	0.009	8	6.33	<0.001
Time x Day	0.0004	716	4.34	<0.001
Time x Day x Treatment	0.0002	1432	2.87	<0.001

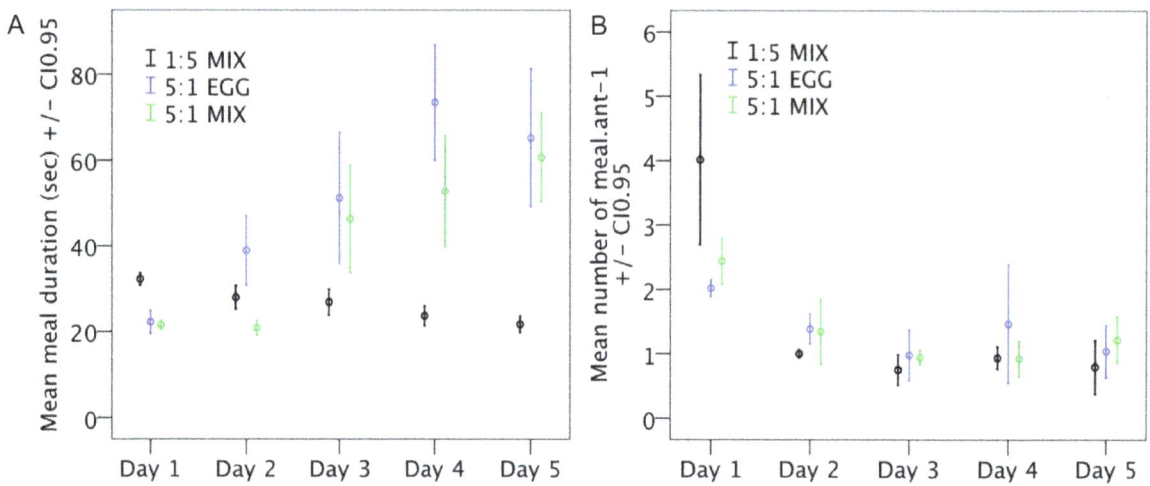

Figure 4. A. Meal duration (in seconds) according to the diet. **B.** Relative number of meals per ant (total number of meals divided by colony size). N = 4 experimental colonies of 200 individuals per treatment. Colony sizes were adjusted to account for ant mortality.

For both high protein diets, the difference observed in proportion of ant feeding was due to a difference in meal duration (Figure 4A) and not to a difference in the absolute number of meals (Figure 4B), while it was the opposite for the high carbohydrate diet. The relative number of meals was higher on the first day especially on the high carbohydrate diet (1:5 MIX) reflecting food deprivation experienced the day before (Figure 4B, diet effect $F_{2,9} = 0.09$ P = 0.919, day effect $F_{4,36} = 94.67$ P<0.001, Interaction diet x day $F_{8,36} = 13.63$ P<0.001). Meal duration increased across days only for the high protein diets (5:1 MIX and 5:1 EGG, diet effect Wald = 157.73 P<0.001, day effect Wald = 52.35 P<0.001, Interaction diet x day Wald = 332.46.14 P<0.001, Figure 4A, Figure S1 in File S1), by contrast it remained constant for the high carbohydrate diet (1:5 MIX).

From day 2, the proportion of inactive ants in the foraging arena was significantly higher for both high protein diets (5:1 MIX

and 5:1 EGG) than for the high carbohydrate diet (1:5 MIX) (Figure 5A, Table 3). The time spent inactive increased for both high protein diets (5:1 MIX and 5:1 EGG) while it remained constant for the high carbohydrate diet (diet effect Wald = 70.51 P<0.001, day effect Wald = 23.95 P<0.001, Interaction diet x day Wald = 82.32 P<0.001, Figure 5B, Figure S2 in File S1). On both high protein diets we observed about ten percent of ants that were inactive for long periods of time, from 10 min to 3 hours (Figure S2 in File S1). In contrast, on high carbohydrate diet no ants were observed inactive for more than 10 min (Figure S2 in File S1). Ants observed motionless for long periods of time looked moribund (slight curling of the antennae and legs).

Experiment 3: Behavior effects of protein type

Ants were observed fighting almost only on the 5:1 EGG diet and the occurrence of this aggressive behavior increased across

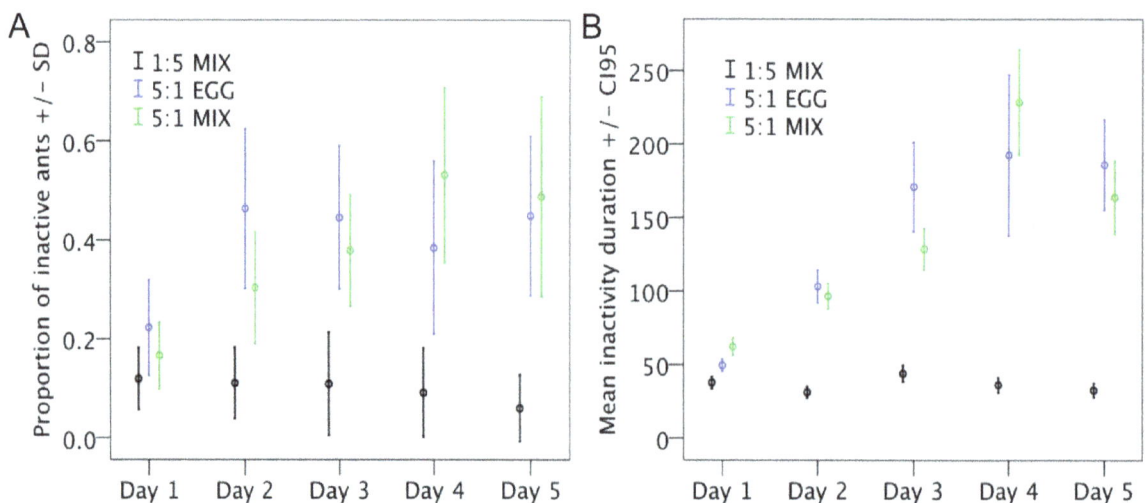

Figure 5. A. Proportion of inactive ants in the foraging arena according to the diet (total number of inactive ants observed divided by the number of ants observed in the foraging arena). **B.** Inactivity duration (in seconds) according to the diet. N = 4 experimental colonies of 200 individuals per treatment. Colony sizes were adjusted to account for ant mortality.

Table 3. Results of a three-way ANOVA to test for the effect of the treatment, the time (1 min to 180 min) and the day (N = 5) at which the measures were done on the proportion of inactive ants.

Source of variation	Mean squares	df	F	p
Between colonies				
Treatment	898.51	2	42.64	<0.001
Within colonies				
Time	0.070	179	10.95	<0.001
Time x Treatment	0.029	358	4.58	<0.001
Day	10.16	4	10.74	<0.001
Day x Treatment	6.42	8	6.79	<0.001
Time x Day	0.022	716	3.26	<0.001
Time x Day x Treatment	0.022	1432	3.36	<0.001

subsequent days (Diet effect $F_{1,6} = 78.72$, P<0.001, Day effect $F_{4,24} = 20.96$ P<0.001, interaction diet x day $F_{4,24} = 19.09$ P<0.001).

Ants fed the 5:1 EGG diet were much more likely to fight with nestmates than ants fed the 5:1 MIX diet, which were never observed to be aggressive towards a nestmate (Table 4, Figure 6B). This effect was slightly more pronounced if the nestmate was also fed the 5:1 EGG diet (Table 4, Figure 6B). As expected in *Lasius niger* [27], all ants responded aggressively towards non-nestmates, no matter the diet they were fed (Table 4, Figure 6B).

Experiment 4: Survival and behavior effects of biotin supplementation

Life expectancy was increased when we added biotin to the 5:1 EGG diet (Biotin effect, Wald = 133.82, P<0.001, mean lifespan±$CI_{0.95}$:19.78±0.19 and 9.53±0.19 for 5:1 EGG+Biotin and 5:1 EGG, Figure 7A) but not when we added biotin to the 5:1 MIX diet (interaction Treatment x Biotin, Wald = 94.21, P<0.001, mean lifespan±$CI_{0.95}$:25.29±0.70 and 23.34±0.44 for

MIX+biotin and MIX, Figure 7A). Nevertheless, ants lived longer on the 5:1 MIX diet supplemented with biotin than on the 5:1 EGG diet supplemented with biotin (Wald = 36.19, P<0.001, Figure 7A). Ants fed the 5:1 EGG+Biotin or the 5:1 MIX+Biotin diet were rarely observed to be aggressive towards a nestmate (Biotin effect $F_{1,12} = 39.15$, P<0.001, Interaction Biotin x diet $F_{1,12} = 37.58$ P<0.001, Figure 7B).

Discussion

The main aim of this study was to investigate how protein quality and quantity affects longevity and behavior in ants. First we showed that lifespan was reduced when ants were fed with high-protein, low- carbohydrate diets. Second, we found that when we modified the protein source of our diets, egg white protein was more deleterious than a mix of various proteins. Third, we showed that being fed egg white protein in large quantity profoundly affects behavior, triggering intra nest aggression. Lastly, we revealed that being fed biotin in addition to egg white protein nearly doubled longevity and decreased intra-nest

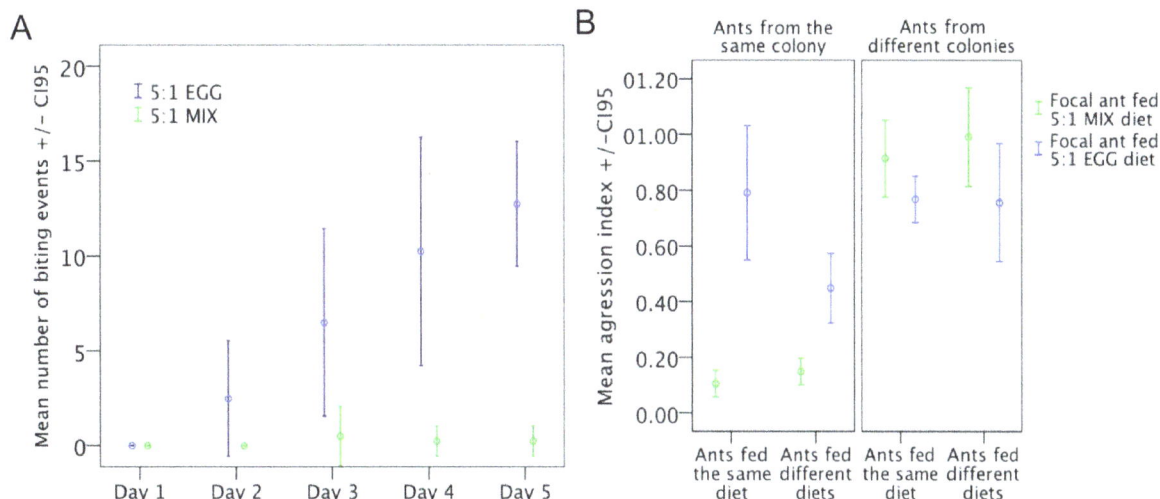

Figure 6. A Mean number of biting events observed. Nests were scanned 60 times (every 3 min for 3 hours) for 5 days. **B** Mean aggression index (see text for definition) of workers fed with a mix of protein (5:1 MIX, green error bars) or egg protein (5:1 EGG, Blue error bars) encountering workers coming from the same or a different colony fed the same or a different diet. N = 4 experimental colonies of 200 individuals per treatment.

Table 4. Linear model testing how an ant's behaviors during an encounter with another ant (aggression index) was affected by the colony of the ant encountered (same or different), the diet of the ant encountered (same or different) and the ant's own food regime (treatment: protein mix or egg protein).

Source of variation	Mean squares	df	F	p
Colony (same or different)	11.21	1, 184	88.42	<0.001
Diet (same or different)	0.16	1, 184	1.30	0.255
Treatment (MIX or EGG)	1.09	1, 184	8.60	0.004
Colony * Diet	0.40	1, 184	3.13	0.079
Colony * Treatment	5.60	1, 184	44.17	<0.001
Diet * Treatment	0.68	1, 184	5.33	0.022
Colony * Diet * Treatment	0.26	1, 184	2.08	0.150

aggression. This latest results suggests that avidin an anti-nutrient present in large quantity in egg white, binds biotin and as a consequence ants suffers from biotin deficiency.

Five hypotheses could be proposed to explain the reduction in survival observed on the 5:1 EGG diet when compared to the 5:1 MIX diet.

First, such a longevity decrease could result from a deficiency or imbalance of amino acids in the food. However a mixture of egg white powder and yolk contain all the essential and non-essential amino acids in ratios and concentrations that make this unlikely as an explanation (Figure S3 in File S1).

Second, were there an increase in amount of food ingested on 5:1 EGG diet, this might explain the decrease in survival probability [18]. The key determinant of the relationship between diet and longevity in ants and other insects is the balance of protein and non-protein energy ingested [1,17–21]. Lifespan is reduced when ants ingest high quantity of protein. However we did not observe any difference between 5:1 EGG and 5:1 MIX in the colony intake. The only differences were between the high carbohydrate diet (1:5 MIX) and the high protein diets (5:1 MIX and 5:1 EGG). Colonies fed with the high-protein diets collected substantial excesses of food relative to the amount collected on the

high carbohydrate diet. The only exception was the very high intake on the high carbohydrate diet seen on day 1. It reflected a combination of food deprivation experienced the day before [17–18,28] and the high phagostimulant efficiency of the high carbohydrate diet in comparison to the high protein diets [29]. After day 1, the intake on the high carbohydrate intake remained constant while the intake on high protein diets increased. In previous studies, we have established that colonies are able to switch between two nutritionally imbalanced but complementary foods to maintain the supply of protein and carbohydrate [17–18]. In this study, we explored the responses of colonies when confined to a single diet containing an excess of nutrient relative to the other. Therefore, the ants were forced to ingest foods that were imbalanced and confronted the situation wherein there is conflict between meeting their requirements for protein and carbohydrates. In all studies published so far in ants, it was shown that colonies without larvae collect substantial excesses of protein when fed with high protein diet in an effort to maintain a constant carbohydrate intake [17–21].

Third, were there an increase in effort spent in foraging activities on 5:1 EGG diet, this might explain the decrease in survival probability [30]. However we did not observe any

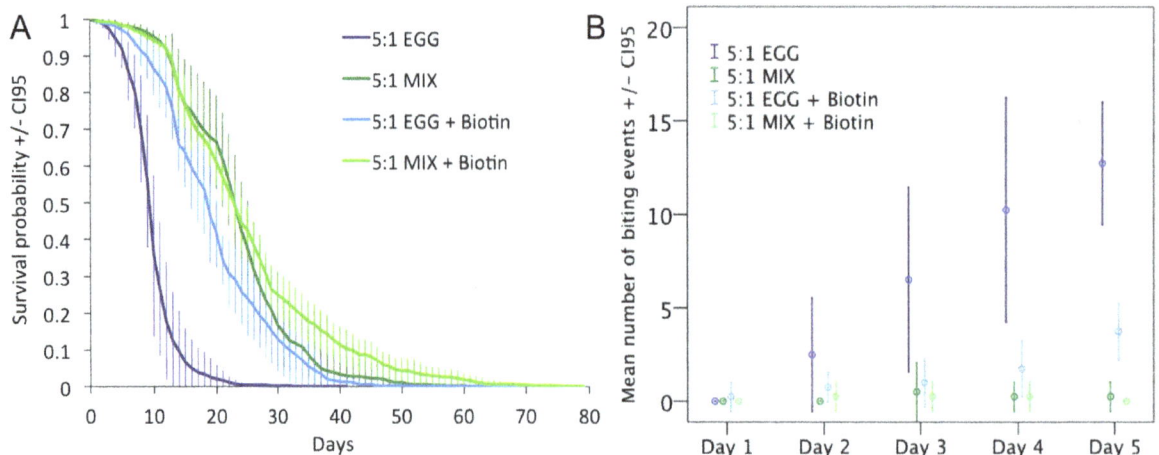

Figure 7. A Ant survival according to the protein type used to prepare the high-protein diet and the supplementation in biotin. N = 8 experimental colonies of 200 individuals per treatment. Mortality dynamics were consistent between colonies of the same treatment. **B** Mean number of biting events observed. Nests were scanned 60 times (every 3 min for 3 hours) for 5 days. N = 4 experimental colonies of 200 individuals per treatment

differences between 5:1 EGG and 5:1 MIX in the number of ants exploring or feeding. The only differences were again between the high carbohydrate diet (1:5 MIX) and the high protein diets (5:1 MIX and 5:1 EGG). On the first day of the experiment, after two days without food, the number of ants exploring and feeding at the high carbohydrate diet increased exponentially during the first hour due to food deprivation experience beforehand, but decreased thereafter, indicating that the colony reached satiety [31]. This pattern was not observed the following days as food was renewed every day. Consequently the foraging activity on the high carbohydrate diet was relatively high on day 1, decreased on day 2 and remained stable. In contrast the number of ants in the foraging arena remained relatively high throughout the days on both high protein diets. Interestingly, these ants were not observed feeding on day 1, but instead appeared to be engaged in 'exploratory' behavior. The number of ant feeding increased from day 2, reflecting again an attempt for compensating the lack of carbohydrate in the high protein diet [17,18].

Fourth, a difference in protein toxicity may explain the reduction in survival observed on the 5:1 EGG diet when compared to the 5:1 MIX. The level of toxicity can be expressed in term of level of sickness. It has been shown that moribund or sick ants [32,33] and honeybees [34] leave their nest and die in isolation. In our experiment, on both high protein diets, numerous ants were observed motionless and moribund close to the edge of the foraging arena, from day 2 for the 5:1 EGG diet and from day 3 for the 5:1 MIX diet. In contrast we did not observe any moribund ant on the high carbohydrate diet (1:5 MIX). It has been shown before that high-protein diets increase mortality rates in ants and other insect [1,17–21], but it remains unclear how. Several explanations have been posed to explain protein toxicity. First, elevated levels of proteins increase the target of rapamycin (TOR) pathway signaling which in turn activate aging processes in mice [35], but it is not yet proven in insect. Second, protein toxicity could result from the limited ability to digest proteinaceous foods in ants [36–39]. Lastly, life-shortening effects of protein could be linked to the elimination of nitrogenous waste. Nevertheless, this does not explain why the 5:1 EGG diet affected the ant earlier than the 5:1 MIX diet. The toxicity of the EGG diet might have been increased by the presence of avidin – an antinutrient in egg, which irreversibly binds biotin and makes it unavailable. Following this hypothesis, ants were dying earlier on an EGG diet in part as result of biotin deficiency or its consequences. The last experiment where ants were fed with EGG diet supplemented with biotin supports this hypothesis. Dietary studies that have added extra biotin back to the diet have allowed insects to overcome the presence of avidin [40]. Interestingly, in honeybees, another social hymenoptera, avidin ingestion had no significant impacts on longevity [41,42]. However the concentration in avidin used in [41,42] was 5 times lower than the concentration in our EGG diet.

Lastly, decreased longevity might have resulted from behavioral modification during interactions. It has been shown that sick ants increase their level of aggression [33]. Following this hypothesis, ants were dying in part as result of elevated level of aggression or its consequences. The third experiment where ants were observed biting each other only on the 5:1 EGG diet supports this hypothesis. As mentioned in Bos et al [33] "self- removal and increased aggression are reminiscent of what is also found in humans, where sick individuals become reclusive and irritable, isolating themselves from other individuals". Interestingly, biotin supplementation limited ants aggression. Two hypotheses could be advanced to explain the link between biotin deficiency and aggression level. First, in insects, certain cuticular hydrocarbons

are synthesized from fatty acid via the elongation-decarboxylation pathway [review in 43]. Because biosynthesis of fatty acid depends in part on biotin [44], the hydrocarbon profiles of ants fed on an egg white diet might have been altered, compromising ant recognition and explaining the aggressive behaviour observed between congeners. However we did not find any evidence for a link between biotin and hydrocarbons synthesis in the literature. Analysis of hydrocarbons profile will be needed to corroborate this hypothesis. Since the brain is quite vulnerable to biotin deficiency [45], the second hypothesis is that biotin deficiency might have led to cognitive impairment and limited ant recognition performance. In rats for example, biotin deficiency produces neurological symptoms that range from ataxia to sensory loss [46].

The question we might ask is why ants ate such large quantities of toxic high protein diet. To this point of the paper we have focused our interpretations on the macronutrients – protein and carbohydrates and stated that ants collected excesses of protein in an effort to acquire a certain carbohydrate intake. As the literature shows [1], macronutrients can explain a good deal of the variation in the behavioral, physiological and performance responses of animals. Macronutrients are, however, clearly not the only functionally important nutritional components of foods: vitamins and minerals are essential to health and also play a critical role in an animal's nutritional strategies. Being able to determine the presence and concentrations of nutrients in foods by taste is clearly advantageous, but not all nutrients in food are detected by specialized taste receptors [1]. This is particularly the case for micronutrients such as vitamins, which are essential to health but required in very small amounts relative to the macronutrients protein and carbohydrate [1]. If the nutritional consequences of eating a food are not apparent until after a meal is processed and absorbed. The animal has to learn from the experience of having eaten a food by associating post-ingestive nutritional consequences (biotin deficiency) with properties of the food, however this take some times as biotin deficiency is not instantaneous but is steadily rising from day to day [13]. In our study, ants confined to an egg white diet, might increase their intake from day to day to also provide limiting biotin, constraining them as a result to ingest more avidin and enter a deadly loop. Here, we have shown that in ants, the effects of biotin deficiency are not limited to health and performance, but extend to behavior modification and by extension to social organization. Knowing that egg white protein is a common component of ant diets in many laboratories, it might be important to add biotin to synthetic diets to improve ant husbandry.

Biotin is a coenzyme required for all forms of life, feeding avidin or streptavidin causes a biotin deficiency that leads to stunted growth and mortality in numerous insect [16,47] to which we can now add ants. Proteins that bind to vitamins, such as biotin, had been shown to represent potential pest-resistance transgene products [40]. The avidin gene has recently been incorporated into genetically modified crop plants, which are then insecticidal to a variety of insects [13,16] and aphids [40]. However, many aphid pests of major plant crops are attended or attacked by ants. In light of our results, it might be important to look at the ecological significance of aphids as carriers of avidin from transgenic plants to ant colonies.

Supporting Information

File S1 Includes Figures S1–S3. **Figure S1:** Natural logarithm of the fraction of ants that are still feeding as a function of time from day 1 to day 5. Note that each plot corresponds to the survival curve of more than 600 meal durations. If the probability for an ant to stop feeding was constant over time then the log-survival curve of the number of ants still feeding should fit a straight line (Haccou and Meelis, 1992). However, from day 2 to day 5, for the high protein diets (5:1 MIX and 5:1 EGG) the curves suggest that the duration of a meal was either short or long. **Figure S2:** Natural logarithm of the fraction of ants that are still inactive as a function of time from day 1 to day 5. Note that each plot corresponds to the survival curve of more than 800 stop durations. If the probability for an ant to initiate a new displacement was constant over time then the log-survival curve of the number of ants still inactive should fit a straight line (Haccou and Meelis, 1992). However, from day 2 to day 5, for the high protein diets (5:1 MIX and 5:1 EGG) the curves suggest that the duration of a stop was either short or long. **Figure S3**: Amino acid

profile of both the MIX diet and the EGG diet. The MIX diet and the EGG diet were prepared with the following protein sources: whey protein 18.3%, casein 73.2%, egg yolk 3.1% and egg white 5.4% (MIX); egg yolk 3.1% and egg white 96.9% (EGG).

Acknowledgments

We thank Jacques Gautrais for his precious scripts.

Author Contributions

Conceived and designed the experiments: AD. Performed the experiments: AD LAP. Analyzed the data: AD LAP. Contributed to the writing of the manuscript: AD SJS.

References

1. Simpson SJ, Raubenheimer D (2012) The nature of nutrition. Princeton University Press. Princeton, New Jersey.
2. Altaye SZ, Pirk CW, Crewe RM, Nicolson SW (2010) Convergence of carbohydrate-biased intake targets in caged worker honeybees fed different protein sources. J Exp Biol 213: 3311–3318.
3. Pirk CWW, Boodhoo C, Human H, Nicolson SW (2010) The importance of protein type and protein to carbohydrate ratio for survival and ovarian activation of caged honeybees (Apis mellifera scutellata). Apidologie. 41: 62–72.
4. Lee KP (2007) The interactive effects of protein quality and macronutrient imbalance on nutrient balancing in an insect herbivore. J Exp Biol 210: 3236–3244.
5. Cooper RA, Schal C (1992) Effects of protein type and concentration on development and reproduction of the German cockroach, Blattella germanica. Entomol Exp Appl 63: 123–134.
6. Wu G (2013) Amino acids: biochemistry and nutrition. CRC Press.
7. Mann J, Truswell AS (2002) Essentials of human nutrition. Oxford University Press.
8. Sarwar Gilani G, Wu Xiao C, Cockell KA (2012) Impact of Antinutritional Factors in Food Proteins on the Digestibility of Protein and the Bioavailability of Amino Acids and on Protein Quality British J Nutr 108: S315–S332.
9. Makkar HPS (1993) Antinutritional factors in foods for livestock. In: Gill M, Owen E, Pollot GE, Lawrence TLJ. (Eds.) Animal Production in Developing Countries. Occasional publication No. 16. British Society of Animal Production, 69–85.
10. Bhatkar A, Whitcomb WH (1970) Artificial diet for rearing various species of ants. Florida Entomol 229–232.
11. Dussutour A, Simpson SJ (2008) Description of a simple synthetic diet for studying nutritional responses in ants. Insectes Soc 55: 329–333.
12. Cohen AC (1999) Artifical media for rearing entomophages comprising sticky, cooked whole egg. US Patent 5945271.
13. Christeller JT, Markwick NP, Burgess EPJ, Malone LA (2010) The use of biotin-binding proteins for insect control. J Econ Entomol 103, 497–508.
14. Yossa R, Sarker PK, Mock DM, Lall SP, Vandenberg GW (2013) Current knowledge on biotin nutrition in fish and research perspectives. Reviews in Aquaculture. DOI: 10.1111/raq.12053.
15. Levinson HZ, Barelkovsky J, Bar Ilan AR (1967) Nutritional effects of vitamin omission and antivitamin administration on development and longevity of the hide beetle Dermestes maculatus (Coleoptera: Dermestidae). J Stored Prod Res 3, 345–352.
16. Kramer KJ, Morgan TD, Throne JE, Dowell FE, Bailey M et al. (2000) Transgenic avidin maize is resistant to storage insect pests. Nature Biotechnology, 18(6): 670–674.
17. Dussutour A, Simpson SJ (2009) Communal nutrition in ants. Curr Biol 19: 740–744.
18. Dussutour A, Simpson SJ (2012) Ant workers die young and colonies collapse when fed a high-protein diet. Proc Roy Soc Lond B 279: 2402–2408.
19. Christensen KL, Gallacher AP, Martin L, Tong D, Elgar MA (2010) Nutrient compensatory foraging in a free-living social insect. Naturwissenschaften. 10: 941–944.
20. Cook SC, Eubanks MD, Gold RE, Behmer ST (2010) Colony-level macronutrient regulation in ants: mechanisms, hoarding and associated costs. Anim Behav 79: 429–437.
21. Cook SC, Behmer ST (2010) Macronutrient regulation in the tropical terrestrial ant, Ectatomma ruidum: a field study. Biotropica. 42: 135–139.
22. Chapuisat M, Keller L (2002) Division of labour influences the rate of ageing in weaver ant workers. Proc. R. Soc. Lond. B 269, 909–913. doi:10.1098/rspb.2002.1962.
23. Lenoir A, Ataya H (1983) Polyéthisme et répartition des niveaux d'activité chez la fourmi Lasius niger L. Zeitschrift für Tierpsychologie, 63(2-3), 213–232.

24. Green NM (1975) Avidin. Adv Protein Chem 29 (1975): 85–133.
25. Simpson SJ, Raubenheimer D (2000) The hungry locust. Adv Stud Behav, 29, 1–44.
26. Green NM, Toms EJ (1970) Purification and crystallization of avidin. Biochem J, 118: 67–70.
27. Lenoir A, Depickère S, Devers S, Christidès JP, Detrain C (2009) Hydrocarbons in the ant Lasius niger: from the cuticle to the nest and home range marking. J Chem Ecol 35(8), 913–921.
28. Dussutour A, Simpson SJ (2008) Carbohydrate regulation in relation to colony growth in ants. J Exp Biol 211(14), 2224–2232.
29. Arganda S, Nicolis SC, Perochain A, Péchabadens C, Latil G, et al. (2014) Collective choice in ants: The role of protein and carbohydrates ratios. Journal of insect physiology. In press (available online).
30. Schmid-Hempel P, Kacelnik A, Houston AI (1985) Honeybees maximize efficiency by not filling their crop. Behav Ecol Sociobiol 17: 61–66.
31. Pasteels JM, Deneubourg JL, Goss S (1987) Self-organization mechanisms in ant societies (i): Trail recruitment to newly discovered food sources. From individual to collective behavior in social insects, 155–175.
32. Heinze J, Walter B (2010) Moribund ants leave their nests to die in social isolation. Current Biology, 20(3), 249–252.
33. Bos N, Lefevre T, Jensen AB, D'ettorre P (2012) Sick ants become unsociable. Journal of evolutionary biology, 25(2), 342–351.
34. Rueppell O, Hayworth MK, Ross NP (2010) Altruistic self- removal of health-compromised honey bee workers from their hive. J. Evol. Biol. 23: 1538–1546.
35. Solon-Biet SM, McMahon AC, Ballard JWO, Ruohonen K, Wu LE, et al (2014) The ratio of macronutrients, not caloric intake, dictates cardiometabolic health, aging, and longevity in ad libitum-fed mice. Cell metabolism, 19(3), 418–430.
36. Holldobler B, Wilson EO (1990) The ants. Cambridge, MA: Harvard University Press.
37. Ricks BL, Vinson SB (1972) Digestive enzymes of the imported fire ant Solenopsis richteri (Hymenoptera: Formicidae). Entomol. Exp. Appl. 15, 329–334.
38. Glancey BM, Vander Meer RK, Glover A, Lofgren CS, Vinson SB (1981) Filtration of micro-particles from liquids ingested by the red imported fire ant Solenopsis invicta Buren. Insect Soc. 28, 395–401.
39. Petralia RS, Sorensen AA, Vinson SB (1980) The labial gland system of larvae of the imported fire ant, Solenopsis invicta Buren: ultrastructure and enzyme analysis. Cell Tissue Res. 206, 145–156.
40. Morgan TD, Oppert B, Czapla TH, Kramer KJ (1993) Avidin and streptavidin as insecticidal and growth inhibiting dietary proteins*. Entomologia experimentalis et applicata, 69(2), 97–108.
41. Malone LA, Tregidga EL, Todd JH, Burgess EP, Philip BA, et al. (2002) Effects of ingestion of a biotin-binding protein on adult and larval honey bees. Apidologie, 33(5), 447–458.
42. Malone LA, Todd JH, Burgess EP, Christeller JT (2004) Development of hypopharyngeal glands in adult honey bees fed with a Bt toxin, a biotin-binding protein and a protease inhibitor. Apidologie, 35(6), 655–664.
43. Blomquist GJ, Bagnéres AG (2010) Insect hydrocarbons: biology, biochemistry and chemical ecology. (Cambridge University Press, Cambridge).
44. Dadd RH (1973) Insect nutrition: current developments and metabolic implications. Ann Rev Entomol 18: 381–420.
45. Zempleni J, Wijeratne SS, Hassan YI (2009). Biotin. Biofactors, 35(1), 36–46.
46. McKay BE, Molineux ML, Turner RW (2004) Biotin is endogenously expressed in select regions of the rat central nervous system. Journal of Comparative Neurology, 473(1), 86–96.
47. Hinchliffe G, Bown DP, Gatehouse JA, Fitches E (2010) Insecticidal activity of recombinant avidin produced in yeast. J Insect Physiol 56, 629–639.

21

Effects of Long-Term Feeding of the Polyphenols Resveratrol and Kaempferol in Obese Mice

Mayte Montero, Sergio de la Fuente, Rosalba I. Fonteriz, Alfredo Moreno, Javier Alvarez*

Instituto de Biología y Genética Molecular (IBGM), Departamento de Bioquímica y Biología Molecular y Fisiología, Facultad de Medicina, Universidad de Valladolid and Consejo Superior de Investigaciones Científicas (CSIC), Valladolid, Spain

Abstract

The effect of the intake of antioxidant polyphenols such as resveratrol and others on survival and different parameters of life quality has been a matter of debate in the last years. We have studied here the effects of the polyphenols resveratrol and kaempferol added to the diet in a murine model undergoing long-term hypercaloric diet. Using 50 mice for each condition, we have monitored weight, survival, biochemical parameters such as blood glucose, insulin, cholesterol, triglycerides and aspartate aminotransferase, neuromuscular coordination measured with the rotarod test and morphological aspect of stained sections of liver and heart histological samples. Our data show that mice fed since they are 3-months-old with hypercaloric diet supplemented with any of these polyphenols reduced their weight by about 5–7% with respect to the controls fed only with hypercaloric diet. We also observed that mice fed with any of the polyphenols had reduced levels of glucose, insulin and cholesterol, and better marks in the rotarod test, but only after 1 year of treatment, that is, during senescence. No effect was observed in the rest of the parameters studied. Furthermore, although treatment with hypercaloric diets induced large changes in the pattern of gene expression in liver, we found no significant changes in gene expression induced by the presence of any of the polyphenols. Thus, our data indicate that addition of resveratrol or kaempferol to mice food produces an initial decrease in weight in mice subjected to hypercaloric diet, but beneficial effects in other parameters such as blood glucose, insulin and cholesterol, and neuromuscular coordination, only appear after prolonged treatments.

Editor: Hemachandra Reddy, Texas Tech University Health Science Centers, United States of America

Funding: This work was supported by grants from Ministerio de Ciencia e Innovación [BFU2008-01871 and BFU 2011-25763; (http://www.idi.mineco.gob.es/portal/site/MICINN/)] and from Junta de Castilla y León [VA029A12-1 and GR105, (http://www.educa.jcyl.es/educacyl/cm/universidad)]. Sergio de la Fuente had an FPI (Formación de Personal Investigador) fellowship from the Spanish Government. The funders had no role in study design, data collection and analysis, decision to publish, or preparation of the manuscript.

Competing Interests: The authors have declared that no competing interests exist.

* Email: jalvarez@ibgm.uva.es

Introduction

The importance of the presence of vegetables in human food has been widely debated, and in the last years considerable scientific literature has been published in relation with the effects on health of diets rich in vegetables or in particular types of polyphenols. Most of these studies are epidemiological, and associate the amount of polyphenols present in the diet with a reduction in the risk for cardiovascular diseases and cancer [1–8]. Some studies have also described other positive effects in neurodegenerative diseases [9,10] or inflammatory diseases [11], and they have also been shown to activate lipolysis [12,13]. All these potentially beneficial effects have led to proposals for their addition in foods as a nutritional supplement.

Moreover, some studies have investigated the effects of some of these individual molecules in cellular or animal models. The more striking example is resveratrol, a polyphenol abundant in red wine, which has been shown to improve the survival and several plasmatic, neurological and genomic markers in mice fed with hypercaloric diets [14]. Previously, it had been shown that this compound was able to prolong considerably the lifespan in several species, including Saccharomyces cerevisiae, Caenorhabditis elegans, Drosophila melanogaster and a short-lived fish [15–19]. More recently, it has also been shown to protect pancreatic islets against oxidative stress in db/db mice [20]. These effects have been attributed not only to the antioxidant properties of this compound, but rather to the specific activation by resveratrol of sirtuins, a family of NAD^+-dependent deacetylases and mono ADP-rybosiltransferases that mediate the physiological effects of caloric restriction. From these data an entire industry has emerged to provide food supplements, cosmetics, and other products based on resveratrol. However, recent data has questioned the results obtained with resveratrol in C. elegans and D. melanogaster [21] and this point is now a matter of strong debate [22–24].

Although at a smaller scale, similar studies have also been carried out with other related polyphenols, but significant differences appear among the reported effects. We were particularly interested in the flavonoid kaempferol, as we showed several years ago that this compound was a potent stimulator of the mitochondrial Ca^{2+} uniporter, a Ca^{2+} channel essential to modulate mitochondrial energy production [25]. It has been shown that kaempferol increases oxygen consumption in cell cultures and isolated mitochondria, potently activates the synthesis of thyroid hormones and modulates a whole series of enzymes and proteins related to metabolism [26]. In addition, a 30 days administration of a kaempferol glucoside to Wistar rats feed with hypercaloric diet reduced the weight and the plasmatic levels of triglycerides [27]. Kaempferol, however, was much less effective as activator of sirtuins [15], and resveratrol, instead, was ineffective as stimulator of thyroid hormone synthesis [14] and little effective as activator of mitochondrial Ca^{2+} uptake [25].

Both compounds (resveratrol and kaempferol) thus seem to act through different signaling pathways, and we thought it would be very interesting to compare their effects on several biochemical, neurological and genomic parameters during a lifetime treatment with these compounds in mice. We have then decided to make a large and complete study of this point by using a very large number (250) 3-months old C57BL mice and carrying out the treatment for the entire life of the mice (24 months). As in previous reports [14,27], we have also studied the effects of these polyphenols in mice fed with hypercaloric diet, periodically monitoring parameters such as food intake, weight, blood glucose, insulin, triglycerides, cholesterol and transaminase activity. Motor coordination was estimated using the rotarod test [28,29], and we have also compared mortality rates, analyzed histological sections of liver and heart, and looked for changes in gene expression patterns in liver using Affymetrix gene expression arrays.

Methods and Materials

Ethics statement

The study was approved by the Ethical Committee for Animal Research of the School of Medicine of the University of Valladolid (Permit Number: 7-2007), and was carried out in strict accordance with the recommendations in the Guide for the Care and Use of Laboratory Animals of the National Institutes of Health.

Mice treatment and diets

250 C57BL male mice were obtained from Janvier (53947 Saint Berthevin Cedex, France) and fed with control diet (Diet D12450B from Research Diets Inc., New Brunswick, NJ 08901 USA) until they were 3-months old and at that moment the different diets were started. This diet contains 3.85 kcal/g. Mice were maintained in standard housing conditions at 22°C, with a 12 h light/dark cycle and with free access to tap water. They were distributed in 5 groups with 50 mice in each of them, and placed in cages containing 10 mice in each for all their life. Group Control (C) continued with the same control diet. Group Hypercaloric Control (HC) was fed with a standard hypercaloric diet (Diet D12492 from Research Diets Inc.), a diet containing a larger proportion of lard and providing 5.24 kcal/g. Control diet contains 1.9% lard, providing 4.44% of the total kcal, while hypercaloric diet contains 31.66% lard, providing 54.35% of the total kcal. Group Hypercaloric Resveratrol (HR) was fed with hypercaloric diet (D12492) supplemented with 250 mg/kg-diet (0,025%) of resveratrol. Group Hypercaloric low Kaempferol (Hk) was fed with hypercaloric diet (D12492) supplemented with 50 mg/kg-diet (0,005%) of kaempferol. Group Hypercaloric high Kaempferol (HK) was fed with hypercaloric diet (D12492) supplemented with 250 mg/kg-diet (0,025%) of kaempferol. Food was stored at 4°C, sealed and at dark. Both resveratrol and kaempferol were obtained from Apin Chemicals Limited, Abingdon, Oxon, OX14 4RU, U.K., and sent to Research Diets to prepare the different types of foods, which were identified by color codes.

The amount of food ingested was measured by putting known amounts of food in every cage and weighting the remaining food

every two days. As shown in Fig. 1 and Table S1, the amount of food ingested kept quite constant along the whole study for all the conditions, although some small changes among the groups could be detected (see Results). The initial weight of the mice was around 25 g, and so the initial dose of the compounds was 25 mg·kg^{-1}·day^{-1} in groups HR (as in ref. 14) and HK, and 5 mg·kg^{-1}·day^{-1} in group Hk. Mice fed hypercaloric diets increased progressively weight along the study (Fig. 2a), reaching values of 55–60 g after 1 year of treatment.

Once the treatment was started, we monitored a series of parameters along the whole life of the mice: mice weight every 15 days, blood glucose, insulin, triglycerides, cholesterol and aspartate aminotransferase every three months, neurological performance with the rotarod test every 3 months, histological studies of liver and heart after 6 and 12 months of hypercaloric diet, and gene expression studies with Affymetrix arrays in liver extracts after 12 months.

Blood biochemistry

Glucose was measured using the Ascensia Breeze 2 Blood Glucose Meter (Bayer HealthCare AG, Wuppertal, Germany).

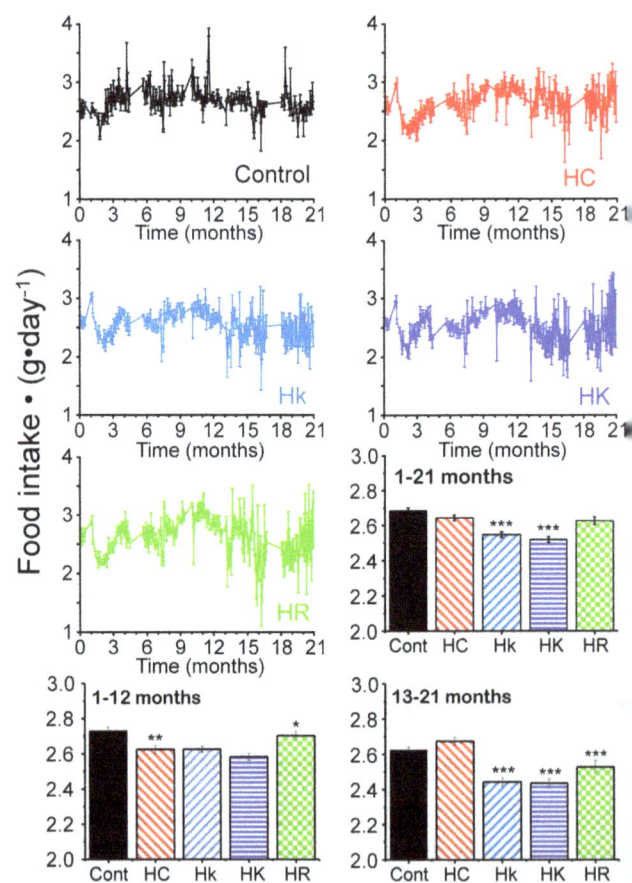

Figure 1. Variation of the mean amount of food ingested by each group of mice along their lifespan. Data are calculated as food intake per mice and per day. Each data was obtained as mean±s.e. n = 5 (5 mice cages containing 10 mice each). The bar diagrams show the mean±s.e. of all the measurements obtained for each group of mice (n = 197 measurements for whole period, n = 112 for the first 12 months, n = 85 for 13–21 months) and the significance (*, p<0.05; **, p<0.005; ***, p<0.001) of the differences (HC group, compared with the control; Hk, HK and HR groups, compared with the HC group).

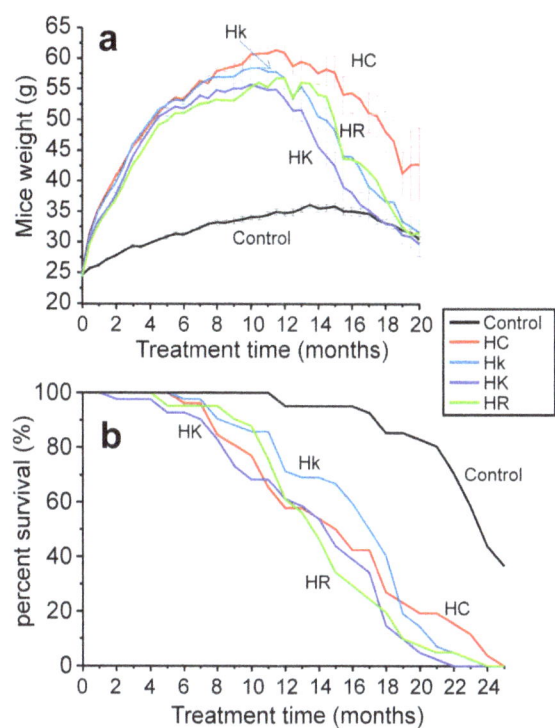

Figure 2. Evolution with time of mice weight and survival.
Panel a. Evolution of mice weight along the study from the start of the hypercaloric diets (time 0 in the figure). Data are mean±s.e, n = 50 data for each value at the beginning, the number of data then decreasing with time according to the survival. **Panel b.** Evolution of mice survival with time in all the groups, expressed as percentage of the starting mice number.

Insulin was measured by ELISA using the kit for mouse insulin from DRG Diagnostics, Marburg, Germany (Ref. EIA-3439). Triglycerides, total cholesterol and alanine aminotransferase were measured using the corresponding kits from BioSystems, Barcelona, Spain (Refs. 11528, 11505 and 11533, respectively). All these measurements were made using blood taken from the tail after overnight fast.

Rotarod test

The Ugo Basile (Comerio, Italy) Mouse Rota Rod was used for the tests. Starting at 0 rpm, the rotarod was accelerated at constant rates of 4–40 rpm. All the mice in the same cage were first trained on the rotarod the day before at a constant rate of 4 rpm until they were able to stay on for one minute. The next day, 5 mice were randomly chosen in that cage and labeled. Then each mice was subjected to three rotarod trials, with 30 min intervals, consisting in a constant acceleration along 5 min from 4 rpm to 40 rpm. The final time to fall of each mouse was obtained as the mean±s.e. of all the measurements for that mouse. Rotarod tests were carried out every 3 months in all the mice groups.

Histological studies

Mice were sacrificed by cervical dislocation. The heart and liver were removed and washed several times in 4°C cold PBS. Tissues were cut in small pieces and fixed for 24 hours in 4% formalin at 4°C. After fixation, tissues were washed 3 times for 30 min in cold PBS, once in distilled water and finally once in 70% ethanol. The tissue sections were then embedded in paraffin in a carousel tissue

processor cut in 4 μm sections and mounted on Superfrost slides. They were then let dry overnight at 37°C before they were finally stained with Haematoxyline and Eosin and imaged in a Nikon Eclipse 80i upright microscope at 20× and 40×.

Gene expression

Three mice of each group (3×5) were sacrificed and their livers were separately minced, immersed in RNAlater RNA stabilization Reagent (Quiagen) and frozen. For RNA extraction, liver chunks were homogenized using the Minibeadbeater-8, and RNA was then extracted using the RNeasy mini kit (Qiagen). RNA concentrations were measured using a NanoDrop spectrophotometer (NanoDrop Technologies, Wilmington, DE, USA), and samples were analyzed using 15 Mouse Gene 1.0 ST arrays, one for each mice, in the Genomic facility of the Cancer Research Center, Salamanca, Spain. Background correction, intra- and inter-microarray normalization, and expression signal calculation was made using a robust microarray analysis algorithm [30–32]. This algorithm calculates the absolute expression signal for each gene in each microarray. Then, a method known as significance analysis of microarray [33] was applied to calculate significant differential expression. This method provides an estimation of the error using the false discovery rate algorithm (FDR) [34]. A FDR of less than 0.1 (10% false positives) was used as standard for all the differential expression calculations, although in some cases it was increased to 0.15 (15% false positives), as indicated. All these methods were applied using R and Bioconductor packages.

Statistics

Data are expressed as mean±s.e. Significance analysis was made with the ANOVA test.

Results

Food ingestion

The mean amount ingested by the group treated with HC diet was significantly smaller in the first 12 months, although its caloric value was still much higher (see Fig. 1 and Table S1). Food intake was little modified in the groups treated with the drugs compared with the HC group in the first 12 months, but it was significantly decreased (by 5–9%) in the senescence period (13–21 months).

Mice weight

Control mice (C) started with a weight of about 25 g, which increased to stabilize at around 30–35 g at one year, and then slowly declined with senescence (see figure 2a). The weight of mice fed high-calorie diet (HC) increased rapidly, reaching around 60 g after one year of treatment with this diet. Then it started to decrease with senescence, tending to return to the original values after 2 years of treatment. Mice treated with HC diet with resveratrol (HR, 25 mg/kg body weight per day) increased weight 7.0±0.4% less (mean±s.e. obtained from the first 20 weight measurements, obtained every 15 days) than HC mice during the first year, and this effect was statistically significant for every measurement (see Table S2). Then, weight also began a rapid decline with senescence, in parallel with the weight loss observed in the HC mice, but starting about 3 months earlier. In the case of mice treated with kaempferol at low concentration (Hk, 5 mg/kg body weight per day), there were no significant differences in weight during most of the first year of treatment (Table S2), but the subsequent drop in weight associated with senescence also occurred about 3 months earlier than in the case of the controls with HC diet. In the case of mice treated with HC diet and kaempferol at high concentration (HK, 25 mg/kg body weight per day), there was also an initial decrease in weight of a $5.0 \pm 0.5\%$ (mean$\pm$s.e. obtained from the first 20 weight measurements, obtained every 15 days), which was statistically significant for most of the measurements, but not all of them (Table S2). Then, the subsequent decline in weight associated with senescence began even earlier, around 4 months earlier than in control mice treated with HC diet, so that the differences in weight with the HC group remained highly significant for most of the senescence period (Table S2). In summary, kaempferol and resveratrol treatments produced significant decreases (5–7%) in weight in mice fed with HC diet during the first year of treatment, and induced an early start of the decrease in weight associated with senescence.

Survival study

In control mice, mortality was very small before they were 18 months old. Feeding with HC diet significantly reduced survival (Fig. 2b), so that 50% mortality was already reached by 15 months of life (compared with only 5% mortality in controls of the same age). The administration of dietary kaempferol or resveratrol at high concentrations did not significantly change the mortality curve with respect to that of the HC mice. In the case of kaempferol at low concentration, 50% mortality was obtained 2 months later, at 17 months, but the subsequent decrease of the survival curve was faster, equaling with the group of mice with HC diet at 19 months. In summary, our data show that treatment with kaempferol or resveratrol did not modify survival in mice fed with hypercaloric diet.

Glucose

Blood glucose levels in control mice were stable around 80–120 mg/dl throughout all the study. In all the other conditions having HC diet, there was a rapid increase in the blood glucose levels, which doubled the previous glucose levels immediately after establishment of the HC diet treatment (Fig. 3a). These high levels were not significantly modified by any of the polyphenols used throughout the first year of treatment. Then, during senescence, there was a slight decrease in the glucose levels, which was more rapid and pronounced in mice treated with polyphenols. In particular, kaempferol at high doses produced a significant decrease with respect to the HC mice ($p < 0.005$, ANOVA test) in blood glucose levels at 15 months of treatment (see red arrows), and both kaempferol and resveratrol produced a decrease ($p < 0.05$) in these values at 18 months. Therefore, significant effects of the polyphenols were only observed after more than 1 year of treatment.

Insulin

Insulin levels in control mice remained around 0.2 to 0.3 μg/l throughout the study. In contrast, in all conditions of HC diet-fed mice there was a large increase in blood insulin levels to values 10–15 times higher (Fig. 3b). Treatment with polyphenols did not modify this increase in the initial measurements. However, there was a statistically significant decrease in the insulin values obtained in mice treated with the high dose of kaempferol between 9 and 15 months (see red arrows, $p < 0.05$ at 9 months, $p < 0.005$ at 15 months). Resveratrol and the low dose of kaempferol also produced statistically significant decreases in insulin levels at 15 months of treatment ($p < 0.05$ and $p < 0.005$, respectively). Thus, as in the case of the glucose levels, a reduction in insulin levels was only apparent after about 1 year of treatment with the polyphenols.

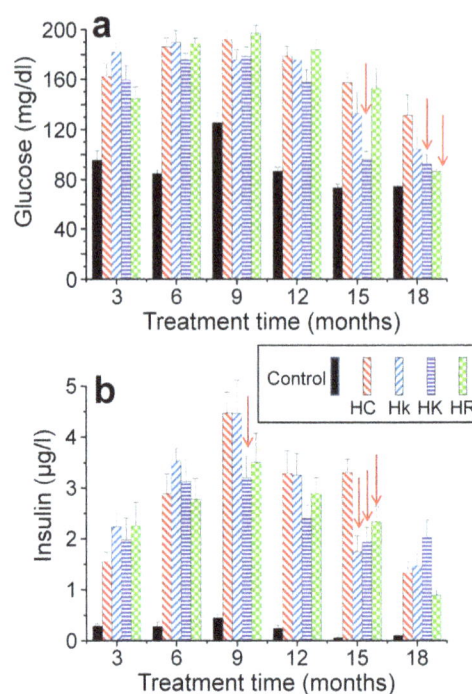

Figure 3. Mean blood glucose (panel a) and insulin (panel b) values obtained every 3 months along the study, from the start of the hypercaloric diet. Arrows indicate the presence of differences statistically significant when the HC condition is compared to the Hk, HK and HR ones (see the text). Data are mean$\pm$s.e, n = 10 data for each value.

Cholesterol

Cholesterol levels in blood were maintained at around 80–100 mg/dl in control mice throughout the study. The administration of HC diet increased these values to levels around 200 mg/dl, which were insensitive to treatment with polyphenols during the first year of treatment (Fig. 4a). In the second year there was a decrease in cholesterol levels in all the conditions. However, this decrease occurred earlier in mice treated with polyphenols. Thus, we observed a significant decrease in cholesterol levels at 15 months of treatment in mice treated with the high dose of kaempferol ($p < 0.005$) or resveratrol ($p < 0.05$) with respect to HC mice. Then, after 18 months of treatment, cholesterol levels were also reduced in the HC mice and the differences disappeared. In the case of mice treated with the high dose of kaempferol, there was also a significant decrease in cholesterol values ($p < 0.05$) at 12 months of treatment. In summary, treatment with polyphenols produced little effects on cholesterol levels in the first year but accelerated the reduction of its levels associated with senescence.

Triglycerides

Triglyceride levels in control mice stayed around 40–60 mg/dl throughout the entire study. Mice fed HC diet showed an increase of these values of about 25–30%, which was homogeneous in mice treated or not with polyphenols (Fig. 4b). This increase was maintained during the first year of treatment and thereafter the difference became smaller and tended to disappear. In any case, there were no significant differences between the data obtained in mice treated with HC diet with or without polyphenols.

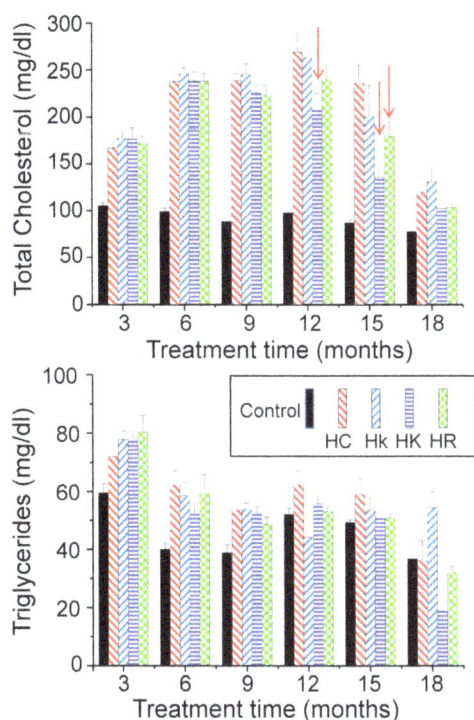

Figure 4. Mean blood total cholesterol (panel a) and triglycerides (panel b) values obtained every 3 months along the study, from the start of the hypercaloric diet. Arrows indicate the presence of differences statistically significant when the HC condition is compared to the HK and HR ones (see the text). Data are mean±s.e, n = 10 data for each value.

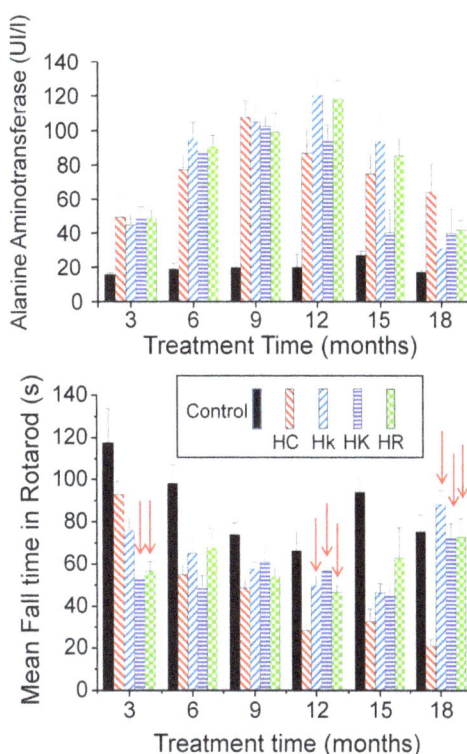

Figure 5. Mean blood alanine aminotransferase (panel a) and rotarod fall time (panel b) values obtained every 3 months along the study. Arrows indicate the presence of differences statistically significant when the HC condition is compared to the Hk, HK and HR ones (see the text). Data are mean±s.e, n = 10 data for each value.

Alanine aminotransferase

Alanine aminotransferase values in blood maintained at around 20–30 IU/L throughout the entire study in control mice. In mice treated with HC diet, there was a progressive increase in these values that reached a maximum of around 100 IU/L after 9 months of treatment (Fig. 5a). Again here, there were no significant differences between the data obtained in mice treated only with HC diet and in those that had also received polyphenols.

Motor coordination study with rotarod

In these studies, the time before fall of the mice from a rotating tube is a measure of the degree of coordination. In control mice, the time before fall decreased slightly with age, but remained between 80 and 120 seconds on average throughout the entire study. In mice treated with HC diet, there was a progressive decrease in that time, reaching values of only 30 seconds after a year of treatment (Fig. 5b). The effect of polyphenols was here quite surprising. In the first measurement, obtained after 3 months of treatment, mice treated with HC diet with kaempferol or resveratrol performed significantly worse (shorter times, p<0.005) than those treated only with diet HC. This effect disappeared at 6 and 9 months, because the values obtained for all mice with high-fat diet were here similar. Interestingly, after 12 months, data obtained by the HC diet treated mice continued to worsen, whereas those treated also with polyphenols remained stable or even improved. In fact, at 12 and 18 months, all the mice groups fed with diets containing polyphenols obtained significantly better marks (p<0.001) than those fed with HC diet.

Histological analysis

Histological studies of liver and heart samples were made at 6 and 12 months after the start of the hypercaloric diet. Fig. 6 shows typical Haematoxyline and Eosin stained sections from livers and hearts obtained 6 months after the start of the hypercaloric diet. Images obtained after 12 months provided similar results. The liver stained sections (Fig. 6a) show that all the mice under hypercaloric diet had clearly enlarged hepatocytes and a large degree of steatosis represented by the vacuolation of hepatocytes. However, after careful examination of a series of sections of each type we did not find any significant difference among the samples from mice fed only with plain hypercaloric diet and those from mice fed with hypercaloric diet plus any of the polyphenolic compounds. Regarding the heart stained sections (Fig. 6b), the hypercaloric diet induced little changes with respect to the controls, and they were not modified either by the presence of the polyphenols.

Gene expression

We have conducted an analysis of whole genome expression arrays in the 5 groups of mice after 1 year of treatment with the different diets. Treatment with HC diet produced a significant change in the pattern of gene expression of multiple genes with metabolic importance, as has been previously described. The list of genes showing a larger change, using a 10% false positives, is in the Table S3.

Concerning the effect of polyphenols, and this is a significant discrepancy with previous data [14], the group of genes that were

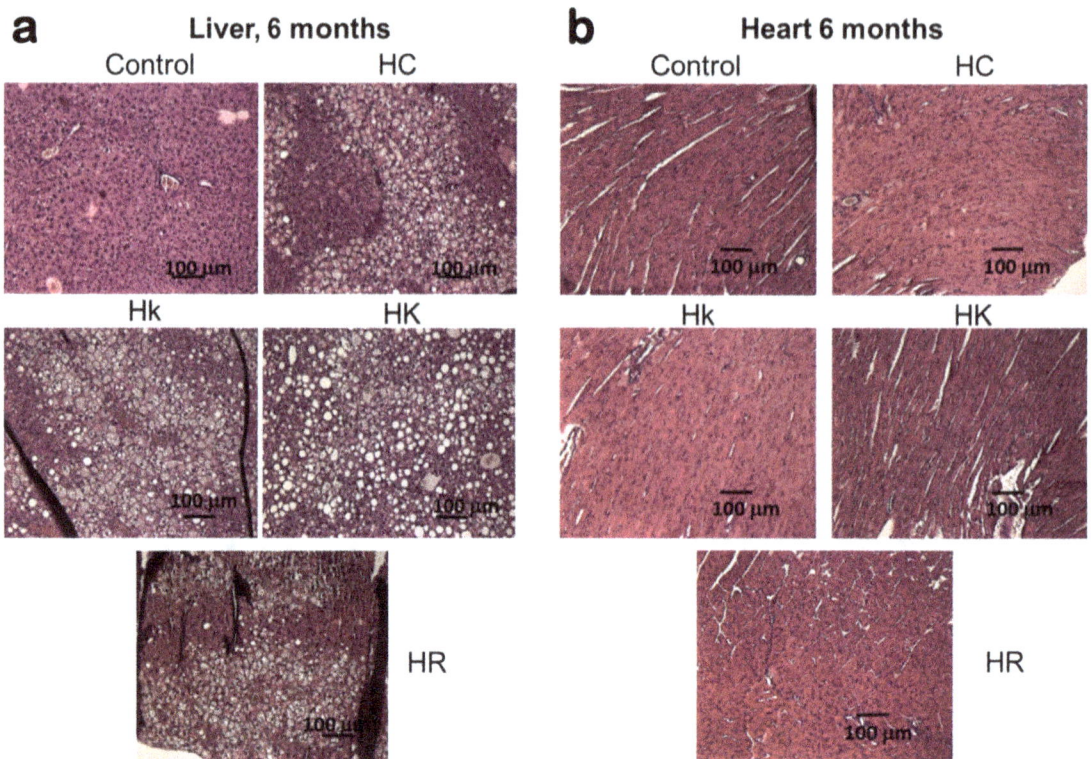

Figure 6. Haematoxylin and Eosin stained sections of liver (panel a) and heart (panel b), obtained from the five mice groups 6 months after the start of the hypercaloric diets and imaged in a Nikon Eclipse 80i upright microscope at 20×.

overexpressed or silenced was very similar in the group treated with hypercaloric diet alone and in the group fed with hypercaloric diet containing resveratrol. In fact, a direct comparison of the gene expression pattern obtained in both mice groups yielded no gene expression change between the two conditions (even using a maximum of 15% false positives).

Regarding the group fed with hypercaloric diet plus low dose of kaempferol, it was also not possible to find any gene changing expression when it was compared with the group fed only with hypercaloric diet. Similarly, in the case of the group fed with the high dose of kaempferol, just a slightly reduced expression of 6 genes by 20–30% was obtained, and only when the percentage of false positives was increased to 15% (see table S4). Some of them are mitochondrial proteins, so it may be interesting to study if they play any role in mitochondrial Ca^{2+} dynamics. However, the main conclusion of the gene expression studies is that the polyphenols added to the HC diet did not induce significant changes in the gene expression pattern generated by the HC diet.

Discussion

Our study has sought to shed light on the controversy over the role of polyphenols on health and survival in mice. We have used resveratrol, a potent activator of sirtuins [15], and kaempferol, which is a poor activator of sirtuins [15] but a potent antioxidant with strong effects on mitochondrial metabolism activation [25–27]. Our macro-study covers the entire lifetime of the mice (24 months) and includes a very high number of C57BL wild-type mice, 50 mice for each condition. This makes the results highly reliable.

Our data show that the main effect of mice treatment with hypercaloric diet containing polyphenols is an initial weight loss of

5–7%, when kaempferol or resveratrol are used at a dose of 25 mg/kg body weight per day. This decrease, which requires using the high dose of kaempferol, can be seen clearly from the beginning of the treatment and is maintained throughout the study. Food ingestion during the first year was similar or higher in the drug-treated groups than in the HC control, so that the weight loss cannot be attributed to a decreased food intake. The decrease in weight induced by resveratrol in hypercaloric diet-fed mice has been described previously [14]. However, the fact that the same effect is obtained with kaempferol is important, because this flavonoid has been shown to be a poor activator of sirtuins [15]. Its effects could rather be attributed to metabolic effects such as increase in oxygen consumption or thyroid hormones synthesis [25–27].

Regarding the weight loss associated with senescence, it occurred 3–4 months earlier in mice treated with polyphenols compared to controls with HC diet, even with the low dose of kaempferol. This decrease was not associated with increased toxicity or mortality because the survival curves of mice treated with HC diet with or without polyphenols were similar. It could be due in part to the 5–9% reduction in food intake observed in the senescence period in all the groups treated with either kaempferol or resveratrol.

As for the study of motor coordination in rotarod, the effects of polyphenols were variable in time: negative after 3 months of treatment, indifferent between 6 and 9 months and finally positive at longer times. Notably, however, the low doses of kaempferol increased rotarod times in the last months as much as the high doses, but hardly produced any alteration at short times. The improvement in rotarod performance after long times of treatment with resveratrol has been reported before [14]. Our data here show that the same findings were also obtained with kaempferol suggesting again that mechanisms additional to sirtuin activation may be involved.

Regarding survival, our data do not confirm the previous report that resveratrol prolongs survival of mice fed with HC diet and makes it similar to that of control mice [14]. Instead, we find that survival in the groups of mice treated with HC diet and resveratrol or kaempferol was similar to that of the group fed HC diet only.

As for the biochemical parameters, it has been reported that treatment with HC diet produces a 30% increase in glucose levels in blood and a four times increase in blood insulin levels, and that both changes were reversed in the presence of resveratrol [14]. Our data, which track more completely the glucose levels at different times of treatment, show instead that glucose levels were doubled by the HC diet and that resveratrol and kaempferol treatments were only able to reduce in part that increase during senescence, after at least 1 year of treatment. As for the insulin levels, our data show a much higher increase, between 10 and 15 times the control values, in the groups treated with HC diet. Again in this case, the polyphenols were only able to partially reverse this effect after 1 year of treatment. The origin of these partial and very delayed effects is unclear, but it could be related to the reduction in food intake and faster decrease in weight we observe during senescence in the drug-treated groups. The lack of metabolic effects of resveratrol is consistent with recent data obtained in obese men [35].

Regarding gene expression, our data show that mice fed with hypercaloric diet show a large number of changes in gene expression, confirming data previously reported by other authors [36,37]. However, the presence of polyphenols in the hypercaloric diet hardly induced any change in gene expression with respect to the mice fed only with hypercaloric diet. These results are again contradictory with previous reports [14], which showed large changes in gene expression induced by resveratrol.

Conclusions

In summary, we have performed here a systematic macro-study of the effect of the polyphenols resveratrol and kaempferol

included in a hypercaloric diet in groups of 50 mice for each condition, monitoring weight, survival and biochemical, neurological, histological and genetic parameters along the whole life of the mice. Our results indicate that these compounds reduced mice weight by 5–7% and prolonged treatments partially reversed some of the biochemical changes induced by the hypercaloric diet. However, the polyphenols did not significant modify survival or liver gene expression.

Supporting Information

Table S1 Food ingestion in the different groups measured in both $g \cdot mice^{-1} \cdot day^{-1}$ or $kcal \cdot mice^{-1} \cdot day^{-1}$.

Table S2 Significance of mice weight differences between drug treated (Hk, HK and HR) and hypercaloric control HC (*, $p < 0.05$; **, $p < 0.005$; ***, $p < 0.001$).

Table S3 Differential gene expression obtained by comparing gene expression arrays for HC mice versus controls.

Table S4 Differential gene expression obtained by comparing gene expression arrays for HK versus HC mice.

Acknowledgments

We thank Pilar Alvarez and Laura González for excellent technical assistance, Dr. Encarna Fermiñán and Diego Alonso, Cancer Research Center (CSIC-University of Salamanca) for help with processing and analysis of the gene expression arrays, and Dr. Angel Barcia, the veterinary of the School of Medicine, for help with animal care.

Author Contributions

Conceived and designed the experiments: MM RIF AM JA. Performed the experiments: MM RIF AM SF. Analyzed the data: MM SF RIF AM JA. Wrote the paper: MM JA.

References

1. Middleton E Jr, Kandaswami C, Theoharides TC (2000) The effects of plant flavonoids on mammalian cells: implications for inflammation, heart disease, and cancer. Pharmacol Rev 52: 673–751.
2. Havsteen BH (2002) The biochemistry and medical significance of the flavonoids. Pharmacol Therap 96: 67–202.
3. Birt DF, Hendrich S, Wang W (2001) Dietary agents in cancer prevention: flavonoids and isoflavonoids. Pharmacol Ther 90: 157–177.
4. Nestel P (2003) Isoflavones: their effects on cardiovascular risk and functions. Curr Opin Lipidol 14: 3–8.
5. Scalbert A, Manach C, Morand C, Remesy C, Jimenez L (2005) Dietary polyphenols and the prevention of diseases. Crit Rev Food Sci Nutr 45: 287–306.
6. Grassi D, Necozione S, Lippi C, Croce G, Valeri L, et al. (2005) Cocoa reduces blood pressure and insulin resistance and improves endothelium-dependent vasodilation in hypertensives. Hypertension 46: 398–405.
7. Manach C, Mazur A, Scalbert A (2005) Polyphenols and prevention of cardiovascular diseases. Curr Opin Lipidol 16: 77–84.
8. Gates MA, Tworoger SS, Hecht JL, De Vivo I, Rosner B, et al. (2007) A prospective study of dietary flavonoid intake and incidence of epithelial ovarian cancer. Int J Cancer 121: 2225–2232.
9. Watanabe CM, Wolffram S, Ader P, Rimbach G, Packer L, et al. (2001) The in vivo neuromodulatory effects of the herbal medicine ginkgo biloba. Proc Natl Acad Sci U S A 98: 6577–6580.
10. Luo Y, Smith JV, Paramasivam V, Burdick A, Curry KJ, et al. (2002) Inhibition of amyloid-beta aggregation and caspase-3 activation by the Ginkgo biloba extract EGb761. Proc Natl Acad Sci U S A 99: 12197–12202.
11. Garcia-Mediavilla V, Crespo I, Collado PS, Esteller A, Sanchez-Campos S, et al. (2007) The anti-inflammatory flavones quercetin and kaempferol cause inhibition of inducible nitric oxide synthase, cyclooxygenase-2 and reactive C-protein, and down-regulation of the nuclear factor kappaB pathway in Chang Liver cells. Eur J Pharmacol 557: 221–229.
12. Kuppusamy UR, Das NP (1992) Effects of flavonoids on cyclic AMP phosphodiesterase and lipid mobilization in rat adipocytes. Biochem Pharmacol 44: 1307–1315.
13. Kuppusamy UR, Das NP (1994) Potentiation of beta-adrenoceptor agonist-mediated lipolysis by quercetin and fisetin in isolated rat adipocytes. Biochem Pharmacol 47: 521–529.
14. Baur JA, Pearson KJ, Price NL, Jamieson HA, Lerin C, et al. (2006) Resveratrol improves health and survival of mice on a high-calorie diet. Nature 444: 337–342.
15. Howitz KT, Bitterman KJ, Cohen HY, Lamming DW, Lavu S, et al. (2003) Small molecule activators of sirtuins extend Saccharomyces cerevisiae lifespan. Nature 425: 191–196.
16. Wood JG, Rogina B, Lavu S, Howitz K, Helfand SL, et al. (2004) Sirtuin activators mimic caloric restriction and delay ageing in metazoans. Nature 430: 686–689.
17. Viswanathan M, Kim SK, Berdichevsky A, Guarente L (2005) A role for SIR-2.1 regulation of ER stress response genes in determining C. elegans life span. Dev Cell 9: 605–615.
18. Jarolim S, Millen J, Heeren G, Laun P, Goldfarb DS, et al. (2004) A novel assay for replicative lifespan in Saccharomyces cerevisiae. FEMS Yeast Res 5: 169–177.
19. Valenzano DR, Terzibasi E, Genade T, Cattaneo A, Domenici L, et al. (2006) Resveratrol prolongs lifespan and retards the onset of age-related markers in a short-lived vertebrate. Curr Biol 16: 296–300.
20. Lee YE, Kim JW, Lee EM, Ahn YB, Song KH, et al. (2012) Chronic Resveratrol Treatment Protects Pancreatic Islets against Oxidative Stress in db/db Mice. PLoS One 7: e50412.
21. Burnett C, Valentini S, Cabreiro F, Goss M, Somogyvári M, et al. (2011) Absence of effects of Sir2 overexpression on lifespan in C. elegans and Drosophila. Nature 477: 482–485.
22. Canto C, Auwerx J (2011) Don't write sirtuins off. Nature 477: 411.

23. Lombard DB, Pletcher SD (2011) A valuable background check. Nature 477: 410.

24. Couzin-Frankel J (2011) Ageing genes: the sirtuin story unravels. Science 334: 1194–1198.

25. Montero M, Lobaton CD, Hernandez-SanMiguel E, SantoDomingo J, Vay L, et al. (2004) Direct activation of the mitochondrial calcium uniporter by natural plant flavonoids. Biochem J 384: 19–24.

26. da-Silva WS, Harney JW, Kim BW, Li J, Bianco SD, et al. (2007) The small polyphenolic molecule kaempferol increases cellular energy expenditure and thyroid hormone activation. Diabetes 56: 767–776.

27. Yu SF, Shun CT, Chen TM, Chen YH (2006) 3-O-beta-D-glucosyl-(1→6)-beta-D-glucosylkaempferol isolated from Sauropus androgenus reduces body weight gain in Wistar rats. Biol Pharm Bull 29: 2510–2513.

28. Lalonde R, Bensoula AN, Filali M (1995) Rotarod sensorimotor learning in cerebellar mutant mice. Neurosci Res 22: 423–426.

29. Shiotsukia H, Yoshimib K, Shimoa Y, Funayamaa M, Takamatsuc Y, et al. (2010) A rotarod test for evaluation of motor skill learning. J Neurosci Meth 189: 180–185.

30. Irizarry RA, Hobbs B, Collin F, Beazer-Barclay YD, Antonellis KJ, et al. (2003) Exploration, normalization, and summaries of high density oligonucleotide array probe level data. Biostatistics 4: 249–264.

31. Irizarry RA, Bolstad BM, Collin F, Cope LM, Hobbs B, et al. (2003) Summaries of Affymetrix GeneChip probe level data. Nucleic Acids Res 31: e15.

32. Bolstad BM, Irizarry RA, Astrand M, Speed TP (2003) A comparison of normalization methods for high density oligonucleotide array data based on variance and bias. Bioinformatics 19: 185–193.

33. Tusher VG, Tibshirani R, Chu G (2001) Significance analysis of microarrays applied to the ionizing radiation response. Proc Natl Acad Sci U S A 98: 5116–5121.

34. Benjamini Y, Drai D, Elmer G, Kafkafi N, Golani I (2001) Controlling the false discovery rate in behavior genetics research. Behav Brain Res 125: 279–284.

35. Poulsen MM, Vestergaard PF, Clasen BF, Radko Y, Christensen LP, et al. (2013) High-dose resveratrol supplementation in obese men: an investigator-initiated, randomized, placebo-controlled clinical trial of substrate metabolism, insulin sensitivity, and body composition. Diabetes 62: 1186–1195.

36. Shockley KR, Witmer D, Burgess-Herbert SL, Paigen B, Churchill GA (2009) Effects of atherogenic diet on hepatic gene expression across mouse strains. Physiol Genomics 39: 172–182.

37. Do GM, Oh HY, Kwon EY, Cho YY, Shin SK, et al. (2011) Long-term adaptation of global transcription and metabolism in the liver of high-fat diet-fed C57BL/6J mice. Mol Nutr Food Res 55: S173–S185.

PERMISSIONS

LIST OF CONTRIBUTORS

Kazutaka Ota and Masanori Kohda
Department of Biology and Geosciences, Osaka City University, Sumiyoshi, Osaka, Japan

Alison M. Roark
Department of Biology, Furman University, Greenville, South Carolina, United States of America

Karen A. Bjorndal
Department of Biology, University of Florida, Gainesville, Florida, United States of America

Thomas J. O'Grady, A. Gregory DiRienzo and Margaret A. Gates
University at Albany, School of Public Health, Rensselaer, New York, United States of America

Cari M. Kitahara
Division of Cancer Epidemiology and Genetics, National Cancer Institute, National Institutes of Health, Rockville, Maryland, United States of America

Noelia Díaz, Laia Ribas and Francesc Piferrer
Institut de Ciències del Mar, Consejo Superior de Investigaciones Científicas (CSIC), Barcelona, Spain

Hae Dong Woo and Jeongseon Kim
Molecular Epidemiology Branch, National Cancer Center, Goyang-si, Korea

Aesun Shin
Molecular Epidemiology Branch, National Cancer Center, Goyang-si, Korea
Department of Preventive Medicine, Seoul National University College of Medicine, Seoul, Republic of Korea

Sara Schärrer and Patrick Presi
Veterinary Public Health Institute/University of Berne, Berne, Switzerland

Jan Hattendorf and Jakob Zinsstag
Swiss Tropical and Public Health Institute/University of Basel, Basel, Switzerland

Nakul Chitnis
Swiss Tropical and Public Health Institute/University of Basel, Basel, Switzerland
Fogarty International Center, National Institutes of Health, Bethesda, Maryland, United States of America

Martin Reist
Federal Food Safety and Veterinary Office, Bern, Switzerland

Theodore A. Evans and Michael J. Beran
Language Research Center, Georgia State University, Atlanta, GA, United States of America

Bonnie Perdue
Department of Psychology, Agnes Scott College, Decatur, GA, United States of America

Toshiki Nakamura, Goro Ishikawa, Hiroyuki Ito and Mika Saito
NARO Tohoku Agricultural Research Center, Morioka, Iwate, Japan

Kazuhiro Nakamura
NARO Tohoku Agricultural Research Center, Morioka, Iwate, Japan
NARO Kyusyu Okinawa Agricultural Research Center, Chikugo, Fukuoka, Japan

Hironobu Jinno and Yasuhiro Yoshimura
Kitami Agricultural Experiment Station, Hokkaido Research Organization, Tokoro-gun, Hokkaido, Japan

Mikako Sato
Kitami Agricultural Experiment Station, Hokkaido Research Organization, Tokoro-gun, Hokkaido, Japan
Central Agricultural Experiment Station, Hokkaido Research Organization, Yubarigun, Hokkaido, Japan

Tsutomu Nishimura
Kitami Agricultural Experiment Station, Hokkaido Research Organization, Tokoro-gun, Hokkaido, Japan
Kamikawa Agricultural Experiment Station, Hokkaido Research Organization, Kamikawa-gun, Hokkaido Japan

Hidekazu Maejima and Yasushi Uehara
Nagano Agricultural Experiment Station, Suzaka Nagano, Japan

Fuminori Kobayashi
National Institute of Agrobiological Sciences Kannondai, Tsukuba, Japan

Emma C. Cancelliere
Department of Anthropology, Graduate Center of the City University of New York, New York, New York United States of America
New York Consortium in Evolutionary Primatology New York, New York, United States of America

Nicole DeAngelis
Department of Animal Science, Cornell University, Ithaca, New York, United States of America

John Bosco Nkurunungi
Department of Biology, Mbarara University of Science and Technology, Mbarara, Uganda

David Raubenheimer
Charles Perkins Centre, Faculty of Veterinary Sciences, School of Biological Sciences, The University of Sydney, Sydney, NSW, Australia

Jessica M. Rothman
Department of Anthropology, Graduate Center of the City University of New York, New York, New York, United States of America
New York Consortium in Evolutionary Primatology, New York, New York, United States of America
Department of Anthropology, Hunter College of the City University of New York, New York City, New York, United States of America

Debbie L. Humphries
Department of Epidemiology of Microbial Disease, Yale School of Public Health, New Haven, Connecticut, United States of America

Jere R. Behrman and Whitney Schott
Department of Economics, University of Pennsylvania, Philadelphia, Pennsylvania, United States of America

Benjamin T. Crookston
Department of Health Science, Brigham Young University, Provo, Utah, United States of America

Kirk A. Dearden
Department of Global Health, Boston University School of Public Health, Boston, Massachusetts, United States of America

Mary E. Penny
Instituto de Investigación Nutricional, Lima, Peru

Xiaolin You
History of Science, Nanjing Agricultural University, Nanjing, P.R. China

Yibo Li
Department of Sociology, Nanjing Agricultural University, Nanjing, P.R. China

Min Zhang
Department of Laws, Nanjing Agricultural University, Nanjing, P.R. China

Huoqi Yan
Philosophy of Science and Technology, Nanjing Agricultural University, Nanjing, P.R. China

Ruqian Zhao
Key Laboratory of Animal Physiology and Biochemistry, Nanjing Agricultural University, Nanjing, P.R. China

Chenwen Cai, Jun Shen, Di Zhao, Yuqi Qiao, Antao Xu, Shuang Jin, Zhihua Ran and Qing Zheng
Key Laboratory of Gastroenterology & Hepatology, Ministry of Health, Division of Gastroenterology and Hepatology, Ren Ji Hospital, School of Medicine, Shanghai Jiao,Tong University, Shanghai Institute of Digestive Diseases, 145 Middle Shandong Road, Shanghai 200001, China

Adele Costabile, Sara Santarelli, Sandrine P. Claus and Glenn R. Gibson
Department of Food and Nutritional Sciences, The University of Reading, Reading, United Kingdom

Jeremy Sanderson, Barry N. Hudspith and Jonathan Brostoff
King's College London, Biomedical & Health Sciences, Dept. of Nutrition and Dietetics, London, United Kingdom

Jane L. Ward3, Alison Lovegrove
Rothamsted Research, Harpenden, Hertfordshire, United Kingdom

Peter R. Shewry
Rothamsted Research, Harpenden, Hertfordshire, United Kingdom
School of Agriculture, Policy and Development, Earley Gate, Reading, United Kingdom

Hannah E. Jones
School of Agriculture, Policy and Development, Earley Gate, Reading, United Kingdom

Andrew M. Whitley
Bread Matters Limited, Macbiehill Farmhouse, Lamancha, West Linton, Peeblesshire, Scotland

Jana Foerster and Heiner Boeing
Department of Epidemiology, German Institute of Human Nutrition Potsdam-Rehbruecke, Nuthetal, Germany

Gertraud Maskarinec
Department of Epidemiology, German Institute of Human Nutrition Potsdam-Rehbruecke, Nuthetal, Germany
University of Hawaii Cancer Center, Honolulu, Hawaii, United States of America

Nicole Reichardt, Adrian Tett and Arjan Narbad
Gut Health and Food Safety, Institute of Food Research, Norwich Research Park, Colney, United Kingdom

Michael Blaut
Department of Gastrointestinal Microbiology, German Institute of Human Nutrition, Potsdam-Rehbruecke, Nuthetal, Germany

Jun Mei, Qizhen Guo, Yan Wu and Yunfei Li
Department of Food Science and Technology, School of Agriculture and Biology, Shanghai Jiao Tong University, Shanghai, P.R. China

Coralie Picoche, Romain Le Gendre, Sylvaine Françoise, Frank Maheux, Benjamin Simon and Aline Gangnery
Laboratoire Environnement Ressources de Normandie, IFREMER, Port en Bessin, France

Jonathan Flye-Sainte-Marie
Université de Bretagne Occidentale, Institut Universitaire Européen de la Mer, Laboratoire des sciences de l'Environnement Marin (LEMAR), UMR 6539 CNRS/UBO/IRD/IFREMER, Plouzané, France

Shane Golden and Reuven Dukas
Animal Behaviour Group, Department of Psychology, Neuroscience & Behaviour, McMaster University, Hamilton, Ontario, Canada

Suzanne Spence, Jennifer Delve and Elaine Stamp
Institute of Health and Society, Newcastle University, Newcastle upon Tyne, England
Human Nutrition Research Centre, Newcastle University, Newcastle upon Tyne, England

Ashley J. Adamson and Martin White
Institute of Health and Society, Newcastle University, Newcastle upon Tyne, England
Human Nutrition Research Centre, Newcastle University, Newcastle upon Tyne, England
Fuse, UKCRC Centre for Translational Research in Public Health, Newcastle upon Tyne, England

John N. S. Matthews
School of Mathematics and Statistics, Newcastle University, Newcastle upon Tyne, England

Hedvig Hogfors, Nisha H. Motwani and Susanna Hajdu
Department of Ecology, Environment and Plant Sciences, Stockholm University, Stockholm, Sweden

Rehab El-Shehawy
IMDEA Agua, Alcalá de Henares, Madrid, Spain

Towe Holmborn
Calluna AB, Stockholm, Sweden

Jonna Engström-Öst
ARONIA Coastal Zone Research Team, Novia University of Applied Sciences & Åbo Akademi University, Ekenäs, Finland

Andreas Brutemark and Anu Vehmaa
ARONIA Coastal Zone Research Team, Novia University of Applied Sciences & Åbo Akademi University, Ekenäs, Finland
Tvärminne Zoological Station, University of Helsinki Hangö, Finland

Elena Gorokhova
Department of Ecology, Environment and Plant Sciences, Stockholm University, Stockholm, Sweden
Department of Applied Environmental Science Stockholm University, Stockholm, Sweden

Laure-Anne Poissonnier and Audrey Dussutour
Research Center on Animal Cognition, The National Center for Scientific Research and Toulouse University Toulouse, France

Stephen J. Simpson
Charles Perkins Centre, The University of Sydney Sydney, New South of Wales, Australia

Mayte Montero, Sergio de la Fuente, Rosalba I Fonteriz, Alfredo Moreno and Javier Alvarez
Instituto de Biología y Genética Molecular (IBGM) Departamento de Bioquímica y Biología Molecula y Fisiología, Facultad de Medicina, Universidad d Valladolid and Consejo Superior de Investigacione Científicas (CSIC), Valladolid, Spain

Index